symmetry

I0049600

Harmonic Oscillators and Two-by-two Matrices in Symmetry Problems in Physics

Edited by
Young Suh Kim

Printed Edition of the Special Issue Published in *Symmetry*

MDPI

Harmonic Oscillators and Two-By-Two Matrices in Symmetry Problems in Physics

Special Issue Editor
Young Suh Kim

MDPI • Basel • Beijing • Wuhan • Barcelona • Belgrade

MDPI

Special Issue Editor

Young Suh Kim
University of Maryland
USA

Editorial Office
MDPI AG
St. Alban-Anlage 66
Basel, Switzerland

This edition is a reprint of the Special Issue published online in the open access journal *Symmetry* (ISSN 2073-8994) from 2014–2017 (available at: http://www.mdpi.com/journal/symmetry/special_issues/physics-matrices).

For citation purposes, cite each article independently as indicated on the article page online and as indicated below:

Author 1; Author 2. Article title. *Journal Name* **Year**, *Article number*, page range.

First Edition 2017

ISBN 978-3-03842-500-7 (Pbk)
ISBN 978-3-03842-501-4 (PDF)

Articles in this volume are Open Access and distributed under the Creative Commons Attribution license (CC BY), which allows users to download, copy and build upon published articles even for commercial purposes, as long as the author and publisher are properly credited, which ensures maximum dissemination and a wider impact of our publications. The book taken as a whole is © 2017 MDPI, Basel, Switzerland, distributed under the terms and conditions of the Creative Commons license CC BY-NC-ND (http://creativecommons.org/licenses/by-nc-nd/4.0/).

Table of Contents

Chapter 1

Chapter 2

Chapter 3

About the Special Issue Editor

Young Suh Kim Dr. Kim came to the United States from South Korea in 1954 after high school graduation, to become a freshman at the Carnegie Institute of Technology (now called Cernegie Mellon University) in Pittsburgh. In 1958, he went to Princeton University to pursue graduate studies in Physics and received his PhD degree in 1961. In 1962, he became an assistant professor of Physics at the University of Maryland at College Park near Washington, DC. In 2007, Dr. Kim became a professor emeritus at the same university and thus became a full-time physicist. Dr. Kim's thesis advisor at Princeton was Sam Treiman, but he had to go to Eugene Wigner when faced with fundamental problems in physics. During this process, he became interested in Wigner's 1939 paper on internal space-time symmetries of physics. Since 1978, his publications have been based primarily on constructing mathematical formulas for understanding this paper. In 1988, Dr. Kim noted that the same set of mathematical devices is applicable to squeezed states in quantum optics. Since then, he has also been publishing papers on optical and information sciences.

Preface to "Harmonic Oscillators and Two-by-two Matrices in Symmetry Problems in Physics"

This book consists of articles published in the two Special Issues entitled "Physics Based on Two-By-Two Matrices" and "Harmonic Oscillators in Modern Physics", in addition to the articles published by the issue editor that are not in those Special Issues.

With a degree of exaggeration, modern physics is the physics of harmonic oscillators and two-by-two matrices. Indeed, they constitute the basic language for the symmetry problems in physics, and thus the main theme of this journal. There is nothing special about the articles published in these Special Issues. In one way or another, most of the articles published in this *Symmetry* journal are based on these two mathematical instruments.

What is special is that the authors of these two Special Issues were able to recognize this aspect of the symmetry problems in physics. They are not the first to do this. In 1963, Eugene Wigner was awarded the Nobel prize for introducing group theoretical methods to physical problems. Wigner's basic scientific language consisted of two-by-two matrices.

Paul A. M. Dirac's four-by-four matrices are two-by-two matrices of two-by-two matrices. In addition, Dirac had another scientific language. He was quite fond of harmonic oscillators. He used the oscillator formalism for the Fock space which is essential to second quantification and quantum field theory. The role of Gaussian functions in coherent and squeezed states in quantum optics is well known. In addition, the oscillator wave functions are used as approximations for many complicated wave functions in physics.

Needless to say, spacial relativity and quantum mechanics are two of the greatest achievements in physics of the past century. Dirac devoted lifelong efforts to making quantum mechanics compatible with Einstein's spacial relativity. He was interested in oscillator wave functions that can be Lorentz-boosted.

This journal will be publishing many interesting papers based on two-by-two matrices and harmonic oscillators. The authors will be very happy to acknowledge that they are following the examples of Dirac and Wigner. We all respect them.

Young Suh Kim
Special Issue Editor

Chapter 1:
Two-By-Two Matrices

symmetry

MDPI

Article

Pseudo Hermitian Interactions in the Dirac Equation

Orlando Panella [1,*] **and Pinaki Roy** [2]

[1] INFN—Istituto Nazionale di Fisica Nucleare, Sezione di Perugia, Via A. Pascoli, Perugia 06123, Italy
[2] Physics and Applied Mathematics Unit, Indian Statistical Institute, 203 Barrackpur Trunck Road Kolkata 700108, India; E-Mail: pinaki@isical.ac.in
* E-Mail: orlando.panella@pg.infn.it; Tel.: +39-075-585-2762; Fax: +39-075-584-7296.

Received: 31 July 2013; in revised form: 18 December 2013 / Accepted: 23 December 2013 / Published: 17 March 2014

Abstract: We consider a $(2+1)$-dimensional massless Dirac equation in the presence of complex vector potentials. It is shown that such vector potentials (leading to complex magnetic fields) can produce bound states, and the Dirac Hamiltonians are η-pseudo Hermitian. Some examples have been explicitly worked out.

Keywords: pseudo Herimitian Hamiltonians; two-dimensional Dirac Equation; complex magnetic fields

1. Introduction

In recent years, the massless Dirac equation in $(2+1)$ dimensions has drawn a lot of attention, primarily because of its similarity to the equation governing the motion of charge carriers in graphene [1,2]. In view of the fact that electrostatic fields alone cannot provide confinement of the electrons, there have been quite a number of works on exact solutions of the relevant Dirac equation with different magnetic field configurations, for example, square well magnetic barriers [3–5], non-zero magnetic fields in dots [6], decaying magnetic fields [7], solvable magnetic field configurations [8], *etc.* On the other hand, at the same time, there have been some investigations into the possible role of non-Hermiticity and $\mathcal{PT}$ symmetry [9] in graphene [10–12], optical analogues of relativistic quantum mechanics [13] and relativistic non-Hermitian quantum mechanics [14], photonic honeycomb lattice [15], *etc.* Furthermore, the $(2+1)$-dimensional Dirac equation with non-Hermitian Rashba and scalar interaction was studied [16]. Here, our objective is to widen the scope of incorporating non-Hermitian interactions in the $(2+1)$-dimensional Dirac equation. We shall introduce η pseudo Hermitian interactions by using imaginary vector potentials. It may be noted that imaginary vector potentials have been studied previously in connection with the localization/delocalization problem [17,18], as well as $\mathcal{PT}$ phase transition in higher dimensions [19]. Furthermore, in the case of the Dirac equation, there are the possibilities of transforming real electric fields to complex magnetic fields and *vice versa* by the application of a complex Lorentz boost [20]. To be more specific, we shall consider η-pseudo Hermitian interactions [21] within the framework of the $(2+1)$-dimensional massless Dirac equation. In particular, we shall examine the exact bound state solutions in the presence of imaginary magnetic fields arising out of imaginary vector potentials. We shall also obtain the η operator, and it will be shown that the Dirac Hamiltonians are η-pseudo Hermitian.

2. The Model

The $(2+1)$-dimensional massless Dirac equation is given by:

$$H\psi = E\psi, \quad H = c\boldsymbol{\sigma}\cdot\boldsymbol{P} = c\begin{pmatrix} 0 & P_- \\ P_+ & 0 \end{pmatrix}, \quad \psi = \begin{pmatrix} \psi_1 \\ \psi_2 \end{pmatrix} \tag{1}$$

where c is the velocity of light and:

$$P_{\pm} = (P_x \pm iP_y) = (p_x + A_x) \pm i(p_y + A_y) \qquad (2)$$

In order to solve Equation (1), it is necessary to decouple the spinor components. Applying the operator, H, from the left in Equation (1), we find:

$$c^2 \begin{pmatrix} P_- P_+ & 0 \\ 0 & P_+ P_- \end{pmatrix} \psi = E^2 \psi \qquad (3)$$

Let us now consider the vector potential to be:

$$A_x = 0, \quad A_y = f(x) \qquad (4)$$

so that the magnetic field is given by:

$$B_z(x) = f'(x) \qquad (5)$$

For the above choice of vector potentials, the component wave functions can be taken of the form:

$$\psi_{1,2}(x,y) = e^{ik_y y} \phi_{1,2}(x) \qquad (6)$$

Then, from (3), the equations for the components are found to be (in units of $\hbar = 1$):

$$\left[-\frac{d^2}{dx^2} + W^2(x) + W'(x) \right] \phi_1(x) = \epsilon^2 \phi_1(x)$$

$$\left[-\frac{d^2}{dx^2} + W^2(x) - W'(x) \right] \phi_2(x) = \epsilon^2 \phi_2(x) \qquad (7)$$

where $\epsilon = (E/c)$, and the function, $W(x)$, is given by:

$$W(x) = k_y + f(x) \qquad (8)$$

2.1. Complex Decaying Magnetic Field

It is now necessary to choose the function, $f(x)$. Our first choice for this function is:

$$f(x) = -(A + iB)\, e^{-x}, \quad -\infty < x < \infty \qquad (9)$$

where $A > 0$ and B are constants. This leads to a complex exponentially decaying magnetic field:

$$B_z(x) = (A + iB)e^{-x} \qquad (10)$$

For $B = 0$ or a purely imaginary number (such that $(A + iB) > 0$), the magnetic field is an exponentially decreasing one, and we recover the case considered in [7,8].

Now, from the second of Equation (7), we obtain:

$$\left[-\frac{d^2}{dx^2} + V_2(x) \right] \phi_2 = \left(\epsilon^2 - k_y^2 \right) \phi_2 \qquad (11)$$

where:

$$V_2(x) = k_y^2 + (A + iB)^2\, e^{-2x} - (2k_y + 1)(A + iB)\, e^{-x} \qquad (12)$$

Symmetry **2014**, *6*, 103–110

It is not difficult to recognize $V_2(x)$ in Equation (12) as the complex analogue of the Morse potential whose solutions are well known [22,23]. Using these results, we find:

$$E_{2,n} = \pm c\sqrt{k_y^2 - (k_y - n)^2}$$

$$\phi_{2,n} = t^{k_y - n} e^{-t/2} L_n^{(2k_y - 2n)}(t), \quad n = 0, 1, 2, \dots < [k_y]$$

(13)

where $t = 2(A + iB)e^{-x}$ and $L_n^{(a)}(t)$ denote generalized Laguerre polynomials. The first point to note here is that for the energy levels to be real, it follows from Equation (13) that the corresponding eigenfunctions are normalizable when the condition $k_y \geq 0$ holds. For $k_y < 0$, the wave functions are not normalizable, *i.e.*, no bound states are possible.

Let us now examine the upper component, ϕ_1. Since ϕ_2 is known, one can always use the intertwining relation:

$$c P_- \psi_2 = E \psi_1$$

(14)

to obtain ϕ_1. Nevertheless, for the sake of completeness, we present the explicit results for ϕ_1. In this case, the potential analogous to Equation (12) reads:

$$V_1(x) = k_y^2 + (A + iB)^2 e^{-2x} - (2k_y - 1)(A + iB) e^{-x}$$

(15)

Clearly, $V_1(x)$ can be obtained from $V_2(x)$ by the replacement $k_y \rightarrow k_y - 1$, and so, the solutions can be obtained from Equation (13) as:

$$E_{1,n} = \pm c\sqrt{k_y^2 - (k_y - n - 1)^2}$$

$$\phi_{1,n} = t^{k_y - n - 1} e^{-t/2} L_n^{(2k_y - 2n - 2)}(t), \quad n = 1, 2, \dots < [k_y - 1]$$

(16)

Note that the $n = 0$ state is missing from the spectrum Equation (16), so that it is a singlet state. Furthermore, $E_{2,n+1} = E_{1,n}$, so that the ground state is a singlet, while the excited ones are doubly degenerate. Similarly, the negative energy states are also paired. In this connection, we would like to note that $\{H, \sigma_3\} = 0$, and consequently, except for the ground state, there is particle hole symmetry. The wave functions for the holes are given by $\sigma_3 \psi$. The precise structure of the wave functions of the original Dirac equation are as follows (we present only the positive energy solutions):

$$E_0 = 0, \quad \psi_0 = \begin{pmatrix} 0 \\ \phi_{2,0} \end{pmatrix}$$

$$E_{n+1} = c\sqrt{k_y^2 - (k_y - n - 1)^2}, \quad \psi_{n+1} = \begin{pmatrix} \phi_{1,n} \\ \phi_{2,n+1} \end{pmatrix}, \quad n = 0, 1, 2, \cdots$$

(17)

It is interesting to note that the spectrum does not depend on the magnetic field. Furthermore, the dispersion relation is no longer linear, as it should be in the presence of interactions. It is also easily checked that when the magnetic field is reversed, *i.e.*, $A \rightarrow -A$ and $B \rightarrow -B$ with the simultaneous change of $k_y \rightarrow -k_y$, the two potentials $V_{1,2}(x) = W(x) \pm W'(x)$ go one into each other, $V_1(x) \leftrightarrow V_2(x)$. Therefore, the solutions are correspondingly interchanged, $\phi_{1,n} \leftrightarrow \phi_{2,n}$ and $E_{1,n} \leftrightarrow E_{2,n}$, but retain the same functional form as in Equations (13) and (16).

Therefore, we find that it is indeed possible to create bound states with an imaginary vector potential. We shall now demonstrate the above results for a second example.

2.2. Complex Hyperbolic Magnetic Field

Here, we choose $f(x)$, which leads to an effective potential of the complex hyperbolic Rosen–Morse type:

$$f(x) = A \tanh(x - i\alpha), \quad -\infty < x < \infty, \quad A \text{ and } \alpha \text{ are real constants} \tag{18}$$

In this case, the complex magnetic field is given by:

$$\mathcal{B}_z(x) = A \operatorname{sech}^2(x - i\alpha) \tag{19}$$

Note that for $\alpha = 0$, we get back the results of [8,24]. Using Equation (18) in the second half of Equation (7), we find:

$$\left[-\frac{d^2}{dx^2} + U_2(x) \right] \phi_2 = (\epsilon^2 - k_y^2 - A^2)\phi_2 \tag{20}$$

where

$$U_2(x) = k_y^2 - A(A+1) \operatorname{sech}^2(x - i\alpha) + 2Ak_y \tanh(x - i\alpha) \tag{21}$$

This is the Hyperbolic Rosen–Morse potential with known energy values and eigenfunctions. In the present case, the eigenvalues and the corresponding eigenfunctions are given by [23,25]:

$$E_{2,n} = \pm c\sqrt{A^2 + k_y^2 - (A - n)^2 - \frac{A^2 k_y^2}{(A-n)^2}}, \quad n = 0, 1, 2, \ldots < [A - \sqrt{Ak_y}] \tag{22}$$

$$\phi_{2,n} = (1 - t)^{s_1/2}(1 + t)^{s_2/2} P_n^{(s_1, s_2)}(t)$$

where $P_n^{(a,b)}(z)$ denotes Jacobi polynomials and:

$$t = \tanh x, \quad s_{1,2} = A - n \pm \frac{Ak_y}{A - n} \tag{23}$$

The energy values corresponding to the upper component of the spinor can be found out by replacing A by $(A - 1)$, and ϕ_1 can be found out using relation Equation (14).

3. η-Pseudo Hermiticity

Let us recall that a Hamiltonian is η-pseudo Hermitian if [21]:

$$\eta H \eta^{-1} = H^\dagger \tag{24}$$

where η is a Hermitian operator. It is known that eigenvalues of a η-pseudo Hermitian Hamiltonian are either all real or are complex conjugate pairs [21]. In view of the fact that in the present examples, the eigenvalues are all real, one is tempted to conclude that the interactions are η pseudo Hermitian. To this end, we first consider case 1, and following [26], let us consider the Hermitian operator:

$$\eta = e^{-\theta p_x}, \quad \theta = \arctan \frac{B}{A} \tag{25}$$

Then, it follows that:

$$\eta c \eta^{-1} = c, \quad \eta p_x \eta^{-1} = p_x, \quad \eta V(x) \eta^{-1} = V(x + i\theta) \tag{26}$$

6

We recall that in both the cases considered here, the Hamiltonian is of the form:

$$H = c\boldsymbol{\sigma} \cdot \boldsymbol{P} = c \begin{pmatrix} 0 & P_- \\ P_+ & 0 \end{pmatrix} \tag{27}$$

where, for the first example:

$$P_{\pm} = p_x \pm i p_y \pm i(A + iB)e^{-x} \tag{28}$$

Then:

$$H^{\dagger} = c \begin{pmatrix} 0 & P_+^{\dagger} \\ P_-^{\dagger} & 0 \end{pmatrix} \tag{29}$$

Now, from Equation (28), it follows that:

$$P_+^{\dagger} = p_x - i p_y - i(A - iB)e^{-x}, \quad P_-^{\dagger} = p_x + i p_y + i(A - iB)e^{-x} \tag{30}$$

and using Equation (26), it can be shown that:

$$\eta P_+ \eta^{-1} = p_x + i p_y + i(A - iB)e^{-x} = P_-^{\dagger}, \quad \eta P_- \eta^{-1} = p_x - i p_y - i(A - iB)e^{-x} = P_+^{\dagger} \tag{31}$$

Next, to demonstrate the pseudo Hermiticity of the Dirac Hamiltonian Equation (27), let us consider the operator $\eta' = \eta \cdot \mathcal{I}_2$, where $\mathcal{I}_2$ is the (2×2) unit matrix. Then, it can be shown that:

$$\eta' H \eta'^{-1} = H^{\dagger} \tag{32}$$

Thus, the Dirac Hamiltonian with a complex decaying magnetic field Equation (10) is η-pseudo Hermitian.

For the magnetic field given by Equation (19), the operator, η, can be found by using relations Equation (26). After a straightforward calculation, it can be shown that the η operator is given by:

$$\eta = e^{-2\alpha p_x} \tag{33}$$

so that, in this second example, also, the Dirac Hamiltonian is η-pseudo Hermitian.

4. Conclusions

Here, we have studied the $(2 + 1)$-dimensional massless Dirac equation (we note that if a massive particle of mass m is considered, the energy spectrum in the first example would become $E_n = c\sqrt{k_y^2 + m^2 c^2 - (k_y - n)^2}$. Similar changes will occur in the second example, too). in the presence of complex magnetic fields, and it has been shown that such magnetic fields can create bound states. It has also been shown that Dirac Hamiltonians in the presence of such magnetic fields are η-pseudo Hermitian. We feel it would be of interest to study the generation of bound states using other types of magnetic fields, e.g., periodic magnetic fields.

Acknowledgments: One of us (P. R.) wishes to thank INFN Sezione di Perugia for supporting a visit during which part of this work was carried out. He would also like to thank the Physics Department of the University of Perugia for its hospitality.

Conflicts of Interest: The authors declare no conflict of interest.

References

1. Novoselov, K.S.; Geim, A.K.; Morozov, S.V.; Jiang, D.; Zhang, Y.; Dubonos, S.V.; Grigorieva, I.V.; Firsov, A.A. Electric field effect in atomically thin carbon films. *Science* **2004**, *306*, 666–669.
2. Novoselov, K.S.; Geim, A.K.; Morozov, S.V.; Jiang, D.; Katsnelson, M.I.; Grigorieva, I.V.; Dubonos, S.V.; Firsov, A.A. Two-dimensional gas of massless Dirac fermions in graphene. *Nature* **2005**, *438*, 197–200.

3. De Martino, A.; Dell'Anna, L.; Egger, R. Magnetic confinement of massless dirac fermions in graphene. *Phys. Rev. Lett.* **2007**, *98*, 066802:1–066802:4.

4. De Martino, A.; Dell'Anna L.; Eggert, R. Magnetic barriers and confinement of Dirac-Weyl quasiparticles in graphene. *Solid State Commun.* **2007**, *144*, 547–550.

5. Dell'Anna, L.; de Martino, A. Multiple magnetic barriers in graphene. *Phys. Rev. B* **2009**, *79*, 045420:1–045420:9.

6. Wang, D.; Jin, G. Bound states of Dirac electrons in a graphene-based magnetic quantum dot. *Phys. Lett. A* **2009**, *373*, 4082–4085.

7. Ghosh, T.K. Exact solutions for a Dirac electron in an exponentially decaying magnetic field. *J. Phys. Condens. Matter* **2009**, *21*, doi:10.1088/0953-8984/21/4/045505.

8. Kuru, S; Negro, J.M.; Nieto, L.M. Exact analytic solutions for a Dirac electron moving in graphene under magnetic fields. *J. Phys. Condens. Matter* **2009**, *21*, doi:10.1088/0953-8984/21/45/455305.

9. Bender, C.M.; Boettcher, S. Real spectra in non-hermitian hamiltonians having PT symmetry. *Phys. Rev. Lett.* **1988**, *80*, 5243–5246.

10. Fagotti, M; Bonati, C.; Logoteta, D.; Marconcini, P.; Macucci, M. Armchair graphene nanoribbons: PT-symmetry breaking and exceptional points without dissipation. *Phys. Rev. B* **2011**, *83*, 241406:1–241406:4.

11. Szameit, A.; Rechtsman, M.C.; Bahat-Treidel, O.; Segev, M. PT-Symmetry in heoneycomeb photonic lattices. *Phys. Rev. A* **2011**, *84*, 021806(R):1–021806(R):5.

12. Esaki, K.; Sato, M.; Hasebe, K.; Kohmoto, M. Edge states and topological phases in non-Hermitian systems. *Phys. Rev. B* **2011**, *84*, 205128:1–205128:19.

13. Longhi, S. Classical simulation of relativistic quantum mechanics in periodic optical structures. *Appl. Phys. B* **2011**, *104*, 453–468.

14. Longhi, S. Optical realization of relativistic non-hermitian quantum mechanics. *Phys. Rev. Lett.* **2010**, *105*, 013903:1–013903:4.

15. Ramezani, H.; Kottos, T.; Kovanis, V.; Christodoulides, D.N. Exceptional-point dynamics in photoni honeycomb lattices with PT-symmetry. *Phys. Rev. A* **2012**, *85*, 013818:1–013818:6.

16. Mandal, B.P.; Gupta, S. Pseudo-hermitian interactions in Dirac theory: Examples. *Mod. Phys. Lett. A* **2010**, *25*, 1723–1732.

17. Hatano, N.; Nelson, D. Localization transitions in non-hermitian quantum mechanics. *Phys. Rev. Lett.* **1996**, *77*, 570–573.

18. Feinberg, J.; Zee, A. Non-Hermitian localization and delocalization. *Phys. Rev. E* **1999**, *59*, 6433–6443.

19. Mandal, B.P.; Mourya, B.K.; Yadav, R.K. PT phase transition in higher-dimensional quantum systems. *Phys. Lett. A* **2013**, *377*, 1043–1046.

20. Tan, L.Z.; Park, C.-H.; Louie, S.G. Graphene Dirac fermions in one dimensional field profiles; Tansforming magnetic to electric field. *Phys. Rev. B* **2010**, *81*, 195426:1–195426:8.

21. Mostafazadeh, A. Pseudo-hermiticity versus PT-symmetry III: Equivalence of pseudo-Hermiticity and the presence of antilinear symmetries. *J. Math. Phys.* **2002**, *43*, 3944–3951.

22. Flügge, S. *Practical Quantum Mechanics*; Springer-Verlag: Berlin, Germany, 1974.

23. Cooper, F.; Khare, A; Sukhatme, U. *Supersymmetry in Quantum Mechanics*; World Scientific Publishing Co. Pte. Ltd.: Singapore, 2001.

24. Milpas, E.; Torres, M.; Murguía, G. Magnetic field barriers in graphene: An analytically solvable model. *J. Phys. Condens. Matter* **2011**, *23*, 245304:1–245304:7.

25. Rosen, N.; Morse, P.M. On the vibrations of polyatomic molecules. *Phys. Rev.* **1932**, *42*, 210–217.

26. Ahmed, Z. Pseudo-hermiticity of hamiltonians under imaginary shift of the coordinate: Real spectrum of complex potentials. *Phys. Lett. A* **2001**, *290*, 19–22.

© 2014 by the authors. Licensee MDPI, Basel, Switzerland. This article is an open access article distributed under the terms and conditions of the Creative Commons Attribution (CC BY) license (http://creativecommons.org/licenses/by/4.0/).

symmetry

MDPI

Article

Spacetime Metrics from Gauge Potentials

Ettore Minguzzi

Dipartimento di Matematica e Informatica "U. Dini", Università degli Studi di Firenze, Via S. Marta 3, I-50139 Firenze, Italy; E-Mail: ettore.minguzzi@unifi.it; Tel./Fax: +39-055-4796-253

Received: 27 January 2014; in revised form: 21 March 2014 / Accepted: 24 March 2014 / Published: 27 March 2014

Abstract: I present an approach to gravity in which the spacetime metric is constructed from a non-Abelian gauge potential with values in the Lie algebra of the group $U(2)$ (or the Lie algebra of quaternions). If the curvature of this potential vanishes, the metric reduces to a canonical curved background form reminiscent of the Friedmann S^3 cosmological metric.

Keywords: gauge theory; G-structure; teleparallel theory

1. Introduction

The observational evidence in favor of Einstein's general theory of relativity has clarified that the spacetime manifold is not flat, and hence that it can be approximated by the flat Minkowski spacetime only over limited regions. Quantum Field Theory, and in particular the perturbative approach through the Feynman's integral, has shown the importance of expanding near a "classical" background configuration. Although we do not have at our disposal a quantum theory of gravity, it would be natural to take a background configuration which approximates as much as possible the homogeneous curved background that is expected to take place over cosmological scales accordingly to the cosmological principle. Therefore, it is somewhat surprising that most classical approaches to quantum gravity start from a perturbation of Minkowski's metric in the form $g_{\mu\nu} = \eta_{\mu\nu} + h_{\mu\nu}$. This approach is ill defined in general unless the manifold is asymptotically flat. Indeed, the expansion depends on the chosen coordinate system, a fact which is at odds with the principle of general covariance.

Expanding over the flat metric is like Taylor expanding a function by taking the first linear approximation near a point. It is clear that the approximation cannot be good far from the point and that no firm global conclusion can be drawn from similar approaches. A good global expansion should be performed in a different way, taking into account the domain of definition of the function. So, a function defined over an interval would be better approximated with a Fourier series than with a Taylor expansion. Despite of these simple analogies, much research has been devoted to quantum gravity by means of expansions of the form $g = \eta + h$, possibly because of the lack of alternatives.

Actually, some years ago [1] I proposed a gauge approach to gravity that solves this problem in a quite simple way and which, I believe, deserves to be better known.

To start with let us observe that general relativity seems to privilege in its very formalism the flat background. Indeed, the Riemann curvature $\mathcal{R}$ measures the extent by which the spacetime is far from flat, namely far from the background

$$\mathcal{R} = 0 \quad \Leftrightarrow \quad (M, g) \text{ is flat.}$$

If the true background is not the flat Minkowski space then as a first step one would have to construct a different curvature F with the property that

$$F = 0 \Leftrightarrow (M, g) \text{ takes the canonical background shape.}$$

It is indeed possible to accomplish this result. Let us first introduce some notations.

Symmetry **2014**, *6*, 164–170

2. Some Notations from Gauge Theory

Gauge theories were axiomatized in the fifties by Ehresmann [2] as connections over principal bundles. Since I need to fix the notation, here I shortly review that setting. A principal bundle is given by a differentiable manifold (the bundle) P, a differentiable manifold (the base) M, a projection

$$\pi : P \to M \tag{1}$$

a Lie group G, and a right action of G on P

$$p \to pg \qquad p \in P, \quad g \in G \tag{2}$$

such that $M = P/G$, *i.e.*, M is the orbit space. Moreover, the fiber bundle P is locally the product $P = M \times G$. To be more precise, given a point $m \in M$ there is an open set U of m, such that $\pi^{-1}(U)$ is diffeomorphic to $U \times G$ and the diffeomorphism preserves the right action. If this property holds also globally the principal bundle is called trivial. The set $\pi^{-1}(m)$ is the fiber of m and it is diffeomorphic to G. Let $\mathcal{G}$ be the Lie algebra of G, and let τ_a be a base of generators

$$[\tau_a, \tau_b] = f_{ab}^c \tau_c \tag{3}$$

Let $p \in P$ be a point of the principal bundle; it can be considered as an application $p : G \to P$ which acts as $g \to pg$. The fundamental fields (We follow mostly the conventions of Kobayashi-Nomizu. The upper star * indicates the pull-back when applied to a function, the fundamental field when applied to a generator, and the horizontal lift when applied to a curve or a tangent vector on the base.) τ_a^* over P are defined in p as the push-forward of the group generators: $\tau_a^* = p_*\tau_a$. They are vertical fields in the sense that they are in the ker of π: $\pi_*(\tau_a^*) = 0$. They form a base of the vertical tangent space at p.

A connection over P is a 1-form $\omega : P \to \mathcal{G}$ with the following properties

(a) $\omega(X^*) = X \qquad X \in \mathcal{G}$
(b) $R_g^*\omega = g^{-1}\omega g$

The tangent space at p is split into the sum of two subspaces: the vertical space, that is the ker of π, and the horizontal space, that is the ker of ω

$$T_pP = H_p \oplus V_p \tag{4}$$

Let U be an open set of M. A section σ is a function $\sigma : U \to \pi^{-1}(U)$ such that $\pi \circ \sigma = I_U$. The gauge potential depends on the section and is defined by

$$A = \tau_a A_\mu^a dx^\mu = \sigma^*\omega \tag{5}$$

where $\{x^\mu\}$ are coordinates on the base. A change of section is sometimes called gauge transformation. The curvature is defined by (The exterior product is defined through $\alpha \wedge \beta = \alpha \otimes \beta - \beta \otimes \alpha$ where α and β are 1-forms. As a consequence $\omega \wedge \omega = [\omega, \omega]$)

$$\Omega = d\omega h = d\omega + \omega \wedge \omega \tag{6}$$

where h projects the vector arguments to the horizontal space [2]. The field strength is defined by $F = \tau_a F_{\mu\nu}^a dx^\mu dx^\nu = \sigma^*\Omega$. In other words

$$F_{\mu\nu}^a = \partial_\mu A_\nu^a - \partial_\nu A_\mu^a + f_{bc}^a A_\mu^b A_\nu^c \tag{7}$$

Given a section one can construct a system of coordinates over P in a canonical way. Simply let (x, g) be the coordinates of the point $p = \sigma(x)g$. In this coordinates the connection can be rewritten

$$\omega = g^{-1}dg + g^{-1}Ag \tag{8}$$

and the curvature can be rewritten

$$\Omega = g^{-1}Fg \tag{9}$$

indeed the form of the connection given here satisfies both the requirements above and $A = \sigma^*\omega$. From these last equations one easily recovers the gauge transformation rules after a change of section $\sigma' = \sigma u(x)$ $(g' = u^{-1}(x)g)$, that is

$$A'_\mu = u^{-1}A_\mu u + u^{-1}\partial_\mu u \tag{10}$$

$$F'_{\mu\nu} = u^{-1}F_{\mu\nu}u \tag{11}$$

3. The Background Metric

We are used to define a manifold through charts $\phi : U \to \mathbb{R}^4$, $U \subset M$, taking values on $\mathbb{R}^4$. Let us instead take them with value in a four-dimensional canonical manifold with enough structure to admit some natural metric. We shall use a matrix Lie group G, but we do not really want to give any special role to the identity of G. We shall see later how to solve this problem. The metric g has to be constructed as a small departure from that naturally present in G and which plays the role of background metric.

We take as background metric the expression

$$g_B = I_g(\theta, \theta) \tag{12}$$

where θ is the Maurer-Cartan form of the group [2], that is $\theta = g^{-1}dg$, and I_g is an adjoint invariant quadratic form on the Lie algebra $\mathcal{G}$, which might depend on $g \in G$. The Maurer-Cartan form has the effect of mapping an element $v \in T_gG$ to the Lie algebra element whose fundamental vector field at g is v.

Of course, we demand that g_B be a Lorentzian metric in a four-dimensional Lie group, and furthermore we want it to represent an isotropic cosmological background, thus G has to contain the $SO(3)$ subgroup. We are lead to the Abelian group of translations T_4 or to the group $U(2)$ (or equivalently the group of quaternions since it shares with $U(2)$ the Lie algebra). In what follows we shall only consider the latter group, the case of the Abelian translation group being simpler.

Thus let us consider the group $U(2)$. Every matrix of this group reads $u = e^{i\lambda}r$ with $0 \le \lambda \le \pi$ where $r \in SU(2)$ (while a quaternion reads $e^\lambda r$, $\lambda \in \mathbb{R}$)

$$r = \begin{pmatrix} r_0 + ir_3 & r_2 + ir_1 \\ -r_2 + ir_1 & r_0 - ir_3 \end{pmatrix}, \qquad \sum_{\mu=0}^{3} r_\mu^2 = 1 \tag{13}$$

The Lie algebra of $U(2)$ is that of anti-hermitian matrices A which read

$$A = i \begin{pmatrix} a^0 + a^3 & a^1 - ia^2 \\ a^1 + ia^2 & a^0 - a^3 \end{pmatrix} \tag{14}$$

By adjoint invariance of I_g we mean $I_{u'gu'^\dagger}(uAu^\dagger, uAu^\dagger) = I_g(A, A)$, for any $u, u' \in U(2)$. Clearly, the adjoint invariance for the Abelian subgroup $U(1)$ is guaranteed because for $u \in U(1)$, $uAu^\dagger = A$, $u'gu'^\dagger = g$. The expressions that satisfy this invariance property are

$$I_g(A, A) = \frac{\alpha(\lambda)}{2}(\operatorname{tr} A)^2 - \frac{\beta(\lambda)}{2}\operatorname{tr}(A^2) \tag{15}$$

$$I_g(A, A) = -2\alpha(\lambda)(a^0)^2 + \beta(\lambda)[(a^0)^2 + (a^1)^2 + (a^2)^2 + (a^3)^2] \tag{16}$$

where α and β are functions of the phase of $g = e^{i\lambda}r$, $r \in SU(2)$ (which is left invariant under adjoint transformations). We get a Lorentzian metric for $2\alpha > \beta$ and $\beta > 0$. With the simple choice $\alpha = \beta = 1$ we get

$$I_g(A, A) = \det A = -(a^0)^2 + (a^1)^2 + (a^2)^2 + (a^3)^2 \tag{17}$$

Notice that $\mathrm{tr}(r^\dagger dr) = 0$ and

$$\mathrm{tr}(r^\dagger dr\, r^\dagger dr) = -\mathrm{tr}(dr^\dagger dr) = -2\det(r^\dagger dr) = -2\sum_{\mu=0}^{3} dr_\mu^2 \tag{18}$$

Let us recall that $\theta = \phi^\dagger d\phi$ where the group element ϕ reads $\phi = re^{i\lambda}$. Thus using $\mathrm{tr}(r^\dagger dr) = 0$ we find for the background metric

$$\begin{aligned}
g_B = I_g(\theta, \theta) &= I\left(r^\dagger dr + id\lambda, r^\dagger dr + id\lambda\right) = \\
&= -I(d\lambda, d\lambda) + I(r^\dagger dr, r^\dagger dr) = -(2\alpha - \beta)d\lambda^2 - \frac{\beta}{2}\mathrm{tr}(r^\dagger dr\, r^\dagger dr) = \\
&= -(2\alpha - \beta)d\lambda^2 + \beta(dr_0^2 + dr_1^2 + dr_2^2 + dr_3^2)
\end{aligned}$$

Recalling the constraint $\sum_{\mu=0}^{3} r_\mu^2 = 1$ we find a background metric which coincides with Friedmann's with a S^3 section.

More specifically, let $\sigma_0 = I$, and let σ_i, $i = 1, 2, 3$, be the Pauli matrices. Let $\tau_\mu = i\sigma_\mu$ be a base for the Lie algebra of $U(2)$. Let us parametrize $\phi \in U(2)$ through

$$\phi = e^{i\lambda\sigma_0}r = e^{i\lambda\sigma_0}e^{i\chi(\tau_1 \sin\theta\cos\varphi + \tau_2 \sin\theta\sin\varphi + \tau_3 \cos\theta)} \tag{19}$$

then the background metric reads

$$g_B = -dt^2 + a^2(t)\left(d\chi^2 + \sin^2\chi(d\theta^2 + \sin^2\theta\, d\varphi^2)\right) \tag{20}$$

where

$$t = \int_0^\lambda d\lambda' \sqrt{2\alpha(\lambda') - \beta(\lambda')} \tag{21}$$

and

$$a^2(t) = \beta(\lambda(t)) \tag{22}$$

These calculations, first presented in [1], show that the Friedmann metric appears rather naturally from the study of the $U(2)$ group. Of course, since this argument depends only on the Lie algebra rather that the group structure, it can be repeated for the group of quaternions [3].

4. Perturbing the Background

In this section we shall suppose that I_g does not depend on g, namely that α and β are constants, this means that we ignore the time dependence of the cosmological background.

We mentioned that we wish to use charts $\phi: U \to G$, $U \subset M$, with value in a group G but that we do not want to assign to the identity of G any special role. To that end, let us assume for simplicity that M is simply connected, and let us introduce a trivial bundle P endowed with a flat connection $\tilde\omega$. The connection being flat is integrable, thus given an horizontal section $\tilde\sigma: M \to P$, and parametrizing every point of P through $p(x, g) = \tilde\sigma(x)g$, we obtain a splitting $P \sim M \times G$. In this way the identity of G does not play any special role since it refers to different points of P depending on the choice of section $\tilde\sigma$.

A second section $\sigma\colon M \to P$ is now related to the former by $\sigma(x)\phi^{-1}(x) = \tilde{\sigma}(x)$, where $\phi\colon M \to G$ is the chart we were looking for. In order to be interpreted as a chart, ϕ has to be injective. The idea is to define the metric

$$g = I(\tilde{A} - A, \tilde{A} - A)$$

where $\tilde{A} = \sigma^*\tilde{\omega}$ is the potential of the flat connection and $A = \sigma^*\omega$ is the potential of a possibly non-trivial connection. From the transformation rule for the potential (10) we obtain

$$\tilde{A} = \phi^{-1}(x)\,\mathrm{d}\phi(x)$$

Let us show that the metric so defined satisfies the property $F = 0 \Rightarrow$ background metric. Suppose that $F = 0$ then σ can be chosen in such a way that $A = 0$, thus the metric becomes

$$F = 0 \quad \Rightarrow \quad g = I(\phi^{-1}(x)\mathrm{d}\phi(x), \phi^{-1}(x)\mathrm{d}\phi(x)) = I(\phi^*\theta, \phi^*\theta) = \phi^* g_B \tag{23}$$

that is, up to a coordinate change the metric coincides with the background metric.

We observe that $A = \tau_a A^a_\mu \mathrm{d}x^\mu$ has 16 components, namely the same number of components as the metric. However, we have an additional degree of freedom given by $\phi(x)$. This function can be completely removed using the invertibility of this map, namely using the coordinates ϕ^μ on the Lie group to parametrize M. In this way the metric reads

$$g = I(\phi^{-1}\mathrm{d}\phi - \tau_a A^a_\mu(\phi)\mathrm{d}\phi^\mu, \phi^{-1}\mathrm{d}\phi - \tau_a A^a_\mu(\phi)\mathrm{d}\phi^\mu)$$

these coordinates are referred as *internal coordinates*. In internal coordinates any gauge transformation induces a coordinate transformation. For instance, the gauge potential transforms as

$$\tau_a A'^a_c = \{u^{-1}\tau_a A^a_b u + u^{-1}\partial_b u\}\frac{\partial\phi^b}{\partial\phi'^c} \tag{24}$$

and the transformation law for the curvature becomes

$$F'_{ab} = u^{-1}F_{cd}\,u\,\frac{\partial\phi^c}{\partial\phi'^a}\frac{\partial\phi^d}{\partial\phi'^b} \tag{25}$$

where $\sigma' = \sigma u$ and the matrix $u(\phi)$ is related to the transformation $\phi'^a(\phi^b)$ by the product $\phi' = \phi\,u(\phi)$. In the same way it can be shown, for example, that the spacetime metric transforms as a tensor under (24).

One can further ask whether the Einstein equations can be rephrased as dynamical equations for the potential A. The answer is affirmative and passes through the vierbein reformulation of the Einstein-Hilbert Lagrangian.

We recall that a tetrad field (vierbein) $e_a = e^\mu_a\partial_\mu$, is a set of four vector fields e_a such that $g_{\mu\nu} = \eta_{ab}e^a_\mu e^b_\nu$. The inverse e^a_μ is defined through $e^\mu_a e^a_\nu = \delta^\mu_\nu$. The Einstein Lagrangian can be rewritten

$$-\frac{\sqrt{-g}}{16\pi}R = \frac{1}{8\pi}v^\nu_{,\nu} + \frac{\sqrt{-g}}{16\pi}\left\{\frac{1}{4}C^{abc}C_{abc} - C^a_{ac}C^{b\ c}_{\ b} + \frac{1}{2}C^{abc}C_{bac}\right\} \tag{26}$$

where the first term on the right-hand side is a total divergence and

$$C^c_{ab} = e^c_\alpha(\partial_a e^\alpha_b - \partial_b e^\alpha_a) = e^\mu_a e^\nu_b(\partial_\nu e^c_\mu - \partial_\mu e^c_\nu) \tag{27}$$

In order to obtain a dynamics for A we select a base τ_a for the Lie algebra such that

$$I(\tau_a, \tau_b) = \eta_{ab}$$

13

Symmetry **2014**, *6*, 164–170

where η_{ab} is the Minkowski metric. Then we make a gauge transformation so as to send the flat potential $\tilde{A}$ to zero. This gauge is called the *OT gauge*. Since $g = I(\tau_a A_\mu^a dx^\mu, \tau_a A_\mu^a dx^\mu)$, the vierbein becomes coincident with the potential

$$e_\mu^a = A_\mu^a$$

so the field equations can be ultimately expressed in terms of A_μ^a. We have observed above that with $F = 0$ the metric becomes that of the Einstein static Universe which is not a solution of the dynamical equations (without cosmological constant). One could wish to obtain a realistic cosmological solution for $F = 0$. At the moment I do not known how to modify the theory so as to accomplish this result (but observe that we never changed the dynamics which is always that given by the Einstein's equations). However, our framework might not need any modification. It can be shown [1] that the scale factor a in front of the Einstein static Universe metric is actually the coupling constant for this theory so the expansion of the Universe could be an effect related to the renormalization of the theory.

In the Abelian case T_4 (not in the $U(2)$ case) the Lagrangian can also be expressed in terms of the curvature (7). Indeed, since $f_{ab}^c = 0$ the curvature becomes coincident with the tensors C_{bc}^a entering the above expression of the Lagrangian (however, observe that the potential still enters the metric and the vierbeins which are used to raise the indices of the curvature). The final expression is quadratic in the curvature F and is related to the teleparallel approach to general relativity [4–7]. Issues related to the renormalizability of the dynamics determined by (26) have yet to be fully studied.

The *OT gauge* approach has been used to infer the dynamics and is complementary to the *internal coordinates* approach mentioned above. Indeed, while the latter allows us to interpret the map $\phi : U \to G, U \subset M$, as a chart with values in G, the *OT frame* approach sends ϕ to the identity, so in the new gauge the non-injective map ϕ cannot be interpreted as a chart. Thus, after having developed the dynamics in the *OT gauge* we would have to make a last gauge transformation to reformulate it in internal coordinates.

Acknowledgments: This work has been partially supported by GNFM of INDAM.

Conflicts of Interest: The author declares no conflicts of interest.

References

1. Minguzzi, E. Gauge invariance in teleparallel gravity theories: A solution to the background structure problem. *Phys. Rev. D* **2002**, *65*, 084048. doi:10.1103/PhysRevD.65.084048.
2. Kobayashi, S.; Nomizu, K. Foundations of Differential Geometry. In *Interscience Tracts in Pure and Applied Mathematics*; Interscience Publishers: New York, NY, USA, 1963; Volume I.
3. Trifonov, V. Natural Geometry of Nonzero Quaternions. *Int. J. Theor. Phys.* **2007**, *46*, 251–257.
4. Cho, Y.M. Einstein Lagrangian as the translational Yang-Mills Lagrangian. *Phys. Rev. D* **1976**, *14*, 2521–2525.
5. Hayashi, K.; Shirafuji, T. New general relativity. *Phys. Rev. D* **1979**, *19*, 3524–3553.
6. Rodrigues, W.A., Jr.; de Souza, Q.A.G.; da Rocha, R. Conservation Laws on Riemann-Cartan, Lorentzian and Teleparallel Spacetimes. *Bull. Soc. Sci. Lett. Lodz. Ser. Rech. Deform.* **2007**, *52*, 37–65, 66–77.
7. Aldrovandi, R.; Pereira, J.G. Teleparallel Gravity. In *Fundamental Theories of Physics*; Springer: Berlin, Germany, 2013; Volume 173.

© 2014 by the author. Licensee MDPI, Basel, Switzerland. This article is an open access article distributed under the terms and conditions of the Creative Commons Attribution (CC BY) license (http://creativecommons.org/licenses/by/4.0/).

symmetry

MDPI

Review

Quantum Local Symmetry of the *D*-Dimensional Non-Linear Sigma Model: A Functional Approach

Andrea Quadri [1,2]

1   Istituto Nazionale di Fisica Nucleare (INFN), Sezione di Milano, via Celoria 16, I-20133 Milano, Italy;
     E-Mail: andrea.quadri@mi.infn.it; Tel.: +39-2-5031-7287; Fax: +39-2-5031-7480
2   Dipartimento di Fisica, Università di Milano, via Celoria 16, I-20133 Milano, Italy

Received: 27 February 2014; in revised form: 31 March 2014 / Accepted: 11 April 2014 /
Published: 17 April 2014

Abstract: We summarize recent progress on the symmetric subtraction of the Non-Linear Sigma Model in *D* dimensions, based on the validity of a certain Local Functional Equation (LFE) encoding the invariance of the SU(2) Haar measure under local left transformations. The deformation of the classical non-linearly realized symmetry at the quantum level is analyzed by cohomological tools. It is shown that all the divergences of the one-particle irreducible (1-PI) amplitudes (both on-shell and off-shell) can be classified according to the solutions of the LFE. Applications to the non-linearly realized Yang-Mills theory and to the electroweak theory, which is directly relevant to the model-independent analysis of LHC data, are briefly addressed.

Keywords: Non-Linear   Sigma   Model;   quantum   symmetries;   renormalization;
Becchi–Rouet–Stora–Tyutin (BRST)

---

1. Introduction

   The purpose of this paper is to provide an introduction to the recent advances in the study of the renormalization properties of the SU(2) Non-Linear Sigma Model (NLSM) and of the quantum deformation of the underlying non-linearly realized classical SU(2) local symmetry. The results reviewed here are based mainly on References [1–19].

   The linear sigma model was originally proposed a long time ago in [20] in the context of elementary particle physics. In this model the pseudoscalar pion fields $\vec{\phi}$ form a chiral multiplet together with a scalar field σ, with $(\sigma, \vec{\phi})$ transforming linearly as a vector under O(4) ~ SU(2) × SU(2)/Z_2. If one considers instead the model on the manifold defined by

$$\sigma^2 + \vec{\phi}^2 = f_\pi^2, \quad \sigma > 0 \tag{1}$$

one obtains a theory where the chiral group SO(4) ~ SU(2) × SU(2) (with SO(4) selected by the positivity condition on σ) is spontaneously broken down to the isotopic spin group SU(2). The composite field σ has a non-vanishing expectation value f_π (to be identified with the pion decay constant), while the pions are massless. Despite the fact that this is only an approximate description (since in reality the pions are massive and chiral SU(2) × SU(2) is not exact, even before being spontaneously broken), the approach turned out to be phenomenologically quite successful and paved the way to the systematic use of effective field theories as a low energy expansion.

   The first step in this direction was to obtain a phenomenological lagrangian directly, by making use of a pion field with non-linear transformation properties dictated by chiral symmetry from the beginning. After the seminal work of Reference [21] for the chiral SU(2) × SU(2) group, non-linearly realized symmetries were soon generalized to arbitrary groups in [22,23] and have since then become a very popular tool [24].

Symmetry **2014**, *6*, 234–255

Modern applications involve, e.g., Chiral Perturbation Theory [25–28], low energy electroweak theories [29] as well as gravity [30].

Effective field theories usually exhibit an infinite number of interaction terms, that can be organized according to the increasing number of derivatives. By dimensional arguments, the interaction terms must then be suppressed by some large mass scale M (so that one expects that the theory is reliable at energies well below M) (For a modern introduction to the problem, see e.g., [31]). In the spirit of the phenomenological lagrangians, the tree-level effective action is used to compute physical quantities up to a given order in the momentum expansion. Only a finite number of derivative interaction vertices contribute to that order, thus allowing to express the physical observables one is interested in through a finite number of parameters (to be eventually fixed by comparison with experimental data). Then the theory can be used to make predictions at the given order of accuracy in the low-energy expansion.

The problem of the mathematically consistent evaluation of quantum corrections in this class of models has a very long history. On general grounds, the derivative couplings tend to worsen the ultraviolet (UV) behavior of the theory, since UV divergent contributions arise in the Feynman amplitudes that cannot be compensated by a multiplicative renormalization of the fields and a redefinition of the mass parameters and the coupling constants in the classical action (truncated at some given order in the momentum expansion). Under these circumstances, one says that the theory is non-renormalizable (A compact introduction to renormalization theory is given in [32]).

It should be stressed that the key point here is the instability of the classical action: no matter how many terms are kept in the derivative expansion of the tree-level action, there exists a sufficiently high loop order where UV divergences appear that cannot be reabsorbed into the classical action. On the other hand, if in a non-anomalous and non-renormalizable gauge theory one allows for *infinitely many* terms in the classical action (all those compatible with the symmetries of the theory), then UV divergences can indeed be reabsorbed by preserving the Batalin-Vilkovisky master equation [33] and the model is said to be renormalizable in the modern sense [34].

Sometimes symmetries are so powerful in constraining the UV divergences that the non-linear theory proves to be indeed renormalizable (although not by power-counting), like for instance the NLSM in two dimensions [35,36] (For a more recent introduction to the subject, see e.g., [37]).

In four dimensions the situation is much less favorable. It has been found many years ago that already at one loop level in the four-dimensional NLSM there exists an infinite number of one-particle irreducible (1-PI) divergent pion amplitudes. Many attempts were then made in the literature in order to classify such divergent terms. Global SU(2) chiral symmetry is not preserved already at one loop level [38–40]. Moreover it turns out that some of the non-symmetric terms can be reabsorbed by a redefinition of the fields [40–43], however in the off-shell four-point ϕ_a amplitudes some divergent parts arise that cannot be reabsorbed by field redefinitions unless derivatives are allowed [40]. These technical difficulties prevented such attempts to evolve into a mathematically consistent subtraction procedure.

More recently it has been pointed out [1] that one can get the full control on the ultraviolet divergences of the ϕ's-amplitudes by exploiting the constraints stemming from the presence of a certain local symmetry, associated with the introduction of a SU(2) background field connection into the theory. This symmetry in encoded in functional form in the so-called Local Functional Equation (LFE) [1]. It turns out that the fundamental divergent amplitudes are not those associated with the quantum fields of the theory, namely the pions, but those corresponding to the background connection and to the composite operator implementing the non-linear constraint [1,2]. These amplitudes are named ancestor amplitudes.

At every order in the loop expansion there is only a finite number of divergent ancestor amplitudes. They uniquely fix the divergent amplitudes involving the pions. Moreover, non-renormalizability of this theory in four dimensions can be traced back to the instability of the classical non-linear local symmetry, that gets deformed by quantum corrections. These results hold for the full off-shell amplitudes [3].

A comment is in order here. In Reference [4] it has been argued that Minimal Subtraction is a symmetric scheme, fulfilling all the symmetries of the NLSM in the LFE approach. This in particular entails that all finite parts of the needed higher order counterterms are consistently set to zero. It should be stressed that this is not the most general solution compatible with the symmetries and the WPC, that is commonly used in the spirit of the most popular effective field theory point of view. Indeed, these finite parts are constrained neither by the LFE nor by the WPC and thus, mathematically, they can be freely chosen, as far as they are introduced at the order prescribed by the WPC and without violating the LFE.

The four dimensional SU(2) NLSM provides a relatively simple playground where to test the approach based on the LFE, that can be further generalized to the SU(N) case (and possibly even to a more general Lie group).

Moreover, when the background vector field becomes dynamical, the SU(2) NLSM action allows one to generate a mass term for the gauge field *à la* Stückelberg [44,45]. The resulting non-linear implementation of the spontaneous symmetry breaking mechanism (as opposed to the linear Higgs mechanism) is widely used in the context of electroweak low energy effective field theories, that are a very important tool in the model-independent analysis of LHC data [46–49].

2. The Classical Non-Linear Sigma Model

The classical SU(2) NLSM in D dimensions is defined by the action

$$S_0 = \int d^D x \, \frac{m_D^2}{4} \text{Tr} \left(\partial_\mu \Omega^\dagger \partial^\mu \Omega \right) \tag{2}$$

where the matrix Ω is a SU(2) group element given by

$$\Omega = \frac{1}{m_D} \left(\phi_0 + i \phi_a \tau_a \right), \quad \Omega^\dagger \Omega = 1, \quad \det \Omega = 1, \quad \phi_0^2 + \phi_a^2 = m_D^2 \tag{3}$$

In the above equation τ_a, $a = 1, 2, 3$ are the Pauli matrices and $m_D = m^{D/2-1}$ is the mass scale of the theory. m has mass dimension 1. ϕ_a are the three independent fields parameterizing the matrix Ω, while we choose the positive solution of the non-linear constraint, yielding

$$\phi_0 = \sqrt{m_D^2 - \phi_a^2} \tag{4}$$

In components one finds

$$S_0 = \int d^D x \left(\frac{1}{2} \partial_\mu \phi_a \partial^\mu \phi_a + \frac{1}{2} \frac{\phi_a \partial_\mu \phi_a \phi_b \partial^\mu \phi_b}{\phi_0^2} \right) \tag{5}$$

The model therefore contains non-polynomial, derivative interactions for the massless scalars ϕ_a.

Equation (2) is invariant under a *global* $SU(2)_L \times SU(2)_R$ chiral transformation

$$\Omega' = U \Omega V^\dagger, \quad U \in SU(2)_L, \quad V \in SU(2)_R \tag{6}$$

We notice that such a global transformation is non-linearly realized, as can be easily seen by looking at its infinitesimal version. E.g., for the left transformation one finds:

$$\delta \phi_a = \frac{1}{2} \alpha \phi_0(x) + \frac{1}{2} \epsilon_{abc} \phi_b(x) \alpha_c, \quad \delta \phi_0(x) = -\frac{1}{2} \alpha \phi_a(x) \tag{7}$$

Since ϕ_0 is given by Equation (4), the first term in the r.h.s. of $\delta \phi_a$ is non-linear (and even non-polynomial) in the quantum fields.

Perturbative quantization of the NLSM requires to carry out the path-integral

$$Z[J] = \int \mathcal{D}\phi_a \, \exp\left(iS_0[\phi_a] + i\int d^D x \, J_a \phi_a\right) \tag{8}$$

by expanding around the free theory and by treating the second term in the r.h.s. of Equation (5) as an interaction. Notice that in Equation (8) the sources J_a are coupled to the fields ϕ_a over which the path-integral is performed. In momentum space the propagator for the ϕ_a fields is

$$\Delta_{\phi_a\phi_b} = i\frac{\delta_{ab}}{p^2} \tag{9}$$

The mass dimension of the ϕ_a is therefore $D/2 - 1$, in agreement with Equation (3).

The presence of two derivatives in the interaction term is the cause (in dimensions greater than 2) of severe UV divergences, leading to the non-renormalizability of the theory.

3. The Approach based on the Local Functional Equation

Some years ago it was recognized that the most effective classification of the UV divergences (both for on-shell and off-shell amplitudes) of the NLSM cannot be achieved in terms of the quantized fields ϕ_a, as it usually happens in power-counting renormalizable theories, but rather through the so-called ancestor amplitudes, *i.e.*, the Green's functions of certain composite operators, whose knowledge completely determines the amplitudes involving at least one ϕ_a-leg. This property follows as a consequence of the existence of an additional local functional identity, the so-called Local Functional Equation (LFE) [1].

The LFE stems from the *local* $SU(2)_L$-symmetry that can be established from the gauge transformation of the flat connection F_μ associated with the matrix Ω:

$$F_\mu = i\Omega\partial_\mu\Omega^\dagger = \frac{1}{2}F_{a\mu}\tau^a \tag{10}$$

i.e., the local $SU_L(2)$-transformation of Ω

$$\Omega' = U\Omega \tag{11}$$

induces a gauge transformation of the flat connection, namely

$$F'_\mu = UF_\mu U^\dagger + iU\partial_\mu U^\dagger \tag{12}$$

S_0 in Equation (2) is not invariant under local $SU(2)_L$ transformations; however it is easy to made it invariant, once one realizes that it can be written as

$$S_0 = \int d^D x \, \frac{m_D^2}{4} Tr(F_\mu^2) \tag{13}$$

Since F_μ transforms as a gauge connection, one can introduce an additional external classical vector source $\tilde{J}_\mu = \frac{1}{2}\tilde{J}_{a\mu}\tau^a$ and replace S_0 with

$$S = \int d^D x \, \frac{m_D^2}{4} Tr(F_\mu - \tilde{J}_\mu)^2 \tag{14}$$

If one requires that $\tilde{J}_{a\mu}$ transforms as a gauge connection under the local SU(2)$_L$ group, S in Equation (14) is invariant under a local SU(2)$_L$ symmetry given by

$$\delta\phi_a = \frac{1}{2}\alpha_a\phi_0 + \frac{1}{2}\epsilon_{abc}\phi_b\alpha_c\,, \qquad \delta\phi_0 = -\frac{1}{2}\alpha_a\phi_a$$
$$\delta\tilde{J}_{a\mu} = \partial_\mu\alpha_a + \epsilon_{abc}\tilde{J}_{b\mu}\alpha_c \tag{15}$$

Notice that in the above equation α_a is a local parameter.

In order to implement the classical local SU(2)$_L$ invariance at the quantum level, one needs to define the composite operator ϕ_0 in Equation (4) by coupling it in the classical action to an external source K_0 through the term

$$S_{ext} = \int d^D x\, K_0\phi_0 \tag{16}$$

K_0 is invariant under δ.

The important observation now is that the variation of full one-particle irreducible (1-PI) vertex functional $\Gamma^{(0)} = S + S_{ext}$ is linear in the quantized fields ϕ_a, i.e.,

$$\delta\Gamma^{(0)} = -\frac{1}{2}\int d^D x\, \alpha_a(x)K_0(x)\phi_a(x) \tag{17}$$

By taking a derivative of both sides of the above equation w.r.t. $\alpha_a(x)$ one obtains the LFE for the tree-level vertex functional $\Gamma^{(0)}$:

$$\mathcal{W}_a(\Gamma^{(0)}) = -\partial_\mu \frac{\delta\Gamma^{(0)}}{\delta\tilde{J}_{a\mu}} + \epsilon_{acb}\tilde{J}_{c\mu}\frac{\delta\Gamma^{(0)}}{\delta\tilde{J}_{b\mu}} + \frac{1}{2}\frac{\delta\Gamma^{(0)}}{\delta K_0(x)}\frac{\delta\Gamma^{(0)}}{\delta\phi_a(x)} + \frac{1}{2}\epsilon_{abc}\phi_c(x)\frac{\delta\Gamma^{(0)}}{\delta\phi_b(x)} = -\frac{1}{2}K_0(x)\phi_a(x) \tag{18}$$

Notice that the ϕ_0-term, entering in the variation of the ϕ_a field, is generated by $\frac{\delta\Gamma^{(0)}}{\delta K_0(x)}$. The advantage of this formulation resides in the fact that it is suitable to be promoted at the quantum level. Indeed by defining the composite operator ϕ_0 by taking functional derivatives w.r.t. its source K_0, one is able to control its renormalization, once radiative corrections are included [50].

In the following Section we are going to give a compact and self-contained presentation of the algebraic techniques used to deal with bilinear functional equations like the LFE in Equation (18).

4. Ancestor Amplitudes and the Weak Power-Counting

We are going to discuss in this Section the consequences of the LFE for the full vertex functional. The imposition of a quantum symmetry in a non-power-counting renormalizable theory is a subtle problem, since in general there is no control on the dimensions of the possible breaking terms as strong as the one guaranteed by the Quantum Action Principle (QAP) in the renormalizable case. Let us discuss the latter case first.

4.1. Renormalizable Theories and the Quantum Action Principle

If the tree-level functional $\Gamma^{(0)}$ is power-counting renormalizable, the renormalization procedure [51] provides a way to compute all higher-order terms in the loop expansion of the full vertex functional $\Gamma[\Phi,\chi] = \sum_{n=0}^{\infty}\hbar^n\Gamma^{(n)}[\Phi,\chi]$, depending on the set of quantized fields Φ and external sources collectively denoted by χ, by fixing order by order only a finite set of action-like normalization conditions. One says that the classical action is therefore stable under radiative corrections, namely the number of free parameters does not increase with the loop order.

This procedure is a recursive one, since it allows to construct $\Gamma^{(n)}$ once $\Gamma^{(j)}$, $j < n$ are known. From a combinatorial point of view, it turns out that Γ is the generating functional of the 1-PI renormalized Feynman amplitudes.

19

A desirable feature of power-counting renormalizable theories is that the dependence of 1-PI Green's functions under an infinitesimal variations of the quantized fields and of the parameters of the model is controlled by the so-called Quantum Action Principle (QAP) [52–55] and can be expressed as the insertion of certain *local* operators with UV dimensions determined by their tree-level approximation (*i.e.*, a polynomial in the fields, the external sources and derivatives thereof).

Let us now consider a certain symmetry δ of the tree-level $\Gamma^{(0)}$ classical action. Under the condition that the symmetry δ is non-anomalous [56], it can be extended to the full vertex functional Γ. In many cases of physical interest the proof that the symmetry is non-anomalous can be performed by making use of cohomological tools. Namely one writes the functional equation associated with the δ-invariance of the tree-level vertex functional as follows:

$$\mathcal{S}(\Gamma^{(0)}) \equiv \int d^D x \sum_{\Phi} \frac{\delta\Gamma^{(0)}}{\delta\Phi(x)} \frac{\delta\Gamma^{(0)}}{\delta\Phi^*(x)} = 0 \tag{19}$$

where Φ^* is an external source coupled in the tree-level vertex functional to the δ-transformation of Φ and the sum is over the quantized fields. Φ^* are known as antifields [33]. If δ is nilpotent (as it happens, e.g., for the Becchi–Rouet–Stora–Tyutin (BRST) operator [57–59] in gauge theories), the recursive proof of the absence of obstructions to the fulfillment of Equation (19) works as follows. Suppose that Equation (19) is satisfied up to order $n - 1$ in the loop expansion. Then by the QAP the n-th order breaking

$$\Delta^{(n)} \equiv \int d^D x \sum_{\Phi} \left(\frac{\delta\Gamma^{(0)}}{\delta\Phi(x)} \frac{\delta\Gamma^{(n)}}{\delta\Phi^*(x)} + \frac{\delta\Gamma^{(n)}}{\delta\Phi(x)} \frac{\delta\Gamma^{(0)}}{\delta\Phi^*(x)} + \sum_{j=1}^{n-1} \frac{\delta\Gamma^{(j)}}{\delta\Phi(x)} \frac{\delta\Gamma^{(n-j)}}{\delta\Phi^*(x)} \right) \tag{20}$$

is a polynomial in the fields, the external sources and their derivatives. The term involving $\Gamma^{(n)}$ in Equation (20) allows to define the linearized operator $\mathcal{S}_0$ according to

$$\mathcal{S}_0(\Gamma^{(n)}) \equiv \int d^D x \sum_{\Phi} \left(\frac{\delta\Gamma^{(0)}}{\delta\Phi(x)} \frac{\delta\Gamma^{(n)}}{\delta\Phi^*(x)} + \frac{\delta\Gamma^{(n)}}{\delta\Phi(x)} \frac{\delta\Gamma^{(0)}}{\delta\Phi^*(x)} \right) \tag{21}$$

$\mathcal{S}_0$ is also nilpotent, as a consequence of the nilpotency of δ and of the tree-level invariance in Equation (19). By exploiting this fact and by applying $\mathcal{S}_0$ on both sides of Equation (20) one finds

$$\mathcal{S}_0(\Delta^{(n)}) = 0 \tag{22}$$

provided that the Wess-Zumino consistency condition [60]

$$\mathcal{S}_0\left(\sum_{j=1}^{n-1} \frac{\delta\Gamma^{(j)}}{\delta\Phi(x)} \frac{\delta\Gamma^{(n-j)}}{\delta\Phi^*(x)} \right) = 0 \tag{23}$$

holds. This is the case, e.g., for the BRST symmetry and the associated master Equation (19), since Equation (23) turns out to be a consequence of a generalized Jacobi identity for the Batalin-Vilkovisky bracket for the conjugated variables (Φ, Φ^*) [33].

The problem of establishing whether the functional identity

$$\mathcal{S}(\Gamma) = 0 \tag{24}$$

holds at order n then boils down to prove that the most general solution to Equation (22) is of the form

$$\Delta^{(n)} = -\mathcal{S}_0(\Xi^{(n)}) \tag{25}$$

since then $\Gamma'^{(n)} \equiv \Gamma^{(n)} + \Xi^{(n)}$ will fulfill Equation (24) at order n in the loop expansion. *I.e.*, the problem reduces to the computation of the cohomology $H(\mathcal{S}_0)$ of the operator $\mathcal{S}_0$ in the space of integrated local polynomials in the fields, the external sources and their derivatives. Two $\mathcal{S}_0$-invariant integrated local polynomials $\mathcal{J}_1$ and $\mathcal{J}_2$ belong to the same cohomology class in $H(\mathcal{S}_0)$ if and only if

$$\mathcal{J}_1 = \mathcal{J}_2 + \mathcal{S}_0(\mathcal{K}) \tag{26}$$

for some integrated local polynomial $\mathcal{K}$. In particular, $H(\mathcal{S}_0)$ is empty if the only cohomology class is the one of the zero element, so that the condition that $\mathcal{J}_1$ is $\mathcal{S}_0$-invariant implies that

$$\mathcal{J}_1 = \mathcal{S}_0(\mathcal{K}) \tag{27}$$

for some $\mathcal{K}$. Hence if one can prove that the cohomology of the operator $\mathcal{S}_0$ is empty in the space of breaking terms, then Equation (25) must be fulfilled by some choice of the functional $\Xi^{(n)}$. Moreover it must be checked that the UV dimensions of the possible counterterms $\Xi^{(n)}$ are compatible with the action-like condition, so that renormalizability of the theory is not violated. An extensive review of BRST cohomologies for gauge theories is given in [61].

4.2. Non-Renormalizable Theories

The QAP does not in general hold for non-renormalizable theories. This does not come as a surprise, since the appearance of UV divergences with higher and higher degree, as one goes up with the loop order, prevents to characterize the induced breaking of a functional identity in terms of a polynomial of a given finite degree (independent of the loop order).

Moreover for the NLSM another important difference must be stressed: the basic Green's functions of the theory are not those of the quantized fields ϕ_a, but those of the flat connection coupled to the external vector source $\tilde{J}_{a\mu}$ and of the non-linear constraint ϕ_0 (coupled to K_0). This result follows from the invertibility of

$$\frac{\delta\Gamma}{\delta K_0} = \phi_0 + O(\hbar)$$

as a formal power series in $\hbar$ (since $\phi_0|_{\phi_a=0} = m_D$). Then the LFE for the vertex functional Γ

$$\mathcal{W}_a(\Gamma) = -\frac{1}{2}K_0(x)\phi_a(x) \tag{28}$$

can be seen as a first-order functional differential equation controlling the dependence of Γ on the fields ϕ_a. Provided that a solution exists (as will be proven in Section 5), Equation (28) determines all the amplitudes involving at least one external ϕ_a-leg in terms of the boundary condition provided by the functional $\Gamma[\tilde{J}, K_0] = \Gamma[\phi, \tilde{J}, K_0]\big|_{\phi_a=0}$.

$\Gamma[\tilde{J}, K_0]$ is the generating functional of the so called ancestor amplitudes, *i.e.*, the 1-PI amplitudes involving only external $\tilde{J}$ and K_0 legs.

It is therefore reasonable to assume the LFE in Equation (28) as the starting point for the quantization of the theory.

From a path-integral point of view, Equation (28) implies that one is performing an integration over the SU(2)-invariant Haar measure of the group, namely one is computing

$$Z[J, \tilde{J}_\mu, K_0] = \int \mathcal{D}\Omega(\phi)\exp\left(i\Gamma^{(0)}[\phi, \tilde{J}_\mu, K_0] + i\int d^D x\, J_a\phi_a\right) \tag{29}$$

where we denote by $\mathcal{D}\Omega(\phi)$ the SU(2) Haar measure (in the coordinate representation spanned by the fields ϕ_a). This clarifies the geometrical meaning of the LFE.

4.3. Weak Power-Counting

As we have already noticed, in four dimensions the NLSM is non power-counting renormalizable, since already at one loop level an infinite number of divergent ϕ-amplitudes exists. One may wonder whether the UV behavior of the ancestor amplitudes (the boundary conditions to the LFE) is better. It turns out that this is indeed the case and one finds that in D dimensions a n-th loop Feynman amplitude $\mathcal{G}$ with N_{K_0} external K_0-legs and $N_{\tilde{J}}$ external $\tilde{J}$-legs has superficial degree of divergence given by [2]

$$d(\mathcal{G}) \leq (D-2)n + 2 - N_{\tilde{J}} - 2N_{K_0} \tag{30}$$

The proof is straightforward although somehow lengthy and will not be reported here. It can be found in [2]. Equation (30) establishes the Weak Power-Counting (WPC) condition: at every loop order only a finite number of superficially divergent ancestor amplitudes exist.

For instance, in $D = 4$ and at one loop order, Equation (30) reduces to

$$d(\mathcal{G}) \leq 4 - N_{\tilde{J}} - 2N_{K_0} \tag{31}$$

i.e., UV divergent amplitudes involve only up to four external $\tilde{J}_\mu$ legs or two K_0-legs.

By taking into account Lorentz-invariance and global $\mathrm{SU}(2)_R$ symmetry, the list of UV divergent amplitudes reduces to

$$\int d^4x\, \partial_\mu \tilde{J}_{av} \partial^\mu \tilde{J}_a^\nu \,, \qquad \int d^4x\, (\partial \tilde{J}_a)^2 \,, \qquad \int d^4x\, \epsilon_{abc} \partial_\mu \tilde{J}_{av} \tilde{J}_b^\mu \tilde{J}_c^\nu \,, \qquad \int d^4x\, (\tilde{J}_a)^2 (\tilde{J}_b)^2$$
$$\int d^4x\, \tilde{J}_{a\mu} \tilde{J}_b^\mu \tilde{J}_{av} \tilde{J}_b^\nu \,, \qquad \int d^4x\, \tilde{J}_{a\mu}^2 \,, \qquad \int d^4x\, K_0^2 \,, \qquad \int d^4x\, K_0 \tilde{J}_a^2 \tag{32}$$

Notice that the counterterms are local.

It should be emphasized that the model is not power-counting renormalizable, even when ancestor amplitudes are considered, since according to Equation (30) the number of UV divergent amplitudes increases as the loop order n grows.

A special case is the 2-dimensional NLSM. For $D = 2$ Equation (30) yields

$$d(\mathcal{G}) \leq 2 - N_{\tilde{J}} - 2N_{K_0} \tag{33}$$

i.e., at every loop order there can be only two UV divergent ancestor amplitudes, namely

$$\int d^2x\, \tilde{J}^2 \qquad \text{and} \qquad \int d^2x\, K_0$$

These are precisely of the same functional form as the ancestor amplitudes entering in the tree-level vertex functional and, in this sense, the model shares the stability property of the classical action typical of power-counting renormalizable models. Renormalizability of the 2-dimensional NLSM can also be established by relying on the Ward identity of global SU(2) symmetry (see e.g., [37]).

A comment is in order here. In References [24,25] the external fields are the sources of connected Green's functions of certain quark-antiquark currents. The ancestor amplitudes in the NLSM, in the approach based on the LFE, do not have a direct physical interpretation of this type, however they have a very clear geometrical meaning. First of all, $\tilde{J}_\mu$ is the source coupled to the flat connection naturally associated with the group element Ω. On the other hand, K_0 is the unique scalar source required, in the special case of the SU(2) group, in order to control the renormalization of the non-linear classical SU(2) transformation of the ϕ_a's and thus plays the role of the so-called antifields [33,50]. The extension to a general Lie group G is addressed at the end of Section 5.

5. Cohomological Analysis of the LFE

In order to study the properties of the LFE, it is very convenient to introduce a fictious BRST operator s by promoting the gauge parameters $\alpha_a(x)$ to classical anticommuting ghosts $\omega_a(x)$. *I.e.*, one sets

$$s\tilde{J}_{a\mu} = \partial_\mu \omega_a + \epsilon_{abc}\tilde{J}_{b\mu}\omega_c \,, \qquad s\phi_a = \frac{1}{2}\omega_a\phi_0 + \frac{1}{2}\epsilon_{abc}\phi_b\omega_c \,, \qquad s\phi_0 = -\frac{1}{2}\omega_a\phi_a$$

$$sK_0 = \frac{1}{2}\omega_a\frac{\delta\Gamma^{(0)}}{\delta\phi_a(x)} \,, \qquad s\omega_a = -\frac{1}{2}\epsilon_{abc}\omega_b\omega_c \tag{34}$$

Some comments are in order here. First of all the BRST operator s acts also on the external source K_0. Moreover, the BRST transformation of ω_a is fixed by nilpotency, namely $s^2 = 0$.

The introduction of the ghosts allows to define a grading w.r.t. the conserved ghost number. ω has ghost number $+1$, while all the other fields and sources have ghost number zero. (The ghost number was called the Faddeev-Popov (ФП) charge in [2].)

In terms of the operator s we can write the n-th order projection ($n \geq 1$) of the LFE in Equation (28) as follows:

$$\left[\int d^D x\, \omega_a \mathcal{W}_a(\Gamma)\right]^{(n)} = s\Gamma^{(n)} + \sum_{j=1}^{n-1}\int d^D x \frac{1}{2}\omega_a\frac{\delta\Gamma^{(j)}}{\delta K_0}\frac{\delta\Gamma^{(n-j)}}{\delta\phi_a} = 0 \tag{35}$$

Notice that the bilinear term in the LFE manifests itself into the presence of the mixed $\frac{\delta\Gamma^{(j)}}{\delta K_0}\frac{\delta\Gamma^{(n-j)}}{\delta\phi_a}$ contribution. Moreover in the r.h.s. there is no contribution from the breaking term linear in ϕ_a in Equation (18) since the latter remains classical.

Suppose now that all divergences have been recursively subtracted up to order $n-1$. At the n-th order the UV divergent part can only come from the term involving $\Gamma^{(n)}$ in Equation (35) and therefore, if the LFE holds, one gets a condition on the UV divergent part $\Gamma^{(n)}_{pol}$ of $\Gamma^{(n)}$:

$$s\Gamma^{(n)}_{pol} = 0 \tag{36}$$

To be specific, one can use Dimensional Regularization and subtract only the pole part of the ancestor amplitudes (after the proper normalization of the ancestor background connection amplitudes

$$\frac{m}{m_D}\frac{\delta^{(n)}\Gamma}{\delta\tilde{J}^{\mu_1}_{a_1}\ldots\delta\tilde{J}^{\mu_n}_{a_n}}$$

The LFE then fixes the correct factor for the normalization of amplitudes involving K_0). This subtraction procedure has been shown to be symmetric [2,4], *i.e.*, to preserve the LFE. The pole parts before subtraction obey the condition in Equation (36).

By the nilpotency of s, solving Equation (36) is equivalent to computing the cohomology of the BRST operator s in the space of local functionals in $\tilde{J}$, ϕ, K_0 and their derivatives with ghost number zero. This can be achieved by using the techniques developed in [62].

One first builds invariant combinations in one-to-one correspondence with the ancestor variables $\tilde{J}_{a\mu}$ and K_0. For that purpose it is more convenient to switch back to matrix notation. The difference $I_\mu \equiv F_\mu - \tilde{J}_\mu$ transforms in the adjoint representation of SU(2), being the difference of two gauge connections. Thus the conjugate of such a difference w.r.t. Ω

$$j_\mu = j_{a\mu}\frac{\tau_a}{2} = \Omega^\dagger I_\mu\Omega \tag{37}$$

is invariant under s. By direct computation one finds

$$
\begin{aligned}
m_D^2 j_{a\mu} &= m_D^2 I_{a\mu} - 2\phi_b^2 I_{a\mu} + 2\phi_b I_{b\mu}\phi_a + 2\phi_0 \epsilon_{abc}\phi_b I_{c\mu} \\
&\equiv m_D^2 R_{ba} I_{b\mu}
\end{aligned}
\tag{38}
$$

The matrix R_{ba} is an element of the adjoint representation of SU(2) and therefore the mapping $\tilde{J}_{a\mu} \to j_{a\mu}$ is invertible.

One can also prove that the following combination

$$
\overline{K}_0 \equiv \frac{m_D^2 K_0}{\phi_0} - \phi_a \frac{\delta S}{\delta \phi_a}
\tag{39}
$$

is invariant [2]. At $\phi_a = 0$ one gets

$$
\overline{K}_0\big|_{\phi_a=0} = m_D K_0
\tag{40}
$$

and therefore the transformation $K_0 \to \overline{K}_0$ is also invertible.

In terms of the new variables $\overline{K}_0$ and j_μ and by differentiating Equation (36) w.r.t. ω_a one gets

$$
\Theta_{ab} \frac{\delta \Gamma_{pol}^{(n)}[j, \overline{K}, \phi]}{\delta \phi_b} = 0
\tag{41}
$$

where $s\phi_b = \omega_a \Theta_{ab}$, i.e.,

$$
\Theta_{ab} = \frac{1}{2}\phi_0 \delta_{ab} + \frac{1}{2}\epsilon_{abc}\phi_c
\tag{42}
$$

Θ_{ab} is invertible and thus Equation (41) yields

$$
\frac{\delta \Gamma_{pol}^{(n)}[j, \overline{K}_0, \phi]}{\delta \phi_b} = 0
\tag{43}
$$

This equation is a very powerful one. It states that the n-th order divergences (after the theory has been made finite up to order $n-1$) of the ϕ-fields can only appear through the invariant combinations $\overline{K}_0$ and $j_{a\mu}$. These invariant variables have been called bleached variables and they are in one-to-one correspondence with the ancestor variables K_0 and $\tilde{J}_{a\mu}$.

The subtraction strategy is thus the following. One computes the divergent part of the properly normalized ancestor amplitudes that are superficially divergent at a given loop order according to the WPC formula in Equation (30). Then the replacement $\tilde{J}_{a\mu} \to j_{a\mu}$ and $K_0 \to \overline{K}_0$ is carried out. This gives the full set of counterterms required to make the theory finite at order n in the loop expansion.

As an example, we give here the explicit form of the one-loop divergent counterterms for the NLSM in $D = 4$ [2] (notice that we have set $g = 1$ according to our conventions in this paper):

$$
\begin{aligned}
\hat{\Gamma}^{(1)} = \frac{1}{D-4}\Big[&-\frac{1}{12}\frac{1}{(4\pi)^2}\frac{m_D^2}{m^2}\left(\mathcal{I}_1 - \mathcal{I}_2 - \mathcal{I}_3\right) + \frac{1}{(4\pi)^2}\frac{1}{48}\frac{m_D^2}{m^2}\left(\mathcal{I}_6 + 2\mathcal{I}_7\right) \\
&+ \frac{1}{(4\pi)^2}\frac{3}{2}\frac{1}{m^2 m_D^2}\mathcal{I}_4 + \frac{1}{(4\pi)^2}\frac{1}{2}\frac{1}{m^2}\mathcal{I}_5\Big]
\end{aligned}
\tag{44}
$$

By projecting the above equation on the relevant monomial in the ϕ_a fields one can get the divergences of the descendant amplitudes. As an example, for the four point ϕ_a function one gets by explicit computation that the contribution from the combination $\mathcal{I}_1 - \mathcal{I}_2 - \mathcal{I}_3$ is zero, while the remaining invariants give

$$
\begin{aligned}
\hat{\Gamma}^{(1)}[\phi\phi\phi] \;=\; & -\frac{1}{D-4}\frac{1}{m_D^2 m^2 (4\pi)^2} \\
& \int d^D x \left(-\frac{1}{3}\partial_\mu\phi_a\partial^\mu\phi_a\partial_\nu\phi_b\partial^\nu\phi_b - \frac{2}{3}\partial_\mu\phi_a\partial_\nu\phi_a\partial^\mu\phi_b\partial^\nu\phi_b \right. \\
& \left. -\frac{3}{2}\phi_a\Box\phi_a\phi_b\Box\phi_b - 2\phi_a\Box\phi_a\partial_\mu\phi_b\partial^\mu\phi_b \right)
\end{aligned}
\tag{45}
$$

The invariants in the combination $\mathcal{I}_6 + 2\mathcal{I}_7$ generate the counterterms in the first line between square brackets; these counterterms are globally SU(2) invariant. The other terms are generated by invariants involving the source K_0. In [39,40] they were constructed by means of a (non-locally invertible) field redefinition of ϕ_a. The full set of mixed four point amplitudes involving at least one ϕ_a legs and the external sources $\tilde{J}_\mu$ and K_0 can be found in [2].

The correspondence with the linear sigma model in the large coupling limit has been studied in [5].

The massive NLSM in the LFE formulation has been studied in [15], while the symmetric subtraction procedure for the LFE associated with polar coordinates in the simplest case of the free complex scalar field has been given in [16].

In the SU(2) NLSM just one scalar source K_0 is sufficient in order to formulate the LFE. For an arbitrary Lie group G the LFE can always be written if one introduces a full set of antifields ϕ_I^*, as follows. Let us denote by $\Omega(\phi_I)$ the group element belonging to G, parameterized by local coordinates ϕ_I. Then under an infinitesimal left G-transformation of parameters α_J

$$
\delta\Omega = i\alpha_J T_J \Omega
\tag{46}
$$

where T_J are the generators of the group G, one has

$$
\delta\phi_I = S_{IJ}(\phi)\alpha_J
\tag{47}
$$

It is convenient to promote the local left invariance to a BRST symmetry by upgrading the parameters α_J to local classical anticommuting ghosts C_J. Then one can introduce in the usual way the couplings with the antifields ϕ_I^* through

$$
S_{ext} = \int d^D x\, \phi_I^* S_{IJ}(\phi) C_J
\tag{48}
$$

and then write the corresponding BV master equation [33]. This is the generalization of the LFE valid for the group G. The cohomology of the linearized BV operator (which is the main tool for identifying the bleached variables, as shown above) has been studied for any Lie group G in [62].

6. Higher Loops

At orders $n > 1$ the LFE for $\Gamma^{(n)}$ is an inhomogeneous equation

$$
s\Gamma^{(n)} = \Delta^{(n)} \equiv -\frac{1}{2}\int d^D x\, \omega_a \sum_{j=1}^{n-1} \frac{\delta\Gamma^{(j)}}{\delta K_0}\frac{\delta\Gamma^{(n-j)}}{\delta\phi_a}
\tag{49}
$$

The above equation can be explicitly integrated by using the techniques of the Slavnov-Taylor (ST) parameterization of the effective action [63–65] (originally developed in order to provide a strategy for the restoration of the ST identity of non-anomalous gauge theories in the absence of a symmetric regularization).

For that purpose it is convenient to redefine the ghost according to

$$\overline{\omega}_a = \Theta_{ab}\omega_b \tag{50}$$

where Θ_{ab} is given in Equation (42). The action of s then reduces to

$$s\overline{K}_0 = sj_{a\mu} = 0, \qquad s\phi_a = \overline{\omega}_a, \qquad s\overline{\omega}_a = 0 \tag{51}$$

This means that the variables $\overline{K}_0$ and $j_{a\mu}$ are invariant, while the pair $(\phi_a, \overline{\omega}_a)$ is a BRST doublet (*i.e.*, a pair of variables u, v such that $s\,u = v, s\,v = 0$) [33,66].

By the nilpotency of s the following consistency condition must hold for $\Delta^{(n)}$:

$$s\Delta^{(n)} = 0 \tag{52}$$

The fulfillment of the above equation as a consequence of the validity of the LFE up to order $n-1$ is proven in [63]. In terms of the new variables Equation (49) reads

$$\int d^D x\, \overline{\omega}_a \frac{\delta\Gamma^{(n)}}{\delta\phi_a} = \Delta^{(n)}[\overline{\omega}_a, \phi_a, \overline{K}_0, j_{a\mu}] \tag{53}$$

By noticing that $\Delta^{(n)}$ is linear in $\overline{\omega}_a$ and by differentiating Equation (53) w.r.t. $\overline{\omega}_a$ we arrive at

$$\frac{\delta\Gamma^{(n)}}{\delta\phi_a(x)} = \frac{\delta\Delta^{(n)}}{\delta\overline{\omega}_a(x)} \tag{54}$$

The above equation controls the explicit dependence of the n-th order vertex functional on ϕ_a (there is in addition an implicit dependence on ϕ_a through the variables $j_{a\mu}$ and $\overline{K}_0$).

The explicit dependence on ϕ_a only appears through lower order terms. Hence it does not influence the n-th order ancestor amplitudes.

The solution of Equation (49) can be written in compact form by using a homotopy operator. Indeed $\Gamma^{(n)}$ will be the sum of a n-th order contribution $\mathcal{A}^{(n)}$, depending only on $j_{a\mu}$ and $\overline{K}_0$, plus a lower order term:

$$\begin{aligned}
\Gamma^{(n)}[\phi_a, K_0, \check{J}_{a\mu}] &= \mathcal{A}^{(n)}[\overline{K}_0, j_{a\mu}] \\
&\quad + \int d^D x \int_0^1 dt\, \phi_a(x)\lambda_t \frac{\delta\Delta^{(n)}}{\delta\overline{\omega}_a(x)}
\end{aligned} \tag{55}$$

The operator λ_t acts as follows on a generic functional $X[\phi_a, \overline{\omega}_a, \overline{K}_0, j_{a\mu}]$:

$$\lambda_t X[\phi_a, \overline{\omega}_a, \overline{K}_0, j_{a\mu}] = X[t\phi_a, t\overline{\omega}_a, \overline{K}_0, j_{a\mu}] \tag{56}$$

The homotopy operator κ for the BRST differential s in the second line of Equation (55) is therefore given by

$$\kappa = \int d^D x \int_0^1 dt\, \phi_a(x)\lambda_t \frac{\delta}{\delta\overline{\omega}_a(x)} \tag{57}$$

and satisfies the condition

$$\{s, \kappa\} = \mathbf{1} \tag{58}$$

where $\mathbf{1}$ denotes the identity on the space of functionals spanned by $\overline{\omega}_a, \phi_a$.

An important remark is in order here. The theory remains finite and respects the LFE if one adds to $\Gamma^{(n)}$ some integrated local monomials in $j_{a\mu}$ and $\overline{K}_0$ and ordinary derivatives thereof (with finite coefficients), compatible with Lorentz symmetry and global SU(2) invariance, while respecting the WPC condition in Equation (30):

$$\Gamma^{(n)}_{finite} = \sum_j \int d^D x \, \mathcal{M}_j(j_{a\mu}, \overline{K}_0) \tag{59}$$

This is a consequence of the non power-counting renormalizability of the theory: one can introduce order by order in the loop expansion an increasing number of finite parameters that do not appear in the classical action. Notice that they cannot be inserted back at tree-level: if one performs such an operation, the WPC condition is lost.

This observation suggests that these finite parameters cannot be easily understood as physical free parameters of the theory, since they cannot appear in the tree-level action. It was then proposed to define the model by choosing the symmetric subtraction scheme discussed in Section 5 and by considering as physical parameters only those present in the classical action plus the scale of the radiative corrections Λ [4]. While acceptable on physical grounds, from the mathematical point of view one may wonder whether there is some deeper reason justifying such a strategy. We will comment briefly on this point in the Conclusions.

7. Applications to Yang-Mills and the Electroweak Theory

When the vector source $\tilde{J}_{a\mu}$ becomes a dynamical gauge field, the NLSM action gives rise to the Stückelberg mass term [67].

The subtraction procedure based on the LFE has been used to implement a mathematically consistent formulation of non-linearly realized massive Yang-Mills theory. SU(2) Yang-Mills in the LFE formalism has been formulated in [6]. The pseudo-Goldstone fields take over the role of the ϕ_a fields of the NLSM. Their Green's functions are fixed by the LFE. The WPC proves to be very restrictive, since by imposing the WPC condition it turns out that the only allowed classical solution is the usual Yang-Mills theory plus the Stückelberg mass term.

This is a very powerful (and somehow surprising) result. Indeed all possible monomials constructed out of $j_{a\mu}$ and ordinary derivatives thereof are gauge-invariant and therefore they could be used as interaction vertices in the classical action.

Otherwise said, the peculiar structure of the Yang-Mills action

$$S_{\text{YM}} = -\int d^4 x \, \frac{1}{4} G_{a\mu\nu} G_a^{\mu\nu} \tag{60}$$

where $G_{a\mu\nu}$ denotes the field strength of the gauge field $A_{a\mu}$

$$G_{a\mu\nu} = \partial_\mu A_{a\nu} - \partial_\nu A_{a\mu} + f_{abc} A_{b\mu} A_{c\nu}$$

is not automatically enforced by the requirement of gauge invariance if the gauge group is non-linearly realized. However if the WPC condition is satisfied, the only admissible solution becomes Yang-Mills theory plus the Stückelberg mass term:

$$S_{\text{nlYM}} = S_{\text{YM}} + \int d^4 x \, \frac{M^2}{2} (A_{a\mu} - F_{a\mu})^2 \tag{61}$$

Massive Yang-Mills theory in the presence of a non-linearly realized gauge group is physically unitary [67] (despite the fact that it violates the Froissart bound [68–74] at tree-level). The counterterms in the Landau gauge have been computed at one loop level in [7]. The formulation of the theory in a general 't Hooft gauge has been given in [8].

The approach based on the LFE can also be used for non-perturbative studies of Yang-Mills theory on the lattice. The phase diagram of SU(2) Yang-Mills has been considered in [17]. Emerging evidence is being accumulated about the formation of isospin scalar bound states [18] in the supposedly confined phase of the theory [19].

An analytic approach based on the massless bound-state formalism for the implementation of the Schwinger mechanism in non-Abelian gauge theories has been presented in [75–77].

A very important physical application of non-linearly realized gauge theories is the formulation of a non-linearly realized electroweak theory, based on the group $SU(2) \times U(1)$. The set of gauge fields comprises the SU(2) fields $A_{a\mu}$ and the hypercharge U(1) gauge connection B_μ. By using the technique of bleached variables one can first construct SU(2) invariant variables in one-to-one correspondence with $A_\mu = A_{a\mu} \frac{\tau_a}{2}$ [8]:

$$w_\mu = \Omega^\dagger g A_\mu \Omega - g' \frac{\tau_3}{2} B_\mu + i\Omega^\dagger \partial_\mu \Omega \equiv w_{a\mu} \frac{\tau_a}{2} \tag{62}$$

In the above equation we have reinserted back for later convenience the SU(2) and U(1) coupling constants g and g'. Since w_μ is SU(2) invariant, the hypercharge generator coincides with the electric charge generator. $w_{3\mu}$ is then the bleached counterpart of the Z_μ field, since

$$Z_\mu = \left. \frac{1}{\sqrt{g^2 + g'^2}} w_{3\mu} \right|_{\phi_a = 0} = c_W A_{3\mu} - s_W B_\mu \tag{63}$$

where s_W and c_W are the sine and cosine of the Weinberg angle

$$s_W = \frac{g'}{\sqrt{g^2 + g'^2}}, \qquad c_W = \frac{g}{\sqrt{g^2 + g'^2}} \tag{64}$$

The photon A_μ is described by the combination orthogonal to Z_μ, namely

$$A_\mu = s_W A_{3\mu} + c_W B_\mu \tag{65}$$

One can built out of $A_{1\mu}$ and $A_{2\mu}$ the charged $W^\pm$ field

$$W_\mu^\pm = \frac{1}{\sqrt{2}} (A_{1\mu} \mp i A_{2\mu}) \tag{66}$$

whose bleached counterpart is simply

$$w_\mu^\pm = \frac{1}{\sqrt{2}} (w_{1\mu} \mp i w_{2\mu}) \tag{67}$$

The WPC allows for the same symmetric couplings of the Standard Model and for two independent mass invariants [9–11]

$$M_W^2 w^+ w^- + \frac{M_Z^2}{2} w_{3\mu}^2 \tag{68}$$

where the mass of the Z and W bosons are not related by the Weinberg relation

$$M_Z = \frac{M_W}{c_W}$$

This is a peculiar signature of the mass generation mechanism *à la* Stückelberg, that is not present in the linearly realized theory *à la* Brout-Englert-Higgs [78–80] (even if one discards the condition of power-counting renormalizability in favour of the WPC) [12].

The inclusion of physical scalar resonances in the non-linearly realized electroweak model, while respecting the WPC, yields some definite prediction for the Beyond the Standard Model (BSM) sector. Indeed it turns out that it is impossible to add a scalar singlet without breaking the WPC condition. The minimal solution requires a SU(2) doublet of scalars, leading to a CP-even physical field (to be identified with the recently discovered scalar resonance at 125.6 GeV) and to three additional heavier physical states, one CP-odd and neutral and two charged ones [13]. The proof of the WPC in this model and the BRST identification of physical states has been given in [14].

The WPC and the symmetries of the theory select uniquely the tree-level action of the non-linearly realized electroweak model. As in the NLSM case, mathematically additional finite counterterms are allowed at higher orders in the loop expansion. In [4] it has been argued that they cannot be interpreted as additional physical parameters (unlike in the effective field theory approach), on the basis of the observation that they are forbidden at tree-level by the WPC, and this strategy has been consistently applied in [7,11].

The question remains open of whether a Renormalization Group equation exists, involving a finite change in the higher order subtractions, in such a way to compensate the change in the sliding scale Λ of the radiative corrections. We notice that in this case the finite higher order counterterms would be a function of the tree-level parameters only (unlike in the conventional effective field theory approach, where they are treated as independent extra free parameters). This issue deserves further investigation, since obviously the possibility of running the scale Λ in a mathematically consistent way would allow to obtain physical predictions of the same observables applicable in different energy regimes.

8. Conclusions

The LFE makes it apparent that the independent amplitudes of the NLSM are not those of the quantum fields, over which the path-integral is carried out, but rather those of the background connection $\tilde{J}_\mu$ and of the source K_0, coupled to the solution of the non-linear constraint ϕ_0. The WPC can be formulated only for these ancestor amplitudes; the LFE in turn fixes the descendant amplitudes, involving at least one pion external leg. Within this formulation, the minimal symmetric subtraction discussed in Section 5 is natural, since it provides a way to implement the idea that the number of ancestor interaction vertices, appearing in the classical action and compatible with the WPC, must be finite.

However, it should be stressed that the most general solution to the LFE, compatible with the WPC, does not forbid to choose different finite parts of the higher order symmetric counterterms (as in the most standard view of effective field theories, where such arbitrariness is associated with extra free parameters of the non-renormalizable theory), as far as they are introduced at the order prescribed by the WPC condition and without violating the LFE.

In this connection it should be noticed that the addition of the symmetric finite renormalizations in Equation (59), that are allowed by the symmetries of the theory, is equivalent to a change in the Hopf algebra [81,82] of the model. This is because the finite counterterms in Equation (59) modify the set of 1-PI Feynman diagrams on which the Hopf algebra is constructed, as a dual of the enveloping algebra of the Lie algebra of Feynman graphs. The approach to renormalization based on Hopf algebras is known to be equivalent [83] to the traditional approach based on the Bogoliubov recursive formula and its explicit solution through the Zimmermann's forest formula [84]. For models endowed with a WPC it might provide new insights into the structure of the UV divergences of the theory. This connection seems to deserve further investigations.

Acknowledgments: It is a pleasure to acknowledge many enlightening discussions with R. Ferrari. Useful comments and a careful reading of the manuscript by D. Bettinelli are also gratefully acknowledged.

Appendix

One-Loop Invariants

We report here the invariants controlling the one-loop divergences of the NLSM in $D = 4$ [2].

$$\mathcal{I}_1 = \int d^D x \left[D_\mu (F - \tilde{J})_\nu \right]_a \left[D^\mu (F - \tilde{J})^\nu \right]_a ,$$

$$\mathcal{I}_2 = \int d^D x \left[D_\mu (F - \tilde{J})^\mu \right]_a \left[D_\nu (F - \tilde{J})^\nu \right]_a ,$$

$$\mathcal{I}_3 = \int d^D x \, \epsilon_{abc} \left[D_\mu (F - \tilde{J})_\nu \right]_a \left(F_b^\mu - \tilde{J}_b^\mu \right) \left(F_c^\nu - \tilde{J}_c^\nu \right) ,$$

$$\mathcal{I}_4 = \int d^D x \left(\frac{m_D^2 K_0}{\phi_0} - \phi_a \frac{\delta S}{\delta \phi_a} \right)^2 ,$$

$$\mathcal{I}_5 = \int d^D x \left(\frac{m_D^2 K_0}{\phi_0} - \phi_a \frac{\delta S}{\delta \phi_a} \right) \left(F_b^\mu - \tilde{J}_b^\mu \right)^2 ,$$

$$\mathcal{I}_6 = \int d^D x \left(F_a^\mu - \tilde{J}_a^\mu \right)^2 \left(F_b^\nu - \tilde{J}_b^\nu \right)^2 ,$$

$$\mathcal{I}_7 = \int d^D x \left(F_a^\mu - \tilde{J}_a^\mu \right) \left(F_a^\nu - \tilde{J}_a^\nu \right) \left(F_{b\mu} - \tilde{J}_{b\mu} \right) \left(F_{b\nu} - \tilde{J}_{b\nu} \right) \tag{A1}$$

In the above equation $D_\mu[F]$ stands for the covariant derivative w.r.t. $F_{a\mu}$

$$D_\mu[F]_{ab} = \delta_{ab} \partial_\mu + \epsilon_{acb} F_{c\mu} \tag{A2}$$

Conflicts of Interest: The author declares no conflict of interest.

References

1. Ferrari, R. Endowing the nonlinear sigma model with a flat connection structure: A way to renormalization. *JHEP* **2005**, doi:10.1088/1126-6708/2005/08/048.
2. Ferrari, R.; Quadri, A. A Weak power-counting theorem for the renormalization of the non-linear sigma model in four dimensions. *Int. J. Theor. Phys.* **2006**, *45*, 2497–2515.
3. Bettinelli, D.; Ferrari, R.; Quadri, A. Path-integral over non-linearly realized groups and Hierarchy solutions. *JHEP* **2007**, doi:10.1088/1126-6708/2007/03/065.
4. Bettinelli, D.; Ferrari, R.; Quadri, A. Further Comments on the Symmetric Subtraction of the Nonlinear Sigma Model. *Int. J. Mod. Phys.* **2008**, *A23*, 211–232.
5. Bettinelli, D.; Ferrari, R.; Quadri, A. The Hierarchy principle and the large mass limit of the linear sigma model. *Int. J. Theor. Phys.* **2007**, *46*, 2560–2590.
6. Bettinelli, D.; Ferrari, R.; Quadri, A. A Massive Yang-Mills Theory based on the Nonlinearly Realized Gauge Group. *Phys. Rev. D* **2008**, *77*, doi:10.1103/PhysRevD.77.045021.
7. Bettinelli, D.; Ferrari, R.; Quadri, A. One-loop self-energy and counterterms in a massive Yang-Mills theory based on the nonlinearly realized gauge group. *Phys. Rev. D* **2008**, *7*, doi:10.1103/PhysRevD.77.105012.
8. Bettinelli, D.; Ferrari, R.; Quadri, A. Gauge Dependence in the Nonlinearly Realized Massive SU(2) Gauge Theory. *J. General. Lie Theor. Appl.* **2008**, *2*, 122–126.
9. Bettinelli, D.; Ferrari, R.; Quadri, A. The SU(2) × U(1) Electroweak Model based on the Nonlinearly Realized Gauge Group. *Int. J. Mod. Phys.* **2009**, *A24*, 2639–2654.
10. Bettinelli, D.; Ferrari, R.; Quadri, A. The SU(2) × U(1) Electroweak Model based on the Nonlinearly Realized Gauge Group. II. Functional Equations and the Weak Power-Counting. *Acta Phys. Polon.* **2010**, *B41*, 597–628.
11. Bettinelli, D.; Ferrari, R.; Quadri, A. One-loop Self-energies in the Electroweak Model with Nonlinearly Realized Gauge Group. *Phys. Rev. D* **2009**, *79*, doi:10.1103/PhysRevD.79.125028.

12. Quadri, A. The Algebra of Physical Observables in Nonlinearly Realized Gauge Theories. *Eur. Phys. J.* **2010**, *C70*, 479–489.

13. Binosi, D.; Quadri, A. Scalar Resonances in the Non-linearly Realized Electroweak Theory. *JHEP* **2013**, *1302*, doi:10.1007/JHEP02(2013)020.

14. Bettinelli, D.; Quadri, A. The Stueckelberg Mechanism in the presence of Physical Scalar Resonances. *Phys. Rev. D* **2013**, *88*, doi:10.1103/PhysRevD.88.065023.

15. Ferrari, R. A Symmetric Approach to the Massive Nonlinear Sigma Model. *J. Math. Phys.* **2011**, *52*, 092303:1–092303:16.

16. Ferrari, R. On the Renormalization of the Complex Scalar Free Field Theory. *J. Math. Phys.* **2010**, *51*, 032305:1–032305:20.

17. Ferrari, R. On the Phase Diagram of Massive Yang-Mills. *Acta Phys. Polon.* **2012**, *B43*, 1965–1980.

18. Ferrari, R. On the Spectrum of Lattice Massive SU(2) Yang–Mills. *Acta Phys. Polon.* **2013**, *B44*, 1871–1885.

19. Ferrari, R. Metamorphosis versus Decoupling in Nonabelian Gauge Theories at Very High Energies. *Acta Phys. Polon.* **2012**, *B43*, 1735–1767.

20. Gell-Mann, M.; Levy, M. The axial vector current in beta decay. *Nuovo Cim.* **1960**, *16*, 705–726.

21. Weinberg, S. Nonlinear realizations of chiral symmetry. *Phys. Rev.* **1968**, *166*, 1568–1577.

22. Coleman, S.R.; Wess, J.; Zumino, B. Structure of phenomenological Lagrangians. 1. *Phys. Rev.* **1969**, *177*, 2239–2247.

23. Callan, C.G., Jr.; Coleman, S.R.; Wess, J.; Zumino, B. Structure of phenomenological Lagrangians. 2. *Phys. Rev.* **1969**, *177*, 2247–2250.

24. Weinberg, S. Phenomenological Lagrangians. *Physica* **1979**, *A96*, 327–340.

25. Gasser, J.; Leutwyler, H. Chiral Perturbation Theory to One Loop. *Ann. Phys.* **1984**, *158*, 142–210.

26. Gasser, J.; Leutwyler, H. Chiral Perturbation Theory: Expansions in the Mass of the Strange Quark. *Nucl. Phys. B* **1985**, *250*, 465–516.

27. Bijnens, J.; Colangelo, G.; Ecker, G. Renormalization of chiral perturbation theory to order p**6. *Ann. Phys.* **2000**, *280*, 100–139.

28. Ecker, G.; Gasser, J.; Leutwyler, H.; Pich, A.; de Rafael, E. Chiral Lagrangians for Massive Spin 1 Fields. *Phys. Lett. B* **1989**, *223*, 425–432.

29. Buchmuller, W.; Wyler, D. Effective Lagrangian Analysis of New Interactions and Flavor Conservation. *Nucl. Phys. B* **1986**, *268*, 621–653.

30. Donoghue, J.F. Introduction to the effective field theory description of gravity. Available online: http://arxiv.org/abs/grqc/9512024 (accessed on 15 April 2014).

31. Weinberg, S. *The Quantum Theory of Fields. Vol. 2: Modern Applications*; Cambridge University Press: Cambridge, UK, 1996.

32. Itzykson, C.; Zuber, J. *Quantum Field Theory*; McGraw-Hill: New York, NY, USA, 1980.

33. Gomis, J.; Paris, J.; Samuel, S. Antibracket, antifields and gauge theory quantization. *Phys. Rep.* **1995**, *259*, 1–145.

34. Gomis, J.; Weinberg, S. Are nonrenormalizable gauge theories renormalizable? *Nucl. Phys. B* **1996**, *469*, 473–487.

35. Brezin, E.; Zinn-Justin, J.; Le Guillou, J. Renormalization of the Nonlinear Sigma Model in (Two + Epsilon) Dimension. *Phys. Rev. D* **1976**, *14*, 2615–2621.

36. Becchi, C.; Piguet, O. On the Renormalization of Two-dimensional Chiral Models. *Nucl. Phys. B* **1989**, *315*, 153–165.

37. Zinn-Justin, J. *Quantum Field Theory and Critical Phenomena*; International Series of Monographs on Physics; Oxford University Press: Oxford, UK, 2002.

38. Ecker, G.; Honerkamp, J. Application of invariant renormalization to the nonlinear chiral invariant pion lagrangian in the one-loop approximation. *Nucl. Phys. B* **1971**, *35*, 481–492.

39. Appelquist, T.; Bernard, C.W. The Nonlinear σ Model in the Loop Expansion. *Phys. Rev. D* **1981**, *23*, doi:10.1103/PhysRevD.23.425.

40. Tataru, L. One Loop Divergences of the Nonlinear Chiral Theory. *Phys. Rev. D* **1975**, *12*, 3351–3352.

41. Gerstein, I.; Jackiw, R.; Weinberg, S.; Lee, B. Chiral loops. *Phys. Rev. D* **1971**, *3*, 2486–2492.

42. Charap, J. Closed-loop calculations using a chiral-invariant lagrangian. *Phys. Rev. D* **1970**, *2*, 1554–1561.

43. Honerkamp, J.; Meetz, K. Chiral-invariant perturbation theory. *Phys. Rev. D* **1971**, *3*, 1996–1998.

44. Stueckelberg, E. Interaction forces in electrodynamics and in the field theory of nuclear forces. *Helv. Phys. Acta* **1938**, *11*, 299–328.

45. Ruegg, H.; Ruiz-Altaba, M. The Stueckelberg field. *Int. J. Mod. Phys.* **2004**, *A19*, 3265–3348.

46. Altarelli, G.; Mangano, M.L. Electroweak Physics. In Proceedings of CERN Workshop on Standard Model Physics (and More) at the LHC, CERN, Geneva, Switzerland, 25–26 May 1999.

47. Azatov, A.; Contino, R.; Galloway, J. Model-Independent Bounds on a Light Higgs. *JHEP* **2012**, *1204*, doi:10.1007/JHEP04(2012)127.

48. Contino, R. The Higgs as a Composite Nambu-Goldstone Boson. Available online: http://arxiv.org/abs/1005.4269 (accessed on 15 April 2014).

49. Espinosa, J.; Grojean, C.; Muhlleitner, M.; Trott, M. First Glimpses at Higgs' face. *JHEP* **2012**, *1212*, doi:DOI:10.1007/JHEP12(2012)045.

50. Zinn-Justin, J. Renormalization of Gauge Theories—Unbroken and broken. *Phys. Rev. D* **1974**, *9*, 933–946.

51. Velo, G.; Wightman, A. Renormalization Theory. In Proceedings of the NATO Advanced Study Institute, Erice, Sicily, Italy, 17–31 August 1975.

52. Breitenlohner, P.; Maison, D. Dimensional Renormalization and the Action Principle. *Commun. Math. Phys.* **1977**, *52*, 11–38.

53. Lam, Y.M.P. Perturbation Lagrangian theory for scalar fields: Ward-Takahasi identity and current algebra. *Phys. Rev. D* **1972**, *6*, 2145–2161.

54. Lam, Y.M.P. Perturbation lagrangian theory for Dirac fields—Ward-Takahashi identity and current algebra. *Phys. Rev. D* **1972**, *6*, 2161–2167.

55. Lowenstein, J. Normal product quantization of currents in Lagrangian field theory. *Phys. Rev. D* **1971**, *4*, 2281–2290.

56. Piguet, O.; Sorella, S. Algebraic renormalization: Perturbative renormalization, symmetries and anomalies. *Lect. Notes Phys.* **1995**, *M28*, 1–134.

57. Becchi, C.; Rouet, A.; Stora, R. Renormalization of Gauge Theories. *Ann. Phys.* **1976**, *98*, 287–321.

58. Becchi, C.; Rouet, A.; Stora, R. Renormalization of the Abelian Higgs-Kibble Model. *Commun. Math. Phys.* **1975**, *42*, 127–162.

59. Becchi, C.; Rouet, A.; Stora, R. The Abelian Higgs-Kibble Model. Unitarity of the S Operator. *Phys. Lett. B* **1974**, *52*, 344–346.

60. Wess, J.; Zumino, B. Consequences of anomalous Ward identities. *Phys. Lett. B* **1971**, *37*, 95–97.

61. Barnich, G.; Brandt, F.; Henneaux, M. Local BRST cohomology in gauge theories. *Phys. Rep.* **2000**, *338*, 439–569.

62. Henneaux, M.; Wilch, A. Local BRST cohomology of the gauged principal nonlinear sigma model. *Phys. Rev. D* **1998**, *58*, 025017:1–025017:14.

63. Quadri, A. Slavnov-Taylor parameterization of Yang-Mills theory with massive fermions in the presence of singlet axial-vector currents. *JHEP* **2005**, *0506*, doi:10.1088/1126-6708/2005/06/068.

64. Quadri, A. Higher order nonsymmetric counterterms in pure Yang-Mills theory. *J. Phys. G* **2004**, *30*, 677–689.

65. Quadri, A. Slavnov-Taylor parameterization for the quantum restoration of BRST symmetries in anomaly free gauge theories. *JHEP* **2003**, *0304*, doi:10.1088/1126-6708/2003/04/017.

66. Quadri, A. Algebraic properties of BRST coupled doublets. *JHEP* **2002**, *0205*, doi:10.1088/1126-6708/2002/05/051.

67. Ferrari, R.; Quadri, A. Physical unitarity for massive non-Abelian gauge theories in the Landau gauge: Stueckelberg and Higgs. *JHEP* **2004**, *0411*, doi:10.1088/1126-6708/2004/11/019.

68. Froissart, M. Asymptotic behavior and subtractions in the Mandelstam representation. *Phys. Rev.* **1961**, *123*, 1053–1057.

69. Cornwall, J.M.; Levin, D.N.; Tiktopoulos, G. Derivation of Gauge Invariance from High-Energy Unitarity Bounds on the s Matrix. *Phys. Rev. D* **1974**, *10*, 1145–1167.

70. Lee, B.W.; Quigg, C.; Thacker, H. Weak Interactions at Very High-Energies: The Role of the Higgs Boson Mass. *Phys. Rev. D* **1977**, *16*, 1519–1531.

71. Weldon, H.A. The Effects of Multiple Higgs Bosons on Tree Unitarity. *Phys. Rev. D* **1984**, *30*, 1547–1558.

72. Chanowitz, M.S.; Gaillard, M.K. The TeV Physics of Strongly Interacting W's and Z's. *Nucl. Phys. B* **1985**, *261*, 379–431.

73. Gounaris, G.; Kogerler, R.; Neufeld, H. Relationship Between Longitudinally Polarized Vector Bosons and their Unphysical Scalar Partners. *Phys. Rev. D* **1986**, *34*, 3257–3259.

74. Bettinelli, D.; Ferrari, R.; Quadri, A. Of Higgs, Unitarity and other Questions. *Proc. Steklov Inst. Math.* **2011**, *272*, 22–38.

75. Aguilar, A.; Ibanez, D.; Mathieu, V.; Papavassiliou, J. Massless bound-state excitations and the Schwinger mechanism in QCD. *Phys. Rev. D* **2012**, *85*, doi:10.1103/PhysRevD.85.014018 .

76. Aguilar, A.; Binosi, D.; Papavassiliou, J. The dynamical equation of the effective gluon mass. *Phys. Rev. D* **2011**, *84*, doi:10.1103/PhysRevD.84.085026.

77. Ibañez, D.; Papavassiliou, J. Gluon mass generation in the massless bound-state formalism. *Phys. Rev. D* **2013**, *87*, doi:10.1103/PhysRevD.87.034008.

78. Higgs, P.W. Broken symmetries, massless particles and gauge fields. *Phys. Lett.* **1964**, *12*, 132–133.

79. Higgs, P.W. Broken Symmetries and the Masses of Gauge Bosons. *Phys. Rev. Lett.* **1964**, *13*, 508–509.

80. Englert, F.; Brout, R. Broken Symmetry and the Mass of Gauge Vector Mesons. *Phys. Rev. Lett.* **1964**, *13*, 321–323.

81. Connes, A.; Kreimer, D. Renormalization in quantum field theory and the Riemann-Hilbert problem. 1. The Hopf algebra structure of graphs and the main theorem. *Commun. Math. Phys.* **2000**, *210*, 249–273.

82. Connes, A.; Kreimer, D. Renormalization in quantum field theory and the Riemann-Hilbert problem. 2. The beta function, diffeomorphisms and the renormalization group. *Commun. Math. Phys.* **2001**, *216*, 215–241.

83. Ebrahimi-Fard, K.; Patras, F. Exponential renormalization. *Ann. Henri Poincare* **2010**, *11*, 943–971.

84. Zimmermann, W. Convergence of Bogolyubov's method of renormalization in momentum space. *Commun. Math. Phys.* **1969**, *15*, 208–234.

© 2014 by the author. Licensee MDPI, Basel, Switzerland. This article is an open access article distributed under the terms and conditions of the Creative Commons Attribution (CC BY) license (http://creativecommons.org/licenses/by/4.0/).

symmetry

MDPI

Article

Dynamical Relation between Quantum Squeezing and Entanglement in Coupled Harmonic Oscillator System

Lock Yue Chew [1],* and Ning Ning Chung [2]

[1] Division of Physics and Applied Physics, School of Physical and Mathematical Sciences, Nanyang Technological University, Singapore 637371, Singapore

[2] Department of Physics, National University of Singapore, Singapore 117542, Singapore;
   E-Mail: phycnn@nus.edu.sg

* E-Mail: lockyue@ntu.edu.sg; Tel.: +65-6316-2968; +65-6316-6984.

Received: 27 February 2014; in revised form: 14 April 2014 / Accepted: 18 April 2014 / Published: 23 April 2014

Abstract: In this paper, we investigate into the numerical and analytical relationship between the dynamically generated quadrature squeezing and entanglement within a coupled harmonic oscillator system. The dynamical relation between these two quantum features is observed to vary monotically, such that an enhancement in entanglement is attained at a fixed squeezing for a larger coupling constant. Surprisingly, the maximum attainable values of these two quantum entities are found to consistently equal to the squeezing and entanglement of the system ground state. In addition, we demonstrate that the inclusion of a small anharmonic perturbation has the effect of modifying the squeezing *versus* entanglement relation into a nonunique form and also extending the maximum squeezing to a value beyond the system ground state.

Keywords: quantum entanglement; squeezed state; coupled harmonic oscillators

PACS: 03.65.Ge, 31.15.MD

1. Introduction

Entanglement is a fundamental resource for non-classical tasks in the field of quantum information [1]. It has been shown to improve communication and computation capabilities via the notion of quantum dense coding [2], quantum teleportation [3], unconditionally secured quantum cryptographic protocols [4,5], and quantum algorithms for integer factorization [6]. For any quantum algorithm operating on pure states, it has been proven that the presence of multi-partite entanglement is necessary if the quantum algorithm is to offer an exponential speed-up over classical computation [7]. Note, however, that a non-zero value of entanglement might not be the necessary condition for quantum computational speed up of algorithm operating on mixed states [8]. In addition, in order to achieve these goals practically, it is necessary to maintain the entanglement within the quantum states which are fragile against the decohering environment. An approach would be to employ an entangled state with as large an entanglement as possible, and the idea is that the production of such entangled state could be tuned through the operation of quantum squeezing.

Indeed, the relation between quantum squeezing and quantum entanglement has been actively pursued in recent years [9–18]. Notably, the creation of entanglement is shown experimentally to be able to induce spin squeezing [9,10]. Such entanglement-induced squeezing has the important outcome of producing measuring instruments that go beyond the precision of current models. In addition, quantum squeezing is found to be able to induce, enhance and even preserve entanglement in decohering environments [11–13]. Previously, we have investigated the relation between the squeezing and entanglement of the ground state of the coupled harmonic oscillator system [16,17]. The ground state entanglement entropy was found to increase monotonically with an increase in quadrature squeezing within this system. When a small anharmonic perturbing potential is added to the system, a further enhancement in quadrature squeezing is observed. While the entropy-squeezing curve shifts to the right in this case, we realized that the entanglement entropy is still a monotonically increasing function in terms of quadrature squeezing.

In this paper, we have extended our earlier work discussed above by investigating into the dynamical relation between quadrature squeezing and entanglement entropy of the coupled harmonic oscillator system. Coupled harmonic oscillator system has served as useful paradigm for many physical systems, such as the field modes of electromagnetic radiation [19–21], the vibrations in molecular systems [22], and the formulation of the Lee model in quantum field theory [23]. It was shown that the coupled harmonic oscillator system possesses the symmetry of the Lorentz group $O(3,3)$ or $SL(4,r)$ classically, and that of the symmetry $O(3,2)$ or $Sp(4)$ quantum mechanically [24]. In addition, the physics of coupled harmonic oscillator system can be conveniently represented by the mathematics of two-by-two matrices, which have played a role in clarifying the physical basis of entanglement [25]. In Section 2 of this paper, we first described the coupled harmonic oscillator model. It is then followed by a discussion on the relation between the dynamically generated squeezing and entanglement of the coupled oscillator systems, which we have determined quantitatively via numerical computation. In Section 3 of the paper, we present analytical results in support of the numerical results obtained in Section 2. Here, we illustrate how the problem can be solved in terms of two-by-two matrices. Then, in Section 4 of the paper, we study how the inclusion of anharmonicity can influence the relation between the dynamically generated squeezing and entanglement. Finally, we give our conclusion in Section 5 of the paper.

2. Dynamical Relation of Quantum Squeezing and Entanglement in Coupled Harmonic Oscillator System

The Hamiltonian of the coupled harmonic oscillator system is described as follow:

$$H = \frac{p_1^2}{2m_1} + \frac{1}{2}m_1\omega_1^2 x_1^2 + \frac{p_2^2}{2m_2} + \frac{1}{2}m_2\omega_2^2 x_2^2 + \lambda(x_2 - x_1)^2 \tag{1}$$

where x_1 and x_2 are the position co-ordinates, while p_1 and p_2 are the momenta of the oscillators. The interaction potential between the two oscillators is assumed to depend quadratically on the distance between the oscillators, and is proportional to the coupling constant λ. For simplicity, we have set $m_1 = m_2 = m$ and $\omega_1 = \omega_2 = \omega$. This Hamiltonian is commonly used to model physical systems such as the vibrating molecules or the squeezed modes of electromagnetic field. In fact, the model has been widely explored [26–28] and is commonly used to elucidate the properties of quantum entanglement in continuous variable systems [29–35].

Next, let us discuss on the relation between the squeezing and entanglement of the lowest energy eigenstate of this coupled harmonic oscillator system. Note that

$$H|g\rangle = E_0|g\rangle \tag{2}$$

with $|g\rangle$ being the ground state and E_0 being the lowest eigen-energy of the coupled oscillator system with Hamiltonian given by Equation (1). Entanglement between the two oscillators can be quantified by the von Neumann entropy:

$$S_{vN} = -\text{Tr}\left[\rho_l \ln \rho_l\right] \tag{3}$$

where ρ_l is the reduced density matrix. For squeezing parameter, we shall adopt the dimensionless definition:

$$S_x = -\ln \frac{\sigma_{x_1}}{\sigma_{x_1}^{(0)}} \tag{4}$$

with $\sigma_{x_1} = \sqrt{\langle x_1^2 \rangle - \langle x_1 \rangle^2}$ being the uncertainty associated with the first oscillator's position and the normalization constant $\sigma_{x_1}^{(0)} = \sqrt{\hbar/2m\omega}$ being the uncertainty associated with the harmonic oscillator's position. For simplicity, we shall evaluate only the position squeezing in the first oscillator.

Indeed, the position uncertainty squeezing and the entanglement entropy of the ground state of this oscillator have been solved analytically by previous studies [36,37] as follows:

$$S_x = -\ln \frac{\sqrt{\frac{\hbar}{2m\omega} \frac{1+\gamma}{2}}}{\sqrt{\frac{\hbar}{2m\omega}}} = -\ln \sqrt{\frac{1+\gamma}{2}} \tag{5}$$

where $\gamma = 1/\sqrt{1 + 4\lambda/m\omega^2}$; and

$$S_{vN} = \cosh^2\left(\frac{\ln\gamma}{4}\right) \ln\left[\cosh^2\left(\frac{\ln\gamma}{4}\right)\right] - \sinh^2\left(\frac{\ln\gamma}{4}\right) \ln\left[\sinh^2\left(\frac{\ln\gamma}{4}\right)\right] \tag{6}$$

As shown in Reference [17], by eliminating γ between Equations (5) and (6), the relation between the squeezing parameter and the von Neumann entropy of the ground state of the coupled harmonic oscillators is obtained as follow:

$$S_{vN} = \frac{(\xi+1)^2}{4\xi} \ln\left(\frac{(\xi+1)^2}{4\xi}\right) - \frac{(\xi-1)^2}{4\xi} \ln\left(\frac{(\xi-1)^2}{4\xi}\right) \tag{7}$$

with

$$\xi = \sqrt{2e^{-2S_x} - 1} \tag{8}$$

This relation is shown as a solid line in Figure 1.

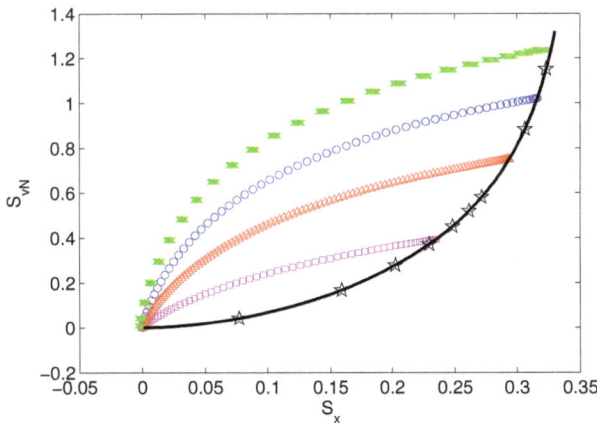

Figure 1. A plot on the dynamical relation between entanglement and squeezing obtained numerically for coupled harmonic oscillator system with the coupling constant $\lambda = 0.75$ (squares), 2 (triangles), 3.75 (circles) and 6 (crosses). Note that the ground state entanglement-squeezing curve given by Equation (7) is plotted as a solid curve for comparison. In addition, the values of the maximum attainable squeezing and entanglement for various λ have been plotted as stars.

In this paper, we have gone beyond the static relation between squeezing and entanglement based on the stationary ground state. In particular, we have explored numerically into the dynamical generation of squeezing and entanglement via the quantum time evolution, with the initial state being the tensor product of the vacuum states ($|0,0\rangle$) of the oscillators. Note that the obtained results hold true for any initial coherent states ($|\alpha_1, \alpha_2\rangle$) since the entanglement dynamics of the coupled harmonic oscillator system is independent of initial states [38]. In general, the system dynamics is either two-frequency periodic or quasi-periodic depending on whether the ratio of the two frequencies, $f_1 = 1$ and $f_2 = \sqrt{1 + 4\lambda}$, are rational or irrational. By yielding the values of the squeezing parameter and the entanglement entropy at the same time point within their respective dynamical evolution, we obtained the dynamical relations between the squeezing and entanglement for different coupling constants $\lambda = 0.75, 2, 3.75$ and 6, as shown in Figure 1. Interestingly, the results show a smooth monotonic increase of the dynamically generated entanglement entropy as the quadrature squeezing increases for each λ. In addition, the dynamically generated entanglement entropy is observed to be larger for a fixed squeezing as λ increases. It is surprising that the maximum attainable values of these two quantum entities determined dynamically are found to fall consistently on the system ground states' squeezing and entanglement relation as given by Equations (7) and (8) for all values of λ. More importantly, this relation also serve as a bound to the entanglement entropy and squeezing that are generated dynamically.

3. Analytical Derivation on the Dynamical Relation between Quantum Squeezing and Entanglement

In this section, we shall perform an analytical study on the dynamical relationship between quantum squeezing and the associated entanglement production. We first yield the second quantized form of the Hamiltonian of the coupled harmonic oscillator system as follow:

$$H = a_1^\dagger a_1 + a_2^\dagger a_2 + 1 + \frac{\lambda}{2}\{(a_1^\dagger + a_1) - (a_2^\dagger + a_2)\}^2 \tag{9}$$

Then, the time evolution of the annihilation operator a_j (as well as the creation operator $a_j^\dagger$) can be determined according to the following Heisenberg equation of motion:

$$\frac{d}{dt}a_j = \frac{1}{i}[a_j, H] \tag{10}$$

From this, we obtain:

$$\frac{d}{dt}\tilde{a} = A\tilde{a} \tag{11}$$

with $\tilde{a} = (a_1 \, a_1^\dagger \, a_2 \, a_2^\dagger)^T$ and

$$A = \begin{pmatrix} B & C \\ C & B \end{pmatrix} \tag{12}$$

Note that

$$B = i\begin{pmatrix} -(1+\lambda) & -\lambda \\ \lambda & 1+\lambda \end{pmatrix} \tag{13}$$

and

$$C = i\begin{pmatrix} \lambda & \lambda \\ -\lambda & -\lambda \end{pmatrix} \tag{14}$$

Due to the symmetry in the coupled oscillator system, the matrix A is symmetric in the form of a two-by-two matrix although it is not symmetric in its full four-by-four matrix form. This symmetric property enables a simple evaluation of the time dependent annihilation and creation operators of the oscillators:

$$\tilde{a}(t) = F\tilde{a}(0) \tag{15}$$

where

$$F = \frac{1}{2}\begin{pmatrix} Je^{D_1t}J - Ke^{D_2t}K^{-1} & Je^{D_1t}J + Ke^{D_2t}K^{-1} \\ Je^{D_1t}J + Ke^{D_2t}K^{-1} & Je^{D_1t}J - Ke^{D_2t}K^{-1} \end{pmatrix} \tag{16}$$

$$J = \begin{pmatrix} 0 & 1 \\ 1 & 0 \end{pmatrix} \tag{17}$$

$$D_1 = \begin{pmatrix} i & 0 \\ 0 & -i \end{pmatrix} \tag{18}$$

$$D_2 = \begin{pmatrix} i\Omega & 0 \\ 0 & -i\Omega \end{pmatrix} \tag{19}$$

and

$$K = \begin{pmatrix} 1 & \beta \\ \beta & 1 \end{pmatrix} \tag{20}$$

with $\Omega = f_2 = \sqrt{1+4\lambda}$ and $\beta = (1+\Omega)/(1-\Omega)$. We then have:

$$a_1(t) = \left(\frac{1}{2}e^{-it} - \eta_1 + \eta_2\right) a_1(0) + \eta_3\, a_1^\dagger(0) + \left(\frac{1}{2}e^{-it} + \eta_1 - \eta_2\right) a_2(0) - \eta_3\, a_2^\dagger(0) \tag{21}$$

$$a_1^\dagger(t) = -\eta_3\, a_1(0) + \left(\frac{1}{2}e^{it} - \eta_1^* + \eta_2^*\right) a_1^\dagger(0) + \eta_3\, a_2(0) + \left(\frac{1}{2}e^{it} + \eta_1^* - \eta_2^*\right) a_2^\dagger(0) \tag{22}$$

$$a_2(t) = \left(\frac{1}{2}e^{-it} + \eta_1 - \eta_2\right) a_1(0) - \eta_3\, a_1^\dagger(0) + \left(\frac{1}{2}e^{-it} - \eta_1 + \eta_2\right) a_2(0) + \eta_3\, a_2^\dagger(0) \tag{23}$$

$$a_2^\dagger(t) = \eta_3\, a_1(0) + \left(\frac{1}{2}e^{it} + \eta_1^* - \eta_2^*\right) a_1^\dagger(0) - \eta_3\, a_2(0) + \left(\frac{1}{2}e^{it} - \eta_1^* + \eta_2^*\right) a_2^\dagger(0) \tag{24}$$

where

$$\eta_1 = \frac{(1-\Omega)^2}{8\Omega} e^{i\Omega t}$$

$$\eta_2 = \frac{(1+\Omega)^2}{8\Omega} e^{-i\Omega t}$$

$$\eta_3 = \frac{i(1-\Omega)(1+\Omega)}{4\Omega} \sin(\Omega t)$$

With these results, we are now ready to determine the analytical expressions of both the quantum entanglement and squeezing against time. For entanglement, we shall employ the criterion developed by Duan *et al.* [39] for quantification since it leads to simplification of the analytical expression while remaining valid as a measure of entanglement in coupled harmonic oscillator systems. According to this criterion, as long as

$$S_D = 2 - (\Delta u)^2 - (\Delta v)^2 > 0 \tag{25}$$

the state of the quantum system is entangled. Note that $u = x_1 + x_2$ and $v = p_1 - p_2$ are two EPR-type operators, whereas Δu and Δv are the corresponding quantum fluctuation. This allows us to express the entanglement measure S_D as follow:

$$S_D(t) = 2(\langle a_1^\dagger a_1 \rangle - \langle a_1^\dagger \rangle \langle a_1 \rangle + \langle a_2^\dagger a_2 \rangle - \langle a_2^\dagger \rangle \langle a_2 \rangle + \langle a_1^\dagger a_2^\dagger \rangle - \langle a_1^\dagger \rangle \langle a_2^\dagger \rangle + \langle a_1 a_2 \rangle - \langle a_1 \rangle \langle a_2 \rangle) \tag{26}$$

Note that the short form $\langle O \rangle$ used in Equation (26) implies $\langle \alpha_1, \alpha_2 | O(t) | \alpha_1, \alpha_2 \rangle$, where $|\alpha_1, \alpha_2\rangle$ represents a tensor product of arbitrary initial coherent states. Recall that the subsequent results are indepedent of the initial states as mentioned in the last section. After substituting Equations (21)–(24) into Equation (26), we obtain the analytical expression of entanglement against time:

$$S_D(t) = (\Omega^2 - 1) \sin^2 \Omega t \tag{27}$$

In coupled harmonic oscillator systems, S_D has a unique monotonic relation with S_{vN} (see Figure 2). For squeezing, we have

$$
\begin{aligned}
S_x(t) &= -\ln \sqrt{\frac{\langle x_1^2 \rangle - \langle x_1 \rangle^2}{0.5}} \\
&= -\ln \sqrt{\langle a_1^{\dagger 2} \rangle - \langle a_1^\dagger \rangle^2 + \langle a_1^2 \rangle - \langle a_1 \rangle^2 + \langle a_1^\dagger a_1 \rangle - \langle a_1^\dagger \rangle \langle a_1 \rangle + \langle a_1 a_1^\dagger \rangle - \langle a_1 \rangle \langle a_1^\dagger \rangle}
\end{aligned}
\tag{28}
$$

Then, by substituting Equations (21)–(24) into Equation (28) as before, we obtain the analytical expression of squeezing against time:

$$S_x(t) = -\ln \sqrt{1 - \frac{\Omega^2 - 1}{2\Omega^2} \sin^2 \Omega t} \tag{29}$$

We can also obtain an analytical expression between S_D and S_x by substituting Equation (27) into Equation (29) with some rearrangement:

$$S_D = 2\Omega^2 \left(1 - e^{-2S_x}\right) \tag{30}$$

It is important to note that S_x can only span a range of values $0 \leq S_x \leq S_x^{(m)}$, where $S_x^{(m)} = -\ln \sqrt{(\Omega^2 + 1)/2\Omega^2}$. Furthermore, for a coupled harmonic oscillator system with a fixed value of λ, the dynamically generated squeezing can be higher than the squeezing in the system's ground state. The analytical result given by Equation (30) is plotted in Figure 3 for $\lambda = 0.75, 2, 3.75, 6$ and 10, with each curve begins at $S_x = 0$, $S_D = 0$ and ends at $S_x = S_x^{(m)}$, $S_D = S_D^{(m)} = \Omega^2 - 1$. In fact, the set of end points given by $S_x = S_x^{(m)}$, $S_D = S_D^{(m)}$ gives rise to the solid curve in Figure 3. Specifically, the maximum entanglement and the maximum squeezing parameter relates as follow:

$$S_D^{(m)} = \frac{1 - \xi^2}{\xi^2} \tag{31}$$

with

$$\xi = \sqrt{2 e^{-2S_x^{(m)}} - 1} \tag{32}$$

Note that Equation (32) is the same as Equation (8), and Equation (31) corresponds to the ground state solid curve of Figure 1. This allows us to deduce the monotonic relation between S_D and S_{vN}, which is performed by evaluating the relation between S_D of the maximum entangled state and S_{vN} of the ground state at equal amount of squeezing. Indeed, the resulting derived relationship shown as solid line in Figure 2 is valid due to the fact that the link between $S_D(t)$ and $S_{vN}(t)$ is found to be expressible by precisely the same curve. Thus, we have concretely affirmed the one to one correspondence between S_D and S_{vN} through this relationship. More importantly, we have clearly demonstrated that the maximum entanglement attained dynamically is the same as the degree of entanglement of a ground state with the same squeezing.

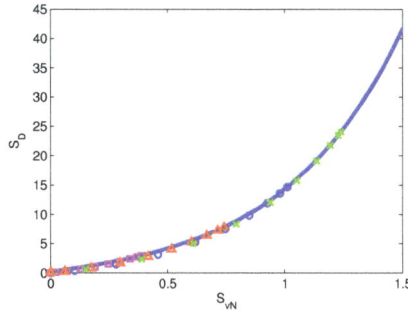

Figure 2. This plot shows the monotonic relation between S_D and S_{vN} in coupled harmonic oscillator systems. $S_D(t)$ and $S_{vN}(t)$ are plotted as squares ($\lambda = 0.75$), triangles ($\lambda = 2$), circles ($\lambda = 3.75$) and crosses ($\lambda = 6$). The relation between the ground state von Neuman entropy given by $S_{vN} = \frac{(\zeta+1)^2}{4\zeta} \ln\left(\frac{(\zeta+1)^2}{4\zeta}\right) - \frac{(\zeta-1)^2}{4\zeta} \ln\left(\frac{(\zeta-1)^2}{4\zeta}\right)$ and the maximum dynamically generated entanglement given by $S_D^{(m)} = \frac{1-\zeta^2}{\zeta^2}$ is plotted as solid curve. Note that both S_{vN} and $S_D^{(m)}$ are functions of the squeezing parameter S_x and $\zeta = \sqrt{2e^{-2S_x} - 1}$.

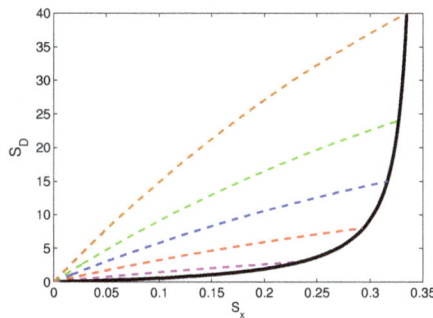

Figure 3. A plot on the dynamical relation between entanglement and squeezing given by Equation (30) for coupled harmonic oscillator system. The relation is dependent on λ and the curves from top to bottom are with respect to $\lambda = 10, 6, 3.75, 2,$ and 0.75 respectively. Note that the thick solid curve represents the values of the maximum attainable squeezing and entanglement for the range $0 < \lambda < 10$.

When projected into the $x_1 - p_2$ or $x_2 - p_1$ plane, the initial coherent state can be represented by a circular distribution with equal uncertainty in both x and p direction. During the time evolution, the circular distribution is being rotated and squeezed. As a result, squeezing and entanglement are generated such that the distribution becomes elliptical in the $x_1 - p_2$ or $x_2 - p_1$ plane with rotation of the ellipse's major axis away from the x- or p-axis which creates entanglement. The generation of squeezing and entanglement reaches their maximum values at the same time when the major axis of the elliptical distribution has rotated $45°$ away from the x- or p-axis. Note that at this point, squeezing is merely in the collective modes. On the other hand, as discussed in Reference [37], the ground state wave function of the coupled harmonic oscillator system is separable in their collective modes. In both cases, entanglement and squeezing relates uniquely as given by Equation (7) and (31).

4. Quantum Squeezing and Entanglement in Coupled Anharmonic Oscillator Systems

Next, let us investigate the effect of including an anharmonic potential on the dynamical relation between squeezing and entanglement through the following Hamiltonian systems:

$$H = \frac{p_1^2}{2m_1} + \frac{1}{2}m_1\omega_1^2 x_1^2 + \frac{p_2^2}{2m_2} + \frac{1}{2}m_2\omega_2^2 x_2^2 + \lambda(x_2 - x_1)^2 + \epsilon(x_1^4 + x_2^4) \tag{33}$$

For simplicity, we consider only the quartic perturbation potential. For previous studies of entanglement in coupled harmonic oscillators with quartic perturbation, see Reference [40] and the references therein. Again, we choose the initial state to be the tensor product of the vacuum states. We then evolve the state numerically through the Hamiltonian given by Equation (33). For the numerical simulation, we consider only a small anharmonic perturbation, *i.e.*, $\epsilon = 0.1$ and 0.2. Note that we have truncated the basis size at $M = 85$ at which the results are found to converge.

With a small anharmonic perturbation, the dynamically generated entanglement entropy is no longer a smooth monotonically increasing function of the quadrature squeezing as before (see Figure 4). This implies that for coupled anharmonic oscillator systems, the dynamically generated degree of entanglement cannot be characterized through a measurement of the squeezing parameter. In addition, when the anharmonic potential is included, the maximum attainable squeezing is much enhanced. This effect is clearly shown in Figure 4, where we observe that the maximum dynamical squeezing extends far beyond the largest squeezing given by the coupled anharmonic oscillator system's ground state at different λ. In addition, as we increase the anharmonic perturbation from 0.1 to 0.2, we found that the maximum attainable squeezing continues to grow with extension going further beyond the largest squeezing given by the ground state of the coupled anharmonic oscillator system.

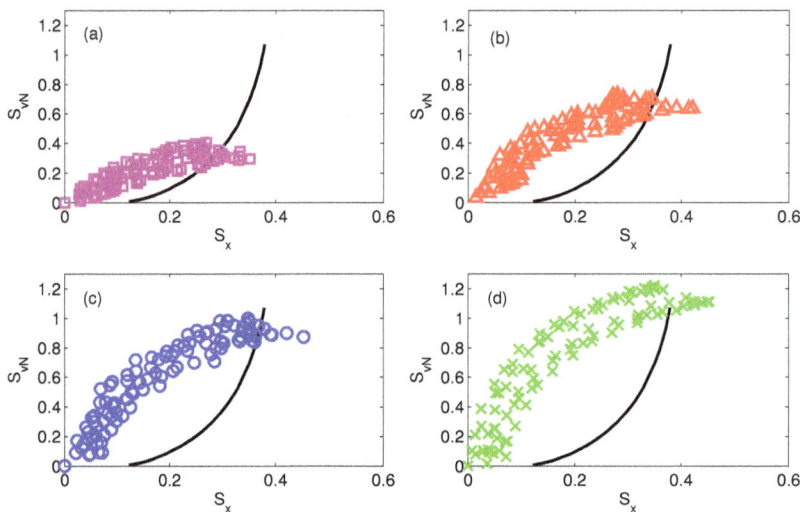

Figure 4. The effect of anharmonicity ($\epsilon = 0.1$) on the dynamical relation between quadrature squeezing and entanglement. Note that we have employed the following parameter: (a) $\lambda = 0.75$; (b) $\lambda = 2$; (c) $\lambda = 3.75$; and (d) $\lambda = 6$. We have plotted the ground state entanglement-squeezing curve of the coupled anharmonic oscillator system with $\epsilon = 0.1$ as solid curve for comparison.

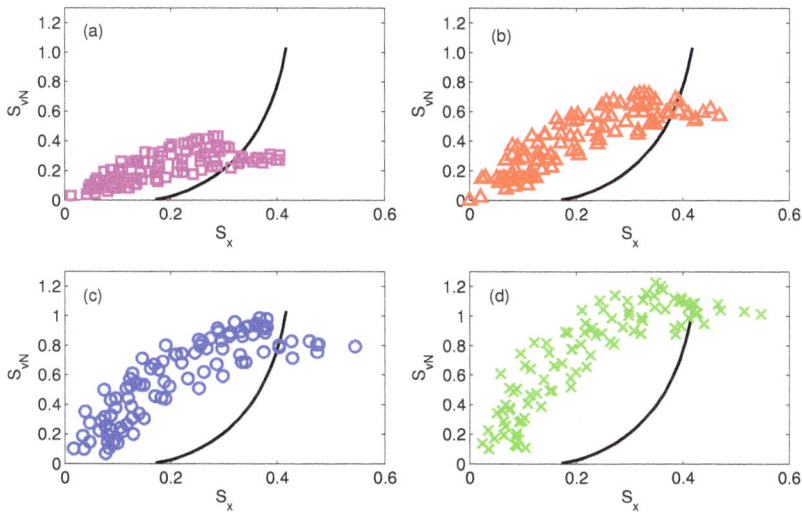

Figure 5. The effect of anharmonicity ($\epsilon = 0.2$) on the dynamical relation between quadrature squeezing and entanglement. Note that we have employed the following parameter: (a) $\lambda = 0.75$, (b) $\lambda = 2$, (c) $\lambda = 3.75$, and (d) $\lambda = 6$. We have plotted the ground state entanglement-squeezing curve of the coupled anharmonic oscillator system with $\epsilon = 0.2$ as solid curve for comparison.

5. Conclusions

We have studied into the dynamical generation of quadrature squeezing and entanglement for both coupled harmonic and anharmonic oscillator systems. Our numerical and analytical results show that the quantitative relation that defines the dynamically generated squeezing and entanglement in coupled harmonic oscillator system is a monotonically increasing function. Such a monotonic relation vanishes, however, when a small anharmonic potential is added to the system. This result implies the possibility of characterizing the dynamically generated entanglement by means of squeezing in the case of coupled harmonic oscillator system. In addition, we have uncovered the unexpected result that the maximum attainable entanglement and squeezing obtained dynamically matches exactly the entanglement-squeezing relation of the system's ground state of the coupled harmonic oscillators. When an anharmonic potential is included, we found that the dynamically generated squeezing can be further enhanced. We percieve that this result may provide important insights to the construction of precision instruments that attempt to beat the quantum noise limit.

Acknowledgments: L. Y. Chew would like to thank Y. S. Kim for the helpful discussion on this work during the ICSSUR 2013 conference held in Nuremberg, Germany.

Author Contributions: All authors contribute equally to the theoretical analysis, numerical computation, and writing of the paper.

Conflicts of Interest: The authors declare no conflict of interest.

References

1. Nielson, M.A.; Chuang, I.L. *Quantum Computation and Quantum Information*; Cambridge University Press: Cambridge, UK, 2000.
2. Bennett, C.H.; Wiesner, S.J. Communication via one- and two-particle operators on Einstein-Podolsky-Rosen states. *Phys. Rev. Lett.* **1992**, *69*, 2881–2884.

3. Bennett, C.H.; Brassard, G.; Crépeau, C.; Jozsa, R.; Peres, A.; Wooters, W.K. Teleporting an unknown quantum state via dual classical and Einstein-Podolsky-Rosen channels. *Phys. Rev. Lett.* **1993**, *70*, 1895–1899.

4. Bennett, C.H.; Brassard, G. Quantum cryptography: Public key distribution and coin tossing. In Proceedings of the IEEE International Conference on Computers, Systems and Signal Processing, IEEE Computer Society, New York, NY, USA, 1984; pp. 175–179.

5. Ekert, A.K. Quantum cryptography based on Bell's theorem. *Phys. Rev. Lett.* **1991**, *67*, 661–663.

6. Shor, P.W. Polynomial-time algorithms for prime factorization and discrete logarithms on a quantum computer. *SIAM J. Comput.* **1997**, *26*, 1484–1509.

7. Jozsa, R.; Linden, N. On the role of entanglement in quantum-computational speed-up. *Proc. R. Soc. Lond. A* **2003**, *459*, 2011–2032.

8. Lanyon, B.P.; Barbieri, M.; Almeida, M.P.; White, A.G. Experimental quantum computing without entanglement. *Phys. Rev. Lett.* **2008**, *101*, 200501:1–200501:4.

9. Sørensen, A.; Duan, L.M.; Cirac, J.I.; Zoller, P. Many-particle entanglement with Bose-Einstein condensates. *Nature* **2001**, *409*, 63–66.

10. Bigelow, N. Squeezing Entanglement. *Nature* **2001**, *409*, 27–28.

11. Furuichi, S.; Mahmoud, A.A. Entanglement in a squeezed two-level atom. *J. Phys. A Math. Gen.* **2001**, *34*, 6851–6857.

12. Xiang, S.; Shao, B.; Song, K. Quantum entanglement and nonlocality properties of two-mode Gaussian squeezed states. *Chin. Phys. B* **2009**, *18*, 418–425.

13. Galve, F.; Pachón, L.A.; Zueco, D. Bringing entanglement to the high temperature limit. *Phys. Rev. Lett.* **2010**, *105*, doi:10.1103/PhysRevLett.105.180501.

14. Ulam-Orgikh, D.; Kitagawa, M. Spin squeezed and decoherence limit in Ramsey spectroscopy. *Phys. Rev. A* **2001**, *64*, doi:10.1103/PhysRevA.64.052106.

15. Wolf, M.M.; Eisert, J.; Plenio, M.B. Entangling power of passive optical elements. *Phys. Rev. Lett.* **2003**, *90*, 047904:1–047904:4.

16. Chung, N.N.; Er, C.H.; Teo, Y.S.; Chew, L.Y. Relation of the entanglement entropy and uncertainty product in ground states of coupled anharmonic oscillators. *Phys. Rev. A* **2010**, *82*, doi:10.1103/PhysRevA.82.014101.

17. Chew, L.Y.; Chung, N.N. Quantum entanglement and squeezing in coupled harmonic and anharmonic oscillators systems. *J. Russ. Laser Res.* **2011**, *32*, 331–337.

18. Er, C.H.; Chung, N.N.; Chew, L.Y. Threshold effect and entanglement enhancement through local squeezing of initial separable states in continuous-variable systems. *Phys. Scripta* **2013**, *87*, doi:10.1088/0031-8949/87/02/025001.

19. Han, D.; Kim, Y.S.; Noz, M.E. Linear canonical transformations of coherent and squeezed states in the Wigner phase space. *Phys. Rev. A* **1988**, *37*, 807–814.

20. Han, D.; Kim, Y.S.; Noz, M.E. Linear canonical transformations of coherent and squeezed states in the Wigner phase space. II. Quantitative analysis. *Phys. Rev. A* **1989**, *40*, 902–912.

21. Han, D.; Kim, Y.S.; Noz, M.E. Linear canonical transformations of coherent and squeezed states in the Wigner phase space. III. Two-mode states. *Phys. Rev. A* **1990**, *41*, 6233–6244.

22. Wilson, E.B.; Decius, J.C.; Cross, P.C. *Molecular Vibration*; McGraw-Hill: New York, NY, USA, 1955.

23. Schweber, S.S. *An Introduction to Relativistic Quantum Field Theory*; Row-Peterson: New York, NY, USA, 1961.

24. Han, D.; Kim, Y.S.; Noz, M.E. O(3,3)-like symmetries of coupled harmonic-oscillators. *J. Math. Phys.* **1995**, *36*, 3940–3954.

25. Kim, Y.S.; Noz, M.E. Coupled oscillators, entangled oscillators, and Lorentz-covariant harmonic oscillators. *J. Opt. B Quantum Semiclass. Opt.* **2005**, *7*, S458–S467.

26. Eisert, J.; Plenio, M.B.; Bose, S.; Hartley, J. Towards quantum entanglement in nanoelectromechanical devices. *Phys. Rev. Lett.* **2004**, *93*, 190402:1–190402:4.

27. Joshi, C.; Jonson, M.; Öhberg, P.; Andersson, E. Constructive role of dissipation for driven coupled bosonic modes. *Phys. Rev. A* **2013**, *87*, 062304:1–062304:4.

28. Joshi, C.; Hutter, A.; Zimmer, F.E.; Jonson, M.; Andersson, E.; Öhberg, P. Quantum entanglement of nanocantilevers. *Phys. Rev. A* **2010**, *82*, doi:10.1103/PhysRevA.82.043846.

29. Ikeda, S.; Fillaux, F. Incoherent elastic-neutron-scattering study of the vibrational dynamics and spin-related symmetry of protons in the $KHCO_3$ crystal. *Phys. Rev. B* **1999**, *59*, 4134–4145.

30. Fillaux, F. Quantum entanglement and nonlocal proton transfer dynamics in dimers of formic acid and analogues. *Chem. Phys. Lett.* **2005**, *408*, 302–306.
31. Audenaert, K.; Eisert, J.; Plenio, M.B.; Werner, R.F. Symmetric qubits from cavity states. *Phys. Rev. A* **2002**, *66*, 042327:1–042327:6.
32. Martina, L.; Soliani, G. Hartree-Fock approximation and entanglement. Available online: http://arxiv.org/abs/0704.3130 (accessed on 18 April 2014).
33. Chung, N.N.; Chew, L.Y. Energy eigenvalues and squeezing properties of general systems of coupled quantum anharmonic oscillators. *Phys. Rev. A* **2007**, *76*, doi:10.1103/ PhysRevA.76.032113.
34. Chung, N.N.; Chew, L.Y. Two-step approach to the dynamics of coupled anharmonic oscillators. *Phys. Rev. A* **2009**, *80*, doi:10.1103/PhysRevA.80.012103.
35. Jellal, A.; Madouri, F.; Merdaci, A. Entanglement in coupled harmonic oscillators studied using a unitary transformation. *J. Stat. Mech.* **2011**, doi:10.1088/1742-5468/2011/09/P09015.
36. McDermott, R.M.; Redmount, I.H. Coupled classical and quantum oscillators. Available online: http://arxiv.org/abs/quant-ph/0403184 (accessed on 18 April 2014).
37. Han, D.; Kim, Y.S.; Noz, M.E. Illustrative example of Feymann's rest of the universe. *Am. J. Phys.* **1999**, *67*, 61–66.
38. Chung, N.N.; Chew, L.Y. Dependence of entanglement dynamics on the global classical dynamical regime. *Phys. Rev. E* **2009**, *80*, 016204:1–016204:7.
39. Duan, L.M.; Giedke, G.; Cirac, J.I.; Zoller, P. Inseparable criterion for continuous variable systems. *Phys. Rev. Lett.* **2000**, *84*, 2722–2725.
40. Joshi, C.; Jonson, M.; Andersson, E.; Öhberg, P. Quantum entanglement of anharmonic oscillators. *J. Phys. B At. Mol. Opt. Phys.* **2011**, *44*, doi:10.1088/0953-4075/44/24/245503.

© 2014 by the authors. Licensee MDPI, Basel, Switzerland. This article is an open access article distributed under the terms and conditions of the Creative Commons Attribution (CC BY) license (http://creativecommons.org/licenses/by/4.0/).

symmetry

MDPI

Article

Closed-Form Expressions for the Matrix Exponential

F. De Zela

Departamento de Ciencias, Sección Física, Pontificia Universidad Católica del Perú, Ap.1761, Lima L32, Peru;
E-Mail: fdezela@pucp.edu.pe; Tel.: +51-1-6262000; Fax: +51-1-6262085

Received: 28 February 2014; in revised form: 16 April 2014 / Accepted: 17 April 2014 / Published: 29 April 2014

Abstract: We discuss a method to obtain closed-form expressions of $f(A)$, where f is an analytic function and A a square, diagonalizable matrix. The method exploits the Cayley–Hamilton theorem and has been previously reported using tools that are perhaps not sufficiently appealing to physicists. Here, we derive the results on which the method is based by using tools most commonly employed by physicists. We show the advantages of the method in comparison with standard approaches, especially when dealing with the exponential of low-dimensional matrices. In contrast to other approaches that require, e.g., solving differential equations, the present method only requires the construction of the inverse of the Vandermonde matrix. We show the advantages of the method by applying it to different cases, mostly restricting the calculational effort to the handling of two-by-two matrices.

Keywords: matrix exponential; Cayley–Hamilton theorem; two-by-two representations; Vandermonde matrices

PACS: 02.30.Tb, 42.25.Ja, 03.65.Fd

1. Introduction

Physicists are quite often faced with the task of calculating $f(A)$, where A is an $n \times n$ matrix and f an analytic function whose series expansion generally contains infinitely many terms. The most prominent example corresponds to $\exp A$. Usual approaches to calculate $f(A)$ consist in either truncating its series expansion, or else finding a way to "re-summate" terms so as to get a closed-form expression. There is yet another option that can be advantageously applied when dealing with an $n \times n$ matrix, and which derives from the Cayley–Hamilton theorem [1]. This theorem states that every square matrix satisfies its characteristic equation. As a consequence of this property, any series expansion can be written in terms of the first n powers of A. While this result is surely very well known among mathematicians, it appears to be not so widespread within the physicists' community [2]. Indeed, most textbooks on quantum mechanics still resort to the Baker–Hausdorff lemma or to special properties of the involved matrices, in order to obtain closed-form expressions of series expansions [3–5]. This happens even when dealing with low-dimensional matrices, *i.e.*, in cases in which exploiting the Cayley–Hamilton theorem would straightforwardly lead to the desired result. Such a state of affairs probably reflects a lack of literature on the subject that is more palatable to physicists than to mathematicians. The present paper aims at dealing with the subject matter by using language and tools that are most familiar to physicists. No claim of priority is made; our purpose is to show how well the derived results fit into the repertoire of tools that physicists routinely employ. To this end, we start addressing the simple, yet rich enough case of 2×2 matrices.

An archetypical example is the Hamiltonian $H = k\boldsymbol{\sigma} \cdot \boldsymbol{B}$ that rules the dynamics of a spin-1/2 particle subjected to a magnetic field $\boldsymbol{B}$. Here, $\boldsymbol{\sigma} = (\sigma_x, \sigma_y, \sigma_z)$ denotes the Pauli spin operator and k is a parameter that provides the above expression with appropriate units. The upsurge of research in several areas of physics—most notably in quantum optics—involving two-level systems, has made a

Hamiltonian of the above type quite ubiquitous. Indeed, the dynamics of any two-level system is ruled by a Hamiltonian that can be written in such a form. Hence, one often requires an explicit, closed-form expression for quantities such as $\exp(i\alpha n \cdot \sigma)$, where n is a unit vector. This closed-form expression can be obtained as a generalization of Euler's formula $\exp i\alpha = \cos\alpha + i\sin\alpha$. It reads

$$\exp(i\alpha n \cdot \sigma) = \cos\alpha I + i\sin\alpha n \cdot \sigma \tag{1}$$

with I denoting the identity operator.

Let us recall how most textbooks of quantum mechanics proceed to demonstrate Equation (1) (see, e.g., [3–5]). The demonstration starts by writing the series expansion $\exp A = \sum_k A^k/k!$ for the case $A = i\alpha n \cdot \sigma$. Next, one invokes the following relationship:

$$(a \cdot \sigma)(b \cdot \sigma) = (a \cdot b)I + i(a \times b) \cdot \sigma \tag{2}$$

whose proof rests on $\sigma_i\sigma_j = \delta_{ij}I + i\epsilon_{ijk}\sigma_k$ (summation over repeated indices being understood). Equation (2) implies that $(n \cdot \sigma)^{2n} = I$, and hence $(n \cdot \sigma)^{2n+1} = n \cdot \sigma$. This allows one to split the power series of $\exp(i\alpha n \cdot \sigma)$ in two parts, one constituted by even and the other by odd powers of $i\alpha n \cdot \sigma$:

$$\exp(i\alpha n \cdot \sigma) = \sum_{n=0}^{\infty} \frac{(i\alpha)^{2n}}{2n!} I + \sum_{n=0}^{\infty} \frac{(i\alpha)^{2n+1}}{(2n+1)!} n \cdot \sigma \tag{3}$$

By similarly splitting Euler's exponential, *i.e.*,

$$\exp i\alpha = \cos\alpha + i\sin\alpha = \sum_{n=0}^{\infty} \frac{(i\alpha)^{2n}}{2n!} + \sum_{n=0}^{\infty} \frac{(i\alpha)^{2n+1}}{(2n+1)!} \tag{4}$$

one sees that Equation (3) is the same as Equation (1).

Although this standard demonstration is a relatively simple one, it seems to be tightly related to the particular properties of the operator $n \cdot \sigma$, as well as to our ability to "re-summate" the series expansion so as to obtain a closed-form expression. There are several other cases [6] in which a relation similar to Equation (1) follows as a consequence of generalizing some properties of the group $SU(2)$ and its algebra to the case $SU(N)$, with $N > 2$. Central to these generalizations and to their associated techniques are both the Cayley–Hamilton theorem and the closure of the Lie algebra $su(N)$ under commutation and anti-commutation of its elements [6]. As already recalled, the Cayley–Hamilton theorem states that any $n \times n$ matrix A satisfies its own characteristic equation $p(A) = 0$, where

$$p(\lambda) = \text{Det}(\lambda I - A) = \lambda^n + c_{n-1}\lambda^{n-1} + \ldots + c_1\lambda + c_0 \tag{5}$$

is A's characteristic polynomial. From $p(A) = 0$ it follows that any power A^k, with $k \geq n$, can be written in terms of the matrices $I = A^0, A, \ldots, A^{n-1}$. Thus, any infinite series, such as the one corresponding to $\exp A$, may be rewritten in terms of the n powers $A^0, A, \ldots, A^{n-1}$. By exploiting this fact one can recover Equation (1). Reciprocally, given A, one can construct a matrix B that satisfies $\exp B = A$, as shown by Dattoli, Mari and Torre [2]. These authors used essentially the same tools as we do here and presented some of the results that we will show below, but leaving them in an implicit form. The aforementioned authors belong to a group that has extensively dealt with our subject matter and beyond it [7], applying the present techniques to cases of current interest [8]. A somewhat different approach was followed by Leonard [9], who related the Cayley–Hamilton theorem to the solution of ordinary differential equations, in order to get closed expressions for the matrix exponential. This technique can be applied to all $n \times n$ matrices, including those that are not diagonalizable. Untidt and Nielsen [10] used this technique when addressing the groups $SU(2)$, $SU(3)$ and $SU(4)$. Now, especially when addressing $SU(2)$, Leonard's approach seems to be unnecessarily involved. This is because there is a trade-off between the wide applicability of the method and its tailoring to a special case. When dealing with diagonalizable matrices, the present approach may prove more useful. Thus, one exploits not only the Cayley–Hamilton theorem, but the diagonalizability of the involved matrices as well. As a result, we are provided with a straightforward way to obtain closed-form expressions for the matrix exponential. There are certainly many other ways that are either more general [9,11] or else better suited to specific cases [12–16], but the present method is especially useful for physical applications.

The rest of the paper is organized as follows. First, we present Leonard's technique in a way that somewhat differs from the approach used in [9]. Thereafter, we show how to obtain Equation (1) by using a technique that can be generalized to diagonalizable $n \times n$ matrices, thereby introducing the method that is the main subject of the present work. As an illustration of this technique, we address some representative cases that were taken from the repertoire of classical mechanics, quantum electrodynamics, quantum optics and from the realm of Lorentz transformations. While the results obtained are known, their derivations should serve to demonstrate the versatility of the method. Let us stress once again that our aim has been to present this method by following an approach that could be appealing to most physicists, rather than to mathematically oriented readers.

2. Closed Form of the Matrix Exponential via the Solution of Differential Equations

Consider the coupled system of differential equations, given by

$$Dx \equiv \frac{dx}{dt} = Ax \tag{6}$$

with $x = (x_1, \ldots, x_n)^T$ and A a constant, $n \times n$ matrix. The matrix exponential appears in the solution of Equation (6), when we write it as $x(t) = e^{At}x(0)$. By successive derivation of this exponential we obtain $D^k e^{At} = A^k e^{At}$. Hence, $p(D)e^{At} \equiv (D^n + c_{n-1}D^{n-1} + \ldots + c_1 D + c_0)e^{At} = p(A)e^{At} = 0$, on account of $p(A) = 0$, *i.e.*, the Cayley–Hamilton theorem. Now, as already noted, this implies that e^{At} can be expressed in terms of $A^0, A, \ldots, A^{n-1}$. Let us consider the matrix $M(t) := \sum_{k=0}^{n-1} y_k(t)A^k$, with the $y_k(t)$ being n independent solutions of the differential equation $p(D)y(t) = 0$. That is, the $y_k(t)$ solve this equation for n different initial conditions that will be conveniently chosen. We have thus that $p(D)M(t) = \sum_{k=0}^{n-1} p(D)y_k(t)A^k = 0$. Our goal is to choose the $y_k(t)$ so that $e^{At} = M(t)$. To this end, we note that $D^k e^{At}|_{t=0} = A^k e^{At}|_{t=0} = A^k$. That is, e^{At} solves $p(D)\Phi(t) = 0$ with the initial conditions $\Phi(0) = A^0, \ldots, D^{n-1}\Phi(0) = A^{n-1}$. It is then clear that we must take the following initial conditions: $D^j y_k(0) = \delta_k^j$, with $j, k \in \{0, \ldots, n-1\}$. In such a case, e^{At} and $M(t)$ satisfy both the same differential equation and the same initial conditions. Hence, $e^{At} = M(t)$.

Summarizing, the method consists in solving the n-th order differential equation $p(D)y(t) = 0$ for n different initial conditions. These conditions read $D^j y_k(0) = \delta_k^j$, with $j, k \in \{0, \ldots, n-1\}$. The matrix exponential is then given by $e^{At} = \sum_{k=0}^{n-1} y_k(t)A^k$. The standard procedure for solving $p(D)y(t) = 0$ requires finding the roots of the characteristic equation $p(\lambda) = 0$. Each root λ with multiplicity m contributes to the general solution with a term $(a_0 + a_1 t + \ldots + a_{m-1}t^{m-1})e^{\lambda t}$, the a_k being fixed by the initial conditions. As already said, this method applies even when the matrix A is not diagonalizable. However, when the eigenvalue problem for A is a solvable one, another approach can be more convenient. We present such an approach in what follows.

3. Closed Form of the Matrix Exponential via the Solution of Algebraic Equations

Let us return to Equation (1). We will derive it anew, this time using standard tools of quantum mechanics. Consider a Hermitian operator A, whose eigenvectors satisfy $A |a_k\rangle = a_k |a_k\rangle$ and span the Hilbert space on which A acts. Thus, the identity operator can be written as $I = \sum_k |a_k\rangle \langle a_k|$. One can also write $A = A \cdot I = \sum_k a_k |a_k\rangle \langle a_k|$. Moreover, $A^m = \sum_k a_k^m |a_k\rangle \langle a_k|$, from which it follows that

$$F(A) = \sum_k F(a_k) |a_k\rangle \langle a_k| \tag{7}$$

Symmetry **2014**, 6, 329–344

for any function $F(A)$ that can be expanded in powers of A.

Let us consider the 2×2 case $A = \mathbf{n} \cdot \boldsymbol{\sigma}$, with $\mathbf{n}$ a unit vector. This matrix has the eigenvalues ± 1 and the corresponding eigenvectors $|n_{\pm}\rangle$. That is, $\mathbf{n} \cdot \boldsymbol{\sigma} |n_{\pm}\rangle = \pm |n_{\pm}\rangle$. We need no more than this to get Equation (1). Indeed, from $\mathbf{n} \cdot \boldsymbol{\sigma} = |n_+\rangle \langle n_+| - |n_-\rangle \langle n_-|$ and $I = |n_+\rangle \langle n_+| + |n_-\rangle \langle n_-|$, it follows that $|n_{\pm}\rangle \langle n_{\pm}| = (I \pm \mathbf{n} \cdot \boldsymbol{\sigma}) /2$. Next, we consider $F(A) = \exp A = \sum_k \exp a_k |a_k\rangle \langle a_k|$, with $A = i\alpha \mathbf{n} \cdot \boldsymbol{\sigma}$. The operator $i\alpha \mathbf{n} \cdot \boldsymbol{\sigma}$ has eigenvectors $|n_{\pm}\rangle$ and eigenvalues $\pm i\alpha$. Thus,

$$\exp(i\alpha\mathbf{n} \cdot \boldsymbol{\sigma}) = e^{i\alpha} |n_+\rangle \langle n_+| + e^{-i\alpha} |n_-\rangle \langle n_-| \tag{8}$$

$$= \frac{1}{2} e^{i\alpha} (I + \mathbf{n} \cdot \boldsymbol{\sigma}) + \frac{1}{2} e^{-i\alpha} (I - \mathbf{n} \cdot \boldsymbol{\sigma}) \tag{9}$$

$$= \left(\frac{e^{i\alpha} + e^{-i\alpha}}{2} \right) I + \left(\frac{e^{i\alpha} - e^{-i\alpha}}{2} \right) \mathbf{n} \cdot \boldsymbol{\sigma} \tag{10}$$

which is Equation (1). Note that it has not been necessary to know the eigenvectors of $A = i\alpha \mathbf{n} \cdot \boldsymbol{\sigma}$. It is a matter of convenience whether one chooses to express $\exp(i\alpha \mathbf{n} \cdot \boldsymbol{\sigma})$ in terms of the projectors $|n_{\pm}\rangle \langle n_{\pm}|$, or in terms of I and $\mathbf{n} \cdot \boldsymbol{\sigma}$.

Let us now see how the above method generalizes when dealing with higher-dimensional spaces. To this end, we keep dealing with rotations. The operator $\exp(i\alpha \mathbf{n} \cdot \boldsymbol{\sigma})$ is a rotation operator acting on spinor space. It is also an element of the group $SU(2)$, whose generators can be taken as $X_i = i\sigma_i/2$, $i = 1, 2, 3$. They satisfy the commutation relations $[X_i, X_j] = \epsilon_{ijk} X_k$ that characterize the rotation algebra. The rotation operator can also act on three-dimensional vectors $\mathbf{r}$. In this case, one often uses the following formula, which gives the rotated vector $\mathbf{r}'$ in terms of the rotation angle θ and the unit vector $\mathbf{n}$ that defines the rotation axis:

$$\mathbf{r}' = \mathbf{r} \cos\theta + \mathbf{n} (\mathbf{n} \cdot \mathbf{r}) [1 - \cos\theta] + (\mathbf{n} \times \mathbf{r}) \sin\theta \tag{11}$$

Equation (11) is usually derived from vector algebra plus some geometrical considerations [17]. We can derive it, alternatively, by the method used above. To this end, we consider the rotation generators X_i for three-dimensional space, which can be read off from the next formula, Equation (12). The rotation matrix is then obtained as $\exp(\theta \mathbf{n} \cdot \mathbf{X})$, with

$$\mathbf{n} \cdot \mathbf{X} = \begin{pmatrix} 0 & -n_3 & n_2 \\ n_3 & 0 & -n_1 \\ -n_2 & n_1 & 0 \end{pmatrix} \equiv M \tag{12}$$

It is straightforward to find the eigenvalues of the non-Hermitian, antisymmetric matrix M. They are 0 and $\pm i$. Let us denote the corresponding eigenvectors as $|n_0\rangle$ and $|n_{\pm}\rangle$, respectively. Similarly to the spin case, we have now

$$I = |n_+\rangle \langle n_+| + |n_-\rangle \langle n_-| + |n_0\rangle \langle n_0| \tag{13}$$

$$M = i |n_+\rangle \langle n_+| - i |n_-\rangle \langle n_-| \tag{14}$$

We need a third equation, if we want to express the three projectors $|n_k\rangle \langle n_k|$, $k = \pm, 0$, in terms of I and M. This equation is obtained by squaring M:

$$M^2 = - |n_+\rangle \langle n_+| - |n_-\rangle \langle n_-| \tag{15}$$

From Equations (13)–(15) we immediately obtain $|n_{\pm}\rangle \langle n_{\pm}| = (\mp iM - M^2) /2$, and $|n_0\rangle \langle n_0| = I + M^2$. Thus, we have

$$\exp(\theta M) = e^{i\theta} |n_+\rangle \langle n_+| + e^{-i\theta} |n_-\rangle \langle n_-| + e^0 |n_0\rangle \langle n_0| \tag{16}$$

$$= I + M \sin\theta + M^2 [1 - \cos\theta] \tag{17}$$

By letting M, as given in Equation (12), act on $r = (x, y, z)^T$, we easily see that $Mr = n \times r$ and $M^2 r = n \times (n \times r) = n (n \cdot r) - r$. Thus, on account of Equation (17), $r' = \exp(\theta M) r$ reads the same as Equation (11).

The general case is now clear. Consider an operator A whose matrix representation is an $N \times N$ matrix. Once the eigenvalues a_k of A (which we assume nondegenerate) have been determined, we can write the N equations: $A^0 = I = \sum_k |a_k\rangle \langle a_k|$, $A = \sum_k a_k |a_k\rangle \langle a_k|$, $A^2 = \sum_{k=1}^{N} a_k^2 |a_k\rangle \langle a_k|, \dots, A^{N-1} = \sum_{k=1}^{N} a_k^{N-1} |a_k\rangle \langle a_k|$, from which it is possible to obtain the N projectors $|a_k\rangle \langle a_k|$ in terms of $I, A, A^2, \dots, A^{N-1}$. To this end, we must solve the system

$$\begin{pmatrix} 1 & 1 & \cdots & 1 \\ a_1 & a_2 & \cdots & a_N \\ a_1^2 & a_2^2 & & a_N^2 \\ \vdots & \vdots & & \vdots \\ a_1^{N-1} & a_2^{N-1} & & a_N^{N-1} \end{pmatrix} \begin{pmatrix} |a_1\rangle \langle a_1| \\ |a_2\rangle \langle a_2| \\ |a_3\rangle \langle a_3| \\ \vdots \\ |a_N\rangle \langle a_N| \end{pmatrix} = \begin{pmatrix} I \\ A \\ A^2 \\ \vdots \\ A^{N-1} \end{pmatrix} \qquad (18)$$

The matrix in Equation (18), with components $V_{k,i} = a_i^{k-1}$ ($k, i \in \{1, \dots, N\}$), is a Vandermonde matrix, whose inverse can be explicitly given [18]. Once we have written the $|a_k\rangle \langle a_k|$ in terms of $I, A, \dots A^{N-1}$, we can express any analytic function of A in terms of these N powers of A, in particular $\exp A = \sum_{k=1}^{N} \exp(a_k) |a_k\rangle \langle a_k|$. For the case $N = 4$, for instance, we have the following result:

$$|a_1\rangle \langle a_1| = \frac{A^3 - A^2(a_2 + a_3 + a_4) + A(a_2 a_3 + a_2 a_4 + a_3 a_4) - a_2 a_3 a_4}{(a_1 - a_2)(a_1 - a_3)(a_1 - a_4)} \qquad (19)$$

$$|a_2\rangle \langle a_2| = \frac{A^3 - A^2(a_1 + a_3 + a_4) + A(a_1 a_3 + a_1 a_4 + a_3 a_4) - a_1 a_3 a_4}{(a_2 - a_1)(a_2 - a_3)(a_2 - a_4)} \qquad (20)$$

$$|a_3\rangle \langle a_3| = \frac{A^3 - A^2(a_1 + a_2 + a_4) + A(a_1 a_2 + a_1 a_4 + a_2 a_4) - a_1 a_2 a_4}{(a_3 - a_1)(a_3 - a_2)(a_3 - a_4)} \qquad (21)$$

$$|a_4\rangle \langle a_4| = \frac{A^3 - A^2(a_1 + a_2 + a_3) + A(a_1 a_2 + a_1 a_3 + a_2 a_3) - a_1 a_3 a_4}{(a_4 - a_1)(a_4 - a_2)(a_4 - a_3)} \qquad (22)$$

The general solution can be written in terms of the inverse of the Vandermonde matrix V. To this end, consider a system of equations that reads like (18), but with the operators entering the column vectors being replaced by numbers, i.e., $|a_j\rangle \langle a_j| \to w_j$, with $j = 1, \dots, N$, and $A^k \to q_{k+1}$, with $k = 0, \dots, N-1$. The solution of this system is given by $w_j = \sum_{k=0}^{N-1} U_{j,k} q_k$, with $U = V^{-1}$, the inverse of the Vandermonde matrix. This matrix inverse can be calculated as follows [18]. Let us define a polynomial $P_j(x)$ of degree $N - 1$ as

$$P_j(x) = \prod_{\substack{n=1 \\ n \neq j}}^{N} \frac{x - a_n}{a_j - a_n} = \sum_{k=1}^{N} U_{j,k} x^{k-1} \qquad (23)$$

The coefficients $U_{j,k}$ of the last equality follow from expanding the preceding expression and collecting equal powers of x. These $U_{j,k}$ are the components of V^{-1}. Indeed, setting $x = a_i$ and observing that $P_j(a_i) = \delta_{ji} = \sum_{k=1}^{N} U_{j,k} a_i^{k-1} = (UV)_{j,i}$, we see that U is the inverse of the Vandermonde matrix. The projectors $|a_j\rangle \langle a_j|$ in Equation (18) can thus be obtained by replacing $x \to A$ in Equation (23). We get in this way the explicit solution

$$|a_j\rangle \langle a_j| = \sum_{k=1}^{N} U_{j,k} A^{k-1} = \prod_{\substack{n=1 \\ n \neq j}}^{N} \frac{A - a_n}{a_j - a_n} \qquad (24)$$

The above expression can be inserted into Equation (7), if one wants to write $F(A)$ in terms of the first N powers of A.

So far, we have assumed that the eigenvalues of A are all nondegenerate. Let us now consider a matrix M with degenerate eigenvalues. As before, we deal with a special case, from which the general formalism can be easily inferred. Let M be of dimension four and with eigenvalues λ_1 and λ_2, which are two-fold degenerate. We can group the projectors as follows:

$$I = (|e_1\rangle \langle e_1| + |e_2\rangle \langle e_2|) + (|e_3\rangle \langle a_3| + |e_4\rangle \langle e_4|) \tag{25}$$

$$M = \lambda_1 (|e_1\rangle \langle e_1| + |e_2\rangle \langle e_2|) + \lambda_2 (|e_3\rangle \langle a_3| + |e_4\rangle \langle e_4|) \tag{26}$$

It is then easy to solve the above equations for the two projectors associated with the two eigenvalues. We obtain

$$|e_1\rangle \langle e_1| + |e_2\rangle \langle e_2| = \frac{\lambda_2 I - M}{\lambda_2 - \lambda_1} \tag{27}$$

$$|e_3\rangle \langle a_3| + |e_4\rangle \langle e_4| = \frac{\lambda_1 I - M}{\lambda_1 - \lambda_2} \tag{28}$$

We can then write

$$e^M = \frac{1}{\lambda_1 - \lambda_2} \left[\left(\lambda_1 e^{\lambda_2} - \lambda_2 e^{\lambda_1} \right) I + \left(e^{\lambda_1} - e^{\lambda_2} \right) M \right] \tag{29}$$

We will need this result for the calculation of the unitary operator that defines the Foldy–Wouthuysen transformation, our next example. It is now clear that in the general case of degenerate eigenvalues, we can proceed similarly to the nondegenerate case, but solving $n < N$ equations.

4. Examples

Let us now see how the method works when applied to some well-known cases. Henceforth, we refer to the method as the Cayley–Hamilton (CH)-method, for short. Our aim is to show the simplicity of the required calculations, as compared with standard techniques.

4.1. The Foldy–Wouthuysen Transformation

The Foldy–Wouthuysen transformation is introduced [19] with the aim of decoupling the upper (φ) and lower (χ) components of a bispinor $\psi = (\varphi, \chi)^T$ that solves the Dirac equation $i\hbar \partial \psi / \partial t = H\psi$, where $H = -i\hbar c \boldsymbol{\alpha} \cdot \nabla + \beta m c^2$. Here, β and $\boldsymbol{\alpha} = (\alpha_x, \alpha_y, \alpha_z)$ are the 4×4 Dirac matrices:

$$\beta = \begin{pmatrix} 1 & 0 \\ 0 & -1 \end{pmatrix}, \quad \boldsymbol{\alpha} = \begin{pmatrix} 0 & \sigma \\ \sigma & 0 \end{pmatrix} \tag{30}$$

The Foldy–Wouthuysen transformation is given by $\psi' = U\psi$, with [19]

$$U = \exp\left(\frac{\theta}{2} \beta \boldsymbol{\alpha} \cdot \boldsymbol{p} \right) \tag{31}$$

We can calculate U by applying Equation (29) for $M = \theta \beta \boldsymbol{\alpha} \cdot \boldsymbol{p}/2 = (\theta |\boldsymbol{p}| /2)\beta \boldsymbol{\alpha} \cdot \boldsymbol{n}$, where $\boldsymbol{n} = \boldsymbol{p}/ |\boldsymbol{p}|$. The eigenvalues of the 4×4 matrix $\beta \boldsymbol{\alpha} \cdot \boldsymbol{n}$ are $\pm i$, each being two-fold degenerate. This follows from noting that the matrices

$$\beta \boldsymbol{\alpha} \cdot \boldsymbol{n} = \begin{pmatrix} 0 & \sigma \cdot \boldsymbol{n} \\ -\sigma \cdot \boldsymbol{n} & 0 \end{pmatrix} \text{ and } \begin{pmatrix} 0 & 1 \\ -1 & 0 \end{pmatrix} \tag{32}$$

have the same eigenvalues. Indeed, because $(\sigma \cdot n)^2 = 1$, the above matrices share the characteristic equation $\lambda^2 + 1 = 0$. Their eigenvalues are thus $\pm i$. The eigenvalues of $M = \theta \beta \alpha \cdot p/2$ are then $\lambda_{1,2} = \pm i\theta |p|/2$. Replacing these values in Equation (29) we obtain

$$\exp\left(\frac{\theta}{2}\beta\alpha \cdot p\right) = \frac{1}{i\theta|p|}\left[\frac{i\theta|p|}{2}\left(e^{-i\theta|p|/2} + e^{i\theta|p|/2}\right)I + \left(e^{i\theta|p|/2} - e^{-i\theta|p|/2}\right)\frac{\theta|p|}{2}\beta\alpha \cdot n\right] \quad (33)$$

$$= \cos\left(|p|\,\theta/2\right) + \sin\left(|p|\,\theta/2\right)\beta\alpha \cdot \frac{p}{|p|} \quad (34)$$

The standard way to get this result requires developing the exponential in a power series. Thereafter, one must exploit the commutation properties of α and β in order to group together odd and even powers of θ. This finally leads to the same closed-form expression that we have arrived at after some few steps.

4.2. Lorentz-Type Equations of Motion

The dynamics of several classical and quantum systems is ruled by equations that can be cast as differential equations for a three-vector S. These equations often contain terms of the form $\Omega\times$. An example of this is the ubiquitous equation

$$\frac{dS}{dt} = \Omega \times S \quad (35)$$

Equation (35) and its variants have been recently addressed by Babusci, Dattoli and Sabia [20], who applied operational methods to deal with them. Instead of writing Equation (35) in matrix form, these authors chose to exploit the properties of the vector product by defining the operator $\hat{\Omega} := \Omega\times$. The solution for the case $\partial\Omega/\partial t = 0$, for instance, was obtained by expanding $\exp(t\hat{\Omega})$ as an infinite series and using the cyclical properties of the vector product in order to get $S(t)$ in closed form. This form is nothing but Equation (11) with the replacements $r' \rightarrow S(t)$, $r \rightarrow S(0)$ and $\theta \rightarrow \Omega t$, where $\Omega := |\Omega|$. We obtained Equation (11) without expanding the exponential and without using any cyclic properties. Our solution follows from writing Equation (35) in matrix form, i.e.,

$$\frac{dS}{dt} = \Omega M S \quad (36)$$

where M is given by Equation (12) with $n = \Omega/\Omega$. The solution $S(t) = \exp(M\Omega t)S(0)$ is then easily written in closed form by applying the CH-method, as in Equation (11). The advantages of this method show up even more sharply when dealing with some extensions of Equation (36). Consider, e.g., the non-homogeneous version of Equation (35):

$$\frac{dS}{dt} = \Omega \times S + N = \Omega M S + N \quad (37)$$

This is the form taken by the Lorentz equation of motion when the electromagnetic field is given by scalar and vector potentials reading $\Phi = -E \cdot r$ and $A = B \times r/2$, respectively [20]. The solution of Equation (37) is easily obtained by acting on both sides with the "integrating (operator-valued) factor" $\exp(-\Omega M t)$. One then readily obtains, for the initial condition $S(0) = S_0$,

$$S(t) = e^{\Omega M t}S_0 + \int_0^t e^{\Omega M(t-s)}N ds \quad (38)$$

The matrix exponentials in Equation (38) can be expressed in their eigenbasis, as in Equation (16). For a time-independent N, the integral in Equation (38) is then trivial. An equivalent solution is given in [20], but written in terms of the evolution operator $\hat{U}(t) = \exp(t\hat{\Omega})$ and its inverse. Inverse operators repeatedly appear within such a framework [20] and are often calculated with the help of the Laplace transform identity: $\hat{A}^{-1} = \int_0^\infty \exp(-s\hat{A})ds$. Depending on $\hat{A}$, this could be not such a straightforward task as it might appear at first sight. Now, while vector notation gives us additional physical insight, vector calculus can rapidly turn into a messy business. Our strategy is therefore to avoid vector calculus and instead rely on the CH-method as much as possible. Only at the end we write down our results, if we wish, in terms of vector products and the like. That is, we use Equations (13)–(17) systematically, in particular Equation (16) when we need to handle $\exp(\theta M)$, e.g., within integrals. The simplification comes about from our working with the eigenbasis of $\exp(\theta M)$, *i.e.*, with the eigenbasis of M. Writing down the final results in three-vector notation amounts to expressing these results in the basis in which M was originally defined, *cf.* Equation (12). Let us denote this basis by $\{|x\rangle, |y\rangle, |z\rangle\}$. The eigenvectors $|n_\pm\rangle$ and $|n_0\rangle$ of M are easily obtained from those of X_3, *cf.* Equation (12). The eigenvectors of X_3 are, in turn, analogous to those of Pauli's σ_y, namely $|\pm\rangle = (|x\rangle \mp i|y\rangle)/\sqrt{2}$, plus a third eigenvector that is orthogonal to the former ones, that is, $|0\rangle = |z\rangle$. In order to obtain the eigenvectors of $n \cdot X$, with $n = (\sin\theta\cos\phi, \sin\theta\sin\phi, \cos\theta)$, we apply the rotation $\exp(\phi X_3)\exp(\theta X_2)$ to the eigenvectors $|\pm\rangle$ and $|0\rangle$, thereby getting $|n_\pm\rangle$ and $|n_0\rangle$, respectively. All these calculations are easily performed using the CH-method.

Once we have $|n_\pm\rangle$ and $|n_0\rangle$, we also have the transformation matrix T that brings M into diagonal form: $T^{-1}MT = M_D = \text{diag}(-i, 0, i)$. Indeed, T's columns are just $|n_-\rangle$, $|n_0\rangle$ and $|n_+\rangle$. After we have carried out all calculations in the eigenbasis of M, by applying T we can express the final result in the basis $\{|x\rangle, |y\rangle, |z\rangle\}$, thereby obtaining the desired expressions in three-vector notation. Let us illustrate this procedure by addressing the evolution equation

$$\frac{dS}{dt} = \Omega \times S + \lambda\Omega \times (\Omega \times S) \tag{39}$$

In matrix form, such an equation reads

$$\frac{dS}{dt} = \Omega MS + \lambda(\Omega M)^2 S = [\Omega M + \lambda(\Omega M)^2]S \equiv AS \tag{40}$$

The solution is given by $S(t) = \exp(At)S_0$. The eigenbasis of A is the same as that of M. We have thus

$$\exp(At) = e^{(i\Omega - \lambda\Omega^2)t}|n_+\rangle\langle n_+| + e^{(-i\Omega - \lambda\Omega^2)t}|n_-\rangle\langle n_-| + |n_0\rangle\langle n_0| \tag{41}$$

The projectors $|n_k\rangle\langle n_k|$ can be written in terms of the powers of A by solving the system

$$I = |n_+\rangle\langle n_+| + |n_-\rangle\langle n_-| + |n_0\rangle\langle n_0| \tag{42}$$

$$A = (i\Omega - \lambda\Omega^2)|n_+\rangle\langle n_+| - (i\Omega + \lambda\Omega^2)|n_-\rangle\langle n_-| \tag{43}$$

$$A^2 = (i\Omega - \lambda\Omega^2)^2|n_+\rangle\langle n_+| + (i\Omega + \lambda\Omega^2)^2|n_-\rangle\langle n_-| \tag{44}$$

Using $A = \Omega M + \lambda(\Omega M)^2$ and $A^2 = -2\lambda\Omega^3 M + (1 - \lambda^2\Omega^2)(\Omega M)^2$, and replacing the solution of the system (42)–(44) in Equation (41) we get

$$\exp(At) = I + e^{-\lambda\Omega^2 t}\sin(\Omega t)M + [1 - e^{-\lambda\Omega^2 t}\cos(\Omega t)]M^2 \tag{45}$$

Finally, we can write the solution $S(t) = \exp(At)S_0$ in the original basis $\{|x\rangle, |y\rangle, |z\rangle\}$, something that in this case amounts to writing $MS_0 = n \times S_0$ and $M^2 S_0 = n(n \cdot S_0) - S_0$. Equation (39) was also addressed in [20], but making use of the operator method. The solution was given in terms of a series expansion for the evolution operator. In order to write this solution in closed form, it is necessary to introduce sin- and cos-like functions [20]. These functions are defined as infinite series involving two-variable Hermite polynomials. The final expression reads like Equation (11), but with sin and cos replaced by the aforementioned functions containing two-variable Hermite polynomials. Now, one can hardly unravel from such an expression the physical features that characterize the system's dynamics.

On the other hand, a solution given as in Equation (45) clearly shows such dynamics, in particular the damping effect stemming from the λ-term in Equation (39), for $\lambda > 0$. Indeed, Equation (45) clearly shows that the state vector $S(t) = \exp(At)S_0$ asymptotically aligns with Ω while performing a damped Larmor precession about the latter.

The case $\partial\Omega/\partial t \neq 0$ is more involved and generally requires resorting to Dyson-like series expansions, e.g., time-ordered exponential integrations. While this subject lies beyond the scope of the present work, it should be mentioned that the CH-method can be advantageously applied also in this context. For instance, time-ordered exponential integrations involving operators of the form $A + B(t)$ do require the evaluation of $\exp A$. Likewise, disentangling techniques make repeated use of matrix exponentials of single operators [21]. In all these cases, the CH-method offers a possible shortcut.

4.3. The Jaynes–Cummings Hamiltonian

We address now a system composed by a two-level atom and a quantized (monochromatic) electromagnetic field. Under the dipole and the rotating-wave approximations, the Hamiltonian of this system reads (in standard notation)

$$H = \frac{\hbar}{2}\omega_0\sigma_z + \hbar\omega a^\dagger a + \hbar g\left(a^\dagger\sigma_- + a\sigma_+\right) \tag{46}$$

Let us denote the upper and lower states of the two-level atom by $|a\rangle$ and $|b\rangle$, respectively, and the Fock states of the photon-field by $|n\rangle$. The Hilbert space of the atom-field system is spanned by the basis $\mathcal{B} = \{|a,n\rangle, |b,n\rangle, n = 0, 1, \ldots\}$. The states $|a,n\rangle$ and $|b,n\rangle$ are eigenstates of the unperturbed Hamiltonian $H_0 = \hbar\omega_0\sigma_z/2 + \hbar\omega a^\dagger a$. The interaction Hamiltonian $V = \hbar g\left(a^\dagger\sigma_- + a\sigma_+\right)$ couples the states $|a,n\rangle$ and $|b,n+1\rangle$ alone. Hence, H can be split into a sum: $H = \sum_n H_n$, with each H_n acting on the subspace $\mathrm{Span}\{|a,n\rangle, |b,n+1\rangle\}$. Within such a subspace, H_n is represented by the 2×2 matrix

$$H_n = \hbar\omega\left(n + \frac{1}{2}\right)I + \hbar\begin{pmatrix} \delta/2 & g\sqrt{n+1} \\ g\sqrt{n+1} & -\delta/2 \end{pmatrix} \tag{47}$$

where $\delta = \omega_0 - \omega$.

A standard way [22] to calculate the evolution operator $U = \exp(-iHt/\hbar)$ goes as follows. One first writes the Hamiltonian in the form $H = H_1 + H_2$, with $H_1 = \hbar\omega\left(a^\dagger a + \sigma_z/2\right)$ and $H_2 = \hbar\delta\sigma_z/2 + \hbar g\left(a^\dagger\sigma_- + a\sigma_+\right)$. Because $[H_1, H_2] = 0$, the evolution operator can be factored as $U = U_1 U_2 = \exp(-iH_1 t/\hbar)\exp(-iH_2 t/\hbar)$. The first factor is diagonal in Span $\mathcal{B}$. The second factor can be expanded in a Taylor series. As it turns out, one can obtain closed-form expressions for the even and the odd powers of the expansion. Thus, a closed-form for U_2 can be obtained as well. As can be seen, this method depends on the realization that Equation (46) can be written in a special form, which renders it possible to factorize U.

Let us now calculate U by the CH-method. We can exploit the fact that H splits as $H = \sum_n H_n$, with $[H_n, H_m] = 0$, and write $U = \prod_n U_n = \prod_n \exp(-iH_n t/\hbar)$. Generally, a 2×2 Hamiltonian H has eigenvalues of the form $E_\pm = \hbar(\lambda_0 \pm \lambda)$. We have thus

$$I = |+\rangle\langle+| + |-\rangle\langle-| \tag{48}$$

$$H/\hbar = (\lambda_0 + \lambda)|+\rangle\langle+| + (\lambda_0 - \lambda)|-\rangle\langle-| \tag{49}$$

so that

$$\exp(-iHt/\hbar) = \exp(-i\lambda_+ t)|+\rangle\langle+| + \exp(-i\lambda_- t)|-\rangle\langle-| \tag{50}$$

$$= \frac{e^{-i\lambda_0 t}}{\lambda}\left[(i\lambda_0\sin\lambda t + \lambda\cos\lambda t)I - i(\sin\lambda t)\frac{H}{\hbar}\right] \tag{51}$$

In our case, H_n has eigenvalues $E_n^{\pm} = \hbar\omega(n+1/2) \pm \hbar\sqrt{\delta^2/4 + g^2(n+1)} \equiv \hbar\omega(n+1/2) \pm \hbar R_n$. Whence,

$$\exp(-iH_n t/\hbar) = \frac{e^{-i\omega(n+1/2)t}}{R_n}\left[\left(i\omega\left(n+\frac{1}{2}\right)\sin\left(R_n t\right) + R_n\cos\left(R_n t\right)\right)I - i\sin(R_n t)\frac{H_n}{\hbar}\right] \tag{52}$$

Replacing H_n from Equation (47) in the above expression we get

$$\exp(-iH_n t/\hbar) = e^{-i\omega(n+1/2)t}\left[\cos\left(R_n t\right)I - \frac{i\sin\left(R_n t\right)}{2R_n}\begin{pmatrix} \delta & 2g\sqrt{n+1} \\ 2g\sqrt{n+1} & -\delta \end{pmatrix}\right] \tag{53}$$

This result enables a straightforward calculation of the evolved state $|\psi(t)\rangle$ out of a general initial state

$$|\psi(0)\rangle = \sum_n C_{a,n}|a,n\rangle + C_{b,n+1}|b,n+1\rangle \tag{54}$$

Equation (53) refers to a matrix representation in the two-dimensional subspace Span$\{|a,n\rangle, |b,n+1\rangle\}$. Let us focus on

$$\cos\left(R_n t\right)I = \begin{pmatrix} \cos\left(R_n t\right) & 0 \\ 0 & \cos\left(R_n t\right) \end{pmatrix} \tag{55}$$

This matrix is a representation in subspace Span$\{|a,n\rangle, |b,n+1\rangle\}$ of the operator

$$\cos\left(t\sqrt{\widehat{\varphi}+g^2}\right)|a\rangle\langle a| + \cos\left(t\sqrt{\widehat{\varphi}}\right)|b\rangle\langle b| \tag{56}$$

where $\widehat{\varphi} := g^2 a^\dagger a + \delta^2/4$. Proceeding similarly with the other operators that enter Equation (53) and observing that $\sin\left(R_n t\right)R_n^{-1}\sqrt{n+1} = \langle n|i\sin(t\sqrt{\widehat{\varphi}+g^2})(\sqrt{\widehat{\varphi}+g^2})^{-1}a|n+1\rangle$, etc., we readily obtain

$$\exp(-iHt/\hbar) = e^{-i\omega(a^\dagger a+\frac{1}{2})t}\begin{pmatrix} \cos\left(t\sqrt{\widehat{\varphi}+g^2}\right) - \dfrac{i\delta\sin\left(t\sqrt{\widehat{\varphi}+g^2}\right)}{2\sqrt{\widehat{\varphi}+g^2}} & -\dfrac{ig\sin\left(t\sqrt{\widehat{\varphi}+g^2}\right)}{\sqrt{\widehat{\varphi}+g^2}}a \\ -\dfrac{ig\sin\left(t\sqrt{\widehat{\varphi}}\right)}{\sqrt{\widehat{\varphi}}}a^\dagger & \cos\left(t\sqrt{\widehat{\varphi}}\right) + \dfrac{i\delta\sin\left(t\sqrt{\widehat{\varphi}}\right)}{2\sqrt{\widehat{\varphi}}} \end{pmatrix} \tag{57}$$

where the 2×2 matrix refers now to the atomic subspace Span$\{|a\rangle, |b\rangle\}$. One can see that the CH-method reduces the amount of calculational effort invested to get Equation (53), as compared with other approaches [22].

4.4. Bispinors and Lorentz Transformations

As a further application, let us consider the representation of Lorentz transformations in the space of bispinors. In coordinate space, Lorentz transformations are given by $\widetilde{x}^\mu = \Lambda^\mu_\nu x^\nu$ (Greek indices run from 0 to 3), with the Λ^μ_ν satisfying $\Lambda^\mu_\nu \Lambda^\tau_\sigma \eta^{\nu\sigma} = \eta^{\mu\tau}$. Here, $\eta^{\mu\nu}$ represents the metric tensor of Minkowsky space ($\eta^{00} = -\eta^{11} = -\eta^{22} = \eta^{33} = 1$, $\eta^{\mu\nu} = 0$ for $\mu \neq \nu$). A bispinor $\psi(x)$ transforms according to [19]

$$\widetilde{\psi}(\widetilde{x}) = \widetilde{\psi}(\Lambda x) = S(\Lambda)\psi(x) \tag{58}$$

with

$$S(\Lambda) = \exp B \tag{59}$$

$$B = -\frac{1}{4}V^{\mu\nu}\gamma_\mu\gamma_\nu \tag{60}$$

Symmetry **2014**, 6, 329–344

The $V^{\mu\nu} = -V^{\nu\mu}$ are the components of an antisymmetric tensor, which has thus six independent components, corresponding to the six parameters defining a Lorentz transformation. The quantities $\gamma_\mu = \eta_{\mu\nu}\gamma^\nu$ satisfy $\gamma^\mu\gamma^\nu + \gamma^\mu\gamma^\nu = 2\eta^{\mu\nu}$. The quantities $\gamma_\mu\gamma_\nu$ are the generators of the Lorentz group. $S(\Lambda)$ is not a unitary transformation, but satisfies

$$S^{-1} = \gamma_0 S^\dagger \gamma_0 \tag{61}$$

For the following, it will be advantageous to define

$$p_i = \gamma_0\gamma_i, \ i = 1, 2, 3 \tag{62}$$

$$q_1 = \gamma_2\gamma_3, \ q_2 = \gamma_3\gamma_1, \ q_3 = \gamma_1\gamma_2 \tag{63}$$

We call the p_i Pauli generators and the q_i quaternion generators. The pseudoscalar $\gamma_5 := \gamma_0\gamma_1\gamma_2\gamma_3$ satisfies $\gamma_5^2 = -1$, $\gamma_5\gamma_\mu = -\gamma_\mu\gamma_5$, so that it commutes with each generator of the Lorentz group:

$$\gamma_5 (\gamma_\mu\gamma_\nu) = (\gamma_\mu\gamma_\nu) \gamma_5 \tag{64}$$

This means that quantities of the form $\alpha + \beta\gamma_5$ ($\alpha, \beta \in \mathbb{R}$) behave like complex numbers upon multiplication with p_i and q_i . We denote the subspace spanned by such quantities as the complex-like subspace C_i and set $i \equiv \gamma_5$. Noting that $i\,p_i = q_i$ and $i\,q_i = -p_i$, the following multiplication rules are easily derived:

$$q_i q_j = \epsilon_{ijk} q_k - \delta_{ij} \tag{65}$$

$$p_i p_j = -\epsilon_{ijk} q_k + \delta_{ij} = -q_i q_j = -i\epsilon_{ijk} p_k + \delta_{ij} \tag{66}$$

$$p_i q_j = \epsilon_{ijk} p_k + i\delta_{ij} = i(-\epsilon_{ijk} q_k + \delta_{ij}) \tag{67}$$

The following commutators can then be straightforwardly obtained:

$$[q_i, q_j] = 2\epsilon_{ijk} q_k \tag{68}$$

$$[p_i, p_j] = -2\epsilon_{ijk} q_k = -2i\epsilon_{ijk} p_k \tag{69}$$

$$[p_i, q_j] = 2\epsilon_{ijk} p_k \tag{70}$$

They make clear why we dubbed the p_i as Pauli generators. Noting that they furthermore satisfy

$$p_i p_j + p_j p_i = 2\delta_{ij} \tag{71}$$

we see the correspondence $i \rightarrow i$, $p_k \rightarrow -\sigma_k$, with i being the imaginary unit and σ_k the Pauli matrices. These matrices, as is well-known, satisfy $[\sigma_i, \sigma_j] = 2i\epsilon_{ijk}\sigma_k$ and the anticommutation relations $\sigma_i\sigma_j + \sigma_j\sigma_i = 2\delta_{ij}$, which follow from $\sigma_i\sigma_j = i\epsilon_{ijk}\sigma_k + \delta_{ij}$.

We can write now $S(\Lambda) = \exp(-\frac{1}{4}V^{\mu\nu}\gamma_\mu\gamma_\nu)$ in terms of p_i and q_i:

$$B = \sum_{i=1}^{3}(\alpha^i p_i + \beta^i q_i) \tag{72}$$

Here, we have set $\alpha^i = -V^{0i}/4$ and $\beta^k\epsilon_{ijk} = -V^{ij}/4$. We can write B in terms of the Pauli-generators alone:

$$B = \sum_{i=1}^{3}(\alpha^i + i\beta^i)p_i \equiv \sum_{i=1}^{3} z^i p_i \tag{73}$$

Considering the isomorphism $p_k \leftrightarrow -\sigma_k$, we could derive the expression for $S(\Lambda) = \exp B$ by splitting the series expansion into even and odd powers of B, and noting that

$$B^2 = (\alpha^2 - \beta^2) + (2\alpha \cdot \beta) \, i \equiv z^2 \in C_i \tag{74}$$

where $\alpha^2 \equiv \alpha \cdot \alpha$, $\beta^2 \equiv \beta \cdot \beta$, and $\alpha \cdot \beta \equiv \sum_{i=1}^{3} \alpha^i \beta^i$. We have then that $B^3 = z^2 B$, $B^4 = z^4$, $B^5 = z^4 B, \ldots$ This allows us to write

$$
\begin{aligned}
\exp(B) &= 1 + B + \frac{z^2}{2!} + \frac{z^2}{3!}B + \frac{z^4}{4!} + \frac{z^4}{5!}B + \ldots = \\
&\left(1 + \sum_{n=1}^{\infty} \frac{z^{2n}}{(2n)!}\right) + B\left(1 + \frac{z^2}{3!} + \frac{z^4}{5!} + \ldots\right) = \cosh z + \frac{\sinh z}{z}B
\end{aligned}
\tag{75}
$$

As in the previous examples, also in this case the above result can be obtained more directly by noting that $B = \sum_{i=1}^{3}(\alpha^i + i\beta^i)p_i \leftrightarrow -\sum_{i=1}^{3}(\alpha^i + i\beta^i)\sigma_i$. This suggests that we consider $\exp(-f \cdot \sigma)$, with $f = \alpha + i\beta \in \mathbb{C}$. The matrix $f \cdot \sigma$ has the (complex) eigenvalues

$$\lambda_{\pm} = \pm\sqrt{\alpha^2 - \beta^2 + 2i\alpha \cdot \beta} \equiv \pm z \tag{76}$$

Writing $|f_{\pm}\rangle$ for the corresponding eigenvectors, i.e., $f \cdot \sigma |f_{\pm}\rangle = \lambda_{\pm} |f_{\pm}\rangle$, we have that

$$I = |f_+\rangle \langle f_+| + |f_-\rangle \langle f_-| \tag{77}$$

$$f \cdot \sigma = \lambda_+ |f_+\rangle \langle f_+| + \lambda_- |f_-\rangle \langle f_-| \tag{78}$$

Solving for $|f_{\pm}\rangle \langle f_{\pm}|$, we get

$$|f_{\pm}\rangle \langle f_{\pm}| = \frac{zI \pm f \cdot \sigma}{2z} \tag{79}$$

We apply now the general decomposition $\exp A = \sum_n \exp a_n |a_n\rangle \langle a_n|$ to the case $A = -f \cdot \sigma$. The operator $\exp(-f \cdot \sigma)$ has eigenvectors $|f_{\pm}\rangle$ and eigenvalues $\exp(\mp z)$. Thus,

$$
\begin{aligned}
\exp(-f \cdot \sigma) &= e^{-z}|f_+\rangle \langle f_+| + e^{z}|f_-\rangle \langle f_-| \tag{80} \\
&= \frac{e^{-z}}{2z}(zI + f \cdot \sigma) + \frac{e^{z}}{2z}(zI - f \cdot \sigma) \tag{81} \\
&= \left(\frac{e^z + e^{-z}}{2}\right)I - \left(\frac{e^z - e^{-z}}{2z}\right)f \cdot \sigma \tag{82} \\
&= \cosh z - \frac{\sinh z}{z}f \cdot \sigma \tag{83}
\end{aligned}
$$

which is equivalent to Equation (75) via the correspondence $\cosh(z) + \sinh(z)B/z \leftrightarrow \cosh(z) - \sinh(z)f \cdot \sigma/z$. We have thus obtained closed-form expressions for $\exp(-f \cdot \sigma)$, with $f = \alpha + i\beta \in \mathbb{C}^3$, i.e., for the elements of $SL(2,\mathbb{C})$, the universal covering group of the Lorentz group. It is interesting to note that the elements of $SL(2,\mathbb{C})$ are related to those of $SU(2)$ by extending the parameters α entering $\exp(i\alpha \cdot n) \in SU(2)$ from the real to the complex domain: $i\alpha \to \alpha + i\beta$. Standard calculations that are carried out with $SU(2)$ elements can be carried out similarly with $SL(2,\mathbb{C})$ elements [15]. A possible realization of $SU(2)$ transformations occurs in optics, by acting on the polarization of light with the help of birefringent elements (waveplates). If we also employ dichroic elements like polarizers, which absorb part of the light, then it is possible to implement $SL(2,\mathbb{C})$ transformations as well. In this way, one can simulate Lorentz transformations in the optical laboratory [23]. The above formalism is of great help for designing the corresponding experimental setup.

5. Conclusions

The method presented in this paper—referred to as the Cayley–Hamilton method—proves advantageous for calculating closed-form expressions of analytic functions $f(A)$ of an $n \times n$ matrix A, in particular matrix exponentials. The matrix A is assumed to be a diagonalizable one, even though only its eigenvalues are needed, not its eigenvectors. We have recovered some known results from classical and quantum mechanics, including Lorentz transformations, by performing the straightforward calculations that the method prescribes. In most cases, the problem at hand was reshaped so as to solve it by dealing with two-by-two matrices only.

Acknowledgments: The author gratefully acknowledges the Research Directorate of the Pontificia Universidad Católica del Perú (DGI-PUCP) for financial support under Grant No. 2014-0064.

Conflicts of Interest: The author declares no conflict of interest.

References

1. Gantmacher, F.R. *The Theory of Matrices*; Chelsea Publishing Company: New York, NY, USA, 1960; p. 83.
2. Dattoli, G.; Mari, C.; Torre, A. A simplified version of the Cayley-Hamilton theorem and exponential forms of the 2×2 and 3×3 matrices. *Il Nuovo Cimento* **1998**, *180*, 61–68.
3. Cohen-Tannoudji, C.; Diu, B.; Laloë, F. *Quantum Mechanics*; John Wiley & Sons: New York, NY, USA, 1977; pp. 983–989.
4. Sakurai, J.J. *Modern Quantum Mechanics*; Addison-Wesley: New York, NY, USA, 1980; pp. 163–168.
5. Greiner, W.; Müller, B. *Quantum Mechanics, Symmetries*; Springer: New York, NY, USA, 1989; p. 68.
6. Weigert, S. Baker-Campbell-Hausdorff relation for special unitary groups $SU(N)$. *J. Phys. A* **1997**, *30*, 8739–8749.
7. Dattoli, G.; Ottaviani, P.L.; Torre, A.; Vásquez, L. Evolution operator equations: Integration with algebraic and finite-difference methods. Applications to physical problems in classical and quantum mechanics and quantum field theory. *Riv. Nuovo Cimento* **1997**, *20*, 1–133.
8. Dattoli, G.; Zhukovsky, K. Quark flavour mixing and the exponential form of the Kobayashi–Maskawa matrix. *Eur. Phys. J. C* **2007**, *50*, 817–821.
9. Leonard, I. The matrix exponential. *SIAM Rev.* **1996**, *38*, 507–512.
10. Untidt, T.S.; Nielsen, N.C. Closed solution to the Baker-Campbell-Hausdorff problem: Exact effective Hamiltonian theory for analysis of nuclear-magnetic-resonance experiments. *Phys. Rev. E* **2002**, *65*, doi:10.1103/PhysRevE.65.021108.
11. Moore, G. Orthogonal polynomial expansions for the matrix exponential. *Linear Algebra Appl.* **2011**, *435*, 537–559.
12. Ding, F. Computation of matrix exponentials of special matrices. *Appl. Math. Comput.* **2013**, *223*, 311–326.
13. Koch, C.T.; Spence, J.C.H. A useful expansion of the exponential of the sum of two non-commuting matrices, one of which is diagonal. *J. Phys. A Math. Gen.* **2003**, *36*, 803–816.
14. Ramakrishna, V.; Zhou, H. On the exponential of matrices in $su(4)$. *J. Phys. A Math. Gen.* **2006**, *39*, 3021–3034.
15. Tudor, T. On the single-exponential closed form of the product of two exponential operators. *J. Phys. A Math. Theor.* **2007**, *40*, 14803–14810.
16. Siminovitch, D.; Untidt, T.S.; Nielsen, N.C. Exact effective Hamiltonian theory. II. Polynomial expansion of matrix functions and entangled unitary exponential operators. *J. Chem. Phys.* **2004**, *120*, 51–66.
17. Goldstein, H. *Classical Mechanics*, 2nd ed.; Addison-Wesley: New York, NY, USA, 1980; pp. 164–174.
18. Press, W.H.; Teukolsky, S.A.; Vetterling, W.T.; Flannery, B.P. *Numerical Recipees in FORTRAN, The Art of Scientific Computing*, 2nd ed.; Cambridge University Press: Cambridge, UK, 1992; pp. 83–84.
19. Bjorken, J.D.; Drell, S.D. *Relativistic Quantum Mechanics*; McGraw-Hill: New York, NY, USA, 1965.
20. Babusci, D.; Dattoli, G.; Sabia, E. Operational methods and Lorentz-type equations of motion. *J. Phys. Math.* **2011**, *3*, 1–17.
21. Puri, R.R. *Mathematical Methods of Quantum Optics*; Springer: New York, NY, USA, 2001; pp. 8–53.

22. Meystre, P.; Sargent, M. *Elements of Quantum Optics*, 2nd ed.; Springer: Berlin, Germany, 1999, pp. 372–373.
23. Kim, Y.S.; Noz, M.E. Symmetries shared by the Poincaré group and the Poincaré sphere. *Symmetry* **2013**, *5*, 233–252.

© 2014 by the author. Licensee MDPI, Basel, Switzerland. This article is an open access article distributed under the terms and conditions of the Creative Commons Attribution (CC BY) license (http://creativecommons.org/licenses/by/4.0/).

symmetry

MDPI

Article

Invisibility and $\mathcal{PT}$ Symmetry: A Simple Geometrical Viewpoint

Luis L. Sánchez-Soto * and Juan J. Monzón

Departamento de Óptica, Facultad de Física, Universidad Complutense, 28040 Madrid, Spain; E-Mail:
jjmonzon@opt.ucm.es
* E-Mail: lsanchez@fis.ucm.es; Tel.: +34-91-3944-680; Fax: +34-91-3944-683.

Received: 24 February 2014; in revised form: 12 May 2014 / Accepted: 14 May 2014 /
Published: 22 May 2014

Abstract: We give a simplified account of the properties of the transfer matrix for a complex one-dimensional potential, paying special attention to the particular instance of unidirectional invisibility. In appropriate variables, invisible potentials appear as performing null rotations, which lead to the helicity-gauge symmetry of massless particles. In hyperbolic geometry, this can be interpreted, via Möbius transformations, as parallel displacements, a geometric action that has no Euclidean analogy.

Keywords: $\mathcal{PT}$ symmetry; SL(2, $\mathbb{C}$); Lorentz group; Hyperbolic geometry

1. Introduction

The work of Bender and coworkers [1–6] has triggered considerable efforts to understand complex potentials that have neither parity ($\mathcal{P}$) nor time-reversal symmetry ($\mathcal{T}$), yet they retain combined $\mathcal{PT}$ invariance. These systems can exhibit real energy eigenvalues, thus suggesting a plausible generalization of quantum mechanics. This speculative concept has motivated an ongoing debate in several forefronts [7,8].

Quite recently, the prospect of realizing $\mathcal{PT}$-symmetric potentials within the framework of optics has been put forward [9,10] and experimentally tested [11]. The complex refractive index takes on here the role of the potential, so they can be realized through a judicious inclusion of index guiding and gain/loss regions. These $\mathcal{PT}$-synthetic materials can exhibit several intriguing features [12–14], one of which will be the main interest of this paper, namely, unidirectional invisibility [15–17].

In all these matters, the time-honored transfer-matrix method is particularly germane [18]. However, a quick look at the literature immediately reveals the different backgrounds and habits in which the transfer matrix is used and the very little cross talk between them.

To remedy this flaw, we have been capitalizing on a number of geometrical concepts to gain further insights into the behavior of one-dimensional scattering [19–26]. Indeed, when one think in a unifying mathematical scenario, geometry immediately comes to mind. Here, we keep going this program and examine the action of the transfer matrices associated to invisible scatterers. Interestingly enough, when viewed in SO(1, 3), they turn to be nothing but parabolic Lorentz transformations, also called null rotations, which play a crucial role in the determination of the little group of massless particles. Furthermore, borrowing elementary techniques of hyperbolic geometry, we reinterpret these matrices as parallel displacements, which are motions without Euclidean counterpart.

We stress that our formulation does not offer any inherent advantage in terms of efficiency in solving practical problems; rather, it furnishes a general and unifying setting to analyze the transfer matrix for complex potentials, which, in our opinion, is more than a curiosity.

2. Basic Concepts on Transfer Matrix

To be as self-contained as possible, we first briefly review some basic facts on the quantum scattering of a particle of mass m by a local complex potential $V(x)$ defined on the real line $\mathbb{R}$ [27–34]. Although much of the renewed interest in this topic has been fuelled by the remarkable case of $\mathcal{PT}$ symmetry, we do not use this extra assumption in this Section.

The problem at hand is governed by the time-independent Schrödinger equation

$$H\Psi(x) = \left[-\frac{d^2}{dx^2} + U(x) \right] \Psi(x) = \varepsilon \, \Psi(x) \tag{1}$$

where $\varepsilon = 2mE/\hbar^2$ and $U(x) = 2mV(x)/\hbar^2$, E being the energy of the particle. We assume that $U(x) \to 0$ fast enough as $x \to \pm\infty$, although the treatment can be adapted, with minor modifications, to cope with potentials for which the limits $U_\pm = \lim_{x \to \pm\infty} U(x)$ are different.

Since $U(x)$ decays rapidly as $|x| \to \infty$, solutions of (1) have the asymptotic behavior

$$\Psi(x) = \begin{cases} A_+ e^{+ikx} + A_- e^{-ikx} & x \to -\infty \\ B_+ e^{+ikx} + B_- e^{-ikx} & x \to \infty \end{cases} \tag{2}$$

Here, $k^2 = \varepsilon$, $A_\pm$ and $B_\pm$ are k-dependent complex coefficients (unspecified, at this stage), and the subscripts $+$ and $-$ distinguish right-moving modes $\exp(+ikx)$ from left-moving modes $\exp(-ikx)$, respectively.

The problem requires to work out the exact solution of (1) and invoke the appropriate boundary conditions, involving not only the continuity of $\Psi(x)$ itself, but also of its derivative. In this way, one has two linear relations among the coefficients $A_\pm$ and $B_\pm$, which can be solved for any amplitude pair in terms of the other two; the result can be expressed as a matrix equation that translates the linearity of the problem. Frequently, it is more advantageous to specify a linear relation between the wave amplitudes on both sides of the scatterer, namely,

$$\begin{pmatrix} B_+ \\ B_- \end{pmatrix} = \mathbf{M} \begin{pmatrix} A_+ \\ A_- \end{pmatrix} \tag{3}$$

$\mathbf{M}$ is the transfer matrix, which depends in a complicated way on the potential $U(x)$. Yet one can extract a good deal of information without explicitly calculating it: let us apply (3) successively to a right-moving $[(A_+ = 1, B_- = 0)]$ and to a left-moving wave $[(A_+ = 0, B_- = 1)]$, both of unit amplitude. The result can be displayed as

$$\begin{pmatrix} T^\ell \\ 0 \end{pmatrix} = \mathbf{M} \begin{pmatrix} 1 \\ R^\ell \end{pmatrix}, \qquad \begin{pmatrix} R^r \\ 1 \end{pmatrix} = \mathbf{M} \begin{pmatrix} 0 \\ T^r \end{pmatrix} \tag{4}$$

where $T^{\ell,r}$ and $R^{\ell,r}$ are the transmission and reflection coefficients for a wave incoming at the potential from the left and from the right, respectively, defined in the standard way as the quotients of the pertinent fluxes [35].

With this in mind, Equation (4) can be thought of as a linear superposition of the two independent solutions

$$\Psi_k^\ell(x) = \begin{cases} e^{+ikx} + R^\ell(k)\, e^{-ikx} & x \to -\infty, \\ T^\ell(k)\, e^{+ikx} & x \to \infty, \end{cases} \quad \Psi_k^r(x) = \begin{cases} T^r(k)\, e^{-ikx} & x \to -\infty \\ e^{-ikx} + R^r(k)\, e^{+ikx} & x \to \infty \end{cases} \tag{5}$$

which is consistent with the fact that, since $\varepsilon > 0$, the spectrum of the Hamiltonian (1) is continuous and there are two linearly independent solutions for a given value of ε. The wave function $\Psi_k^\ell(x)$ represents a wave incident from $-\infty$ $[\exp(+ikx)]$ and the interaction with the potential produces a reflected wave $[R^\ell(k)\exp(-ikx)]$ that escapes to $-\infty$ and a transmitted wave $[T^\ell(k)\exp(+ikx)]$ that moves off to $+\infty$. The solution $\Psi_k^r(x)$ can be interpreted in a similar fashion.

Because of the Wronskian of the solutions (5) is independent of x, we can compute $W(\Psi_k^\ell, \Psi_k^r) = \Psi_k^\ell \Psi_k^{r\prime} - \Psi_k^{\ell\prime} \Psi_k^r$ first for $x \to -\infty$ and then for $x \to \infty$; this gives

$$\frac{i}{2k} W(\Psi_k^\ell, \Psi_k^r) = T^r(k) = T^\ell(k) \equiv T(k) \tag{6}$$

We thus arrive at the important conclusion that, irrespective of the potential, the transmission coefficient is always independent of the input direction.

Taking this constraint into account, we go back to the system (4) and write the solution for $\mathbf{M}$ as

$$M_{11}(k) = T(k) - \frac{R^\ell(k)R^r(k)}{T(k)}, \quad M_{12}(k) = \frac{R^r(k)}{T(k)}, \quad M_{21}(k) = -\frac{R^\ell(k)}{T(k)}, \quad M_{22}(k) = \frac{1}{T(k)} \tag{7}$$

A straightforward check shows that $\det \mathbf{M} = +1$, so $\mathbf{M} \in \mathrm{SL}(2, \mathbb{C})$; a result that can be drawn from a number of alternative and more elaborate arguments [36].

One could also relate outgoing amplitudes to the incoming ones (as they are often the magnitudes one can externally control): this is precisely the scattering matrix, which can be concisely formulated as

$$\begin{pmatrix} B_+ \\ A_- \end{pmatrix} = \mathbf{S} \begin{pmatrix} A_+ \\ B_- \end{pmatrix} \tag{8}$$

with matrix elements

$$S_{11}(k) = T(k), \quad S_{12}(k) = R^r(k), \quad S_{21}(k) = R^\ell(k), \quad S_{22}(k) = T(k) \tag{9}$$

Finally, we stress that transfer matrices are very convenient mathematical objects. Suppose that V_1 and V_2 are potentials with finite support, vanishing outside a pair of adjacent intervals I_1 and I_2. If $\mathbf{M}_1$ and $\mathbf{M}_2$ are the corresponding transfer matrices, the total system (with support $I_1 \cup I_2$) is described by

$$\mathbf{M} = \mathbf{M}_1 \mathbf{M}_2 \tag{10}$$

This property is rather helpful: we can connect simple scatterers to create an intricate potential landscape and determine its transfer matrix by simple multiplication. This is a common instance in optics, where one routinely has to treat multilayer stacks. However, this important property does not seem to carry over into the scattering matrix in any simple way [37,38], because the incoming amplitudes for the overall system cannot be obtained in terms of the incoming amplitudes for every subsystem.

3. Spectral Singularities

The scattering solutions (5) constitute quite an intuitive way to attack the problem and they are widely employed in physical applications. Nevertheless, it is sometimes advantageous to look at the fundamental solutions of (1) in terms of left- and right-moving modes, as we have already used in (2).

Indeed, the two independent solutions of (1) can be formally written down as [39]

$$
\begin{aligned}
\Psi_k^{(+)}(x) &= e^{+ikx} + \int_x^\infty K_+(x, x') e^{+ikx'} dx' \\
\Psi_k^{(-)}(x) &= e^{-ikx} + \int_{-\infty}^x K_-(x, x') e^{-ikx'} dx'
\end{aligned}
\tag{11}
$$

The kernels $K_\pm(x, x')$ enjoy a number of interesting properties. What matters for our purposes is that the resulting $\Psi_k^{(\pm)}(x)$ are analytic with respect to k in $\mathbb{C}_+ = \{z \in \mathbb{C} | \operatorname{Im} z > 0\}$ and continuous on the real axis. In addition, it is clear that

$$\Psi_k^{(+)}(x) = e^{+ikx} \quad x \to \infty, \qquad \Psi_k^{(-)}(x) = e^{-ikx} \quad x \to -\infty \tag{12}$$

that is, they are the Jost functions for this problem [31].

Let us look at the Wronskian of the Jost functions $W(\Psi_k^{(-)}, \Psi_k^{(+)})$, which, as a function of k, is analytical in $\mathbb{C}_+$. A spectral singularity is a point $k_* \in \mathbb{R}_+$ of the continuous spectrum of the Hamiltonian (1) such that

$$W(\Psi_{k_*}^{(-)}, \Psi_{k_*}^{(+)}) = 0 \tag{13}$$

so $\Psi_k^{(\pm)}(x)$ become linearly dependent at k_* and the Hamiltonian is not diagonalizable. In fact, the set of zeros of the Wronskian is bounded, has at most a countable number of elements and its limit points can lie in a bounded subinterval of the real axis [40]. There is an extensive theory of spectral singularities for (1) that was started by Naimark [41]; the interested reader is referred to, e.g., Refs. [42–46] for further details.

The asymptotic behavior of $\Psi_k^\pm(x)$ at the opposite extremes of $\mathbb{R}$ with respect to those in (12) can be easily worked out by a simple application of the transfer matrix (and its inverse); viz,

$$\begin{aligned}
\Psi_k^{(-)}(x) &= M_{12}e^{+ikx} + M_{22}e^{-ikx} \quad x \to \infty \\[1em]
\Psi_k^{(+)}(x) &= M_{22}e^{+ikx} - M_{21}e^{-ikx} \quad x \to -\infty
\end{aligned} \tag{14}$$

Using $\Psi_k^\pm(x)$ in (12) and (14), we can calculate

$$\frac{i}{2k} W(\Psi_k^{(-)}, \Psi_k^{(+)}) = M_{22}(k) \tag{15}$$

Upon comparing with the definition (13), we can reinterpret the spectral singularities as the real zeros of $M_{22}(k)$ and, as a result, the reflection and transmission coefficients diverge therein. The converse holds because $M_{12}(k)$ and $M_{21}(k)$ are entire functions, lacking singularities. This means that, in an optical scenario, spectral singularities correspond to lasing thresholds [47–49].

One could also consider the more general case that the Hamiltonian (1) has, in addition to a continuous spectrum corresponding to $k \in \mathbb{R}_+$, a possibly complex discrete spectrum. The latter corresponds to the square-integrable solutions of that represent bound states. They are also zeros of $M_{22}(k)$, but unlike the zeros associated with the spectral singularities these must have a positive imaginary part [36].

The eigenvalues of **S** are

$$s_\pm = \frac{1}{M_{22}(k)} \left[1 \pm \sqrt{1 - M_{11}(k)M_{22}(k)} \right] \tag{16}$$

At a spectral singularity, s_+ diverges, while $s_- \to M_{11}(k)/2$, which suggests identifying spectral singularities with resonances with a vanishing width.

4. Invisibility and $\mathcal{PT}$ Symmetry

As heralded in the Introduction, unidirectional invisibility has been lately predicted in $\mathcal{PT}$ materials. We shall elaborate on the ideas developed by Mostafazadeh [50] in order to shed light into this intriguing question.

The potential $U(x)$ is called reflectionless from the left (right), if $R^\ell(k) = 0$ and $R^r(k) \neq 0$ [$R^r(k) = 0$ and $R^\ell(k) \neq 0$]. From the explicit matrix elements in (7) and (9), we see that unidirectional reflectionlessness implies the non-diagonalizability of both **M** and **S**. Therefore, the parameters of the potential for which it becomes reflectionless correspond to exceptional points of **M** and **S** [51,52].

The potential is called invisible from the left (right), if it is reflectionless from left (right) and in addition $T(k) = 1$. We can easily express the conditions for the unidirectional invisibility as

$$M_{12}(k) \neq 0, \qquad M_{11}(k) = M_{22}(k) = 1 \qquad \text{(left invisible)}$$

$$M_{21}(k) \neq 0, \qquad M_{11}(k) = M_{22}(k) = 1 \qquad \text{(right invisible)}$$

(17)

Next, we scrutinize the role of $\mathcal{PT}$-symmetry in the invisibility. For that purpose, we first briefly recall that the parity transformation "reflects" the system with respect to the coordinate origin, so that $x \mapsto -x$ and the momentum $p \mapsto -p$. The action on the wave function is

$$\Psi(x) \mapsto (\mathcal{P}\Psi)(x) = \Psi(-x) \tag{18}$$

On the other hand, the time reversal inverts the sense of time evolution, so that $x \mapsto x$, $p \mapsto -p$ and $i \mapsto -i$. This means that the operator $\mathcal{T}$ implementing such a transformation is antiunitary and its action reads

$$\Psi(x) \mapsto (\mathcal{T}\Psi)(x) = \Psi^*(x) \tag{19}$$

Consequently, under a combined $\mathcal{PT}$ transformation, we have

$$\Psi(x) \mapsto (\mathcal{PT}\Psi)(x) = \Psi^*(-x) \tag{20}$$

Let us apply this to a general complex scattering potential. The transfer matrix of the $\mathcal{PT}$-transformed system, we denote by $\mathbf{M}^{(\mathcal{PT})}$, fulfils

$$\begin{pmatrix} A_+^* \\ A_-^* \end{pmatrix} = \mathbf{M}^{(\mathcal{PT})} \begin{pmatrix} B_+^* \\ B_-^* \end{pmatrix} \tag{21}$$

Comparing with (3), we come to the result

$$\mathbf{M}^{(\mathcal{PT})} = (\mathbf{M}^{-1})^* \tag{22}$$

and, because det $\mathbf{M} = 1$, this means

$$M_{11} \xmapsto{\mathcal{PT}} M_{22}^*, \qquad M_{12} \xmapsto{\mathcal{PT}} -M_{12}^*, \qquad M_{21} \xmapsto{\mathcal{PT}} -M_{21}^*, \qquad M_{22} \xmapsto{\mathcal{PT}} M_{11}^* \tag{23}$$

When the system is invariant under this transformation $[\mathbf{M}^{(\mathcal{PT})} = \mathbf{M}]$, it must hold

$$\mathbf{M}^{-1} = \mathbf{M}^* \tag{24}$$

a fact already noticed by Longhi [48] and that can be also recast as [53]

$$\text{Re}\left(\frac{R^\ell}{T}\right) = \text{Re}\left(\frac{R^r}{T}\right) = 0 \tag{25}$$

This can be equivalently restated in the form

$$\rho^\ell - \tau = \pm\pi/2, \qquad \rho^r - \tau = \pm\pi/2 \tag{26}$$

with $\tau = \arg(T)$ and $\rho_{\ell,r} = \arg(R_{\ell,r})$. Hence, if we look at the complex numbers R^ℓ, R^r, and T as phasors, Equation (26) tell us that R^ℓ and R^r are always collinear, while T is simultaneously perpendicular to them. We draw the attention to the fact that the same expressions have been derived for lossless symmetric beam splitters [54]: we have shown that they hold true for any $\mathcal{PT}$-symmetric structure.

A direct consequence of (23) is that there are particular instances of $\mathcal{PT}$-invariant systems that are invisible, although not every invisible potential is $\mathcal{PT}$ invariant. In this respect, it is worth stressing, that even ($\mathcal{P}$-symmetric) potentials do not support unidirectional invisibility and the same holds for real ($\mathcal{T}$-symmetric) potentials.

In optics, beam propagation is governed by the paraxial wave equation, which is equivalent to a Schrödinger-like equation, with the role of the potential played here by the refractive index. Therefore, a necessary condition for a complex refractive index to be $\mathcal{PT}$ invariant is that its real part is an even function of x, while the imaginary component (loss and gain profile) is odd.

5. Relativistic Variables

To move ahead, let us construct the Hermitian matrices

$$\mathbf{X} = \begin{pmatrix} X_+ \\ X_- \end{pmatrix} \otimes \begin{pmatrix} X_+^* & X_-^* \end{pmatrix} = \begin{pmatrix} |X_+|^2 & X_+X_-^* \\ X_+^*X_- & |X_-|^2 \end{pmatrix} \tag{27}$$

where $X_\pm$ refers to either $A_\pm$ or $B_\pm$; *i.e.*, the amplitudes that determine the behavior at each side of the potential. The matrices $\mathbf{X}$ are quite reminiscent of the coherence matrix in optics or the density matrix in quantum mechanics.

One can verify that $\mathbf{M}$ acts on $\mathbf{X}$ by conjugation

$$\mathbf{X}' = \mathbf{M}\mathbf{X}\mathbf{M}^\dagger \tag{28}$$

The matrix $\mathbf{X}'$ is associated with the amplitudes $B_\pm$ and $\mathbf{X}$ with $A_\pm$.

Let us consider the set $\sigma^\mu = (\mathbb{1}, \boldsymbol{\sigma})$, with Greek indices running from 0 to 3. The σ^μ are the identity and the standard Pauli matrices, which constitute a basis of the linear space of 2×2 complex matrices. For the sake of covariance, it is convenient to define $\tilde{\sigma}^\mu \equiv \sigma_\mu = (\mathbb{1}, -\boldsymbol{\sigma})$, so that [55]

$$\mathrm{Tr}(\tilde{\sigma}^\mu \sigma_\nu) = 2\delta^\mu_\nu \tag{29}$$

and δ^μ_ν is the Kronecker delta. To any Hermitian matrix $\mathbf{X}$ we can associate the coordinates

$$x^\mu = \tfrac{1}{2}\mathrm{Tr}(\mathbf{X}\tilde{\sigma}^\mu) \tag{30}$$

The congruence (28) induces in this way a transformation

$$x'^\mu = \Lambda^\mu_\nu(\mathbf{M})\,x^\nu \tag{31}$$

where $\Lambda^\mu_\nu(\mathbf{M})$ can be found to be

$$\Lambda^\mu_\nu(\mathbf{M}) = \tfrac{1}{2}\mathrm{Tr}\left(\tilde{\sigma}^\mu\mathbf{M}\sigma_\nu\mathbf{M}^\dagger\right) \tag{32}$$

This equation can be solved to obtain $\mathbf{M}$ from Λ. The matrices $\mathbf{M}$ and $-\mathbf{M}$ generate the same Λ, so this homomorphism is two-to-one. The variables x^μ are coordinates in a Minkovskian (1+3)-dimensional space and the action of the system can be seen as a Lorentz transformation in SO(1, 3).

Having set the general scenario, let us have a closer look at the transfer matrix corresponding to right invisibility (the left invisibility can be dealt with in an analogous way); namely,

$$\mathbf{M} = \begin{pmatrix} 1 & R \\ 0 & 1 \end{pmatrix} \tag{33}$$

where, for simplicity, we have dropped the superscript from R^r, as there is no risk of confusion. Under the homomorphism (32) this matrix generates the Lorentz transformation

$$\Lambda(\mathbf{M}) = \begin{pmatrix} 1 + |R|^2/2 & \operatorname{Re} R & -\operatorname{Im} R & -|R|^2/2 \\ \operatorname{Re} R & 1 & 0 & -\operatorname{Re} R \\ -\operatorname{Im} R & 0 & 1 & \operatorname{Im} R \\ |R|^2/2 & \operatorname{Re} R & -\operatorname{Im} R & 1 - |R|^2/2 \end{pmatrix} \tag{34}$$

According to Wigner [56], the little group is a subgroup of the Lorentz transformations under which a standard vector s^μ remains invariant. When s^μ is timelike, the little group is the rotation group SO(3). If s^μ is spacelike, the little group are the boosts SO(1, 2). In this context, the matrix (34) is an instance of a null rotation; the little group when s^μ is a lightlike or null vector, which is related to E(2), the symmetry group of the two-dimensional Euclidean space [57].

If we write (34) in the form $\Lambda(\mathbf{M}) = \exp(i\mathbf{N})$, we can easily work out that

$$\mathbf{N} = \begin{pmatrix} 0 & \operatorname{Re} R & -\operatorname{Im} R & 0 \\ \operatorname{Re} R & 0 & 0 & -\operatorname{Re} R \\ -\operatorname{Im} R & 0 & 0 & \operatorname{Im} R \\ 0 & \operatorname{Re} R & -\operatorname{Im} R & 0 \end{pmatrix} \tag{35}$$

This is a nilpotent matrix and the vectors annihilated by $\mathbf{N}$ are invariant by $\Lambda(\mathbf{M})$. In terms of the Lie algebra so(1, 3), $\mathbf{N}$ can be expressed as

$$\mathbf{N} = \operatorname{Re} R \, (\mathbf{K}_1 + \mathbf{J}_2) - \operatorname{Im} R \, (\mathbf{K}_2 + \mathbf{J}_1) \tag{36}$$

where $\mathbf{K}_i$ generate boosts and $\mathbf{J}_i$ rotations ($i = 1, 2, 3$) [58]. Observe that the rapidity of the boost and the angle of the rotation have the same norm. The matrix $\mathbf{N}$ define a two-parameter Abelian subgroup.

Let us take, for the time being, $\operatorname{Re} R = 0$, as it happens for $\mathcal{PT}$-invariant invisibility. We can express $\mathbf{K}_2 + \mathbf{J}_1$ as the differential operator

$$\mathbf{K}_2 + \mathbf{J}_1 \mapsto (x^2 \partial_0 + x^0 \partial_2) + (x^2 \partial_3 - x^3 \partial_2) = x^2 (\partial_0 + \partial_3) + (x^0 - x^3) \partial_2 \tag{37}$$

As we can appreciate, the combinations

$$x^2, \qquad x^0 - x^3, \qquad (x^0)^2 - (x^1)^2 - (x^3)^2 \tag{38}$$

remain invariant. Suppressing the inessential coordinate x^2, the flow lines of the Killing vector (37) is the intersection of a null plane, $x^0 - x^3 = c_2$ with a hyperboloid $(x^0)^2 - (x^1)^2 - (x^3)^2 = c_3$. The case $c_3 = 0$ has the hyperboloid degenerate to a light cone with the orbits becoming parabolas lying in corresponding null planes.

6. Hyperbolic Geometry and Invisibility

Although the relativistic hyperboloid in Minkowski space constitute by itself a model of hyperbolic geometry (understood in a broad sense, as the study of spaces with constant negative curvature), it is not the best suited to display some features.

Let us consider the customary tridimensional hyperbolic space $\mathbb{H}^3$, defined in terms of the upper half-space $\{(x, y, z) \in \mathbb{R}^3 | z > 0\}$, equipped with the metric [59]

$$ds^2 = \frac{\sqrt{dx^2 + dy^2 + dz^2}}{z} \tag{39}$$

The geodesics are the semicircles in $\mathbb{H}^3$ orthogonal to the plane $z = 0$.

We can think of the plane $z = 0$ in $\mathbb{R}^3$ as the complex plane $\mathbb{C}$ with the natural identification $(x, y, z) \mapsto w = x + iy$. We need to add the point at infinity, so that $\hat{\mathbb{C}} = \mathbb{C} \cup \infty$, which is usually referred to as the Riemann sphere and identify $\hat{\mathbb{C}}$ as the boundary of $\mathbb{H}^3$.

Every matrix $\mathbf{M}$ in SL(2, $\mathbb{C}$) induces a natural mapping in $\mathbb{C}$ via Möbius (or bilinear) transformations [60]

$$w' = \frac{M_{11} w + M_{12}}{M_{21} w + M_{22}} \tag{40}$$

Note that any matrix obtained by multiplying $\mathbf{M}$ by a complex scalar λ gives the same transformation, so a Möbius transformation determines its matrix only up to scalar multiples. In other words, we need to quotient out SL(2, $\mathbb{C}$) by its center $\{\mathbb{1}, -\mathbb{1}\}$: the resulting quotient group is known as the projective linear group and is usually denoted PSL(2, $\mathbb{C}$).

Observe that we can break down the action (40) into a composition of maps of the form

$$w \mapsto w + \lambda, \qquad w \mapsto \lambda w, \qquad w \mapsto -1/w \tag{41}$$

with $\lambda \in \mathbb{C}$. Then we can extend the Möbius transformations to all $\mathbb{H}^3$ as follows:

$$(w, z) \mapsto (w + \lambda, z), \qquad (w, z) \mapsto (\lambda w, |\lambda| z), \qquad (w, z) \mapsto \left(-\frac{w^*}{|w^2| + z^2}, \frac{z}{|w^2| + z^2} \right) \tag{42}$$

The expressions above come from decomposing the action on $\hat{\mathbb{C}}$ of each of the elements of PSL(2, $\mathbb{C}$) in question into two inversions (reflections) in circles in $\hat{\mathbb{C}}$. Each such inversion has a unique extension to $\mathbb{H}_3$ as an inversion in the hemisphere spanned by the circle and composing appropriate pairs of inversions gives us these formulas.

In fact, one can show that PSL(2, $\mathbb{C}$) preserves the metric on $\mathbb{H}_3$. Moreover every isometry of $\mathbb{H}_3$ can be seen to be the extension of a conformal map of $\hat{\mathbb{C}}$ to itself, since it must send hemispheres orthogonal to $\hat{\mathbb{C}}$ to hemispheres orthogonal to $\hat{\mathbb{C}}$, hence circles in $\hat{\mathbb{C}}$ to circles in $\hat{\mathbb{C}}$. Thus all orientation-preserving isometries of $\mathbb{H}_3$ are given by elements of PSL(2, $\mathbb{C}$) acting as above.

In the classification of these isometries the notion of fixed points is of utmost importance. These points are defined by the condition $w' = w$ in (40), whose solutions are

$$w_f = \frac{(M_{11} - M_{22}) \pm \sqrt{[\mathrm{Tr}(\mathbf{M})]^2 - 4}}{2M_{21}} \tag{43}$$

So, they are determined by the trace of $\mathbf{M}$. When the trace is a real number, the induced Möbius transformations are called elliptic, hyperbolic, or parabolic, according $[\mathrm{Tr}(\mathbf{M})]^2$ is lesser than, greater than, or equal to 4, respectively. The canonical representatives of those matrices are [61]

$$\underbrace{\begin{pmatrix} e^{i\theta/2} & 0 \\ 0 & e^{-i\theta/2} \end{pmatrix}}_{\text{elliptic}}, \quad \underbrace{\begin{pmatrix} e^{\zeta/2} & 0 \\ 0 & e^{-\zeta/2} \end{pmatrix}}_{\text{hyperbolic}}, \quad \underbrace{\begin{pmatrix} 1 & \lambda \\ 0 & 1 \end{pmatrix}}_{\text{parabolic}} \tag{44}$$

while the induced geometrical actions are

$$w' = w e^{i\theta}, \qquad w' = w e^{\zeta}, \qquad w' = w + \lambda \tag{45}$$

that is, a rotation of angle θ (so fixes the axis z), a squeezing of parameter ζ (it has two fixed points in $\hat{\mathbb{C}}$, no fixed points in $\mathbb{H}_3$, and every hyperplane in $\mathbb{H}_3$ that contains the geodesic joining the two fixed points in $\hat{\mathbb{C}}$ is invariant); and a parallel displacement of magnitude λ, respectively. We emphasize that this later action is the only one without Euclidean analogy. Indeed, in view of (33), this is precisely the action associated to an invisible scatterer. The far-reaching consequences of this geometrical interpretation will be developed elsewhere.

7. Concluding Remarks

We have studied unidirectional invisibility by a complex scattering potential, which is characterized by a set of $\mathcal{PT}$ invariant equations. Consequently, the $\mathcal{PT}$-symmetric invisible configurations are quite special, for they possess the same symmetry as the equations.

We have shown how to cast this phenomenon in term of space-time variables, having in this way a relativistic presentation of invisibility as the set of null rotations. By resorting to elementary notions of hyperbolic geometry, we have interpreted in a natural way the action of the transfer matrix in this case as a parallel displacement.

We think that our results are yet another example of the advantages of these geometrical methods: we have devised a geometrical tool to analyze invisibility in quite a concise way that, in addition, can be closely related to other fields of physics.

Acknowledgments: We acknowledge illuminating discussions with Antonio F. Costa, José F. Cariñena and José María Montesinos. Financial support from the Spanish Research Agency (Grant FIS2011-26786) is gratefully acknowledged.

Author Contributions: Both authors contributed equally to the theoretical analysis, numerical calculations, and writing of the paper.

Conflicts of Interest: The authors declare no conflict of interest.

References

1. Bender, C.M.; Boettcher, S. Real spectra in non-Hermitian Hamiltonians having $\mathcal{PT}$ symmetry. *Phys. Rev. Lett.* **1998**, *80*, 5243–5246.
2. Bender, C.M.; Boettcher, S.; Meisinger, P.N. $\mathcal{PT}$-symmetric quantum mechanics. *J. Math. Phys.* **1999**, *40*, 2201–2229.
3. Bender, C.M.; Brody, D.C.; Jones, H.F. Complex extension of quantum mechanics. *Phys. Rev. Lett.* **2002**, *89*, doi:10.1103/PhysRevLett.89.270401.
4. Bender, C.M.; Brody, D.C.; Jones, H.F. Must a Hamiltonian be Hermitian? *Am. J. Phys.* **2003**, *71*, 1095–1102.
5. Bender, C.M. Making sense of non-Hermitian Hamiltonians. *Rep. Prog. Phys.* **2007**, *70*, 947–1018.
6. Bender, C.M.; Mannheim, P.D. $\mathcal{PT}$ symmetry and necessary and sufficient conditions for the reality of energy eigenvalues. *Phys. Lett. A* **2010**, *374*, 1616–1620.
7. Assis, P. *Non-Hermitian Hamiltonians in Field Theory: $\mathcal{PT}$-symmetry and Applications*; VDM: Saarbrücken, Germany, 2010.
8. Moiseyev, N. *Non-Hermitian Quantum Mechanics*; Cambridge University Press: Cambridge, UK, 2011.
9. El-Ganainy, R.; Makris, K.G.; Christodoulides, D.N.; Musslimani, Z.H. Theory of coupled optical $\mathcal{PT}$-symmetric structures. *Opt. Lett.* **2007**, *32*, 2632–2634.
10. Bendix, O.; Fleischmann, R.; Kottos, T.; Shapiro, B. Exponentially fragile $\mathcal{PT}$ symmetry in lattices with localized eigenmodes. *Phys. Rev. Lett.* **2009**, *103*, doi:10.1103/PhysRevLett.103.030402.
11. Ruter, C.E.; Makris, K.G.; El-Ganainy, R.; Christodoulides, D.N.; Segev, M.; Kip, D. Observation of parity-time symmetry in optics. *Nat. Phys.* **2010**, *6*, 192–195.
12. Makris, K.G.; El-Ganainy, R.; Christodoulides, D.N.; Musslimani, Z.H. Beam dynamics in $\mathcal{PT}$ symmetric optical lattices. *Phys. Rev. Lett.* **2008**, *100*, 103904:1–103904:4.
13. Longhi, S. Bloch oscillations in complex crystals with $\mathcal{PT}$ symmetry. *Phys. Rev. Lett.* **2009**, *103*, 123601:1–123601:4.
14. Sukhorukov, A.A.; Xu, Z.; Kivshar, Y.S. Nonlinear suppression of time reversals in $\mathcal{PT}$-symmetric optical couplers. *Phys. Rev. A* **2010**, *82*, doi:10.1103/PhysRevA.82.043818.
15. Ahmed, Z.; Bender, C.M.; Berry, M.V. Reflectionless potentials and $\mathcal{PT}$ symmetry. *J. Phys. A* **2005**, *38*, L627–L630.
16. Lin, Z.; Ramezani, H.; Eichelkraut, T.; Kottos, T.; Cao, H.; Christodoulides, D.N. Unidirectional invisibility dnduced by $\mathcal{PT}$-symmetric periodic structures. *Phys. Rev. Lett.* **2011**, *106*, doi:10.1103/PhysRevLett.106.213901.
17. Longhi, S. Invisibility in $\mathcal{PT}$-symmetric complex crystals. *J. Phys. A* **2011**, *44*, doi:10.1088/1751-8113/44/48/485302.

18. Sánchez-Soto, L.L.; Monzón, J.J.; Barriuso, A.G.; Cariñena, J. The transfer matrix: A geometrical perspective. *Phys. Rep.* **2012**, *513*, 191–227.

19. Monzón, J.J.; Sánchez-Soto, L.L. Lossles multilayers and Lorentz transformations: More than an analogy. *Opt. Commun.* **1999**, *162*, 1–6.

20. Monzón, J.J.; Sánchez-Soto, L.L. Fullly relativisticlike formulation of multilayer optics. *J. Opt. Soc. Am. A* **1999**, *16*, 2013–2018.

21. Monzón, J.J.; Yonte, T.; Sánchez-Soto, L.L. Basic factorization for multilayers. *Opt. Lett.* **2001**, *26*, 370–372.

22. Yonte, T.; Monzón, J.J.; Sánchez-Soto, L.L.; Cariñena, J.F.; López-Lacasta, C. Understanding multilayers from a geometrical viewpoint. *J. Opt. Soc. Am. A* **2002**, *19*, 603–609.

23. Monzón, J.J.; Yonte, T.; Sánchez-Soto, L.L.; Cariñena, J.F. Geometrical setting for the classification of multilayers. *J. Opt. Soc. Am. A* **2002**, *19*, 985–991.

24. Barriuso, A.G.; Monzón, J.J.; Sánchez-Soto, L.L. General unit-disk representation for periodic multilayers. *Opt. Lett.* **2003**, *28*, 1501–1503.

25. Barriuso, A.G.; Monzón, J.J.; Sánchez-Soto, L.L.; Cariñena, J.F. Vectorlike representation of multilayers. *J. Opt. Soc. Am. A* **2004**, *21*, 2386–2391.

26. Barriuso, A.G.; Monzón, J.J.; Sánchez-Soto, L.L.; Costa, A.F. Escher-like quasiperiodic heterostructures. *J. Phys. A* **2009**, *42*, 192002:1–192002:9.

27. Muga, J.G.; Palao, J.P.; Navarro, B.; Egusquiza, I.L. Complex absorbing potentials. *Phys. Rep.* **2004**, *395*, 357–426.

28. Levai, G.; Znojil, M. Systematic search for $\mathcal{PT}$-symmetric potentials with real spectra. *J. Phys. A* **2000**, *33*, 7165–7180.

29. Ahmed, Z. Schrödinger transmission through one-dimensional complex potentials. *Phys. Rev. A* **2001**, *64*, 042716:1–042716:4.

30. Ahmed, Z. Energy band structure due to a complex, periodic, $\mathcal{PT}$-invariant potential. *Phys. Lett. A* **2001**, *286*, 231–235.

31. Mostafazadeh, A. Spectral singularities of complex scattering potentials and infinite reflection and transmission coefficients at real energies. *Phys. Rev. Lett.* **2009**, *102*, 220402:1–220402:4.

32. Cannata, F.; Dedonder, J.P.; Ventura, A. Scattering in $\mathcal{PT}$-symmetric quantum mechanics. *Ann. Phys.* **2007**, *322*, 397–433.

33. Chong, Y.D.; Ge, L.; Stone, A.D. $\mathcal{PT}$-symmetry breaking and laser-absorber modes in optical scattering systems. *Phys. Rev. Lett.* **2011**, *106*, doi:10.1103/PhysRevLett.106.093902.

34. Ahmed, Z. New features of scattering from a one-dimensional non-Hermitian (complex) potential. *J. Phys. A* **2012**, *45*, doi:10.1088/1751-8113/45/3/032004.

35. Boonserm, P.; Visser, M. One dimensional scattering problems: A pedagogical presentation of the relationship between reflection and transmission amplitudes. *Thai J. Math.* **2010**, *8*, 83–97.

36. Mostafazadeh, A.; Mehri-Dehnavi, H. Spectral singularities, biorthonormal systems and a two-parameter family of complex point interactions. *J. Phys. A* **2009**, *42*, doi:10.1088/1751-8113/ 42/12/125303.

37. Aktosun, T. A factorization of the scattering matrix for the Schrödinger equation and for the wave equation in one dimension. *J. Math. Phys.* **1992**, *33*, 3865–3869.

38. Aktosun, T.; Klaus, M.; van der Mee, C. Factorization of scattering matrices due to partitioning of potentials in one-dimensional Schrödinger-type equations. *J. Math. Phys.* **1996**, *37*, 5897–5915.

39. Marchenko, V.A. *Sturm-Liouville Operators and Their Applications*; AMS Chelsea: Providence, RI, USA, 1986.

40. Tunca, G.; Bairamov, E. Discrete spectrum and principal functions of non-selfadjoint differential operator. *Czech J. Math.* **1999**, *49*, 689–700.

41. Naimark, M.A. Investigation of the spectrum and the expansion in eigenfunctions of a non-selfadjoint operator of the second order on a semi-axis. *AMS Transl.* **1960**, *16*, 103–193.

42. Pavlov, B.S. The nonself-adjoint Schrödinger operators. *Topics Math. Phys.* **1967**, *1*, 87–114.

43. Naimark, M.A. *Linear Differential Operators: Part II*; Ungar: New York, NY, USA, 1968.

44. Samsonov, B.F. SUSY transformations between diagonalizable and non-diagonalizable Hamiltonians. *J. Phys. A* **2005**, *38*, L397–L403.

45. Andrianov, A.A.; Cannata, F.; Sokolov, A.V. Spectral singularities for non-Hermitian one-dimensional Hamiltonians: Puzzles with resolution of identity. *J. Math. Phys.* **2010**, *51*, 052104:1–052104:22.

46. Chaos-Cador, L.; García-Calderón, G. Resonant states for complex potentials and spectral singularities. *Phys. Rev. A* **2013**, *87*, doi:10.1103/PhysRevA.87.042114.

47. Schomerus, H. Quantum noise and self-sustained radiation of $\mathcal{PT}$-symmetric systems. *Phys. Rev. Lett.* **2010**, *104*, doi:10.1103/PhysRevLett.104.233601.

48. Longhi, S. $\mathcal{PT}$-symmetric laser absorber. *Phys. Rev. A* **2010**, *82*, doi:10.1103/ PhysRevA.82.031801.

49. Mostafazadeh, A. Nonlinear spectral singularities of a complex barrier potential and the lasing threshold condition. *Phys. Rev. A* **2013**, *87*, doi:10.1103/PhysRevA.87.063838.

50. Mostafazadeh, A. Invisibility and $\mathcal{PT}$ symmetry. *Phys. Rev. A* **2013**, *87*, doi:10.1103/ PhysRevA.87.012103.

51. Müller, M.; Rotter, I. Exceptional points in open quantum systems. *J. Phys. A* **2008**, *41*, 244018:1–244018:15.

52. Mehri-Dehnavi, H.; Mostafazadeh, A. Geometric phase for non-Hermitian Hamiltonians and its holonomy interpretation. *J. Math. Phys.* **2008**, *49*, 082105:1–082105:17.

53. Monzón, J.J.; Barriuso, A.G.; Montesinos-Amilibia, J.M.; Sánchez-Soto, L.L. Geometrical aspects of $\mathcal{PT}$-invariant transfer matrices. *Phys. Rev. A* **2013**, *87*, doi:10.1103/PhysRevA.87.012111.

54. Mandel, L.; Wolf, E. *Optical Coherence and Quantum Optics*; Cambridge University Press: Cambridge, UK, 1995.

55. Barut, A.O.; Rączka, R. *Theory of Group Representations and Applications*; PWN: Warszaw, Poland, 1977; Section 17.2.

56. Wigner, E. On unitary representations of the inhomogeneous Lorentz group. *Ann. Math.* **1939**, *40*, 149–204.

57. Kim, Y.S.; Noz, M.E. *Theory and Applications of the Poincaré Group*; Reidel: Dordrecht, The Netherlands, 1986.

58. Weinberg, S. *The Quantum Theory of Fields*; Cambridge University Press: Cambridge, UK, 2005; Volume 1.

59. Iversen, B. *Hyperbolic Geometry*; Cambridge University Press: Cambridge, UK, 1992; Chapter VIII.

60. Ratcliffe, J.G. *Foundations of Hyperbolic Manifolds*; Springer: Berlin, Germany, 2006; Section 4.3.

61. Anderson, J.W. *Hyperbolic Geometry*; Springer: New York, NY, USA, 1999; Chapter 3.

© 2014 by the authors. Licensee MDPI, Basel, Switzerland. This article is an open access article distributed under the terms and conditions of the Creative Commons Attribution (CC BY) license (http://creativecommons.org/licenses/by/4.0/).

symmetry

MDPI

Article

Wigner's Space-Time Symmetries Based on the Two-by-Two Matrices of the Damped Harmonic Oscillators and the Poincaré Sphere

Sibel Başkal [1], Young S. Kim [2,*] and Marilyn E. Noz [3]

[1] Department of Physics, Middle East Technical University, Ankara 06800, Turkey;
  E-Mail: baskal@newton.physics.metu.edu.tr
[2] Center for Fundamental Physics, University of Maryland, College Park, MD 20742, USA
[3] Department of Radiology, New York University, New York, NY 10016, USA; E-Mail: marilyne.noz@gmail.com
* E-Mail: yskim@umd.edu; Tel.: +1-301-937-1306.

Received: 28 February 2014; in revised form: 28 May 2014 / Accepted: 9 June 2014 / Published: 25 June 2014

Abstract: The second-order differential equation for a damped harmonic oscillator can be converted to two coupled first-order equations, with two two-by-two matrices leading to the group $Sp(2)$. It is shown that this oscillator system contains the essential features of Wigner's little groups dictating the internal space-time symmetries of particles in the Lorentz-covariant world. The little groups are the subgroups of the Lorentz group whose transformations leave the four-momentum of a given particle invariant. It is shown that the damping modes of the oscillator correspond to the little groups for massive and imaginary-mass particles respectively. When the system makes the transition from the oscillation to damping mode, it corresponds to the little group for massless particles. Rotations around the momentum leave the four-momentum invariant. This degree of freedom extends the $Sp(2)$ symmetry to that of $SL(2,c)$ corresponding to the Lorentz group applicable to the four-dimensional Minkowski space. The Poincaré sphere contains the $SL(2,c)$ symmetry. In addition, it has a non-Lorentzian parameter allowing us to reduce the mass continuously to zero. It is thus possible to construct the little group for massless particles from that of the massive particle by reducing its mass to zero. Spin-1/2 particles and spin-1 particles are discussed in detail.

Keywords: damped harmonic oscillators; coupled first-order equations; unimodular matrices; Wigner's little groups; Poincaré sphere; $Sp(2)$ group; $SL(2,c)$ group; gauge invariance; neutrinos; photons

PACS: 03.65.Fd, 03.67.-a, 05.30.-d

1. Introduction

We are quite familiar with the second-order differential equation

$$m\frac{d^2y}{dt^2} + b\frac{dy}{dt} + Ky = 0 \tag{1}$$

for a damped harmonic oscillator. This equation has the same mathematical form as

$$L\frac{d^2Q}{dt^2} + R\frac{dQ}{dt} + \frac{1}{C}Q = 0 \tag{2}$$

for electrical circuits, where L, R, and C are the inductance, resistance, and capacitance respectively. These two equations play fundamental roles in physical and engineering sciences. Since they start from the same set of mathematical equations, one set of problems can be studied in terms of the other. For instance, many mechanical phenomena can be studied in terms of electrical circuits.

In Equation (1), when $b = 0$, the equation is that of a simple harmonic oscillator with the frequency $\omega = \sqrt{K/m}$. As b increases, the oscillation becomes damped. When b is larger than $2\sqrt{Km}$, the oscillation disappears, as the solution is a damping mode.

Consider that increasing b continuously, while difficult mechanically, can be done electrically using Equation (2) by adjusting the resistance R. The transition from the oscillation mode to the damping mode is a continuous physical process.

This b term leads to energy dissipation, but is not regarded as a fundamental force. It is inconvenient in the Hamiltonian formulation of mechanics and troublesome in transition to quantum mechanics, yet, plays an important role in classical mechanics. In this paper this term will help us understand the fundamental space-time symmetries of elementary particles.

We are interested in constructing the fundamental symmetry group for particles in the Lorentz-covariant world. For this purpose, we transform the second-order differential equation of Equation (1) to two coupled first-order equations using two-by-two matrices. Only two linearly independent matrices are needed. They are the anti-symmetric and symmetric matrices

$$A = \begin{pmatrix} 0 & -i \\ i & 0 \end{pmatrix}, \quad \text{and} \quad S = \begin{pmatrix} 0 & i \\ i & 0 \end{pmatrix} \tag{3}$$

respectively. The anti-symmetric matrix A is Hermitian and corresponds to the oscillation part, while the symmetric S matrix corresponds to the damping.

These two matrices lead to the $Sp(2)$ group consisting of two-by-two unimodular matrices with real elements. This group is isomorphic to the three-dimensional Lorentz group applicable to two space-like and one time-like coordinates. This group is commonly called the $O(2,1)$ group.

This $O(2,1)$ group can explain all the essential features of Wigner's little groups dictating internal space-time symmetries of particles [1]. Wigner defined his little groups as the subgroups of the Lorentz group whose transformations leave the four-momentum of a given particle invariant. He observed that the little groups are different for massive, massless, and imaginary-mass particles. It has been a challenge to design a mathematical model which will combine those three into one formalism, but we show that the damped harmonic oscillator provides the desired mathematical framework.

For the two space-like coordinates, we can assign one of them to the direction of the momentum, and the other to the direction perpendicular to the momentum. Let the direction of the momentum be along the z axis, and let the perpendicular direction be along the x axis. We therefore study the kinematics of the group within the zx plane, then see what happens when we rotate the system around the z axis without changing the momentum [2].

The Poincaré sphere for polarization optics contains the $SL(2,c)$ symmetry isomorphic to the four-dimensional Lorentz group applicable to the Minkowski space [3–7]. Thus, the Poincaré sphere extends Wigner's picture into the three space-like and one time-like coordinates. Specifically, this extension adds rotations around the given momentum which leaves the four-momentum invariant [2].

While the particle mass is a Lorentz-invariant variable, the Poincaré sphere contains an extra variable which allows the mass to change. This variable allows us to take the mass-limit of the symmetry operations. The transverse rotational degrees of freedom collapse into one gauge degree of freedom and polarization of neutrinos is a consequence of the requirement of gauge invariance [8,9].

The $SL(2,c)$ group contains symmetries not seen in the three-dimensional rotation group. While we are familiar with two spinors for a spin-$1/2$ particle in nonrelativistic quantum mechanics, there are two additional spinors due to the reflection properties of the Lorentz group. There are thus 16 bilinear combinations of those four spinors. This leads to two scalars, two four-vectors, and one antisymmetric four-by-four tensor. The Maxwell-type electromagnetic field tensor can be obtained as a massless limit of this tensor [10].

In Section 2, we review the damped harmonic oscillator in classical mechanics, and note that the solution can be either in the oscillation mode or damping mode depending on the magnitude of the damping parameter. The translation of the second order equation into a first order differential equation with two-by-two matrices is possible. This first-order equation is similar to the Schrödinger equation for a spin-1/2 particle in a magnetic field.

Section 3 shows that the two-by-two matrices of Section 2 can be formulated in terms of the $Sp(2)$ group. These matrices can be decomposed into the Bargmann and Wigner decompositions. Furthermore, this group is isomorphic to the three-dimensional Lorentz group with two space and one time-like coordinates.

In Section 4, it is noted that this three-dimensional Lorentz group has all the essential features of Wigner's little groups which dictate the internal space-time symmetries of the particles in the Lorentz-covariant world. Wigner's little groups are the subgroups of the Lorentz group whose transformations leave the four-momentum of a given particle invariant. The Bargmann Wigner decompositions are shown to be useful tools for studying the little groups.

In Section 5, we note that the given momentum is invariant under rotations around it. The addition of this rotational degree of freedom extends the $Sp(2)$ symmetry to the six-parameter $SL(2, c)$ symmetry. In the space-time language, this extends the three dimensional group to the Lorentz group applicable to three space and one time dimensions.

Section 6 shows that the Poincaré sphere contains the symmetries of $SL(2, c)$ group. In addition, it contains an extra variable which allows us to change the mass of the particle, which is not allowed in the Lorentz group.

In Section 7, the symmetries of massless particles are studied in detail. In addition to rotation around the momentum, Wigner's little group generates gauge transformations. While gauge transformations on spin-1 photons are well known, the gauge invariance leads to the polarization of massless spin-1/2 particles, as observed in neutrino polarizations.

In Section 8, it is noted that there are four spinors for spin-1/2 particles in the Lorentz-covariant world. It is thus possible to construct 16 bilinear forms, applicable to two scalars, and two vectors, and one antisymmetric second-rank tensor. The electromagnetic field tensor is derived as the massless limit. This tensor is shown to be gauge-invariant.

2. Classical Damped Oscillators

For convenience, we write Equation (1) as

$$\frac{d^2y}{dt^2} + 2\mu\frac{dy}{dt} + \omega^2 y = 0 \tag{4}$$

with

$$\omega = \sqrt{\frac{K}{m}}, \quad \text{and} \quad \mu = \frac{b}{2m} \tag{5}$$

The damping parameter μ is positive when there are no external forces. When ω is greater than μ, the solution takes the form

$$y = e^{-\mu t}\left[C_1\cos(\omega' t) + C_2\sin(\omega' t)\right] \tag{6}$$

where

$$\omega' = \sqrt{\omega^2 - \mu^2} \tag{7}$$

and C_1 and C_2 are the constants to be determined by the initial conditions. This expression is for a damped harmonic oscillator. Conversely, when μ is greater than ω, the quantity inside the square-root sign is negative, then the solution becomes

$$y = e^{-\mu t}\left[C_3\cosh(\mu' t) + C_4\sinh(\mu' t)\right] \tag{8}$$

with

$$\mu' = \sqrt{\mu^2 - \omega^2} \tag{9}$$

If $\omega = \mu$, both Equations (6) and (8) collapse into one solution

$$y(t) = e^{-\mu t}\left[C_5 + C_6\, t\right] \tag{10}$$

These three different cases are treated separately in textbooks. Here we are interested in the transition from Equation (6) to Equation (8), via Equation (10). For convenience, we start from μ greater than ω with μ' given by Equation (9).

For a given value of μ, the square root becomes zero when ω equals μ. If ω becomes larger, the square root becomes imaginary and divides into two branches.

$$\pm i\sqrt{\omega^2 - \mu^2} \tag{11}$$

This is a continuous transition, but not an analytic continuation. To study this in detail, we translate the second order differential equation of Equation (4) into the first-order equation with two-by-two matrices.

Given the solutions of Equations (6) and (10), it is convenient to use $\psi(t)$ defined as

$$\psi(t) = e^{\mu t} y(t), \quad \text{and} \quad y = e^{-\mu t} \psi(t) \tag{12}$$

Then $\psi(t)$ satisfies the differential equation

$$\frac{d^2 \psi(t)}{dt^2} + (\omega^2 - \mu^2)\psi(t) = 0 \tag{13}$$

2.1. Two-by-Two Matrix Formulation

In order to convert this second-order equation to a first-order system, we introduce $\psi_1(t)$ and $\psi_2(t)$ satisfying two coupled differential equations

$$\frac{d\psi_1(t)}{dt} = (\mu - \omega)\psi_2(t) \tag{14}$$

$$\frac{d\psi_2(t)}{dt} = (\mu + \omega)\psi_1(t) \tag{15}$$

which can be written in matrix form as

$$\frac{d}{dt}\begin{pmatrix} \psi_1 \\ \psi_2 \end{pmatrix} = \begin{pmatrix} 0 & \mu - \omega \\ \mu + \omega & 0 \end{pmatrix}\begin{pmatrix} \psi_1 \\ \psi_2 \end{pmatrix} \tag{16}$$

Using the Hermitian and anti-Hermitian matrices of Equation (3) in Section 1, we construct the linear combination

$$H = \omega \begin{pmatrix} 0 & -i \\ i & 0 \end{pmatrix} + \mu \begin{pmatrix} 0 & i \\ i & 0 \end{pmatrix} \tag{17}$$

We can then consider the first-order differential equation

$$i\frac{\partial}{\partial t}\psi(t) = H\psi(t) \tag{18}$$

While this equation is like the Schrödinger equation for an electron in a magnetic field, the two-by-two matrix is not Hermitian. Its first matrix is Hermitian, but the second matrix is anti-Hermitian. It is of course an interesting problem to give a physical interpretation to this non-Hermitian matrix in connection with quantum dissipation [11], but this is beyond the scope of the present paper. The solution of Equation (18) is

$$\psi(t) = \exp\left\{\begin{pmatrix} 0 & -\omega + \mu \\ \omega + \mu & 0 \end{pmatrix} t\right\}\begin{pmatrix} C_7 \\ C_8 \end{pmatrix} \tag{19}$$

where $C_7 = \psi_1(0)$ and $C_8 = \psi_2(0)$ respectively.

2.2. Transition from the Oscillation Mode to Damping Mode

It appears straight-forward to compute this expression by a Taylor expansion, but it is not. This issue was extensively discussed in the earlier papers by two of us [12,13]. The key idea is to write the matrix

$$\begin{pmatrix} 0 & -\omega + \mu \\ \omega + \mu & 0 \end{pmatrix} \tag{20}$$

as a similarity transformation of

$$\omega'\begin{pmatrix} 0 & -1 \\ 1 & 0 \end{pmatrix} \qquad (\omega > \mu) \tag{21}$$

and as that of

$$\mu'\begin{pmatrix} 0 & 1 \\ 1 & 0 \end{pmatrix} \qquad (\mu > \omega) \tag{22}$$

with ω' and μ' defined in Equations (7) and (9), respectively.

Then the Taylor expansion leads to

$$\begin{pmatrix} \cos(\omega't) & -\sqrt{(\omega-\mu)/(\omega+\mu)}\ \sin(\omega't) \\ \sqrt{(\omega+\mu)/(\omega-\mu)}\ \sin(\omega't) & \cos(\omega't) \end{pmatrix} \tag{23}$$

when ω is greater than μ. The solution $\psi(t)$ takes the form

$$\begin{pmatrix} C_7\cos(\omega't) - C_8\sqrt{(\omega-\mu)/(\omega+\mu)}\ \sin(\omega't) \\ C_7\sqrt{(\omega+\mu)/(\omega-\mu)}\ \sin(\omega't) + C_8\cos(\omega't) \end{pmatrix} \tag{24}$$

If μ is greater than ω, the Taylor expansion becomes

$$\begin{pmatrix} \cosh(\mu't) & \sqrt{(\mu-\omega)/(\mu+\omega)}\ \sinh(\mu't) \\ \sqrt{(\mu+\omega)/(\mu-\omega)}\ \sinh(\mu't) & \cosh(\mu't) \end{pmatrix} \tag{25}$$

When ω is equal to μ, both Equations (23) and (25) become

$$\begin{pmatrix} 1 & 0 \\ 2\omega t & 1 \end{pmatrix} \tag{26}$$

If ω is sufficiently close to but smaller than μ, the matrix of Equation (25) becomes

$$\begin{pmatrix} 1 + (\epsilon/2)(2\omega t)^2 & +\epsilon(2\omega t) \\ (2\omega t) & 1 + (\epsilon/2)(2\omega t)^2 \end{pmatrix} \tag{27}$$

with

$$\epsilon = \frac{\mu - \omega}{\mu + \omega} \tag{28}$$

If ω is sufficiently close to μ, we can let

$$\mu + \omega = 2\omega, \quad \text{and} \quad \mu - \omega = 2\mu\epsilon \tag{29}$$

If ω is greater than μ, ϵ defined in Equation (28) becomes negative, the matrix of Equation (23) becomes

$$\begin{pmatrix} 1 - (-\epsilon/2)(2\omega t)^2 & -(-\epsilon)(2\omega t) \\ 2\omega t & 1 - (-\epsilon/2)(2\omega t)^2 \end{pmatrix} \tag{30}$$

We can rewrite this matrix as

$$\begin{pmatrix} 1 - (1/2) \left[\left(2\omega\sqrt{-\epsilon} \right) t \right]^2 & -\sqrt{-\epsilon} \left[\left(2\omega\sqrt{-\epsilon} \right) t \right] \\ 2\omega t & 1 - (1/2) \left[\left(2\omega\sqrt{-\epsilon} \right) t \right]^2 \end{pmatrix} \tag{31}$$

If ϵ becomes positive, Equation (27) can be written as

$$\begin{pmatrix} 1 + (1/2) \left[\left(2\omega\sqrt{\epsilon} \right) t \right]^2 & \sqrt{\epsilon} \left[\left(2\omega\sqrt{\epsilon} \right) t \right] \\ 2\omega t & 1 + (1/2) \left[\left(2\omega\sqrt{\epsilon} \right) t \right]^2 \end{pmatrix} \tag{32}$$

The transition from Equation (31) to Equation (32) is continuous as they become identical when $\epsilon = 0$. As ϵ changes its sign, the diagonal elements of above matrices tell us how $\cos(\omega' t)$ becomes $\cosh(\mu' t)$. As for the upper-right element element, $-\sin(\omega' t)$ becomes $\sinh(\mu' t)$. This non-analytic continuity is discussed in detail in one of the earlier papers by two of us on lens optics [13]. This type of continuity was called there "tangential continuity." There, the function and its first derivative are continuous while the second derivative is not.

2.3. Mathematical Forms of the Solutions

In this section, we use the Heisenberg approach to the problem, and obtain the solutions in the form of two-by-two matrices. We note that

1. For the oscillation mode, the trace of the matrix is smaller than 2. The solution takes the form of

$$\begin{pmatrix} \cos(x) & -e^{-\eta} \sin(x) \\ e^{\eta} \sin(x) & \cos(x) \end{pmatrix} \tag{33}$$

   with trace $2\cos(x)$. The trace is independent of η.
2. For the damping mode, the trace of the matrix is greater than 2.

$$\begin{pmatrix} \cosh(x) & e^{-\eta} \sinh(x) \\ e^{\eta} \sinh(x) & \cosh(x) \end{pmatrix} \tag{34}$$

   with trace $2\cosh(x)$. Again, the trace is independent of η.
3. For the transition mode, the trace is equal to 2, and the matrix is triangular and takes the form of

$$\begin{pmatrix} 1 & 0 \\ \gamma & 1 \end{pmatrix} \tag{35}$$

When x approaches zero, the Equations (33) and (34) take the form

$$\begin{pmatrix} 1 - x^2/2 & -xe^{-\eta} \\ xe^{\eta} & 1 - x^2/2 \end{pmatrix}, \quad \text{and} \quad \begin{pmatrix} 1 + x^2/2 & xe^{-\eta} \\ xe^{\eta} & 1 + x^2/2 \end{pmatrix} \tag{36}$$

respectively. These two matrices have the same lower-left element. Let us fix this element to be a positive number γ. Then

$$x = \gamma e^{-\eta} \tag{37}$$

Then the matrices of Equation (36) become

$$\begin{pmatrix} 1 - \gamma^2 e^{-2\eta}/2 & -\gamma e^{-2\eta} \\ \gamma & 1 - \gamma^2 e^{-2\eta}/2 \end{pmatrix}, \quad \text{and} \quad \begin{pmatrix} 1 + \gamma^2 e^{-2\eta}/2 & \gamma e^{-2\eta} \\ \gamma & 1 + \gamma^2 e^{-2\eta}/2 \end{pmatrix} \tag{38}$$

If we introduce a small number ϵ defined as

$$\epsilon = \sqrt{\gamma} e^{-\eta} \tag{39}$$

the matrices of Equation (38) become

$$\begin{pmatrix} e^{-\eta/2} & 0 \\ 0 & e^{\eta/2} \end{pmatrix} \begin{pmatrix} 1 - \gamma\epsilon^2/2 & \sqrt{\gamma}\epsilon \\ \sqrt{\gamma}\epsilon & 1 - \gamma\epsilon^2/2 \end{pmatrix} \begin{pmatrix} e^{\eta/2} & 0 \\ 0 & e^{-\eta/2} \end{pmatrix}$$

$$\begin{pmatrix} e^{-\eta/2} & 0 \\ 0 & e^{\eta/2} \end{pmatrix} \begin{pmatrix} 1 + \gamma\epsilon^2/2 & \sqrt{\gamma}\epsilon \\ \sqrt{\gamma}\epsilon & 1 + \gamma\epsilon^2/2 \end{pmatrix} \begin{pmatrix} e^{\eta/2} & 0 \\ 0 & e^{-\eta/2} \end{pmatrix} \tag{40}$$

respectively, with $e^{-\eta} = \epsilon/\sqrt{\gamma}$.

3. Groups of Two-by-Two Matrices

If a two-by-two matrix has four complex elements, it has eight independent parameters. If the determinant of this matrix is one, it is known as an unimodular matrix and the number of independent parameters is reduced to six. The group of two-by-two unimodular matrices is called $SL(2,c)$. This six-parameter group is isomorphic to the Lorentz group applicable to the Minkowski space of three space-like and one time-like dimensions [14].

We can start with two subgroups of $SL(2,c)$.

1. While the matrices of $SL(2,c)$ are not unitary, we can consider the subset consisting of unitary matrices. This subgroup is called $SU(2)$, and is isomorphic to the three-dimensional rotation group. This three-parameter group is the basic scientific language for spin-1/2 particles.

2. We can also consider the subset of matrices with real elements. This three-parameter group is called $Sp(2)$ and is isomorphic to the three-dimensional Lorentz group applicable to two space-like and one time-like coordinates.

In the Lorentz group, there are three space-like dimensions with $x, y,$ and z coordinates. However, for many physical problems, it is more convenient to study the problem in the two-dimensional (x, z) plane first and generalize it to three-dimensional space by rotating the system around the z axis. This process can be called Euler decomposition and Euler generalization [2].

First, we study $Sp(2)$ symmetry in detail, and achieve the generalization by augmenting the two-by-two matrix corresponding to the rotation around the z axis. In this section, we study in detail properties of $Sp(2)$ matrices, then generalize them to $SL(2,c)$ in Section 5.

There are three classes of $Sp(2)$ matrices. Their traces can be smaller or greater than two, or equal to two. While these subjects are already discussed in the literature [15–17] our main interest is what happens as the trace goes from less than two to greater than two. Here we are guided by the model we have discussed in Section 2, which accounts for the transition from the oscillation mode to the damping mode.

3.1. Lie Algebra of Sp(2)

The two linearly independent matrices of Equation (3) can be written as

$$K_1 = \frac{1}{2} \begin{pmatrix} 0 & i \\ i & 0 \end{pmatrix}, \quad \text{and} \quad J_2 = \frac{1}{2} \begin{pmatrix} 0 & -i \\ i & 0 \end{pmatrix} \tag{41}$$

However, the Taylor series expansion of the exponential form of Equation (23) or Equation (25) requires an additional matrix

$$K_3 = \frac{1}{2} \begin{pmatrix} i & 0 \\ 0 & -i \end{pmatrix} \tag{42}$$

These matrices satisfy the following closed set of commutation relations.

$$[K_1, J_2] = iK_3, \qquad [J_2, K_3] = iK_1, \qquad [K_3, K_1] = -iJ_2 \tag{43}$$

These commutation relations remain invariant under Hermitian conjugation, even though K_1 and K_3 are anti-Hermitian. The algebra generated by these three matrices is known in the literature as the group $Sp(2)$ [17]. Furthermore, the closed set of commutation relations is commonly called the Lie algebra. Indeed, Equation (43) is the Lie algebra of the $Sp(2)$ group.

The Hermitian matrix J_2 generates the rotation matrix

$$R(\theta) = \exp\left(-i\theta J_2\right) = \begin{pmatrix} \cos(\theta/2) & -\sin(\theta/2) \\ \sin(\theta/2) & \cos(\theta/2) \end{pmatrix} \tag{44}$$

and the anti-Hermitian matrices K_1 and K_2, generate the following squeeze matrices.

$$S(\lambda) = \exp\left(-i\lambda K_1\right) = \begin{pmatrix} \cosh(\lambda/2) & \sinh(\lambda/2) \\ \sinh(\lambda/2) & \cosh(\lambda/2) \end{pmatrix} \tag{45}$$

and

$$B(\eta) = \exp\left(-i\eta K_3\right) = \begin{pmatrix} \exp\left(\eta/2\right) & 0 \\ 0 & \exp\left(-\eta/2\right) \end{pmatrix} \tag{46}$$

respectively.

Returning to the Lie algebra of Equation (43), since K_1 and K_3 are anti-Hermitian, and J_2 is Hermitian, the set of commutation relation is invariant under the Hermitian conjugation. In other words, the commutation relations remain invariant, even if we change the sign of K_1 and K_3, while keeping that of J_2 invariant. Next, let us take the complex conjugate of the entire system. Then both the J and K matrices change their signs.

3.2. Bargmann and Wigner Decompositions

Since the $Sp(2)$ matrix has three independent parameters, it can be written as [15]

$$\begin{pmatrix} \cos\left(\alpha_1/2\right) & -\sin\left(\alpha_1/2\right) \\ \sin\left(\alpha_1/2\right) & \cos\left(\alpha_1/2\right) \end{pmatrix} \begin{pmatrix} \cosh\chi & \sinh\chi \\ \sinh\chi & \cosh\chi \end{pmatrix} \begin{pmatrix} \cos\left(\alpha_2/2\right) & -\sin\left(\alpha_2/2\right) \\ \sin\left(\alpha_2/2\right) & \cos\left(\alpha_2/2\right) \end{pmatrix} \tag{47}$$

This matrix can be written as

$$\begin{pmatrix} \cos(\delta/2) & -\sin(\delta/2) \\ \sin(\delta/2) & \cos(\delta/2) \end{pmatrix} \begin{pmatrix} a & b \\ c & d \end{pmatrix} \begin{pmatrix} \cos(\delta/2) & \sin(\delta/2) \\ -\sin(\delta/2) & \cos(\delta/2) \end{pmatrix} \tag{48}$$

where

$$\begin{pmatrix} a & b \\ c & d \end{pmatrix} = \begin{pmatrix} \cos(\alpha/2) & -\sin(\alpha/2) \\ \sin(\alpha/2) & \cos(\alpha/2) \end{pmatrix} \begin{pmatrix} \cosh\chi & \sinh\chi \\ \sinh\chi & \cosh\chi \end{pmatrix} \begin{pmatrix} \cos(\alpha/2) & -\sin(\alpha/2) \\ \sin(\alpha/2) & \cos(\alpha/2) \end{pmatrix} \tag{49}$$

with

$$\delta = \frac{1}{2}(\alpha_1 - \alpha_2), \quad \text{and} \quad \alpha = \frac{1}{2}(\alpha_1 + \alpha_2) \tag{50}$$

If we complete the matrix multiplication of Equation (49), the result is

$$\begin{pmatrix} (\cosh\chi)\cos\alpha & \sinh\chi - (\cosh\chi)\sin\alpha \\ \sinh\chi + (\cosh\chi)\sin\alpha & (\cosh\chi)\cos\alpha \end{pmatrix} \tag{51}$$

We shall call hereafter the decomposition of Equation (49) the Bargmann decomposition. This means that every matrix in the $Sp(2)$ group can be brought to the Bargmann decomposition by a similarity transformation of rotation, as given in Equation (48). This decomposition leads to an equidiagonal matrix with two independent parameters.

For the matrix of Equation (49), we can now consider the following three cases. Let us assume that χ is positive, and the angle θ is less than $90°$. Let us look at the upper-right element.

1. If it is negative with $[\sinh\chi < (\cosh\chi)\sin\alpha]$, then the trace of the matrix is smaller than 2, and the matrix can be written as

$$\begin{pmatrix} \cos(\theta/2) & -e^{-\eta}\sin(\theta/2) \\ e^{\eta}\sin(\theta/2) & \cos(\theta/2) \end{pmatrix} \tag{52}$$

with

$$\cos(\theta/2) = (\cosh\chi)\cos\alpha, \quad \text{and} \quad e^{-2\eta} = \frac{(\cosh\chi)\sin\alpha - \sinh\chi}{(\cosh\chi)\sin\alpha + \sinh\chi} \tag{53}$$

2. If it is positive with $[\sinh\chi > (\cosh\chi)\sin\alpha)]$, then the trace is greater than 2, and the matrix can be written as

$$\begin{pmatrix} \cosh(\lambda/2) & e^{-\eta}\sinh(\lambda/2) \\ e^{\eta}\sinh(\lambda/2) & \cosh(\lambda/2) \end{pmatrix} \tag{54}$$

with

$$\cosh(\lambda/2) = (\cosh\chi)\cos\alpha, \quad \text{and} \quad e^{-2\eta} = \frac{\sinh\chi - (\cosh\chi)\sin\alpha}{(\cosh\chi)\sin\alpha + \sinh\chi} \tag{55}$$

3. If it is zero with $[(\sinh\chi = (\cosh\chi)\sin\alpha)]$, then the trace is equal to 2, and the matrix takes the form

$$\begin{pmatrix} 1 & 0 \\ 2\sinh\chi & 1 \end{pmatrix} \tag{56}$$

The above repeats the mathematics given in Section 2.3.

Returning to Equations (52) and (53), they can be decomposed into

$$M(\theta, \eta) = \begin{pmatrix} e^{\eta/2} & 0 \\ 0 & e^{-\eta/2} \end{pmatrix} \begin{pmatrix} \cos(\theta/2) & -\sin(\theta/2) \\ \sin(\theta/2) & \cos(\theta/2) \end{pmatrix} \begin{pmatrix} e^{-\eta/2} & 0 \\ 0 & e^{\eta/2} \end{pmatrix} \tag{57}$$

and

$$M(\lambda, \eta) = \begin{pmatrix} e^{\eta/2} & 0 \\ 0 & e^{-\eta/2} \end{pmatrix} \begin{pmatrix} \cosh(\lambda/2) & \sinh(\lambda/2) \\ \sinh(\lambda/2) & \cos(\lambda/2) \end{pmatrix} \begin{pmatrix} e^{-\eta/2} & 0 \\ 0 & e^{\eta/2} \end{pmatrix} \tag{58}$$

respectively. In view of the physical examples given in Section 6, we shall call this the "Wigner decomposition." Unlike the Bargmann decomposition, the Wigner decomposition is in the form of a similarity transformation.

We note that both Equations (57) and (58) are written as similarity transformations. Thus

$$[M(\theta, \eta)]^n = \begin{pmatrix} \cos(n\theta/2) & -e^{-\eta}\sin(n\theta/2) \\ e^{\eta}\sin(n\theta/2) & \cos(n\theta/2) \end{pmatrix} \tag{59}$$

$$[M(\lambda, \eta)]^n = \begin{pmatrix} \cosh(n\lambda/2) & e^{\eta}\sinh(n\lambda/2) \\ e^{-\eta}\sinh(n\lambda/2) & \cosh(n\lambda/2) \end{pmatrix} \tag{60}$$

$$[M(\gamma)]^n = \begin{pmatrix} 1 & 0 \\ n\gamma & 1 \end{pmatrix} \tag{61}$$

These expressions are useful for studying periodic systems [18].

The question is what physics these decompositions describe in the real world. To address this, we study what the Lorentz group does in the real world, and study isomorphism between the $Sp(2)$ group and the Lorentz group applicable to the three-dimensional space consisting of one time and two space coordinates.

3.3. Isomorphism with the Lorentz Group

The purpose of this section is to give physical interpretations of the mathematical formulas given in Section 3.2. We will interpret these formulae in terms of the Lorentz transformations which are normally described by four-by-four matrices. For this purpose, it is necessary to establish a correspondence between the two-by-two representation of Section 3.2 and the four-by-four representations of the Lorentz group.

Let us consider the Minkowskian space-time four-vector

$$(t, z, x, y) \tag{62}$$

where $(t^2 - z^2 - x^2 - y^2)$ remains invariant under Lorentz transformations. The Lorentz group consists of four-by-four matrices performing Lorentz transformations in the Minkowski space.

In order to give physical interpretations to the three two-by-two matrices given in Equations (44)–(46), we consider rotations around the y axis, boosts along the x axis, and boosts along the z axis. The transformation is restricted in the three-dimensional subspace of (t, z, x). It is then straight-forward to construct those four-by-four transformation matrices where the y coordinate remains invariant. They are given in Table 1. Their generators also given. Those four-by-four generators satisfy the Lie algebra given in Equation (43).

Table 1. Matrices in the two-by-two representation, and their corresponding four-by-four generators and transformation matrices.

Matrices	Generators	Four-by-Four	Transform matrices
$R(\theta)$	$J_2 = \frac{1}{2}\begin{pmatrix} 0 & -i \\ i & 0 \end{pmatrix}$	$\begin{pmatrix} 0 & 0 & 0 & 0 \\ 0 & 0 & -i & 0 \\ 0 & i & 0 & 0 \\ 0 & 0 & 0 & 0 \end{pmatrix}$	$\begin{pmatrix} 1 & 0 & 0 & 0 \\ 0 & \cos\theta & -\sin\theta & 0 \\ 0 & \sin\theta & \cos\theta & 0 \\ 0 & 0 & 0 & 1 \end{pmatrix}$
$B(\eta)$	$K_3 = \frac{1}{2}\begin{pmatrix} i & 0 \\ 0 & -i \end{pmatrix}$	$\begin{pmatrix} 0 & i & 0 & 0 \\ i & 0 & 0 & 0 \\ 0 & 0 & 0 & 0 \\ 0 & 0 & 0 & 0 \end{pmatrix}$	$\begin{pmatrix} \cosh\eta & \sinh\eta & 0 & 0 \\ \sinh\eta & \cosh\eta & 0 & 0 \\ 0 & 0 & 1 & 0 \\ 0 & 0 & 0 & 1 \end{pmatrix}$
$S(\lambda)$	$K_1 = \frac{1}{2}\begin{pmatrix} 0 & i \\ i & 0 \end{pmatrix}$	$\begin{pmatrix} 0 & 0 & i & 0 \\ 0 & 0 & 0 & 0 \\ i & 0 & 0 & 0 \\ 0 & 0 & 0 & 0 \end{pmatrix}$	$\begin{pmatrix} \cosh\lambda & 0 & \sinh\lambda & 0 \\ 0 & 1 & 0 & 0 \\ \sinh\lambda & 0 & \cosh\lambda & 0 \\ 0 & 0 & 0 & 1 \end{pmatrix}$

Symmetry **2014**, 6, 473–515

4. Internal Space-Time Symmetries

We have seen that there corresponds a two-by-two matrix for each four-by-four Lorentz transformation matrix. It is possible to give physical interpretations to those four-by-four matrices. It must thus be possible to attach a physical interpretation to each two-by-two matrix.

Since 1939 [1] when Wigner introduced the concept of the little groups many papers have been published on this subject, but most of them were based on the four-by-four representation. In this section, we shall give the formalism of little groups in the language of two-by-two matrices. In so doing, we provide physical interpretations to the Bargmann and Wigner decompositions introduced in Section 3.2.

4.1. Wigner's Little Groups

In [1], Wigner started with a free relativistic particle with momentum, then constructed subgroups of the Lorentz group whose transformations leave the four-momentum invariant. These subgroups thus define the internal space-time symmetry of the given particle. Without loss of generality, we assume that the particle momentum is along the z direction. Thus rotations around the momentum leave the momentum invariant, and this degree of freedom defines the helicity, or the spin parallel to the momentum.

We shall use the word "Wigner transformation" for the transformation which leaves the four-momentum invariant:

1. For a massive particle, it is possible to find a Lorentz frame where it is at rest with zero momentum. The four-momentum can be written as $m(1,0,0,0)$, where m is the mass. This four-momentum is invariant under rotations in the three-dimensional (z, x, y) space.
2. For an imaginary-mass particle, there is the Lorentz frame where the energy component vanishes. The momentum four-vector can be written as $p(0,1,0,0)$, where p is the magnitude of the momentum.
3. If the particle is massless, its four-momentum becomes $p(1,1,0,0)$. Here the first and second components are equal in magnitude.

The constant factors in these four-momenta do not play any significant roles. Thus we write them as $(1,0,0,0)$, $(0,1,0,0)$, and $(1,1,0,0)$ respectively. Since Wigner worked with these three specific four-momenta [1], we call them Wigner four-vectors.

All of these four-vectors are invariant under rotations around the z axis. The rotation matrix is

$$Z(\phi) = \begin{pmatrix} 1 & 0 & 0 & 0 \\ 0 & 1 & 0 & 0 \\ 0 & 0 & \cos\phi & -\sin\phi \\ 0 & 0 & \sin\phi & \cos\phi \end{pmatrix} \tag{63}$$

In addition, the four-momentum of a massive particle is invariant under the rotation around the y axis, whose four-by-four matrix was given in Table 1. The four-momentum of an imaginary particle is invariant under the boost matrix $S(\lambda)$ given in Table 1. The problem for the massless particle is more complicated, but will be discussed in detail in Section 7. See Table 2.

Table 2. Wigner four-vectors and Wigner transformation matrices applicable to two space-like and one time-like dimensions. Each Wigner four-vector remains invariant under the application of its Wigner matrix.

Mass	Wigner Four-Vector	Wigner Transformation
Massive	$(1,0,0,0)$	$\begin{pmatrix} 1 & 0 & 0 & 0 \\ 0 & \cos\theta & -\sin\theta & 0 \\ 0 & \sin\theta & \cos\theta & 0 \\ 0 & 0 & 0 & 1 \end{pmatrix}$
Massless	$(1,1,0,0)$	$\begin{pmatrix} 1+\gamma^2/2 & -\gamma^2/2 & \gamma & 0 \\ \gamma^2/2 & 1-\gamma^2/2 & \gamma & 0 \\ -\gamma & \gamma & 1 & 0 \\ 0 & 0 & 0 & 1 \end{pmatrix}$
Imaginary mass	$(0,1,0,0)$	$\begin{pmatrix} \cosh\lambda & 0 & \sinh\lambda & 0 \\ 0 & 1 & 0 & 0 \\ \sinh\lambda & 0 & \cosh\lambda & 0 \\ 0 & 0 & 0 & 1 \end{pmatrix}$

4.2. Two-by-Two Formulation of Lorentz Transformations

The Lorentz group is a group of four-by-four matrices performing Lorentz transformations on the Minkowskian vector space of (t, z, x, y), leaving the quantity

$$t^2 - z^2 - x^2 - y^2 \tag{64}$$

invariant. It is possible to perform the same transformation using two-by-two matrices [7,14,19].

In this two-by-two representation, the four-vector is written as

$$X = \begin{pmatrix} t+z & x-iy \\ x+iy & t-z \end{pmatrix} \tag{65}$$

where its determinant is precisely the quantity given in Equation (64) and the Lorentz transformation on this matrix is a determinant-preserving, or unimodular transformation. Let us consider the transformation matrix as [7,19]

$$G = \begin{pmatrix} \alpha & \beta \\ \gamma & \delta \end{pmatrix}, \quad \text{and} \quad G^\dagger = \begin{pmatrix} \alpha^* & \gamma^* \\ \beta^* & \delta^* \end{pmatrix} \tag{66}$$

with

$$\det(G) = 1 \tag{67}$$

and the transformation

$$X' = GXG^\dagger \tag{68}$$

Since G is not a unitary matrix, Equation (68) not a unitary transformation, but rather we call this the "Hermitian transformation". Equation (68) can be written as

$$\begin{pmatrix} t'+z' & x'-iy' \\ x+iy & t'-z' \end{pmatrix} = \begin{pmatrix} \alpha & \beta \\ \gamma & \delta \end{pmatrix} \begin{pmatrix} t+z & x-iy \\ x+iy & t-z \end{pmatrix} \begin{pmatrix} \alpha^* & \gamma^* \\ \beta^* & \delta^* \end{pmatrix} \tag{69}$$

It is still a determinant-preserving unimodular transformation, thus it is possible to write this as a four-by-four transformation matrix applicable to the four-vector (t, z, x, y) [7,14].

Since the G matrix starts with four complex numbers and its determinant is one by Equation (67), it has six independent parameters. The group of these G matrices is known to be locally isomorphic to the group of four-by-four matrices performing Lorentz transformations on the four-vector (t, z, x, y). In other words, for each G matrix there is a corresponding four-by-four Lorentz-transform matrix [7].

The matrix G is not a unitary matrix, because its Hermitian conjugate is not always its inverse. This group has a unitary subgroup called $SU(2)$ and another consisting only of real matrices called $Sp(2)$. For this later subgroup, it is sufficient to work with the three matrices $R(\theta), S(\lambda)$, and $B(\eta)$ given in Equations (44)–(46) respectively. Each of these matrices has its corresponding four-by-four matrix applicable to the (t, z, x, y). These matrices with their four-by-four counterparts are tabulated in Table 1.

The energy-momentum four vector can also be written as a two-by-two matrix. It can be written as

$$P = \begin{pmatrix} p_0 + p_z & p_x - ip_y \\ p_x + ip_y & p_0 - p_z \end{pmatrix} \tag{70}$$

with

$$\det(P) = p_0^2 - p_x^2 - p_y^2 - p_z^2 \tag{71}$$

which means

$$\det(P) = m^2 \tag{72}$$

where m is the particle mass.

The Lorentz transformation can be written explicitly as

$$P' = GPG^\dagger \tag{73}$$

or

$$\begin{pmatrix} p_0' + p_z' & p_x' - ip_y' \\ p_x' + ip_y' & E' - p_z' \end{pmatrix} = \begin{pmatrix} \alpha & \beta \\ \gamma & \delta \end{pmatrix} \begin{pmatrix} p_0 + p_z & p_x - ip_y \\ p_x + ip_y & p_0 - p_z \end{pmatrix} \begin{pmatrix} \alpha^* & \gamma^* \\ \beta^* & \delta^* \end{pmatrix} \tag{74}$$

This is an unimodular transformation, and the mass is a Lorentz-invariant variable. Furthermore, it was shown in [7] that Wigner's little groups for massive, massless, and imaginary-mass particles can be explicitly defined in terms of two-by-two matrices.

Wigner's little group consists of two-by-two matrices satisfying

$$P = WPW^\dagger \tag{75}$$

The two-by-two W matrix is not an identity matrix, but tells about the internal space-time symmetry of a particle with a given energy-momentum four-vector. This aspect was not known when Einstein formulated his special relativity in 1905, hence the internal space-time symmetry was not an issue at that time. We call the two-by-two matrix W the Wigner matrix, and call the condition of Equation (75) the Wigner condition.

If determinant of W is a positive number, then P is proportional to

$$P = \begin{pmatrix} 1 & 0 \\ 0 & 1 \end{pmatrix} \tag{76}$$

corresponding to a massive particle at rest, while if the determinant is negative, it is proportional to

$$P = \begin{pmatrix} 1 & 0 \\ 0 & -1 \end{pmatrix} \tag{77}$$

corresponding to an imaginary-mass particle moving faster than light along the z direction, with a vanishing energy component. If the determinant is zero, P is

$$P = \begin{pmatrix} 1 & 0 \\ 0 & 0 \end{pmatrix} \tag{78}$$

which is proportional to the four-momentum matrix for a massless particle moving along the z direction.

For all three cases, the matrix of the form

$$Z(\phi) = \begin{pmatrix} e^{-i\phi/2} & 0 \\ 0 & e^{i\phi/2} \end{pmatrix} \tag{79}$$

will satisfy the Wigner condition of Equation (75). This matrix corresponds to rotations around the z axis.

For the massive particle with the four-momentum of Equation (76), the transformations with the rotation matrix of Equation (44) leave the P matrix of Equation (76) invariant. Together with the $Z(\phi)$ matrix, this rotation matrix leads to the subgroup consisting of the unitary subset of the G matrices. The unitary subset of G is $SU(2)$ corresponding to the three-dimensional rotation group dictating the spin of the particle [14].

For the massless case, the transformations with the triangular matrix of the form

$$\begin{pmatrix} 1 & \gamma \\ 0 & 1 \end{pmatrix} \tag{80}$$

leave the momentum matrix of Equation (78) invariant. The physics of this matrix has a stormy history, and the variable γ leads to a gauge transformation applicable to massless particles [8,9,20,21].

For a particle with an imaginary mass, a W matrix of the form of Equation (45) leaves the four-momentum of Equation (77) invariant.

Table 3 summarizes the transformation matrices for Wigner's little groups for massive, massless, and imaginary-mass particles. Furthermore, in terms of their traces, the matrices given in this subsection can be compared with those given in Section 2.3 for the damped oscillator. The comparisons are given in Table 4.

Of course, it is a challenging problem to have one expression for all three classes. This problem has been discussed in the literature [12], and the damped oscillator case of Section 2 addresses the continuity problem.

Table 3. Wigner vectors and Wigner matrices in the two-by-two representation. The trace of the matrix tells whether the particle m^2 is positive, zero, or negative.

Particle Mass	Four-Momentum	Transform Matrix	Trace
Massive	$\begin{pmatrix} 1 & 0 \\ 0 & 1 \end{pmatrix}$	$\begin{pmatrix} \cos(\theta/2) & -\sin(\theta/2) \\ \sin(\theta/2) & \cos(\theta/2) \end{pmatrix}$	less than 2
Massless	$\begin{pmatrix} 1 & 0 \\ 0 & 0 \end{pmatrix}$	$\begin{pmatrix} 1 & \gamma \\ 0 & 1 \end{pmatrix}$	equal to 2
Imaginary mass	$\begin{pmatrix} 1 & 0 \\ 0 & -1 \end{pmatrix}$	$\begin{pmatrix} \cosh(\lambda/2) & \sinh(\lambda/2) \\ \sinh(\lambda/2) & \cosh(\lambda/2) \end{pmatrix}$	greater than 2

Table 4. Damped Oscillators and Space-time Symmetries. Both share $Sp(2)$ as their symmetry group.

Trace	Damped Oscillator	Particle Symmetry
Smaller than 2	Oscillation Mode	Massive Particles
Equal to 2	Transition Mode	Massless Particles
Larger than 2	Damping Mode	Imaginary-mass Particles

5. Lorentz Completion of Wigner's Little Groups

So far we have considered transformations applicable only to (t, z, x) space. In order to study the full symmetry, we have to consider rotations around the z axis. As previously stated, when a particle moves along this axis, this rotation defines the helicity of the particle.

In [1], Wigner worked out the little group of a massive particle at rest. When the particle gains a momentum along the z direction, the single particle can reverse the direction of momentum, the spin, or both. What happens to the internal space-time symmetries is discussed in this section.

5.1. Rotation around the z Axis

In Section 3, our kinematics was restricted to the two-dimensional space of z and x, and thus includes rotations around the y axis. We now introduce the four-by-four matrix of Equation (63) performing rotations around the z axis. Its corresponding two-by-two matrix was given in Equation (79). Its generator is

$$J_3 = \frac{1}{2} \begin{pmatrix} 1 & 0 \\ 0 & -1 \end{pmatrix} \tag{81}$$

If we introduce this additional matrix for the three generators we used in Sections 3 and 3.2, we end up the closed set of commutation relations

$$[J_i, J_j] = i\epsilon_{ijk}J_k, \qquad [J_i, K_j] = i\epsilon_{ijk}K_k, \qquad [K_i, K_j] = -i\epsilon_{ijk}J_k \tag{82}$$

with

$$J_i = \frac{1}{2}\sigma_i, \quad \text{and} \quad K_i = \frac{i}{2}\sigma_i \tag{83}$$

where σ_i are the two-by-two Pauli spin matrices.

For each of these two-by-two matrices there is a corresponding four-by-four matrix generating Lorentz transformations on the four-dimensional Lorentz group. When these two-by-two matrices are imaginary, the corresponding four-by-four matrices were given in Table 1. If they are real, the corresponding four-by-four matrices were given in Table 5.

Table 5. Two-by-two and four-by-four generators not included in Table 1. The generators given there and given here constitute the set of six generators for $SL(2, c)$ or of the Lorentz group given in Equation (82).

Generator	Two-by-Two	Four-by-Four
J_3	$\frac{1}{2}\begin{pmatrix} 1 & 0 \\ 0 & -1 \end{pmatrix}$	$\begin{pmatrix} 0 & 0 & 0 & 0 \\ 0 & 0 & 0 & 0 \\ 0 & 0 & 0 & -i \\ 0 & 0 & i & 0 \end{pmatrix}$
J_1	$\frac{1}{2}\begin{pmatrix} 0 & 1 \\ 1 & 0 \end{pmatrix}$	$\begin{pmatrix} 0 & 0 & 0 & 0 \\ 0 & 0 & 0 & i \\ 0 & 0 & 0 & 0 \\ 0 & -i & 0 & 0 \end{pmatrix}$
K_2	$\frac{1}{2}\begin{pmatrix} 0 & 1 \\ -1 & 0 \end{pmatrix}$	$\begin{pmatrix} 0 & 0 & 0 & i \\ 0 & 0 & 0 & 0 \\ 0 & 0 & 0 & 0 \\ i & 0 & 0 & 0 \end{pmatrix}$

This set of commutation relations is known as the Lie algebra for the $SL(2, c)$, namely the group of two-by-two elements with unit determinants. Their elements are complex. This set is also the Lorentz group performing Lorentz transformations on the four-dimensional Minkowski space.

This set has many useful subgroups. For the group $SL(2, c)$, there is a subgroup consisting only of real matrices, generated by the two-by-two matrices given in Table 1. This three-parameter subgroup is precisely the $Sp(2)$ group we used in Sections 3 and 3.2. Their generators satisfy the Lie algebra given in Equation (43).

In addition, this group has the following Wigner subgroups governing the internal space-time symmetries of particles in the Lorentz-covariant world [1]:

1. The J_i matrices form a closed set of commutation relations. The subgroup generated by these Hermitian matrices is $SU(2)$ for electron spins. The corresponding rotation group does not change the four-momentum of the particle at rest. This is Wigner's little group for massive particles.

   If the particle is at rest, the two-by-two form of the four-vector is given by Equation (76). The Lorentz transformation generated by J_3 takes the form

$$\begin{pmatrix} e^{i\phi/2} & 0 \\ 0 & e^{-i\phi/2} \end{pmatrix} \begin{pmatrix} 1 & 0 \\ 0 & 1 \end{pmatrix} \begin{pmatrix} e^{-i\phi/2} & 0 \\ 0 & e^{i\phi/2} \end{pmatrix} = \begin{pmatrix} 1 & 0 \\ 0 & 1 \end{pmatrix} \tag{84}$$

   Similar computations can be carried out for J_1 and J_2.

2. There is another $Sp(2)$ subgroup, generated by $K_1, K_2,$ and J_3. They satisfy the commutation relations

$$[K_1, K_2] = -iJ_3, \qquad [J_3, K_1] = iK_2, \qquad [K_2, J_3] = iK_1. \tag{85}$$

   The Wigner transformation generated by these two-by-two matrices leave the momentum four-vector of Equation (77) invariant. For instance, the transformation matrix generated by K_2 takes the form

$$\exp\left(-i\xi K_2\right) = \begin{pmatrix} \cosh(\xi/2) & i\sinh(\xi/2) \\ i\sinh(\xi/2) & \cosh(\xi/2) \end{pmatrix} \tag{86}$$

   and the Wigner transformation takes the form

$$\begin{pmatrix} \cosh(\xi/2) & i\sinh(\xi/2) \\ -i\sinh(\xi/2) & \cosh(\xi/2) \end{pmatrix} \begin{pmatrix} 1 & 0 \\ 0 & -1 \end{pmatrix} \begin{pmatrix} \cosh(\xi/2) & i\sinh(\xi/2) \\ -i\sinh(\xi/2) & \cosh(\xi/2) \end{pmatrix} = \begin{pmatrix} 1 & 0 \\ 0 & -1 \end{pmatrix} \tag{87}$$

   Computations with K_2 and J_3 lead to the same result.

Since the determinant of the four-momentum matrix is negative, the particle has an imaginary mass. In the language of the four-by-four matrix, the transformation matrices leave the four-momentum of the form $(0, 1, 0, 0)$ invariant.

3. Furthermore, we can consider the following combinations of the generators:

$$N_1 = K_1 - J_2 = \begin{pmatrix} 0 & i \\ 0 & 0 \end{pmatrix}, \quad \text{and} \quad N_2 = K_2 + J_1 = \begin{pmatrix} 0 & 1 \\ 0 & 0 \end{pmatrix} \tag{88}$$

Together with J_3, they satisfy the the following commutation relations.

$$[N_1, N_2] = 0, \qquad [N_1, J_3] = -iN_2, \qquad [N_2, J_3] = iN_1 \tag{89}$$

In order to understand this set of commutation relations, we can consider an $x\, y$ coordinate system in a two-dimensional space. Then rotation around the origin is generated by

$$J_3 = -i \left(x \frac{\partial}{\partial y} - y \frac{\partial}{\partial x} \right) \tag{90}$$

and the two translations are generated by

$$N_1 = -i \frac{\partial}{\partial x}, \quad \text{and} \quad N_2 = -i \frac{\partial}{\partial y} \tag{91}$$

for the x and y directions respectively. These operators satisfy the commutations relations given in Equation (89).

The two-by-two matrices of Equation (88) generate the following transformation matrix.

$$G(\gamma, \phi) = \exp\left[-i\gamma \left(N_1 \cos\phi + N_2 \sin\phi\right)\right] = \begin{pmatrix} 1 & \gamma e^{-i\phi} \\ 0 & 1 \end{pmatrix} \tag{92}$$

The two-by-two form for the four-momentum for the massless particle is given by Equation (78). The computation of the Hermitian transformation using this matrix is

$$\begin{pmatrix} 1 & \gamma e^{-i\phi} \\ 0 & 1 \end{pmatrix} \begin{pmatrix} 1 & 0 \\ 0 & 0 \end{pmatrix} \begin{pmatrix} 1 & 0 \\ \gamma e^{i\phi} & 1 \end{pmatrix} = \begin{pmatrix} 1 & 0 \\ 0 & 0 \end{pmatrix} \tag{93}$$

confirming that N_1 and N_2, together with J_3, are the generators of the $E(2)$-like little group for massless particles in the two-by-two representation. The transformation that does this in the physical world is described in the following section.

5.2. E(2)-Like Symmetry of Massless Particles

From the four-by-four generators of $K_{1,2}$ and $J_{1,2}$, we can write

$$N_1 = \begin{pmatrix} 0 & 0 & i & 0 \\ 0 & 0 & i & 0 \\ i & -i & 0 & 0 \\ 0 & 0 & 0 & 0 \end{pmatrix}, \quad \text{and} \quad N_2 = \begin{pmatrix} 0 & 0 & 0 & i \\ 0 & 0 & 0 & i \\ 0 & 0 & 0 & 0 \\ i & -i & 0 & 0 \end{pmatrix} \tag{94}$$

These matrices lead to the transformation matrix of the form

$$G(\gamma, \phi) = \begin{pmatrix} 1 + \gamma^2/2 & -\gamma^2/2 & \gamma \cos \phi & \gamma \sin \phi \\ \gamma^2/2 & 1 - \gamma^2/2 & \gamma \cos \phi & \gamma \sin \phi \\ -\gamma \cos \phi & \gamma \cos \phi & 1 & 0 \\ -\gamma \sin \phi & \gamma \sin \phi & 0 & 1 \end{pmatrix} \tag{95}$$

This matrix leaves the four-momentum invariant, as we can see from

$$G(\gamma, \phi) \begin{pmatrix} 1 \\ 1 \\ 0 \\ 0 \end{pmatrix} = \begin{pmatrix} 1 \\ 1 \\ 0 \\ 0 \end{pmatrix} \tag{96}$$

When it is applied to the photon four-potential

$$G(\gamma, \phi) \begin{pmatrix} A_0 \\ A_3 \\ A_1 \\ A_2 \end{pmatrix} = \begin{pmatrix} A_0 \\ A_3 \\ A_1 \\ A_2 \end{pmatrix} + \gamma \left(A_1 \cos \phi + A_2 \sin \phi \right) \begin{pmatrix} 1 \\ 1 \\ 0 \\ 0 \end{pmatrix} \tag{97}$$

with the Lorentz condition which leads to $A_3 = A_0$ in the zero mass case. Gauge transformations are well known for electromagnetic fields and photons. Thus Wigner's little group leads to gauge transformations.

In the two-by-two representation, the electromagnetic four-potential takes the form

$$\begin{pmatrix} 2A_0 & A_1 - iA_2 \\ A_1 + iA_2 & 0 \end{pmatrix} \tag{98}$$

with the Lorentz condition $A_3 = A_0$. Then the two-by-two form of Equation (97) is

$$\begin{pmatrix} 1 & \gamma e^{-i\phi} \\ 0 & 1 \end{pmatrix} \begin{pmatrix} 2A_0 & A_1 - iA_2 \\ A_1 + iA_2 & 0 \end{pmatrix} \begin{pmatrix} 1 & 0 \\ \gamma e^{i\phi} & 1 \end{pmatrix} \tag{99}$$

which becomes

$$\begin{pmatrix} A_0 & A_1 - iA_2 \\ A_1 + iA_2 & 0 \end{pmatrix} + \begin{pmatrix} 2\gamma \left(A_1 \cos \phi - A_2 \sin \phi \right) & 0 \\ 0 & 0 \end{pmatrix} \tag{100}$$

This is the two-by-two equivalent of the gauge transformation given in Equation (97).

For massless spin-1/2 particles starting with the two-by-two expression of $G(\gamma, \phi)$ given in Equation (92), and considering the spinors

$$u = \begin{pmatrix} 1 \\ 0 \end{pmatrix}, \quad \text{and} \quad v = \begin{pmatrix} 0 \\ 1 \end{pmatrix} \tag{101}$$

for spin-up and spin-down states respectively,

$$Gu = u, \quad \text{and} \quad Gv = v + \gamma e^{-i\phi} u \tag{102}$$

This means that the spinor u for spin up is invariant under the gauge transformation while v is not. Thus, the polarization of massless spin-1/2 particle, such as neutrinos, is a consequence of the gauge invariance. We shall continue this discussion in Section 7.

5.3. Boosts along the z Axis

In Sections 4.1 and 5.1, we studied Wigner transformations for fixed values of the four-momenta. The next question is what happens when the system is boosted along the z direction, with the transformation

$$\begin{pmatrix} t' \\ z' \end{pmatrix} = \begin{pmatrix} \cosh\eta & \sinh\eta \\ \sinh\eta & \cosh\eta \end{pmatrix} \begin{pmatrix} t \\ z \end{pmatrix} \tag{103}$$

Then the four-momenta become

$$(\cosh\eta, \sinh\eta, 0, 0), \quad (\sinh\eta, \cosh\eta, 0, 0), \quad e^{\eta}(1, 1, 0, 0) \tag{104}$$

respectively for massive, imaginary, and massless particles cases. In the two-by-two representation, the boost matrix is

$$\begin{pmatrix} e^{\eta/2} & 0 \\ 0 & e^{-\eta/2} \end{pmatrix} \tag{105}$$

and the four-momenta of Equation (104) become

$$\begin{pmatrix} e^{\eta} & 0 \\ 0 & e^{-\eta} \end{pmatrix}, \quad \begin{pmatrix} e^{\eta} & 0 \\ 0 & -e^{-\eta} \end{pmatrix}, \quad \begin{pmatrix} e^{\eta} & 0 \\ 0 & 0 \end{pmatrix} \tag{106}$$

respectively. These matrices become Equations (76)–(78) respectively when $\eta = 0$.

We are interested in Lorentz transformations which leave a given non-zero momentum invariant. We can consider a Lorentz boost along the direction preceded and followed by identical rotation matrices, as described in Figure 1 and the transformation matrix as

$$\begin{pmatrix} \cos(\alpha/2) & -\sin(\alpha/2) \\ \sin(\alpha/2) & \cos(\alpha/2) \end{pmatrix} \begin{pmatrix} \cosh\chi & -\sinh\chi \\ -\sinh\chi & \cosh\chi \end{pmatrix} \begin{pmatrix} \cos(\alpha/2) & -\sin(\alpha/2) \\ \sin(\alpha/2) & \cos(\alpha/2) \end{pmatrix} \tag{107}$$

which becomes

$$\begin{pmatrix} (\cos\alpha)\cosh\chi & -\sinh\chi - (\sin\alpha)\cosh\chi \\ -\sinh\chi + (\sin\alpha)\cosh\chi & (\cos\alpha)\cosh\chi \end{pmatrix} \tag{108}$$

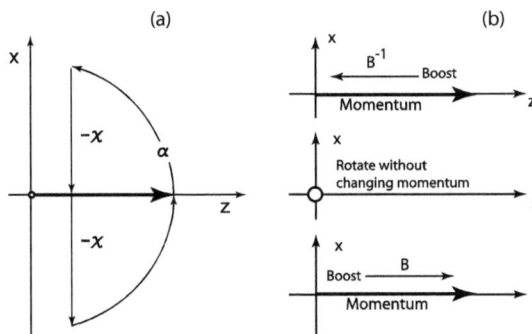

Figure 1. Bargmann and Wigner decompositions. (a) Bargmann decomposition; (b) Wigner decomposition. In the Bargmann decomposition, we start from a momentum along the z direction. We can rotate, boost, and rotate to bring the momentum to the original position. The resulting matrix is the product of one boost and two rotation matrices. In the Wigner decomposition, the particle is boosted back to the frame where the Wigner transformation can be applied. Make a Wigner transformation there and come back to the original state of the momentum. This process also can also be written as the product of three simple matrices.

Except the sign of χ, the two-by-two matrices of Equations (107) and (108) are identical with those given in Section 3.2. The only difference is the sign of the parameter χ. We are thus ready to interpret this expression in terms of physics.

1. If the particle is massive, the off-diagonal elements of Equation (108) have opposite signs, and this matrix can be decomposed into

$$\begin{pmatrix} e^{\eta/2} & 0 \\ 0 & e^{-\eta/2} \end{pmatrix} \begin{pmatrix} \cos(\theta/2) & -\sin(\theta/2) \\ \sin(\theta/2) & \cos(\theta/2) \end{pmatrix} \begin{pmatrix} e^{\eta/2} & 0 \\ 0 & e^{-\eta/2} \end{pmatrix} \tag{109}$$

with

$$\cos(\theta/2) = (\cosh \chi) \cos \alpha, \quad \text{and} \quad e^{2\eta} = \frac{\cosh(\chi) \sin \alpha + \sinh \chi}{\cosh(\chi) \sin \alpha - \sinh \chi} \tag{110}$$

and

$$e^{2\eta} = \frac{p_0 + p_z}{p_0 - p_z} \tag{111}$$

According to Equation (109) the first matrix (far right) reduces the particle momentum to zero. The second matrix rotates the particle without changing the momentum. The third matrix boosts the particle to restore its original momentum. This is the extension of Wigner's original idea to moving particles.

2. If the particle has an imaginary mass, the off-diagonal elements of Equation (108) have the same sign,

$$\begin{pmatrix} e^{\eta/2} & 0 \\ 0 & e^{-\eta/2} \end{pmatrix} \begin{pmatrix} \cosh(\lambda/2) & -\sinh(\lambda/2) \\ \sinh(\lambda/2) & \cosh(\lambda/2) \end{pmatrix} \begin{pmatrix} e^{\eta/2} & 0 \\ 0 & e^{-\eta/2} \end{pmatrix} \tag{112}$$

with

$$\cosh(\lambda/2) = (\cosh \chi) \cos \alpha, \quad \text{and} \quad e^{2\eta} = \frac{\sinh \chi + \cosh(\chi) \sin \alpha}{\cosh(\chi) \sin \alpha - \sinh \chi} \tag{113}$$

and

$$e^{2\eta} = \frac{p_0 + p_z}{p_z - p_0} \tag{114}$$

This is also a three-step operation. The first matrix brings the particle momentum to the zero-energy state with $p_0 = 0$. Boosts along the x or y direction do not change the four-momentum. We can then boost the particle back to restore its momentum. This operation is also an extension of the Wigner's original little group. Thus, it is quite appropriate to call the formulas of Equations (109) and (112) Wigner decompositions.

3. If the particle mass is zero with

$$\sinh \chi = (\cosh \chi) \sin \alpha \tag{115}$$

the η parameter becomes infinite, and the Wigner decomposition does not appear to be useful. We can then go back to the Bargmann decomposition of Equation (107). With the condition of Equations (115) and (108) becomes

$$\begin{pmatrix} 1 & -\gamma \\ 0 & 1 \end{pmatrix} \tag{116}$$

with

$$\gamma = 2 \sinh \chi \tag{117}$$

The decomposition ending with a triangular matrix is called the Iwasawa decomposition [16,22] and its physical interpretation was given in Section 5.2. The γ parameter does not depend on η.

Thus, we have given physical interpretations to the Bargmann and Wigner decompositions given in Section (3.2). Consider what happens when the momentum becomes large. Then η becomes large for nonzero mass cases. All three four-momenta in Equation (106) become

$$e^{\eta} \begin{pmatrix} 1 & 0 \\ 0 & 0 \end{pmatrix} \tag{118}$$

As for the Bargmann-Wigner matrices, they become the triangular matrix of Equation (116), with $\gamma = \sin(\theta/2)e^{\eta}$ and $\gamma = \sinh(\lambda/2)e^{\eta}$, respectively for the massive and imaginary-mass cases.

In Section 5.2, we concluded that the triangular matrix corresponds to gauge transformations. However, particles with imaginary mass are not observed. For massive particles, we can start with the three-dimensional rotation group. The rotation around the z axis is called helicity, and remains invariant under the boost along the z direction. As for the transverse rotations, they become gauge transformation as illustrated in Table 6.

Table 6. Covariance of the energy-momentum relation, and covariance of the internal space-time symmetry. Under the Lorentz boost along the z direction, J_3 remains invariant, and this invariant component of the angular momentum is called the helicity. The transverse component J_1 and J_2 collapse into a gauge transformation. The γ parameter for the massless case has been studied in earlier papers in the four-by-four matrix formulation of Wigner's little groups [8,21].

Massive, Slow	Covariance	Massless, Fast
$E = p^2/2m$	Einstein's $E = mc^2$	$E = cp$
J_3		Helicity
	Wigner's Little Group	
J_1, J_2		Gauge Transformation

5.4. Conjugate Transformations

The most general form of the $SL(2,c)$ matrix is given in Equation (66). Transformation operators for the Lorentz group are given in exponential form as:

$$D = \exp\left\{ -i \sum_{i=1}^{3} (\theta_i J_i + \eta_i K_i) \right\} \tag{119}$$

where the J_i are the generators of rotations and the K_i are the generators of proper Lorentz boosts. They satisfy the Lie algebra given in Equation (43). This set of commutation relations is invariant under the sign change of the boost generators K_i. Thus, we can consider "dot conjugation" defined as

$$\dot{D} = \exp\left\{ -i \sum_{i=1}^{3} (\theta_i J_i - \eta_i K_i) \right\} \tag{120}$$

Since K_i are anti-Hermitian while J_i are Hermitian, the Hermitian conjugate of the above expression is

$$D^{\dagger} = \exp\left\{ -i \sum_{i=1}^{3} (-\theta_i J_i + \eta_i K_i) \right\} \tag{121}$$

while the Hermitian conjugate of G is

$$\dot{D}^{\dagger} = \exp\left\{ -i \sum_{i=1}^{3} (-\theta_i J_i - \eta_i K_i) \right\} \tag{122}$$

Since we understand the rotation around the z axis, we can now restrict the kinematics to the zt plane, and work with the $Sp(2)$ symmetry. Then the D matrices can be considered as Bargmann decompositions. First, D and $\dot{D}$, and their Hermitian conjugates are

$$D(\alpha, \chi) = \begin{pmatrix} (\cos\alpha)\cosh\chi & \sinh\chi - (\sin\alpha)\cosh\chi \\ \sinh\chi + (\sin\alpha)\cosh\chi & (\cos\alpha)\cosh\chi \end{pmatrix} \tag{123}$$

$$\dot{D}(\alpha, \chi) = \begin{pmatrix} (\cos\alpha)\cosh\chi & -\sinh\chi - (\sin\alpha)\cosh\chi \\ -\sinh\chi + (\sin\alpha)\cosh\chi & (\cos\alpha)\cosh\chi \end{pmatrix} \tag{124}$$

These matrices correspond to the "D loops" given in Figure 2a,b respectively. The "dot" conjugation changes the direction of boosts. The dot conjugation leads to the inversion of the space which is called the parity operation.

We can also consider changing the direction of rotations. Then they result in the Hermitian conjugates. We can write their matrices as

$$D^\dagger(\alpha, \chi) = \begin{pmatrix} (\cos\alpha)\cosh\chi & \sinh\chi + (\sin\alpha)\cosh\chi \\ \sinh\chi - (\sin\alpha)\cosh\chi & (\cos\alpha)\cosh\chi \end{pmatrix} \tag{125}$$

$$\dot{D}^\dagger(\alpha, \chi) = \begin{pmatrix} (\cos\alpha)\cosh\chi & -\sinh\chi + (\sin\alpha)\cosh\chi \\ -\sinh\chi - (\sin\alpha)\cosh\chi & (\cos\alpha)\cosh\chi \end{pmatrix} \tag{126}$$

From the exponential expressions from Equation (119) to Equation (122), it is clear that

$$D^\dagger = \dot{D}^{-1}, \quad \text{and} \quad \dot{D}^\dagger = D^{-1} \tag{127}$$

The D loop given in Figure 1 corresponds to $\dot{D}$. We shall return to these loops in Section 7.

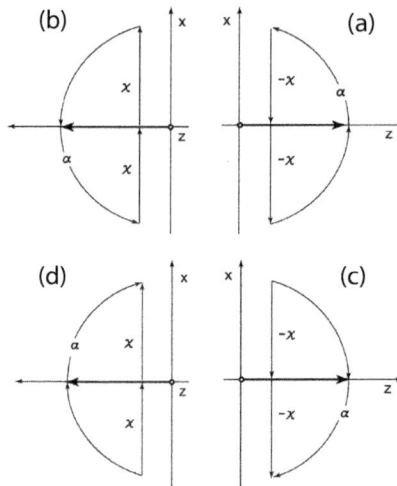

Figure 2. Four D-loops resulting from the Bargmann decomposition. (**a**) Bargmann decomposition from Figure 1; (**b**) Direction of the Lorentz boost is reversed; (**c**) Direction of rotation is reversed; (**d**) Both directions are reversed. These operations correspond to the space-inversion, charge conjugation, and the time reversal respectively.

6. Symmetries Derivable from the Poincaré Sphere

The Poincaré sphere serves as the basic language for polarization physics. Its underlying language is the two-by-two coherency matrix. This coherency matrix contains the symmetry of $SL(2, c)$ isomorphic to the the the Lorentz group applicable to three space-like and one time-like dimensions [4,6,7].

For polarized light propagating along the z direction, the amplitude ratio and phase difference of electric field x and y components traditionally determine the state of polarization. Hence, the polarization can be changed by adjusting the amplitude ratio or the phase difference or both. Usually, the optical device which changes amplitude is called an "attenuator" (or "amplifier") and the device which changes the relative phase a "phase shifter".

Let us start with the Jones vector:

$$\begin{pmatrix} \psi_1(z,t) \\ \psi_2(z,t) \end{pmatrix} = \begin{pmatrix} a \exp\left[i(kz - \omega t)\right] \\ a \exp\left[i(kz - \omega t)\right] \end{pmatrix} \tag{128}$$

To this matrix, we can apply the phase shift matrix of Equation (79) which brings the Jones vector to

$$\begin{pmatrix} \psi_1(z,t) \\ \psi_2(z,t) \end{pmatrix} = \begin{pmatrix} a \exp\left[i(kz - \omega t - i\phi/2)\right] \\ a \exp\left[i(kz - \omega t + i\phi/2)\right] \end{pmatrix} \tag{129}$$

The generator of this phase-shifter is J_3 given Table 5.

The optical beam can be attenuated differently in the two directions. The resulting matrix is

$$e^{-\mu} \begin{pmatrix} e^{\eta/2} & 0 \\ 0 & e^{-\eta/2} \end{pmatrix} \tag{130}$$

with the attenuation factor of $\exp\left(-\mu_0 + \eta/2\right)$ and $\exp\left(-\mu - \eta/2\right)$ for the x and y directions respectively. We are interested only the relative attenuation given in Equation (46) which leads to different amplitudes for the x and y component, and the Jones vector becomes

$$\begin{pmatrix} \psi_1(z,t) \\ \psi_2(z,t) \end{pmatrix} = \begin{pmatrix} ae^{\mu/2} \exp\left[i(kz - \omega t - i\phi/2)\right] \\ ae^{-\mu/2} \exp\left[i(kz - \omega t + i\phi/2)\right] \end{pmatrix} \tag{131}$$

The squeeze matrix of Equation (46) is generated by K_3 given in Table 1.

The polarization is not always along the x and y axes, but can be rotated around the z axis using Equation (79) generated by J_2 given in Table 1.

Among the rotation angles, the angle of $45°$ plays an important role in polarization optics. Indeed, if we rotate the squeeze matrix of Equation (46) by $45°$, we end up with the squeeze matrix of Equation (45) generated by K_1 given also in Table 1.

Each of these four matrices plays an important role in special relativity, as we discussed in Sections 3.2 and 6. Their respective roles in optics and particle physics are given in Table 7.

Table 7. Polarization optics and special relativity share the same mathematics. Each matrix has its clear role in both optics and relativity. The determinant of the Stokes or the four-momentum matrix remains invariant under Lorentz transformations. It is interesting to note that the decoherence parameter (least fundamental) in optics corresponds to the $(mass)^2$ (most fundamental) in particle physics.

Polarization Optics	Transformation Matrix	Particle Symmetry
Phase shift by ϕ	$\begin{pmatrix} e^{-i\phi/2} & 0 \\ 0 & e^{i\phi/2} \end{pmatrix}$	Rotation around z.
Rotation around z	$\begin{pmatrix} \cos(\theta/2) & -\sin(\theta/2) \\ \sin(\theta/2) & \cos(\theta/2) \end{pmatrix}$	Rotation around y.
Squeeze along x and y	$\begin{pmatrix} e^{\eta/2} & 0 \\ 0 & e^{-\eta/2} \end{pmatrix}$	Boost along z.
Squeeze along $45°$	$\begin{pmatrix} \cosh(\lambda/2) & \sinh(\lambda/2) \\ \sinh(\lambda/2) & \cosh(\lambda/2) \end{pmatrix}$	Boost along x.
$a^4 (\sin \zeta)^2$	Determinant	$(mass)^2$

The most general form for the two-by-two matrix applicable to the Jones vector is the G matrix of Equation (66). This matrix is of course a representation of the $SL(2,c)$ group. It brings the simplest Jones vector of Equation (128) to its most general form.

6.1. Coherency Matrix

However, the Jones vector alone cannot tell us whether the two components are coherent with each other. In order to address this important degree of freedom, we use the coherency matrix defined as [3,23]

$$C = \begin{pmatrix} S_{11} & S_{12} \\ S_{21} & S_{22} \end{pmatrix} \tag{132}$$

where

$$< \psi_i^* \psi_j >= \frac{1}{T} \int_0^T \psi_i^*(t + \tau)\psi_j(t)dt \tag{133}$$

where T is a sufficiently long time interval. Then, those four elements become [4]

$$S_{11} =< \psi_1^* \psi_1 >= a^2, \qquad S_{12} =< \psi_1^* \psi_2 >= a^2(\cos \zeta)e^{-i\phi} \tag{134}$$

$$S_{21} =< \psi_2^* \psi_1 >= a^2(\cos \zeta)e^{+i\phi}, \qquad S_{22} =< \psi_2^* \psi_2 >= a^2 \tag{135}$$

The diagonal elements are the absolute values of ψ_1 and ψ_2 respectively. The angle ϕ could be different from the value of the phase-shift angle given in Equation (79), but this difference does not play any role in the reasoning. The off-diagonal elements could be smaller than the product of ψ_1 and ψ_2, if the two polarizations are not completely coherent.

The angle ζ specifies the degree of coherency. If it is zero, the system is fully coherent, while the system is totally incoherent if ζ is $90°$. This can therefore be called the "decoherence angle."

While the most general form of the transformation applicable to the Jones vector is G of Equation (66), the transformation applicable to the coherency matrix is

$$C' = G C G^\dagger \tag{136}$$

The determinant of the coherency matrix is invariant under this transformation, and it is

$$\det(C) = a^4(\sin \zeta)^2 \tag{137}$$

Thus, angle ζ remains invariant. In the language of the Lorentz transformation applicable to the four-vector, the determinant is equivalent to the $(mass)^2$ and is therefore a Lorentz-invariant quantity.

6.2. Two Radii of the Poincaré Sphere

Let us write explicitly the transformation of Equation (136) as

$$
\begin{pmatrix} S'_{11} & S'_{12} \\ S'_{21} & S'_{22} \end{pmatrix} = \begin{pmatrix} \alpha & \beta \\ \gamma & \delta \end{pmatrix} \begin{pmatrix} S_{11} & S_{12} \\ S_{21} & S_{22} \end{pmatrix} \begin{pmatrix} \alpha^* & \gamma^* \\ \beta^* & \delta^* \end{pmatrix}
\tag{138}
$$

It is then possible to construct the following quantities,

$$
S_0 = \frac{S_{11} + S_{22}}{2}, \qquad S_3 = \frac{S_{11} - S_{22}}{2}
\tag{139}
$$

$$
S_1 = \frac{S_{12} + S_{21}}{2}, \qquad S_2 = \frac{S_{12} - S_{21}}{2i}
\tag{140}
$$

These are known as the Stokes parameters, and constitute a four-vector (S_0, S_3, S_1, S_2) under the Lorentz transformation.

In the Jones vector of Equation (128), the amplitudes of the two orthogonal components are equal. Thus, the two diagonal elements of the coherency matrix are equal. This leads to $S_3 = 0$, and the problem is reduced from the sphere to a circle. In the resulting two-dimensional subspace, we can introduce the polar coordinate system with

$$
R = \sqrt{S_1^2 + S_2^2}
\tag{141}
$$

$$
S_1 = R \cos \phi
\tag{142}
$$

$$
S_2 = R \sin \phi
\tag{143}
$$

The radius R is the radius of this circle, and is

$$
R = a^2 \cos \zeta
\tag{144}
$$

The radius R takes its maximum value S_0 when $\zeta = 0°$. It decreases as ζ increases and vanishes when $\zeta = 90°$. This aspect of the radius R is illustrated in Figure 3.

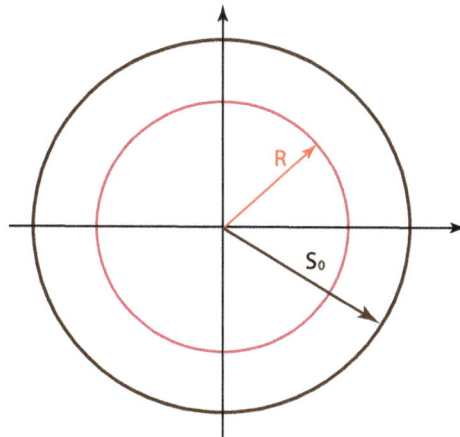

Figure 3. Radius of the Poincaré sphere. The radius R takes its maximum value S_0 when the decoherence angle ζ is zero. It becomes smaller as ζ increases. It becomes zero when the angle reaches $90°$.

In order to see its implications in special relativity, let us go back to the four-momentum matrix of $m(1,0,0,0)$. Its determinant is m^2 and remains invariant. Likewise, the determinant of the coherency matrix of Equation (132) should also remain invariant. The determinant in this case is

$$S_0^2 - R^2 = a^4 \sin^2 \zeta \tag{145}$$

This quantity remains invariant under the Hermitian transformation of Equation (138), which is a Lorentz transformation as discussed in Sections 3.2 and 6. This aspect is shown on the last row of Table 7.

The coherency matrix then becomes

$$C = a^2 \begin{pmatrix} 1 & (\cos\zeta)e^{-i\phi} \\ (\cos\zeta)e^{i\phi} & 1 \end{pmatrix} \tag{146}$$

Since the angle ϕ does not play any essential role, we can let $\phi = 0$, and write the coherency matrix as

$$C = a^2 \begin{pmatrix} 1 & \cos\zeta \\ \cos\zeta & 1 \end{pmatrix} \tag{147}$$

The determinant of the above two-by-two matrix is

$$a^4 \left(1 - \cos^2\zeta\right) = a^4 \sin^2\zeta \tag{148}$$

Since the Lorentz transformation leaves the determinant invariant, the change in this ζ variable is not a Lorentz transformation. It is of course possible to construct a larger group in which this variable plays a role in a group transformation [6], but here we are more interested in its role in a particle gaining a mass from zero or the mass becoming zero.

6.3. Extra-Lorentzian Symmetry

The coherency matrix of Equation (146) can be diagonalized to

$$a^2 \begin{pmatrix} 1 + \cos\zeta & 0 \\ 0 & 1 - \cos\zeta \end{pmatrix} \tag{149}$$

by a rotation. Let us then go back to the four-momentum matrix of Equation (70). If $p_x = p_y = 0$, and $p_z = p_0 \cos\zeta$, we can write this matrix as

$$p_0 \begin{pmatrix} 1 + \cos\zeta & 0 \\ 0 & 1 - \cos\zeta \end{pmatrix} \tag{150}$$

Thus, with this extra variable, it is possible to study the little groups for variable masses, including the small-mass limit and the zero-mass case.

For a fixed value of p_0, the $(mass)^2$ becomes

$$(mass)^2 = (p_0 \sin\zeta)^2, \quad \text{and} \quad (momentum)^2 = (p_0 \cos\zeta)^2 \tag{151}$$

resulting in

$$(energy)^2 = (mass)^2 + (momentum)^2 \tag{152}$$

This transition is illustrated in Figure 4. We are interested in reaching a point on the light cone from mass hyperbola while keeping the energy fixed. According to this figure, we do not have to make an excursion to infinite-momentum limit. If the energy is fixed during this process, Equation (152) tells the mass and momentum relation, and Figure 5 illustrates this relation.

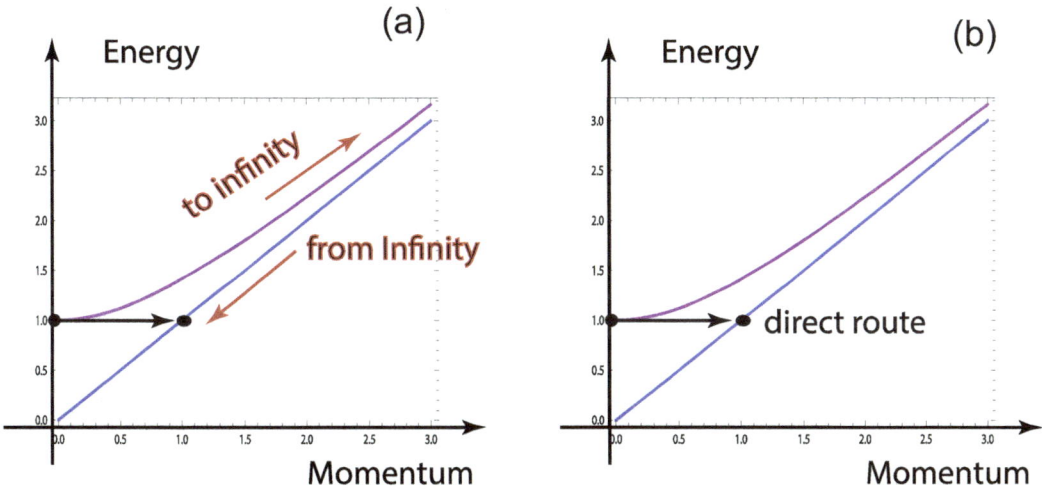

Figure 4. Transition from the massive to massless case. (**a**) Transition within the framework of the Lorentz group; (**b**) Trasnsition allowed in the symmetry of the Poincaré sphere. Within the framework of the Lorentz group, it is not possible to go from the massive to massless case directly, because it requires the change in the mass which is a Lorentz-invariant quantity. The only way is to move to infinite momentum and jump from the hyperbola to the light cone, and come back. The extra symmetry of the Poincaré sphere allows a direct transition

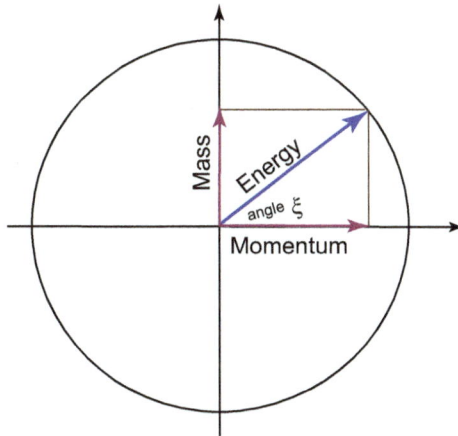

Figure 5. Energy-momentum-mass relation. This circle illustrates the case where the energy is fixed, while the mass and momentum are related according to the triangular rule. The value of the angle ξ changes from zero to $180°$. The particle mass is negative for negative values of this angle. However, in the Lorentz group, only $(mass)^2$ is a relevant variable, and negative masses might play a role for theoretical purposes.

Within the framework of the Lorentz group, it is possible, by making an excursion to infinite momentum where the mass hyperbola coincides with the light cone, to then come back to the desired point. On the other hand, the mass formula of Equation (151) allows us to go there directly. The decoherence mechanism of the coherency matrix makes this possible.

7. Small-Mass and Massless Particles

We now have a mathematical tool to reduce the mass of a massive particle from its positive value to zero. During this process, the Lorentz-boosted rotation matrix becomes a gauge transformation for the spin-1 particle, as discussed Section 5.2. For spin-1/2 particles, there are two issues.

1. It was seen in Section 5.2 that the requirement of gauge invariance lead to a polarization of massless spin-1/2 particle, such as neutrinos. What happens to anti-neutrinos?
2. There are strong experimental indications that neutrinos have a small mass. What happens to the $E(2)$ symmetry?

7.1. Spin-1/2 Particles

Let us go back to the two-by-two matrices of Section 5.4, and the two-by-two D matrix. For a massive particle, its Wigner decomposition leads to

$$D = \begin{pmatrix} \cos(\theta/2) & -e^{-\eta}\sin(\theta/2) \\ e^{\eta}\sin(\theta/2) & \cos(\theta/2) \end{pmatrix} \tag{153}$$

This matrix is applicable to the spinors u and v defined in Equation (101) respectively for the spin-up and spin-down states along the z direction.

Since the Lie algebra of $SL(2,c)$ is invariant under the sign change of the K_i matrices, we can consider the "dotted" representation, where the system is boosted in the opposite direction, while the direction of rotations remain the same. Thus, the Wigner decomposition leads to

$$\dot{D} = \begin{pmatrix} \cos(\theta/2) & -e^{\eta}\sin(\theta/2) \\ e^{-\eta}\sin(\theta/2) & \cos(\theta/2) \end{pmatrix} \tag{154}$$

with its spinors

$$\dot{u} = \begin{pmatrix} 1 \\ 0 \end{pmatrix}, \quad \text{and} \quad \dot{v} = \begin{pmatrix} 0 \\ 1 \end{pmatrix} \tag{155}$$

For anti-neutrinos, the helicity is reversed but the momentum is unchanged. Thus, $D^{\dagger}$ is the appropriate matrix. However, $D^{\dagger} = \dot{D}^{-1}$ as was noted in Section 5.4. Thus, we shall use $\dot{D}$ for anti-neutrinos.

When the particle mass becomes very small,

$$e^{-\eta} = \frac{m}{2p} \tag{156}$$

becomes small. Thus, if we let

$$e^{\eta}\sin(\theta/2) = \gamma, \quad \text{and} \quad e^{-\eta}\sin(\theta/2) = \epsilon^2 \tag{157}$$

then the D matrix of Equation (153) and the $\dot{D}$ of Equation (154) become

$$\begin{pmatrix} 1 - \gamma\epsilon^2/2 & -\epsilon^2 \\ \gamma & 1 - \gamma\epsilon^2 \end{pmatrix}, \quad \text{and} \quad \begin{pmatrix} 1 - \gamma\epsilon^2/2 & -\gamma \\ \epsilon^2 & 1 - \gamma\epsilon^2 \end{pmatrix} \tag{158}$$

respectively where γ is an independent parameter and

$$\epsilon^2 = \gamma \left(\frac{m}{2p} \right)^2 \tag{159}$$

When the particle mass becomes zero, they become

$$\begin{pmatrix} 1 & 0 \\ \gamma & 1 \end{pmatrix}, \quad \text{and} \quad \begin{pmatrix} 1 & -\gamma \\ 0 & 1 \end{pmatrix} \tag{160}$$

respectively, applicable to the spinors (u, v) and $(\dot{u}, \dot{v})$ respectively.

For neutrinos,

$$\begin{pmatrix} 1 & 0 \\ \gamma & 1 \end{pmatrix} \begin{pmatrix} 1 \\ 0 \end{pmatrix} = \begin{pmatrix} 1 \\ \gamma \end{pmatrix}, \quad \text{and} \quad \begin{pmatrix} 1 & 0 \\ \gamma & 1 \end{pmatrix} \begin{pmatrix} 0 \\ 1 \end{pmatrix} = \begin{pmatrix} 0 \\ 1 \end{pmatrix} \tag{161}$$

For anti-neutrinos,

$$\begin{pmatrix} 1 & -\gamma \\ 0 & 1 \end{pmatrix} \begin{pmatrix} 1 \\ 0 \end{pmatrix} = \begin{pmatrix} 1 \\ 0 \end{pmatrix}, \quad \text{and} \quad \begin{pmatrix} 1 & -\gamma \\ 0 & 1 \end{pmatrix} \begin{pmatrix} 0 \\ 1 \end{pmatrix} = \begin{pmatrix} -\gamma \\ 1 \end{pmatrix} \tag{162}$$

It was noted in Section 5.2 that the triangular matrices of Equation (160) perform gauge transformations. Thus, for Equations (161) and (162) the requirement of gauge invariance leads to the polarization of neutrinos. The neutrinos are left-handed while the anti-neutrinos are right-handed. Since, however, nature cannot tell the difference between the dotted and undotted representations, the Lorentz group cannot tell which neutrino is right handed. It can say only that the neutrinos and anti-neutrinos are oppositely polarized.

If the neutrino has a small mass, the gauge invariance is modified to

$$\begin{pmatrix} 1 - \gamma\epsilon^2/2 & -\epsilon^2 \\ \gamma & 1 - \gamma\epsilon^2/2 \end{pmatrix} \begin{pmatrix} 0 \\ 1 \end{pmatrix} = \begin{pmatrix} 0 \\ 1 \end{pmatrix} - \epsilon^2 \begin{pmatrix} 1 \\ \gamma/2 \end{pmatrix} \tag{163}$$

and

$$\begin{pmatrix} 1 - \gamma\epsilon^2/2 & -\gamma \\ \epsilon^2 & 1 - \gamma\epsilon^2 \end{pmatrix} \begin{pmatrix} 1 \\ 0 \end{pmatrix} = \begin{pmatrix} 1 \\ 0 \end{pmatrix} + \epsilon^2 \begin{pmatrix} -\gamma/2 \\ 1 \end{pmatrix} \tag{164}$$

respectively for neutrinos and anti-neutrinos. Thus the violation of the gauge invariance in both cases is proportional to ϵ^2 which is $m^2/4p^2$.

7.2. Small-Mass Neutrinos in the Real World

Whether neutrinos have mass or not and the consequences of this relative to the Standard Model and lepton number is the subject of much theoretical speculation [24,25], and of cosmology [26], nuclear reactors [27], and high energy experimentations [28,29]. Neutrinos are fast becoming an important component of the search for dark matter and dark radiation [30]. Their importance within the Standard Model is reflected by the fact that they are the only particles which seem to exist with only one direction of chirality, *i.e.*, only left-handed neutrinos have been confirmed to exist so far.

It was speculated some time ago that neutrinos in constant electric and magnetic fields would acquire a small mass, and that right-handed neutrinos would be trapped within the interaction field [31]. Solving generalized electroweak models using left- and right-handed neutrinos has been discussed recently [32]. Today these right-handed neutrinos which do not participate in weak interactions are called "sterile" neutrinos [33]. A comprehensive discussion of the place of neutrinos in the scheme of physics has been given by Drewes [30]. We should note also that the three different neutrinos, namely ν_e, ν_μ, and ν_τ, may have different masses [34].

8. Scalars, Four-Vectors, and Four-Tensors

In Sections 5 and 7, our primary interest has been the two-by-two matrices applicable to spinors for spin-1/2 particles. Since we also used four-by-four matrices, we indirectly studied the four-component particle consisting of spin-1 and spin-zero components.

If there are two spin 1/2 states, we are accustomed to construct one spin-zero state, and one spin-one state with three degeneracies.

In this paper, we are confronted with two spinors, but each spinor can also be dotted. For this reason, there are 16 orthogonal states consisting of spin-one and spin-zero states. How many spin-zero states? How many spin-one states?

For particles at rest, it is known that the addition of two one-half spins result in spin-zero and spin-one states. In this paper, we have two different spinors behaving differently under the Lorentz boost. Around the z direction, both spinors are transformed by

$$Z(\phi) = \exp\left(-i\phi J_3\right) = \begin{pmatrix} e^{-i\phi/2} & 0 \\ 0 & e^{i\phi/2} \end{pmatrix} \tag{165}$$

However, they are boosted by

$$B(\eta) = \exp\left(-i\eta K_3\right) = \begin{pmatrix} e^{\eta/2} & 0 \\ 0 & e^{-\eta/2} \end{pmatrix} \tag{166}$$

$$\dot{B}(\eta) = \exp\left(i\eta K_3\right) = \begin{pmatrix} e^{-\eta/2} & 0 \\ 0 & e^{\eta/2} \end{pmatrix} \tag{167}$$

applicable to the undotted and dotted spinors respectively. These two matrices commute with each other, and also with the rotation matrix $Z(\phi)$ of Equation (165). Since K_3 and J_3 commute with each other, we can work with the matrix $Q(\eta, \phi)$ defined as

$$Q(\eta, \phi) = B(\eta)Z(\phi) = \begin{pmatrix} e^{(\eta - i\phi)/2} & 0 \\ 0 & e^{-(\eta - i\phi)/2} \end{pmatrix} \tag{168}$$

$$\dot{Q}(\eta, \phi) = \dot{B}(\eta)\dot{Z}(\phi) = \begin{pmatrix} e^{-(\eta + i\phi)/2} & 0 \\ 0 & e^{(\eta + i\phi)/2} \end{pmatrix} \tag{169}$$

When this combined matrix is applied to the spinors,

$$Q(\eta, \phi)u = e^{(\eta - i\phi)/2}u, \qquad Q(\eta, \phi)v = e^{-(\eta - i\phi)/2}v \tag{170}$$

$$\dot{Q}(\eta, \phi)\dot{u} = e^{-(\eta + i\phi)/2}\dot{u}, \qquad \dot{Q}(\eta, \phi)\dot{v} = e^{(\eta + i\phi)/2}\dot{v} \tag{171}$$

If the particle is at rest, we can construct the combinations

$$uu, \qquad \frac{1}{\sqrt{2}}(uv + vu), \qquad vv \tag{172}$$

to construct the spin-1 state, and

$$\frac{1}{\sqrt{2}}(uv - vu) \tag{173}$$

for the spin-zero state. There are four bilinear states. In the $SL(2, c)$ regime, there are two dotted spinors. If we include both dotted and undotted spinors, there are 16 independent bilinear combinations. They are given in Table 8. This table also gives the effect of the operation of $Q(\eta, \phi)$.

Table 8. Sixteen combinations of the $SL(2,c)$ spinors. In the $SU(2)$ regime, there are two spinors leading to four bilinear forms. In the $SL(2,c)$ world, there are two undotted and two dotted spinors. These four spinors lead to 16 independent bilinear combinations.

Spin 1			Spin 0
$uu,$	$\frac{1}{\sqrt{2}}(uv+vu),$	$vv,$	$\frac{1}{\sqrt{2}}(uv-vu)$
$\dot{u}\dot{u},$	$\frac{1}{\sqrt{2}}(\dot{u}\dot{v}+\dot{v}\dot{u}),$	$\dot{v}\dot{v},$	$\frac{1}{\sqrt{2}}(\dot{u}\dot{v}-\dot{v}\dot{u})$
$u\dot{u},$	$\frac{1}{\sqrt{2}}(u\dot{v}+v\dot{u}),$	$v\dot{v},$	$\frac{1}{\sqrt{2}}(u\dot{v}-v\dot{u})$
$\dot{u}u,$	$\frac{1}{\sqrt{2}}(\dot{u}v+\dot{v}u),$	$\dot{v}v,$	$\frac{1}{\sqrt{2}}(\dot{u}v-\dot{v}u)$
After the Operation of $Q(\eta,\phi)$ and $\dot{Q}(\eta,\phi)$			
$e^{-i\phi}e^{\eta}uu,$	$\frac{1}{\sqrt{2}}(uv+vu),$	$e^{i\phi}e^{-\eta}vv,$	$\frac{1}{\sqrt{2}}(uv-vu)$
$e^{-i\phi}e^{-\eta}\dot{u}\dot{u},$	$\frac{1}{\sqrt{2}}(\dot{u}\dot{v}+\dot{v}\dot{u}),$	$e^{i\phi}e^{\eta}\dot{v}\dot{v},$	$\frac{1}{\sqrt{2}}(\dot{u}\dot{v}-\dot{v}\dot{u})$
$e^{-i\phi}u\dot{u},$	$\frac{1}{\sqrt{2}}(e^{\eta}u\dot{v}+e^{-\eta}v\dot{u}),$	$e^{i\phi}v\dot{v},$	$\frac{1}{\sqrt{2}}(e^{\eta}u\dot{v}-e^{-\eta}v\dot{u})$
$e^{-i\phi}\dot{u}u,$	$\frac{1}{\sqrt{2}}(\dot{u}v+\dot{v}u),$	$e^{i\phi}\dot{v}v,$	$\frac{1}{\sqrt{2}}(e^{-\eta}\dot{u}v-e^{\eta}\dot{v}u)$

Among the bilinear combinations given in Table 8, the following two are invariant under rotations and also under boosts.

$$S = \frac{1}{\sqrt{2}}(uv-vu), \quad \text{and} \quad \dot{S} = -\frac{1}{\sqrt{2}}(\dot{u}\dot{v}-\dot{v}\dot{u}) \tag{174}$$

They are thus scalars in the Lorentz-covariant world. Are they the same or different? Let us consider the following combinations

$$S_+ = \frac{1}{\sqrt{2}}\left(S+\dot{S}\right), \quad \text{and} \quad S_- = \frac{1}{\sqrt{2}}\left(S-\dot{S}\right) \tag{175}$$

Under the dot conjugation, S_+ remains invariant, but S_- changes its sign.

Under the dot conjugation, the boost is performed in the opposite direction. Therefore it is the operation of space inversion, and S_+ is a scalar while S_- is called the pseudo-scalar.

8.1. Four-Vectors

Let us consider the bilinear products of one dotted and one undotted spinor as $u\dot{u}, u\dot{v}, \dot{u}v, v\dot{v},$ and construct the matrix

$$U = \begin{pmatrix} u\dot{v} & v\dot{v} \\ u\dot{u} & v\dot{u} \end{pmatrix} \tag{176}$$

Under the rotation $Z(\phi)$ and the boost $B(\eta)$ they become

$$\begin{pmatrix} e^{\eta}u\dot{v} & e^{-i\phi}v\dot{v} \\ e^{i\phi}u\dot{u} & e^{-\eta}v\dot{u} \end{pmatrix} \tag{177}$$

Indeed, this matrix is consistent with the transformation properties given in Table 8, and transforms like the four-vector

$$\begin{pmatrix} t+z & x-iy \\ x+iy & t-z \end{pmatrix} \tag{178}$$

This form was given in Equation (65), and played the central role throughout this paper. Under the space inversion, this matrix becomes

$$\begin{pmatrix} t-z & -(x-iy) \\ -(x+iy) & t+z \end{pmatrix} \tag{179}$$

This space inversion is known as the parity operation.

The form of Equation (176) for a particle or field with four-components, is given by (V_0, V_z, V_x, V_y). The two-by-two form of this four-vector is

$$U = \begin{pmatrix} V_0 + V_z & V_x - iV_y \\ V_x + iV_y & V_0 - V_z \end{pmatrix} \tag{180}$$

If boosted along the z direction, this matrix becomes

$$\begin{pmatrix} e^\eta \left(V_0 + V_z \right) & V_x - iV_y \\ V_x + iV_y & e^{-\eta} \left(V_0 - V_z \right) \end{pmatrix} \tag{181}$$

In the mass-zero limit, the four-vector matrix of Equation (181) becomes

$$\begin{pmatrix} 2A_0 & A_x - iA_y \\ A_x + iA_y & 0 \end{pmatrix} \tag{182}$$

with the Lorentz condition $A_0 = A_z$. The gauge transformation applicable to the photon four-vector was discussed in detail in Section 5.2.

Let us go back to the matrix of Equation (180), we can construct another matrix $\dot{U}$. Since the dot conjugation leads to the space inversion,

$$\dot{U} = \begin{pmatrix} \dot{u}v & \dot{v}v \\ \dot{u}u & \dot{v}u \end{pmatrix} \tag{183}$$

Then

$$\dot{u}v \simeq (t - z), \qquad \dot{v}u \simeq (t + z) \tag{184}$$

$$\dot{v}v \simeq -(x - iy), \qquad \dot{u}u \simeq -(x + iy) \tag{185}$$

where the symbol $\simeq$ means "transforms like".

Thus, U of Equation (176) and $\dot{U}$ of Equation (183) used up 8 of the 16 bilinear forms. Since there are two bilinear forms in the scalar and pseudo-scalar as given in Equation (175), we have to give interpretations to the six remaining bilinear forms.

8.2. Second-Rank Tensor

In this subsection, we are studying bilinear forms with both spinors dotted and undotted. In Section 8.1, each bilinear spinor consisted of one dotted and one undotted spinor. There are also bilinear spinors which are both dotted or both undotted. We are interested in two sets of three quantities satisfying the $O(3)$ symmetry. They should therefore transform like

$$(x + iy)/\sqrt{2}, \qquad (x - iy)/\sqrt{2}, \qquad z \tag{186}$$

which are like

$$uu, \qquad vv, \qquad (uv + vu)/\sqrt{2} \tag{187}$$

respectively in the $O(3)$ regime. Since the dot conjugation is the parity operation, they are like

$$-\dot{u}\dot{u}, \qquad -\dot{v}\dot{v}, \qquad -(\dot{u}\dot{v} + \dot{v}\dot{u})/\sqrt{2} \tag{188}$$

In other words,

$$(uu\dot{)} = -\dot{u}\dot{u}, \quad \text{and} \quad (vv\dot{)} = -\dot{v}\dot{v} \tag{189}$$

We noticed a similar sign change in Equation (184).

In order to construct the z component in this $O(3)$ space, let us first consider

$$f_z = \frac{1}{2}\left[(uv + vu) - (\dot{u}\dot{v} + \dot{v}\dot{u})\right], \qquad g_z = \frac{1}{2i}\left[(uv + vu) + (\dot{u}\dot{v} + \dot{v}\dot{u})\right] \tag{190}$$

where f_z and g_z are respectively symmetric and anti-symmetric under the dot conjugation or the parity operation. These quantities are invariant under the boost along the z direction. They are also invariant under rotations around this axis, but they are not invariant under boost along or rotations around the x or y axis. They are different from the scalars given in Equation (174).

Next, in order to construct the x and y components, we start with $g_\pm$ as

$$f_+ = \frac{1}{\sqrt{2}}\left(uu - \dot{u}\dot{u}\right) \qquad g_+ = \frac{1}{\sqrt{2}i}\left(uu + \dot{u}\dot{u}\right) \tag{191}$$

$$f_- = \frac{1}{\sqrt{2}}\left(vv - \dot{v}\dot{v}\right) \qquad g_- = \frac{1}{\sqrt{2}i}\left(vv + \dot{v}\dot{v}\right) \tag{192}$$

Then

$$f_x = \frac{1}{\sqrt{2}}\left(f_+ + f_-\right) = \frac{1}{2}\left[(uu - \dot{u}\dot{u}) + (vv - \dot{v}\dot{v})\right] \tag{193}$$

$$f_y = \frac{1}{\sqrt{2}i}\left(f_+ - f_-\right) = \frac{1}{2i}\left[(uu - \dot{u}\dot{u}) - (vv - \dot{v}\dot{v})\right] \tag{194}$$

and

$$g_x = \frac{1}{\sqrt{2}}\left(g_+ + g_-\right) = \frac{1}{2i}\left[(uu + \dot{u}\dot{u}) + (vv + \dot{v}\dot{v})\right] \tag{195}$$

$$g_y = \frac{1}{\sqrt{2}i}\left(g_+ - g_-\right) = -\frac{1}{2}\left[(uu + \dot{u}\dot{u}) - (vv + \dot{v}\dot{v})\right] \tag{196}$$

Here f_x and f_y are symmetric under dot conjugation, while g_x and g_y are anti-symmetric.

Furthermore, f_z, f_x, and f_y of Equations (190) and (193) transform like a three-dimensional vector. The same can be said for g_i of Equations (190) and (195). Thus, they can grouped into the second-rank tensor

$$T = \begin{pmatrix} 0 & -g_z & -g_x & -g_y \\ g_z & 0 & -f_y & f_x \\ g_x & f_y & 0 & -f_z \\ g_y & -f_x & f_z & 0 \end{pmatrix} \tag{197}$$

whose Lorentz-transformation properties are well known. The g_i components change their signs under space inversion, while the f_i components remain invariant. They are like the electric and magnetic fields respectively.

If the system is Lorentz-booted, f_i and g_i can be computed from Table 8. We are now interested in the symmetry of photons by taking the massless limit. According to the procedure developed in Section 6, we can keep only the terms which become larger for larger values of η. Thus,

$$f_x \to \frac{1}{2}\left(uu - \dot{v}\dot{v}\right), \qquad f_y \to \frac{1}{2i}\left(uu + \dot{v}\dot{v}\right) \tag{198}$$

$$g_x \to \frac{1}{2i}\left(uu + \dot{v}\dot{v}\right), \qquad g_y \to -\frac{1}{2}\left(uu - \dot{v}\dot{v}\right) \tag{199}$$

in the massless limit.

Then the tensor of Equation (197) becomes

$$
F = \begin{pmatrix}
0 & 0 & -E_x & -E_y \\
0 & 0 & -B_y & B_x \\
E_x & B_y & 0 & 0 \\
E_y & -B_x & 0 & 0
\end{pmatrix}
\tag{200}
$$

with

$$
B_x \simeq \frac{1}{2}\left(uu - \dot{v}\dot{v}\right), \qquad B_y \simeq \frac{1}{2i}\left(uu + \dot{v}\dot{v}\right)
\tag{201}
$$

$$
E_x = \frac{1}{2i}\left(uu + \dot{v}\dot{v}\right), \qquad E_y = -\frac{1}{2}\left(uu - \dot{v}\dot{v}\right)
\tag{202}
$$

The electric and magnetic field components are perpendicular to each other. Furthermore,

$$
E_x = B_y, \qquad E_y = -B_x
\tag{203}
$$

In order to address this question, let us go back to Equation (191). In the massless limit,

$$
B_+ \simeq E_+ \simeq uu, \qquad B_- \simeq E_- \simeq \dot{v}\dot{v}
\tag{204}
$$

The gauge transformation applicable to u and $\dot{v}$ are the two-by-two matrices

$$
\begin{pmatrix} 1 & -\gamma \\ 0 & 1 \end{pmatrix}, \quad \text{and} \quad \begin{pmatrix} 1 & 0 \\ -\gamma & 1 \end{pmatrix}
\tag{205}
$$

respectively as noted in Sections 5.2 and 7.1. Both u and $\dot{v}$ are invariant under gauge transformations, while $\dot{u}$ and v do not.

The B_+ and E_+ are for the photon spin along the z direction, while B_- and E_- are for the opposite direction. In 1964 [35], Weinberg constructed gauge-invariant state vectors for massless particles starting from Wigner's 1939 paper [1]. The bilinear spinors uu and and $\dot{v}\dot{v}$ correspond to Weinberg's state vectors.

8.3. Possible Symmetry of the Higgs Mechanism

In this section, we discussed how the two-by-two formalism of the group $SL(2,c)$ leads the scalar, four-vector, and tensor representations of the Lorentz group. We discussed in detail how the four-vector for a massive particle can be decomposed into the symmetry of a two-component massless particle and one gauge degree of freedom. This aspect was studied in detail by Kim and Wigner [20,21], and their results are illustrated in Figure 6. This decomposition is known in the literature as the group contraction.

The four-dimensional Lorentz group can be contracted to the Euclidean and cylindrical groups. These contraction processes could transform a four-component massive vector meson into a massless spin-one particle with two spin components, and one gauge degree of freedom.

Since this contraction procedure is spelled out detail in [21], as well as in the present paper, its reverse process is also well understood. We start with one two-component massless particle with one gauge degree of freedom, and end up with a massive vector meson with its four components.

The mathematics of this process is not unlike the Higgs mechanism [36,37], where one massless field with two degrees of freedom absorbs one gauge degree freedom to become a quartet of bosons, namely that of $W, Z^{\pm}$ plus the Higgs boson. As is well known, this mechanism is the basis for the theory of electro-weak interaction formulated by Weinberg and Salam [38,39].

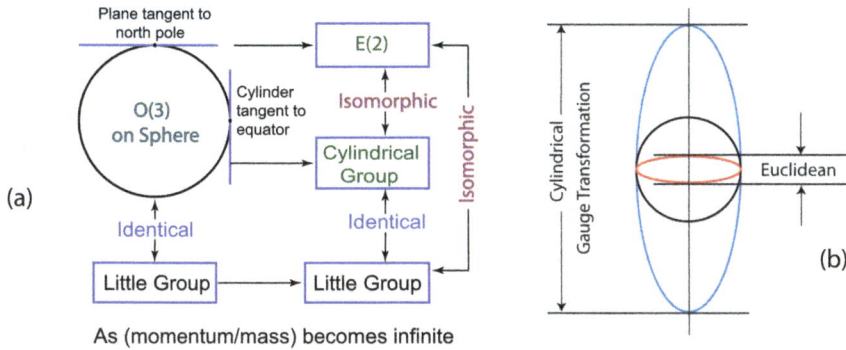

Figure 6. Contractions of the three-dimensional rotation group. (a) Contraction in terms of the tangential plane and the tangential cylinder [20]; (b) Contraction in terms of the expansion and contraction of the longitudinal axis [21]. In both cases, the symmetry ends up with one rotation around the longitudinal direction and one translational degree along the longitudinal axis. The rotation and translation corresponds to the helicity and gauge degrees of freedom.

The word "spontaneous symmetry breaking" is used for the Higgs mechanism. It could be an interesting problem to see that this symmetry breaking for the two Higgs doublet model can be formulated in terms of the Lorentz group and its contractions. In this connection, we note an interesting recent paper by Dée and Ivanov [40].

9. Conclusions

The damped harmonic oscillator, Wigner'e little groups, and the Poincaré sphere belong to the three different branches of physics. In this paper, it was noted that they are based on the same mathematical framework, namely the algebra of two-by-two matrices.

The second-order differential equation for damped harmonic oscillators can be formulated in terms of two-by-two matrices. These matrices produce the algebra of the group $Sp(2)$. While there are three trace classes of the two-by-two matrices of this group, the damped oscillator tells us how to make transitions from one class to another.

It is shown that Wigner's three little groups can be defined in terms of the trace classes of the $Sp(2)$ group. If the trace is smaller than two, the little group is for massive particles. If greater than two, the little group is for imaginary-mass particles. If the trace is equal to two, the little group is for massless particles. Thus, the damped harmonic oscillator provides a procedure for transition from one little group to another.

The Poincaré sphere contains the symmetry of the six-parameter $SL(2, c)$ group. Thus, the sphere provides the procedure for extending the symmetry of the little group defined within the Lorentz group of three-dimensional Minkowski space to its full Lorentz group in the four-dimensional space-time. In addition, the Poincaré sphere offers the variable which allows us to change the symmetry of a massive particle to that of a massless particle by continuously decreasing the mass.

In this paper, we extracted the mathematical properties of Wigner's little groups from the damped harmonic oscillator and the Poincaré sphere. In so doing, we have shown that the transition from one little group to another is tangentially continuous.

This subject was initiated by Inönü and Wigner in 1953 as the group contraction [41]. In their paper, they discussed the contraction of the three-dimensional rotation group becoming contracted to the two-dimensional Euclidean group with one rotational and two translational degrees of freedom. While the $O(3)$ rotation group can be illustrated by a three-dimensional sphere, the plane tangential at

Symmetry **2014**, *6*, 473–515

the north pole is for the $E(2)$ Euclidean group. However, we can also consider a cylinder tangential at the equatorial belt. The resulting cylindrical group is isomorphic to the Euclidean group [20]. While the rotational degree of freedom of this cylinder is for the photon spin, the up and down translations on the surface of the cylinder correspond to the gauge degree of freedom of the photon, as illustrated in Figure 6.

It was noted also that the Bargmann decomposition of two-by-two matrices, as illustrated in Figure 1 and Figure 2, allows us to study more detailed properties of the little groups, including space and time reflection reflection properties. Also in this paper, we have discussed how the scalars, four-vectors, and four-tensors can be constructed from the two-by-two representation in the Lorentz-covariant world.

In addition, it should be noted that the symmetry of the Lorentz group is also contained in the squeezed state of light [14] and the $ABCD$ matrix for optical beam transfers [18]. We also mentioned the possibility of understanding the mathematics of the Higgs mechanism in terms of the Lorentz group and its contractions.

Acknowledgements

In his 1939 paper [1], Wigner worked out the subgroups of the Lorentz group whose transformations leave the four momentum of a given particle invariant. In so doing, he worked out their internal space-time symmetries. In spite of its importance, this paper remains as one of the most difficult papers to understand. Wigner was eager to make his paper understandable to younger physicists.

While he was the pioneer in introducing the mathematics of group theory to physics, he was also quite fond of using two-by-two matrices to explain group theoretical ideas. He asked one of the present authors (Young S. Kim) to rewrite his 1939 paper [1] using the language of those matrices. This is precisely what we did in the present paper.

We are grateful to Eugene Paul Wigner for this valuable suggestion.

Author Contributions

This paper is largely based on the earlier papers by Young S. Kim and Marilyn E. Noz, and those by Sibel Başkal and Young S. Kim. The two-by-two formulation of the damped oscillator in Section 2 was jointly developed by Sibel Başkal and Yound S. Kim during the summer of 2012. Marilyn E. Noz developed the idea of the symmetry of small-mass neutrinos in Section 7. The limiting process in the symmetry of the Poincaré sphere was formulated by Young S. Kim. Sibel Başkal initially constructed the four-by-four tensor representation in Section 8.

The initial organization of this paper was conceived by Young S. Kim in his attempt to follow Wigner's suggestion to translate his 1939 paper into the language of two-by-two matrices. Sibel Başkal and Marilyn E. Noz tightened the organization and filled in the details.

Conflicts of Interest

The authors declare no conflicts of interest.

References

1. Wigner, E. On unitary representations of the inhomogeneous Lorentz Group. *Ann. Math.* **1939**, *40*, 149–204.
2. Han, D.; Kim, Y.S.; Son, D. Eulerian parametrization of Wigner little groups and gauge transformations in terms of rotations in 2-component spinors. *J. Math. Phys.* **1986**, *27*, 2228–2235.
3. Born, M.; Wolf, E. *Principles of Optics*, 6th ed.; Pergamon: Oxford, UK, 1980.

4. Han, D.; Kim, Y.S.; Noz, M.E. Stokes parameters as a Minkowskian four-vector. *Phys. Rev. E* **1997**, *56*, 6065–6076.

5. Brosseau, C. *Fundamentals of Polarized Light: A Statistical Optics Approach*; John Wiley: New York, NY, USA, 1998.

6. Başkal, S.; Kim, Y.S. De Sitter group as a symmetry for optical decoherence. *J. Phys. A* **2006**, *39*, 7775–7788.

7. Kim, Y.S.; Noz, M.E. Symmetries shared by the Poincaré Group and the Poincaré Sphere. *Symmetry* **2013**, *5*, 233–252.

8. Han, D.; Kim, Y.S.; Son, D. E(2)-like little group for massless particles and polarization of neutrinos. *Phys. Rev. D* **1982**, *26*, 3717–3725.

9. Han, D.; Kim, Y.S.; Son, D. Photons, neutrinos and gauge transformations. *Am. J. Phys.* **1986**, *54*, 818–821.

10. Başkal, S.; Kim, Y.S. Little groups and Maxwell-type tensors for massive and massless particles. *Europhys. Lett.* **1997**, *40*, 375–380.

11. Leggett, A.; Chakravarty, S.; Dorsey, A.; Fisher, M.; Garg, A.; Zwerger, W. Dynamics of the dissipative 2-state system. *Rev. Mod. Phys.* **1987**, *59*, 1–85.

12. Başkal, S.; Kim, Y.S. One analytic form for four branches of the ABCD matrix. *J. Mod. Opt.* **2010**, *57*, 1251–1259.

13. Başkal, S.; Kim, Y.S. Lens optics and the continuity problems of the ABCD matrix. *J. Mod. Opt.* **2014**, *61*, 161–166.

14. Kim, Y.S.; Noz, M.E. *Theory and Applications of the Poincaré Group*; Reidel: Dordrecht, The Netherlands, 1986.

15. Bargmann, V. Irreducible unitary representations of the Lorentz group. *Ann. Math.* **1947**, *48*, 568–640.

16. Iwasawa, K. On some types of topological groups. *Ann. Math.* **1949**, *50*, 507–558.

17. Guillemin, V.; Sternberg, S. *Symplectic Techniques in Physics*; Cambridge University Press: Cambridge, UK, 1984.

18. Başkal, S.; Kim, Y.S. Lorentz Group in Ray and Polarization Optics. In *Mathematical Optics: Classical, Quantum and Computational Methods*; Lakshminarayanan, V., Calvo, M.L., Alieva, T., Eds.; CRC Taylor and Francis: New York, NY, USA, 2013; Chapter 9, pp. 303–340.

19. Naimark, M.A. *Linear Representations of the Lorentz Group*; Pergamon: Oxford, UK, 1964.

20. Kim, Y.S.; Wigner, E.P. Cylindrical group and masless particles. *J. Math. Phys.* **1987**, *28*, 1175–1179.

21. Kim, Y.S.; Wigner, E.P. Space-time geometry of relativistic particles. *J. Math. Phys.* **1990**, *31*, 55–60.

22. Georgieva, E.; Kim, Y.S. Iwasawa effects in multilayer optics. *Phys. Rev. E* **2001**, *64*, doi:10.1103/PhysRevE.64.026602.

23. Saleh, B.E.A.; Teich, M.C. *Fundamentals of Photonics*, 2nd ed.; John Wiley: Hoboken, NJ, USA, 2007.

24. Papoulias, D.K.; Kosmas, T.S. Exotic Lepton Flavour Violating Processes in the Presence of Nuclei. *J. Phys.: Conf. Ser.* **2013**, *410*, 012123:1–012123:5.

25. Dinh, D.N.; Petcov, S.T.; Sasao, N.; Tanaka, M.; Yoshimura, M. Observables in neutrino mass spectroscopy using atoms. *Phys. Lett. B.* **2013**, *719*, 154–163.

26. Miramonti, L.; Antonelli, V. Advancements in Solar Neutrino physics. *Int. J. Mod. Phys. E.* **2013**, *22*, 1–16.

27. Li, Y.-F.; Cao, J.; Jun, Y.; Wang, Y.; Zhan, L. Unambiguous determination of the neutrino mass hierarchy using reactor neutrinos. *Phys. Rev. D.* **2013**, *88*, 013008:1–013008:9.

28. Bergstrom, J. Combining and comparing neutrinoless double beta decay experiments using different 584 nuclei. *J. High Energy Phys.* **2013**, *02*, 093:1–093:27.

29. Han, T.; Lewis, I.; Ruiz, R.; Si, Z.-G. Lepton number violation and W' chiral couplings at the LHC. *Phys. Rev. D* **2013**, *87*, 035011:1–035011:25.

30. Drewes, M. The phenomenology of right handed neutrinos. *Int. J. Mod. Phys. E* **2013**, *22*, 1330019:1–1330019:75.

31. Barut, A.O.; McEwan, J. The four states of the massless neutrino with pauli coupling by spin-gauge invariance. *Lett. Math. Phys.* **1986**, *11*, 67–72.

32. Palcu, A. Neutrino Mass as a consequence of the exact solution of 3-3-1 gauge models without exotic electric charges. *Mod. Phys. Lett. A* **2006**, *21*, 1203–1217.

33. Bilenky, S.M. Neutrino. *Phys. Part. Nucl.* **2013**, *44*, 1–46.

34. Alhendi, H. A.; Lashin, E. I.; Mudlej, A. A. Textures with two traceless submatrices of the neutrino mass matrix. *Phys.Rev. D* **2008**, *77*, 013009.1–013009.1-13.

35. Weinberg, S. Photons and gravitons in S-Matrix theory: Derivation of charge conservation and equality of gravitational and inertial mass. *Phys. Rev.* **1964**, *135*, B1049–B1056.

36. Higgs, P.W. Broken symmetries and the masses of gauge bosons. *Phys. Rev. Lett.* **1964**, *13*, 508–509.

Symmetry **2014**, *6*, 473–515

37. Guralnik, G.S.; Hagen, C.R.; Kibble, T.W.B. Global conservation laws and massless particles. *Phys. Rev. Lett.* **1964**, *13*, 585–587.
38. Weinberg, S. A model of leptons. *Phys. Rev. Lett.* **1967**, *19*, 1265–1266.
39. Weinberg, S. *Quantum Theory of Fields, Volume II, Modern Applications*; Cambridge University Press: Cambridge, UK, 1996.
40. Dée, A.; Ivanov, I.P. Higgs boson masses of the general two-Higgs-doublet model in the Minkowski-space formalism. *Phys. Rev. D* **2010**, *81*, 015012:1–015012:8.
41. Inönü, E.; Wigner, E.P. On the contraction of groups and their representations. *Proc. Natl. Acad. Sci. USA* **1953**, *39*, 510–524.

© 2014 by the authors. Licensee MDPI, Basel, Switzerland. This article is an open access article distributed under the terms and conditions of the Creative Commons Attribution (CC BY) license (http://creativecommons.org/licenses/by/4.0/).

Chapter 2:
Harmonic Oscillators in Modern Physics

symmetry

MDPI

Article

Analytical Solutions of Temporal Evolution of Populations in Optically-Pumped Atoms with Circularly Polarized Light

Heung-Ryoul Noh

Department of Physics, Chonnam National University, Gwangju 500-757, Korea; hrnoh@chonnam.ac.kr;
Tel.: +82-62-530-3366

Academic Editor: Young Suh Kim
Received: 10 December 2015; Accepted: 14 March 2016 ; Published: 19 March 2016

Abstract: We present an analytical calculation of temporal evolution of populations for optically pumped atoms under the influence of weak, circularly polarized light. The differential equations for the populations of magnetic sublevels in the excited state, derived from rate equations, are expressed in the form of inhomogeneous second-order differential equations with constant coefficients. We present a general method of analytically solving these differential equations, and obtain explicit analytical forms of the populations of the ground state at the lowest order in the saturation parameter. The obtained populations can be used to calculate lineshapes in various laser spectroscopies, considering transit time relaxation.

Keywords: second-order differential equations; optical pumping; analytical solutions

PACS: 02.30.Hq; 32.80.Xx; 32.30.-r

1. Introduction

When an atom is illuminated by single-mode laser light, the populations of the magnetic sublevels and coherences between them exhibit complicated temporal variations. This phenomenon is called optical pumping, which is widely used in the preparation of internal atomic states of interest [1,2]. It has recently been observed that optical pumping affects the lineshapes in saturated absorption spectroscopy (SAS) [3], electromagnetically induced transparency (EIT) [4], and absorption of cold atoms with a Λ-type three-level scheme [5]. Nonlinear effects in optical pumping have also been investigated [6,7].

The temporal dynamics of the internal states of an atom are accurately described by density matrix equations [8,9]. In some special cases, however, a simpler method can be employed to solve for the dynamics of the internal states of the atom, using rate equations [10,11]. Furthermore, when the intensity of light is weak, the rate equations can be solved analytically [12–15]. These analytical solutions are practically very useful; once they are obtained, it is readily possible to obtain analytically computed quantities such as the absorption coefficient of a probe beam and lineshape functions in nonlinear laser spectroscopy. We have previously reported analytical solutions for SAS [16,17] and polarization spectroscopy (PS) [18].

Interestingly, the equations governing the temporal dynamics of populations at the weak intensity limit are homogeneous or inhomogeneous second-order linear differential equations (DEs) with constant coefficients [12–15]. Unlike the harmonic oscillator in mechanics, where under- or over-damped motions are observed [19], the equations for optical pumping show only over-damped behaviors. However, this system exhibits a variety of inhomogeneous DEs. In a recent publication, we reported the method of solving these equations analytically, in the context of a pedagogical description of the method of solving inhomogeneous DEs [15]. Although the method is straightforward in principle, it is not easy to obtain analytical solutions for complicated atomic structures, such as Cs. Extending the previous study [15], in this paper, we present a general method of analytically solving the DEs for such a complicated atom.

2. Theory

The energy level diagram under consideration is shown in Figure 1. Since alkali-metal atoms are considered, there are two ground states with $F_g = I + 1/2$ and $F_g = I - 1/2$ (I: nuclear spin angular momentum quantum number). We consider a σ^+ polarized weak laser beam, whose Rabi frequency is Ω and optical frequency is $\omega = \omega_0 + \delta$ (ω_0 is the resonance frequency and δ is the laser frequency detuning). We assume that the laser frequency is tuned to the transition from one of the two ground states (in Figure 1, the state $F_g = I + 1/2$). Then, the other ground state (in Figure 1, the state $F_g = I - 1/2$) is not excited by laser light, and can be populated by spontaneous emission from the excited state when the optical transition is not cycling. The populations (and the states themselves) of the magnetic sublevels in the excited, upper ground, and lower ground states are labeled, respectively, as g_i, f_i, and h_i with $i = 1, 2, \cdots$.

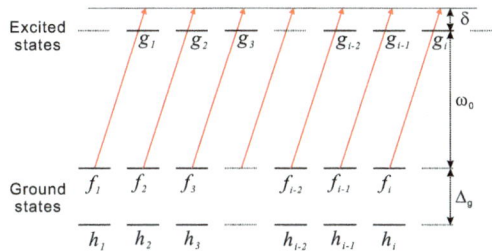

Figure 1. An energy level diagram for an optically pumped atom under the influence of circularly polarized light.

The internal dynamics of the atom can be described by the density matrix equation in the frame rotating with frequency ω:

$$\dot{\rho} = -(i/\hbar)[H, \rho] + \dot{\rho}_{sp}, \tag{1}$$

where ρ is the density operator. In Equation (1), the Hamiltonian, H, is given by

$$H = -\sum_j \hbar\delta |g_j\rangle \langle g_j| - \sum_j \hbar\Delta_g |h_j\rangle \langle h_j| - \frac{\hbar\Omega}{2} \sum_j C_j^j |g_j\rangle \langle f_j| + \text{h.c.}, \tag{2}$$

where Δ_g is the hyperfine splitting between the two ground states and h.c. denotes the harmonic conjugate. In Equation (2), the first two terms in the right-hand side represent the bare atomic Hamiltonian and the rest terms denote the atom-photon interaction Hamiltonian [20]. C_i^j is the normalized transition strength between the states f_i and g_j, and $R_i^j \equiv \left(C_i^j\right)^2$ is given below (Equation (13)). In Equation (1), $\dot{\rho}_{sp}$ represents spontaneous emission term, whose matrix representations are given by:

$$
\begin{aligned}
&\langle g_i| \dot{\rho}_{sp} |g_j\rangle = -\Gamma \langle g_i| \rho |g_j\rangle, \\
&\langle g_i| \dot{\rho}_{sp} |f_j\rangle = -\frac{\Gamma}{2} \langle g_i| \rho |f_j\rangle, \quad \langle g_i| \dot{\rho}_{sp} |h_j\rangle = -\frac{\Gamma}{2} \langle g_i| \rho |h_j\rangle, \\
&\langle f_i| \dot{\rho}_{sp} |f_j\rangle = \Gamma \sum_{\epsilon=-2}^{0} C_i^{i+\epsilon} C_j^{j+\epsilon} \langle g_{i+\epsilon}| \rho |g_{j+\epsilon}\rangle, \\
&\langle h_i| \dot{\rho}_{sp} |h_j\rangle = \Gamma \sum_{\epsilon=-2}^{0} D_i^{i+\epsilon} D_j^{j+\epsilon} \langle g_{i+\epsilon}| \rho |g_{j+\epsilon}\rangle,
\end{aligned}
\tag{3}
$$

and $\langle \mu | \dot{\rho}_{sp} | \nu \rangle = \langle \nu | \dot{\rho}_{sp} | \mu \rangle^*$ when $\mu \neq \nu$, where Γ is the decay rate of the excited state. D_i^j is the normalized transition strength between the states h_i and g_j, and $T_i^j \equiv \left(D_i^j \right)^2$ is also given below (Equation (13)). Inserting Equations (2) and (3) into Equation (1), we can obtain the following differential equations for the optical coherences and populations:

$$\langle g_i | \dot{\rho} | f_i \rangle = \left(i\delta - \frac{\Gamma}{2} \right) \langle g_i | \rho | f_i \rangle + \frac{i}{2} C_i^i \Omega \left(g_i - f_i \right), \tag{4}$$

$$\dot{g}_i = -\Gamma g_i + \frac{i}{2} C_i^i \Omega \left(\langle g_i | \rho | f_i \rangle - \langle f_i | \rho | g_i \rangle \right), \tag{5}$$

$$\dot{f}_i = \Gamma \sum_{j=i-2}^{i} \left(C_i^j \right)^2 g_j - \frac{i}{2} C_i^i \Omega \left(\langle g_i | \rho | f_i \rangle - \langle f_i | \rho | g_i \rangle \right), \tag{6}$$

$$\dot{h}_i = \Gamma \sum_{j=i-2}^{i} \left(D_i^j \right)^2 g_j, \tag{7}$$

where we use simplified expressions for the populations: $\langle g_i | \rho | g_i \rangle = g_i$, $\langle f_i | \rho | f_i \rangle = f_i$, and $\langle h_i | \rho | h_i \rangle = h_i$. In Equations (4)–(7), we assume that $\langle g_i | \rho | h_i \rangle = 0$ because Δ_g is much larger than $|\delta|$ and Γ. We note that, because the polarization of light is σ^+, and therefore the Zeeman coherences between the magnetic sublevels in the excited and ground states disappear.

In Equation (4), the characteristic decay rate of the optical coherence is $\Gamma/2$, which is much larger than the characteristic decay rate of the populations ($\sim s\Gamma$; see Equation (12) below for definition of s). Thus, the optical coherences evolve much faster than the populations, which is called the rate equation approximation [21]. Owing to this rate equation approximation, $\langle g_i | \rho | f_i \rangle$ can be expressed in terms of the populations as follows by letting $\langle g_i | \dot{\rho} | f_i \rangle = 0$:

$$\langle g_i | \rho | f_i \rangle = \frac{C_i^i \Omega}{i\Gamma + 2\delta} \left(f_i - g_i \right). \tag{8}$$

Then, inserting Equation (8) and its complex conjugate into Equations (5)–(7), we can obtain the following rate equations for the populations:

$$\dot{f}_i = -\frac{\Gamma}{2} s R_i^i \left(f_i - g_i \right) + \sum_{j=i-2}^{i} \Gamma R_i^j g_j, \tag{9}$$

$$\dot{g}_i = \frac{\Gamma}{2} s R_i^i \left(f_i - g_i \right) - \Gamma g_i, \tag{10}$$

$$\dot{h}_i = \sum_{j=i-2}^{i} \Gamma T_i^j g_j, \tag{11}$$

for $i = 1, 2, \cdots$. In Equations (9)–(11), s is the saturation parameter, which is given by

$$s = \frac{\Omega^2/2}{\delta^2 + \Gamma^2/4}, \tag{12}$$

and $R_i^j = \left(C_i^j \right)^2$ and $T_i^j = \left(D_i^j \right)^2$. We note that s is a function of both the δ and Rabi frequency. Notably, the reference of the frequency detuning differs, depending on the transition line considered. When i and j refer to the states $|F_g, m_g\rangle$ and $|F_e, m_e\rangle$, respectively, the transition strength (R_i^j) is given by

$$R_{F_g, m_g}^{F_e, m_e} = (2L_e + 1)(2J_e + 1)(2J_g + 1)(2F_e + 1)(2F_g + 1)$$
$$\times \left[\left\{ \begin{matrix} L_e & J_e & S \\ J_g & L_g & 1 \end{matrix} \right\} \left\{ \begin{matrix} J_e & F_e & I \\ F_g & J_g & 1 \end{matrix} \right\} \left(\begin{matrix} F_g & 1 & F_e \\ m_g & m_e - m_g & -m_e \end{matrix} \right) \right]^2, \tag{13}$$

where L and S denote the orbital and electron spin angular momenta, respectively, and the curly (round) brackets represent the 6J (3J) symbol. T_i^j are similarly obtained by using different F_g values in Equation (13).

The explicit form of Equation (9) is given by

$$\dot{f}_i = \frac{\Gamma}{2}sR_i^i(g_i - f_i) + \Gamma\left(R_i^{i-2}g_{i-2} + R_i^{i-1}g_{i-1} + R_i^i g_i\right),\tag{14}$$

and f_i can be expressed in terms of $\dot{g}_i$ and g_i from Equation (10) at the lowest order in s as follows:

$$f_i = \frac{2}{\Gamma s R_i^i}(\dot{g}_i + \Gamma g_i).\tag{15}$$

Insertion of Equations (10) and (15) into Equation (14) yields the following DE for g_i:

$$\ddot{g}_i + \Gamma\left(1 + \frac{s}{2}R_i^i\right)\dot{g}_i + \frac{s}{2}\Gamma^2 R_i^i\left(1 - R_i^i\right)g_i$$
$$= \frac{s}{2}\Gamma^2 R_i^{i-2}R_i^i g_{i-2} + \frac{s}{2}\Gamma^2 R_i^{i-1}R_i^i g_{i-1}.\tag{16}$$

when $i = 1$, the right-hand side of Equation (16) vanishes. Therefore, Equation (16) becomes a homogeneous DE. In contrast, when $i \neq 1$, Equation (16) becomes an inhomogeneous DE because the right-hand side terms are functions of g_i.

We solve Equation (16) from $i = 1$ consecutively. As is well-known, the solution of Equation (16) consists of two parts: a homogeneous solution and a particular solution. We first find the solutions of the homogeneous equation by inserting the equation $g_i \sim e^{\lambda \Gamma t}$ into Equation (16). Then, we have two values $(\lambda_{2i-1}, \lambda_{2i})$ for λ as follows:

$$\lambda_{2i-1(2i)} = \frac{1}{4}\left(-2 - sR_i^i - (+)\sqrt{4 - 4sR_i^i + s(8+s)\left(R_i^i\right)^2}\right),$$

which can be approximated as follows in the weak intensity limit:

$$\lambda_{2i-1} \simeq -1 - \frac{s}{2}\left(R_i^i\right)^2, \quad \lambda_{2i} \simeq -\frac{s}{2}R_i^i\left(1 - R_i^i\right).$$

We consider the case of $i = 1$ in Equation (16). Then, the solution is given by

$$g_1 = C_{1,1}e^{\lambda_1 \Gamma t} + C_{1,2}e^{\lambda_2 \Gamma t},$$

where the coefficients $C_{1,1}$ and $C_{1,2}$ should be determined using the initial conditions. In the case of $i = 2$, the right-hand side in Equation (16) contains the terms of $e^{\lambda_1 \Gamma t}$ and $e^{\lambda_2 \Gamma t}$. Therefore, g_2 has four exponential terms:

$$g_2 = C_{2,1}e^{\lambda_1 \Gamma t} + C_{2,2}e^{\lambda_2 \Gamma t} + C_{2,3}e^{\lambda_3 \Gamma t} + C_{2,4}e^{\lambda_4 \Gamma t},$$

where the coefficients should also be determined. Therefore, we can express g_j generally as follows:

$$g_j = \sum_{k=1}^{2j} C_{j,k}e^{\lambda_k \Gamma t}.\tag{17}$$

We find $C_{j,k}$ with $k = 1, 2, \cdots, 2j$ by means of recursion relations; *i.e.*, $C_{j,k}$ are expressed in terms of $C_{i,l}$ with $i < j$ and $l = 1, 2, \cdots, 2i$. Inserting Equation (17) into Equation (16), we obtain

$$
\begin{aligned}
g_i = {} & C_{i,2i-1} e^{\lambda_{2i-1}\Gamma t} + C_{i,2i} e^{\lambda_{2i}\Gamma t} \\
& + \sum_{k=1}^{2(i-1)} \frac{(s/2) R_i^{i-1} R_i^i C_{i-1,k}}{\lambda_k^2 + \lambda_k + \frac{s}{2} R_i^i \left(1 + \lambda_k - R_i^i\right)} e^{\lambda_k \Gamma t} \\
& + \sum_{k=1}^{2(i-2)} \frac{(s/2) R_i^{i-2} R_i^i C_{i-2,k}}{\lambda_k^2 + \lambda_k + \frac{s}{2} R_i^i \left(1 + \lambda_k - R_i^i\right)} e^{\lambda_k \Gamma t}.
\end{aligned}
\tag{18}
$$

Comparing Equations (17) and (18) gives

$$
C_{i,k} = \frac{(s/2) R_i^i \left(R_i^{i-1} C_{i-1,k} + R_i^{i-2} C_{i-2,k} \right)}{\lambda_k^2 + \lambda_k + \frac{s}{2} R_i^i \left(1 + \lambda_k - R_i^i\right)},
\tag{19}
$$

$$
\text{for } k = 1, 2, \cdots, 2(i-2),
$$

$$
C_{i,k} = \frac{(s/2) R_i^{i-1} R_i^i C_{i-1,k}}{\lambda_k^2 + \lambda_k + \frac{s}{2} R_i^i \left(1 + \lambda_k - R_i^i\right)},
\tag{20}
$$

$$
\text{for } k = 2i - 3 \text{ and } 2(i-1).
$$

The remaining two coefficients, $C_{i,2i-1}$ and $C_{i,2i}$, can be derived from Equation (18) using two initial conditions for $g_i(0)$ and $\dot{g}_i(0)$:

$$
g_i(0) = 0, \quad \dot{g}_i(0) = \frac{s}{2} p_0 R_i^i,
$$

where p_0 is the population of each sublevel in the ground state at equilibrium, which is equal to $1/[2(2I+1)]$. Then, the results are given by

$$
\begin{aligned}
C_{i,2i-1} = {} & \tfrac{1}{2Q_i} \left[2 \left(A_i + 2A_i' + B_i + 2B_i' \right) \right. \\
& \left. + \left(A_i + B_i - 2p_0 \right) s R_i^i \right] - \tfrac{A_i + B_i}{2},
\end{aligned}
\tag{21}
$$

$$
\begin{aligned}
C_{i,2i} = {} & -\tfrac{1}{2Q_i} \left[2 \left(A_i + 2A_i' + B_i + 2B_i' \right) \right. \\
& \left. + \left(A_i + B_i - 2p_0 \right) s R_i^i \right] - \tfrac{A_i + B_i}{2},
\end{aligned}
\tag{22}
$$

where

$$
Q_i = \sqrt{4 + s R_i^i \left(-4 + (8+s) R_i^i \right)},
$$

$$
A_i = \sum_{k=1}^{2(i-1)} \frac{(s/2) R_i^{i-1} R_i^i C_{i-1,k}}{\lambda_k^2 + \lambda_k + \frac{s}{2} R_i^i \left(1 + \lambda_k - R_i^i\right)}, \quad \text{for } i \geq 2
$$

$$
B_i = \sum_{k=1}^{2(i-2)} \frac{(s/2) R_i^{i-2} R_i^i C_{i-2,k}}{\lambda_k^2 + \lambda_k + \frac{s}{2} R_i^i \left(1 + \lambda_k - R_i^i\right)}, \quad \text{for } i \geq 3,
$$

$$
A_i' = \sum_{k=1}^{2(i-1)} \frac{(s/2) R_i^{i-1} R_i^i \lambda_k C_{i-1,k}}{\lambda_k^2 + \lambda_k + \frac{s}{2} R_i^i \left(1 + \lambda_k - R_i^i\right)}, \quad \text{for } i \geq 2
$$

$$
B_i' = \sum_{k=1}^{2(i-2)} \frac{(s/2) R_i^{i-2} R_i^i \lambda_k C_{i-2,k}}{\lambda_k^2 + \lambda_k + \frac{s}{2} R_i^i \left(1 + \lambda_k - R_i^i\right)}, \quad \text{for } i \geq 3,
$$

and

$$
A_1 = 0, \quad A_1' = 0, \quad B_1 = B_2 = 0, \quad \text{and } B_1' = B_2' = 0.
$$

The coefficients in g_i from g_1 can be obtained by successively using the recursion relations in Equations (19)–(22). Once g_i are obtained, f_i can be obtained using Equation (15). Up to the lowest order in s, the result is given by

$$f_i = \sum_{k=1}^{i} \frac{2C_{i,2k}}{sR_i^i} e^{\lambda_{2k}\Gamma t}.$$ (23)

Since $\lambda_k \sim -1$ for odd k, g_i can be expressed as follows:

$$g_i = \sum_{k=1}^{i} \left(C_{i,2k-1} e^{-\Gamma t} + C_{i,2k} e^{\lambda_{2k}\Gamma t} \right).$$ (24)

Taking the derivative of Equation (24) with respect to time and letting $t = 0$, we have

$$\dot{g}_i(0) = -\sum_{k=1}^{i} C_{i,2k-1},$$

up to the first order in s, since λ_{2k} ($k = 1, 2, \cdots, i$) are already in the first order in s. Because one of the initial conditions is $\dot{g}_i(0) = sp_0 R_i^i/2$, and $g_i(0) = \sum_{k=1}^{i} (C_{i,2k-1} + C_{i,2k}) = 0$ from the other initial condition, we obtain the following equations:

$$\sum_{k=1}^{i} C_{i,2k-1} = -\sum_{k=1}^{i} C_{i,2k} = -\frac{s}{2} p_0 R_i^i.$$ (25)

Using the relations in Equations (23) and (25), we find the simplified form of g_i as follows:

$$g_i = \frac{R_i^i s}{2} \left(f_i - p_0 e^{-\Gamma t} \right).$$ (26)

We obtain the populations of the sublevels in the ground state, which are not excited by laser light. The one or two magnetic sublevels with higher magnetic quantum numbers correspond to this case. We can easily obtain analytical populations by integrating the populations spontaneously transferred from the excited state, and the result is given by

$$f_i = p_0 + \sum_{k=1}^{i-2} R_i^{i-2} C_{i-2,2k} \frac{e^{\lambda_{2k}\Gamma t} - 1}{\lambda_{2k}} + \sum_{k=1}^{i-1} R_i^{i-1} C_{i-1,2k} \frac{e^{\lambda_{2k}\Gamma t} - 1}{\lambda_{2k}}.$$ (27)

In several cases of atomic transition systems, λ_k can duplicate, and the method of solving particular solutions given in Equation (18) no longer holds. We may solve for the particular solutions using the method presented in our previous paper [15]. However, it is also possible to solve by intentionally modifying λ_k to satisfy the conditions that all λ_k are unique. One possible method is setting $R_i^j \to R_i^j + \delta_{i,j} j\epsilon$, where ϵ is a constant that is taken as zero at the final stage of the calculation. Although this method is not novel, it is very efficient.

The populations (h_i) of the sublevels in the ground state, which are not excited by laser light, can be easily obtained analytically by integrating the populations spontaneously transferred from the excited state (Equation (11)), and the result is given by

$$h_i = p_0 + \sum_{l=-2}^{0} \sum_{k=1}^{i+l} T_i^{i+l} C_{i+l,2k} \frac{e^{\lambda_{2k}\Gamma t} - 1}{\lambda_{2k}}.$$ (28)

3. Calculated Results

Based on the method developed in Section 2, here we present the calculated results of the populations for the two transition schemes: (i) $F_g = 4 \to F_e = 5$ and (ii) $F_g = 3 \to F_e = 3$ for the D2 line of Cs. The energy level diagram for the Cs-D2 line is shown in Figure 2a, and the energy level diagrams for these two transitions are shown in Figure 2b,c. Owing to the large hyperfine splitting in the excited states, it is justifiable to neglect the off-resonant transitions; *i.e.*, the $F_g = 4 \to F_e = 4$ and $F_g = 4 \to F_e = 3$ transitions can be neglected when the laser light is tuned to the $F_g = 4 \to F_e = 5$ transition line. Although it is in principle possible to include the off-resonant transitions in the analytical calculation of the populations [13], the complicated analytical solutions may not be practically useful.

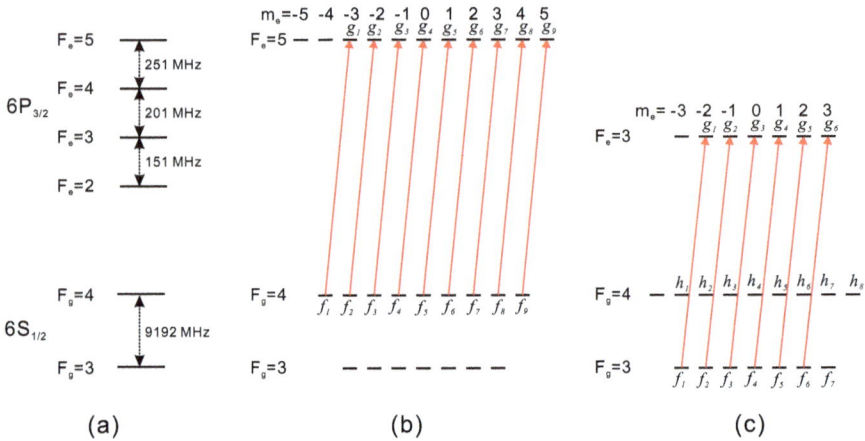

Figure 2. (a) Energy level diagram of the Cs-D2 line. (b) Energy level diagrams for the $F_g = 4 \to F_e = 5$ cycling transition line and (c) for the $F_g = 3 \to F_e = 3$ transition line illuminated by σ^+ polarized laser light.

3.1. Results for the $F_g = 4 \to F_e = 5$ Transition

The $F_g = 4 \to F_e = 5$ transition shown in Figure 2b is cycling, and is used in many experiments, such as laser cooling and trapping [22]. Because σ^+ polarized laser light is illuminated, the sublevels with $m_e = -5$ and -4 are not optically excited. The normalized transition strengths, for the transitions presented in Figure 2b, are given by

$$\left(R_1^1, R_2^2, R_3^3, R_4^4, R_5^5, R_6^6, R_7^7, R_8^8, R_9^9 \right)$$
$$= \left(\frac{1}{45}, \frac{1}{15}, \frac{2}{15}, \frac{2}{9}, \frac{1}{3}, \frac{7}{15}, \frac{28}{45}, \frac{4}{5}, 1 \right).$$

For the transition for $i = 1$, we obtain $\lambda_1 \simeq -1$ and $\lambda_2 \simeq -\frac{22}{2025}s$, and

$$C_{1,1} = -\frac{s}{1440}, \quad C_{1,2} = \frac{s}{1440}.$$

Thus, using Equation (23), we obtain

$$f_1 = \frac{1}{16} e^{-22s\Gamma t/2025}.$$

The λ_4 for the transition for $i = 2$ is approximately given by $-\frac{7}{225}s$, and the coefficients are given by

$$C_{2,1} = \frac{s}{240}, \quad C_{2,2} = \frac{s}{2460},$$
$$C_{2,3} = -\frac{s}{160}, \quad C_{2,4} = \frac{11}{6560}s.$$

Therefore, we have

$$f_2 = \frac{1}{82}e^{-22s\Gamma t/2025} + \frac{33}{656}e^{-7s\Gamma t/225}.$$

The remaining λ_{2k} $(k = 2, \cdots, 9)$ values are given by

$$(\lambda_6, \lambda_8, \lambda_{10}, \lambda_{12}, \lambda_{14}, \lambda_{16}, \lambda_{18})$$
$$= \left(-\frac{13}{225}s, -\frac{7}{81}s, -\frac{s}{9}, -\frac{28}{225}s, -\frac{238}{2025}s, -\frac{2}{25}s, 0\right),$$

and the remaining populations are explicitly given by

$$f_3 = \frac{413}{31\,160}e^{-22\tau/2025} + \frac{77}{2624}e^{-7\tau/225} + \frac{121}{6080}e^{-13\tau/225},$$

$$f_4 = \frac{2317}{264\,860}e^{-22\tau/2025} + \frac{693}{20\,992}e^{-7\tau/225} + \frac{1089}{44\,080}e^{-13\tau/225} - \frac{1001}{252\,416}e^{-7\tau/81},$$

$$f_5 = \frac{25\,577}{3\,072\,376}e^{-22\tau/2025} + \frac{4235}{125\,952}e^{-7\tau/225} + \frac{5203}{141\,056}e^{-13\tau/225}$$
$$- \frac{5005}{504\,832}e^{-7\tau/81} - \frac{143}{22\,272}e^{-\tau/9},$$

$$f_6 = \frac{148\,693}{17\,666\,162}e^{-22\tau/2025} + \frac{1925}{47\,232}e^{-7\tau/225} + \frac{2057}{35\,264}e^{-13\tau/225}$$
$$- \frac{1625}{63\,104}e^{-7\tau/81} - \frac{715}{16\,704}e^{-\tau/9} + \frac{13}{552}e^{-28\tau/225},$$

$$f_7 = \frac{921\,751}{89\,260\,608}e^{-22\tau/2025} + \frac{2519}{41\,984}e^{-7\tau/225} + \frac{891}{7424}e^{-13\tau/225}$$
$$- \frac{49\,075}{504\,832}e^{-7\tau/81} - \frac{5555}{7424}e^{-\tau/9} - \frac{273}{736}e^{-28\tau/225} + \frac{209}{192}e^{-238\tau/2025},$$

$$f_8 = \frac{39\,041\,249}{2\,119\,939\,440}e^{-22\tau/2025} + \frac{1561}{10\,496}e^{-7\tau/225} + \frac{225\,071}{352\,640}e^{-13\tau/225} + \frac{219\,275}{126\,208}e^{-7\tau/81}$$
$$+ \frac{9955}{3712}e^{-\tau/9} + \frac{3367}{3680}e^{-28\tau/225} - \frac{77}{24}e^{-238\tau/2025} - \frac{459}{160}e^{-2\tau/25},$$

$$f_9 = \frac{9}{16} - \frac{1\,205\,666\,281}{8\,479\,757\,760}e^{-22\tau/2025} - \frac{74\,771}{188\,928}e^{-7\tau/225} - \frac{316\,701}{352\,640}e^{-13\tau/225}$$
$$- \frac{404\,009}{252\,416}e^{-7\tau/81} - \frac{62\,953}{33\,408}e^{-\tau/9} - \frac{3133}{5520}e^{-28\tau/225} + \frac{407}{192}e^{-238\tau/2025} + \frac{459}{160}e^{-2\tau/25},$$

where we use a simplified notation: $\tau \equiv s\Gamma t$. Since the $F_g = 4 \rightarrow F_e = 5$ transition is cycling, the populations in the magnetic sublevels in the $F_g = 3$ ground state remain at their equilibrium value, $1/16$. It should be also noted that the sum of the ground state populations is conserved, *i.e.*,

$$\sum_{i=1}^{9} f_i = \frac{9}{16}.$$

From Equation (26), the populations of the sublevels in the excited state can be expressed in terms of the populations in the ground state as follows:

$$g_i = \frac{R_i^i s}{2}\left(f_i - \frac{1}{16}e^{-\Gamma t}\right).$$

The constants in f_9 and g_9 can be accurately calculated using Equation (10). In the steady-state regime, all the populations except f_9 and g_9 vanish, and these satisfy the following equations:

$$\frac{\Gamma}{2}s\left[f_9(\infty) - g_9(\infty)\right] - \Gamma g_9(\infty) = 0, \quad f_9(\infty) + g_9(\infty) = \frac{9}{16},$$

with $R_9^9 = 1$. Then, we have

$$f_9(\infty) = \frac{9(2+s)}{32(1+s)}, \quad g_9(\infty) = \frac{9s}{32(1+s)},$$

which can be used in a more accurate calculation of the SAS spectrum.

3.2. Results for the $F_g = 3 \to F_e = 3$ Transition

Now we present the calculated results of the populations for the $F_g = 3 \to F_e = 3$ transition of the D2 line of Cs. The energy level diagram for the transition is shown in Figure 2c. The sublevel of the excited state with $m_e = -3$ is not optically excited, and thus the sublevel of the upper-ground state with $m_g = -4$ is not filled by spontaneous emission. We also obtain the solutions for the populations in the other ground state ($F_g = 4$). To prevent the duplication of the transition strengths in this transition, we introduce ϵ so that the transition strengths are given explicitly by

$$\left(R_1^1, R_2^2, R_3^3, R_4^4, R_5^5, R_6^6\right)$$
$$= \left(\frac{3}{16} + \epsilon, \frac{5}{16} + 2\epsilon, \frac{3}{8} + 3\epsilon, \frac{3}{8} + 4\epsilon, \frac{5}{16} + 5\epsilon, \frac{3}{16} + 6\epsilon\right).$$

We take $\epsilon \to 0$ at the final stage of the calculation. The λ_{2k} ($k = 1, \cdots, 6$) values at $\epsilon \to 0$ are given by

$$(\lambda_2, \lambda_4, \lambda_6, \lambda_8, \lambda_{10}, \lambda_{12})$$
$$= \left(-\frac{39}{512}s, -\frac{55}{512}s, -\frac{15}{128}s, -\frac{15}{128}s, -\frac{55}{512}s, -\frac{39}{512}s\right).$$

We first find various C_{ik} values using the recursion relations in Equations (19)–(22). For the transition for $i = 1$, we obtain

$$C_{1,1} = -\frac{3}{512}s, \quad C_{1,2} = \frac{3}{512}s;$$

thus, using Equation (23), we obtain

$$f_1 = \frac{1}{16}e^{-39s\Gamma t/512}.$$

Using a similar method, we can obtain f_2 and f_3 as follows:

$$f_2 = \frac{3}{64}e^{-39\tau/512} + \frac{1}{64}e^{-55\tau/512},$$
$$f_3 = \frac{25}{448}e^{-39\tau/512} + \frac{1}{64}e^{-55\tau/512} - \frac{1}{112}e^{-15\tau/128},$$

where the simplified notation, $\tau \equiv s\Gamma t$, is used. In the calculation of f_4, because λ_6 and λ_8 are equal, f_4 may contain the term $\sim \tau e^{-15\tau/128}$. However, because the transition between g_3 and f_4 is prohibited, the particular solution for f_4 does not contain the term $\sim \tau e^{-15\tau/128}$. In contrast, f_5, f_6, and f_7 contain the terms proportional to τ. The results for f_4, f_5, and f_6 are explicitly given by

$$
\begin{aligned}
f_4 &= \frac{15}{224}e^{-39\tau/512} + \frac{3}{32}e^{-55\tau/512} - \frac{11}{112}e^{-15\tau/128}, \\
f_5 &= \frac{135}{896}e^{-39\tau/512} + \left(-\frac{173}{640} + \frac{9\tau}{4096}\right)e^{-55\tau/512} + \frac{51}{280}e^{-15\tau/128}, \\
f_6 &= \left(\frac{269}{12\,544} + \frac{1125\tau}{114\,688}\right)e^{-39\tau/512} \\
&\quad + \left(\frac{19}{256} - \frac{45\tau}{16\,384}\right)e^{-55\tau/512} - \frac{13}{392}e^{-15\tau/128}.
\end{aligned}
$$

Since f_7 is not excited by laser light, using Equation (27) yields,

$$
\begin{aligned}
f_7 &= \frac{68\,971}{327\,184} - \left(\frac{343\,323}{2\,119\,936} + \frac{10\,125\tau}{1\,490\,944}\right)e^{-39\tau/512} \\
&\quad + \left(\frac{1371}{30\,976} + \frac{135\tau}{180\,224}\right)e^{-55\tau/512} - \frac{3}{98}e^{-15\tau/128}.
\end{aligned}
$$

The populations of the sublevels in the excited state, using Equation (26), can be expressed as follows:

$$
g_i = \frac{R_i^i s}{2}\left(f_i - \frac{1}{16}e^{-\Gamma t}\right).
$$

The populations of the sublevels in the ground state $F_g = 4$ can be obtained using Equation (28), and are presented in the appendix.

4. Conclusions

We have presented a general method of solving homogeneous or inhomogeneous second-order DEs corresponding to the optical pumping phenomenon with σ^+ polarized laser light. Unlike the harmonic oscillator in mechanics or electrical circuits, this system only exhibits over-damped behavior. Although the method of solving inhomogeneous DEs with constant coefficients is straightforward in principle, obtaining accurate analytical solutions for the equations related to optically pumped atoms, in particular, those with complicated atomic structures, such as Cs, is cumbersome. Our method of solving the DEs provides an easy way to obtain analytical solutions at the weak intensity limit. This method is general and applicable to most atoms. As stated in Section 1, the obtained analytical form of the populations can be used in the calculation of spectroscopic lineshapes such as in saturated absorption spectroscopy (SAS) [16,17] and polarization spectroscopy (PS) [18]. Calculations of SAS and PS for Cs atoms are in progress.

Acknowledgments: This research was supported by Basic Science Research Program through the National Research Foundation of Korea (NRF) funded by the Ministry of Science, ICT and future Planning (2014R1A2A2A01006654).

Conflicts of Interest: The authors declare no conflict of interest.

Appendix

When the laser frequency is tuned to the $F_g = 3 \rightarrow F_e = 3$ transition (Figure 2c), the populations of the sublevels in the ground state $F_g = 4$ are given by

$$h_1 = \frac{23}{312} - \frac{7}{624}e^{-39\tau/512},$$

$$h_2 = \frac{93}{1144} - \frac{41}{2496}e^{-39\tau/512} - \frac{5}{2112}e^{-55\tau/512},$$

$$h_3 = \frac{895}{10\,296} - \frac{1109}{52\,416}e^{-39\tau/512} - \frac{3}{704}e^{-55\tau/512} + \frac{1}{1008}e^{-15\tau/128},$$

$$h_4 = \frac{235}{2574} - \frac{685}{26\,208}e^{-39\tau/512} - \frac{19}{1760}e^{-55\tau/512} + \frac{41}{5040}e^{-15\tau/128},$$

$$h_5 = \frac{10\,727}{113\,256} - \frac{3475}{104\,832}e^{-39\tau/512}$$
$$- \left(\frac{2641}{232\,320} + \frac{3\tau}{45\,056}\right)e^{-55\tau/512} + \frac{31}{2520}e^{-15\tau/128},$$

$$h_6 = \frac{143\,477}{1\,472\,328} - \left(\frac{843\,497}{19\,079\,424} + \frac{125\tau}{1\,490\,944}\right)e^{-39\tau/512}$$
$$+ \left(\frac{401}{30\,976} - \frac{45\tau}{180\,224}\right)e^{-55\tau/512} - \frac{13}{3528}e^{-15\tau/128},$$

$$h_7 = \frac{293\,731}{2\,944\,656} - \left(\frac{147\,347}{2\,725\,632} + \frac{125\tau}{212\,992}\right)e^{-39\tau/512}$$
$$+ \left(\frac{7889}{154\,880} - \frac{63\tau}{180\,224}\right)e^{-55\tau/512} - \frac{43}{1260}e^{-15\tau/128},$$

$$h_8 = \frac{299\,023}{2\,944\,656} - \left(\frac{24\,497}{681\,408} + \frac{125\tau}{53\,248}\right)e^{-39\tau/512}$$
$$+ \left(-\frac{959}{116\,160} + \frac{21\tau}{45\,056}\right)e^{-55\tau/512} + \frac{13}{2520}e^{-15\tau/128}.$$

Finally, we note that the sum of the populations is conserved, *i.e.*,

$$\frac{1}{16} + \sum_{i=1}^{7} f_i + \sum_{i=1}^{8} h_i = 1,$$

where $1/16$ is the population at the sublevel $m_g = -4$ in the upper ground state.

References

1. Happer, W. Optical pumping. *Rev. Mod. Phys.* **1972**, *44*, 169–249.
2. McClelland, J.J. Optical State Preparation of Atoms. In *Atomic, Molecular, and Optical Physics: Atoms and Molecules*; Dunning, F.B., Hulet, R.G., Eds; Academic Press: San Diego, CA, USA, 1995; pp. 145–170.
3. Smith, D.A.; Hughes, I.G. The role of hyperfine pumping in multilevel systems exhibiting saturated absorption. *Am. J. Phys.* **2004**, *72*, 631–637.
4. Magnus, F.; Boatwright, A.L.; Flodin, A.; Shiell, R.C. Optical pumping and electromagnetically induced transparency in a lithium vapour. *J. Opt. B: Quantum Semiclass. Opt.* **2005**, *7*, 109–118.
5. Han, H.S.; Jeong, J.E.; Cho, D. Line shape of a transition between two levels in a three-level Λ configuration. *Phys. Rev. A* **2011**, *84*, doi:10.1103/PhysRevA.84.032502.
6. Sydoryk, I.; Bezuglov, N.N.; Beterov, I.I.; Miculis, K.; Saks, E.; Janovs, A.; Spels, P.; Ekers, A. Broadening and intensity redistribution in the Na(3p) hyperfine excitation spectra due to optical pumping in the weak excitation limit. *Phys. Rev. A* **2008**, *77*, doi:10.1103/PhysRevA.77.042511.
7. Porfido, N.; Bezuglov, N.N.; Bruvelis, M.; Shayeganrad, G.; Birindelli, S.; Tantussi, F.; Guerri, I.; Viteau, M.; Fioretti, A.; Ciampini, D.; et al. Nonlinear effects in optical pumping of a cold and slow atomic beam. *Phys. Rev. A* **2015**, *92*, doi:10.1103/PhysRevA.92.043408.
8. McClelland, J.J.; Kelley, M.H. Detailed look at aspects of optical pumping in sodium. *Phys. Rev. A* **1985**, *31*, 3704–3710.

9.   Farrell, P.M.; MacGillivary, W.R.; Standage, M.C. Quantum-electrodynamic calculation of hyperfine-state populations in atomic sodium. *Phys. Rev. A* **1988**, *37*, 4240–4251.

10.   Balykin, V.I. Cyclic interaction of Na atoms with circularly polarized laser radiation. *Opt. Commun.* **1980**, *33*, 31–36.

11.   Liu, S.; Zhang, Y.; Fan, D.; Wu, H.; Yuan, P. Selective optical pumping process in Doppler-broadened atoms. *Appl. Opt.* **2011**, *50*, 1620–1624.

12.   Moon, G.; Shin, S.R.; Noh, H.R. Analytic solutions for the populations of an optically-pumped multilevel atom. *J. Korean Phys. Soc.* **2008**, *53*, 552–557.

13.   Moon, G.; Heo, M.S.; Shin, S.R.; Noh, H.R.; Jhe, W. Calculation of analytic populations for a multilevel atom at low laser intensity. *Phys. Rev. A* **2008**, *78*, doi:10.1103/PhysRevA.78.015404.

14.   Won, J.Y.; Jeong, T.; Noh, H.R. Analytical solutions of the time-evolution of the populations for D_1 transition line of the optically-pumped alkali-metal atoms with $I = 3/2$. *Optik* **2013**, *124*, 451–455.

15.   Noh, H.R. Analytical Study of Optical Pumping for the D1 Line of ^{85}Rb Atoms. *J. Korean Phys. Soc.* **2104**, *64*, 1630–1635.

16.   Moon, G.; Noh, H.R. Analytic solutions for the saturated absorption spectra. *J. Opt. Soc. Am. B* **2008**, *25*, 701–711.

17.   Moon, G.; Noh, H.R. Analytic Solutions for the Saturated Absorption Spectrum of the ^{85}Rb Atom with a Linearly Polarized Pump Beam. *J. Korean Phys. Soc.* **2009**, *54*, 13–22.

18.   Do, H.D.; Heo, M.S.; Moon, G.; Noh, H.R.; Jhe, W. Analytic calculation of the lineshapes in polarization spectroscopy of rubidium. *Opt. Commun.* **2008**, *281*, 4042–4047.

19.   Thornton, T.; Marion, J.B. *Classical Dynamics of Particles and Systems*, 5th ed.; Brooks/Cole: New York, NY, USA, 2004.

20.   Cohen-Tannoudji, C.; Dupont-Roc, J.; Grynberg, G. *Atom–Photon Interactions Basic Processes and Applications*; Wiley: New York, NY, USA, 1992.

21.   Meystre, P.; Sargent, M., III. *Elements of Quantum Optics*; Springer: New York, NY, USA, 2007.

22.   Metcalf, H.J.; van der Straten, P. *Laser Cooling and Trapping*; Springer: New York, NY, USA, 1999.

© 2016 by the author. Licensee MDPI, Basel, Switzerland. This article is an open access article distributed under the terms and conditions of the Creative Commons Attribution (CC BY) license (http://creativecommons.org/licenses/by/4.0/).

symmetry

MDPI

Article

Local Dynamics in an Infinite Harmonic Chain

M. Howard Lee

Department of Physics and Astronomy, University of Georgia, Athens, GA 30602, USA; mhlee@uga.edu

Academic Editor: Young Suh Kim
Received: 26 February 2016; Accepted: 6 April 2016; Published: 15 April 2016

Abstract: By the method of recurrence relations, the time evolution in a local variable in a harmonic chain is obtained. In particular, the autocorrelation function is obtained analytically. Using this result, a number of important dynamical quantities are obtained, including the memory function of the generalized Langevin equation. Also studied are the ergodicity and chaos in a local dynamical variable.

Keywords: recurrence relations; harmonic chain; local dynamics; ergodicity; chaos

1. Introduction

A harmonic chain has been a useful model for a variety of dynamical phenomena, such as the lattice vibrations in solids, Brownian motion and diffusion. It has also been a useful model for testing theoretical concepts, such as the thermodynamic limit, irreversibility and ergodicity. One can study these properties in a harmonic chain. In this work, we shall touch on most of these issues analytically.

The dynamics in a chain of nearest-neighbor (nn) coupled monatomic oscillators (defined in Section 3) has been studied in the past almost exclusively by means of normal modes [1]. If there are N oscillators in a chain, the single-particle or individual coordinates of the oscillators q_i, $i = 1, 2, ..N$, are replaced by the total or collective coordinates Q_j, $j = 1, 2, ..N$. In the space of the collective coordinates, the "collective" oscillators are no longer coupled. As a result, their motions are simply periodic. Each collective oscillator would have a unique frequency associated with it (if degeneracy due to symmetry could be ignored).

On the one hand, this collective picture is very helpful in understanding the dynamics of a harmonic chain by avoiding what might be a complicated picture due to a set of motions of coupled single particles. If only the collective behavior is required, this approach is certainly sufficient.

On the other hand, if one wishes to know the dynamics of a single oscillator in a chain, the traditional approach becomes cumbersome. Why would one wish to know the dynamics of one oscillator in a chain? There may be a defect in a chain, for example. It may be a heavier or lighter mass than its neighbors'. Diffusivity is attributed to the motions of single oscillators. For these and other physical reasons that will become apparent, there is a need to study how a single oscillator embedded in a chain evolves in time. We shall term it local dynamics to be distinguished from total dynamics.

In the 1980s, a new method of calculating the time evolution in a Hermitian system was developed, known as the method of recurrence relations [2]. It solves the Heisenberg equation of motion for a dynamical variable of physical interest, which may be the momentum of a single particle, the number or current density. Although it was intended to deal with dynamical variables of quantum origin, *i.e.*, operators, it was found to be applicable to classical variables by replacing commutators with Poisson brackets. During the past three decades, this method has been widely applied to a variety of dynamical issues emanating from the electron gas, lattice spins, lattice vibrations and classical fluids. For reviews, see [3–7]. For a partial list of recent papers, see [8–21].

Formally, this method shows what types of solutions are admissible [22]. It provides a deeper insight into the memory function and the Langevin equation. It has also provided a basis from which to developed the ergometric theory of the ergodic hypothesis.

In Section 2, we will briefly introduce the method of recurrence relations, mostly by assertion, referring the proofs to the original sources and review articles. In Section 3, the dynamics of a local variable (a single particle) in an infinite harmonic chain will be solved by the method of recurrence relations. Some useful physical applications will follow to complete this work.

2. Method of Recurrence Relations

Let A be a dynamical variable, e.g., a spin operator, and $H(A)$ an N-body Hamiltonian. The number of particles N is not restricted initially. The Hamiltonian H must however be Hermitian, which means that there is to be no dissipation in the dynamics of A. The time evolution of A is to be given by the Heisenberg equation of motion:

$$\dot{A}(t) = i[H, A(t)] \tag{1}$$

with $\hbar = 1$ and $[H, A] = HA - AH$. If A is a classical variable, the rhs of Equation (1) is to be replaced by the Poisson brackets.

A formal solution for Equation (1) may be viewed in geometrical terms. Let $A(t)$ be a vector in an inner product space S of d dimensions. This space is spanned by d basis vectors f_k, $k = 0, 1, ..d - 1, d \geq 2$. These basis vectors are mutually orthogonal:

$$(f_k, f_{k'}) = 0 \ if \ k' \neq k \tag{2}$$

where (,) denotes an inner product, which defines the space S. Observe that they are time independent. In terms of these, $A(t)$ may be expressed as:

$$A(t) = \sum_k a_k(t) f_k \tag{3}$$

where a_k, $k = 0, 1, ..d - 1$, is a set of functions or basis functions conjugate to the basis vectors. They carry time dependence.

As t evolves, this vector $A(t)$ evolves in this space S. Its motion in S is governed by Equation (1), so that it is H specific. Since $||A(t)|| = ||A||$, that is $(A(t), A(t)) = (A, A)$, the "length" of $A(t)$ in S is an invariant of time. As t evolves, $A(t)$ may only rotate in S. This means that there is a Bessel equality, which limits what kind of rotation is allowed.

Since both the basis vectors and functions are only formally stated, Equation (3) is not yet useful. One does not know what is d, the dimensionality of S. To make it useful, we need to realize S, an abstract space by defining the inner product in a physically-useful way.

2.1. Kubo Scalar Product

We shall realize S by the Kubo scalar product (KSP) as follows: let X and Y be two vectors in S. The inner product of X and Y is defined as:

$$(X, Y) = 1/\beta \int_0^\beta d\lambda < X(\lambda) Y^* > - < X >< Y^* > \tag{4}$$

where $\beta = 1/k_B T$, T temperature, $< .. >$ means an ensemble average, $*$ means Hermitian conjugation and:

$$X(\lambda) = e^{\lambda H} X e^{-\lambda H} \tag{5}$$

Equation (4) is known as KSP in many body theory [23]. There is a deep physical reason for using KSP to realize S [24]. When realized by KSP, it shall be denoted $\bar{S}$.

2.2. Basis Vectors

We have proved that the basis vectors in $\tilde{S}$ satisfy the following recurrence relation, known as RR I:

$$f_{k+1} = \dot{f}_k + \Delta_k f_{k-1}, \quad k = 0, 1, 2, .., d-1 \tag{6}$$

where $\dot{f}_k = i[H, f_k]$, $\Delta_k = ||f_k||/||f_{k-1}||$, with $f_{-1} = 0$ and $\Delta_0 = 1$.

If $k = 0$ in Equation (6), $f_1 = \dot{f}_0$. With $f_0 = A$ (by choice), f_1 is obtained and, therewith, Δ_1.

Given Δ_1, by setting $k = 1$ in Equation (6), one can calculate f_2, therewith Δ_2. If proceeding in this manner, $f_d = 0$ for some finite value of d giving a finite dimensional $\tilde{S}$ or $f_d \neq 0$ as $d \rightarrow \infty$ giving an infinite dimensional $\tilde{S}$. By RR I, we can determine d and, thus, generate all of the basis vectors needed to span $A(t)$ in $\tilde{S}$ for a particular H. In addition, we can construct the hypersurface σ:

$$\sigma = (\Delta_1, \Delta_2, ..\Delta_{d-1}) \tag{7}$$

As we shall see, the dynamics is governed by σ. The Δ's known as the recurrants are successive ratios of the norms of f_k. They are static quantities, so that they are in principle calculable as a function of parameters, such as temperature, wave vectors, etc., for a given H. They collectively define the shape of $\tilde{S}$, constraining what kind of trajectory is possible for $A(t)$.

2.3. Basis Functions

If RR I is applied to Equation (1), it yields a recurrence relation for the basis functions: with $a_{-1} = 0$,

$$\Delta_{k+1} a_{k+1} = -\dot{a}_k + a_{k-1}, \quad k = 0, 1, ..d-1 \tag{8}$$

where $\dot{a}_k = d/dt \ a_k$. Equation (8) is known as RR II. It is actually composed of two recurrence relations, one for $k = 0$ (because of $a_{-1} = 0$) and another for the rest $k = 1, 2, ..d-1$.

There is an important boundary condition on a_k. By Equation (3), $A(t = 0) = A = f_0$. Thus, $a_0(t = 0) = 1$ and $a_k(t = 0) = 0$, $k \neq 0$. These basis functions are autocorrelation functions. For example, $a_0 = (A(t), A)/(A, A)$, $a_1 = (A(t), f_1)/(f_1, f_1) = (A(t), \dot{A})/(\dot{A}, \dot{A})$, etc. Hence, the static and dynamic information is to be contained in them.

2.4. Continued Fractions

If a_0 is known, the rest of the basis functions can be obtained one by one by RR II. To obtain it, let $L_z a_k(t) = \tilde{a}_k(z)$, $k = 0, 1, ..d-1$, where L_z is the Laplace transform operator. The RR II is transformed to:

$$1 = z\tilde{a}_0 + \Delta_1 \tilde{a}_1 \tag{9}$$

$$\tilde{a}_{k-1} = z\tilde{a}_k + \Delta_{k+1}\tilde{a}_{k+1}, \quad k = 1, 2, ..d-1 \tag{10}$$

From Equation (9), $\tilde{a}_0$ is obtained in terms of $\tilde{b}_1 = \tilde{a}_1/\tilde{a}_0$. By setting $k = 1$ in Equation (10), $\tilde{b}_1$ in terms of $\tilde{b}_2 = \tilde{a}_2/\tilde{a}_1$. Proceeding term by term, we obtain the continued fraction form for $\tilde{a}_0$:

$$\tilde{a}_0(z) = 1/(z + \Delta_1/(z + ... + \Delta_{d-1}/z))..) \tag{11}$$

If the hypersurface is determined, the continued fraction may be summable. By taking L_z^{-1} on Equation (11), we can obtain $a_0(t)$:

$$a_0(t) = 1/2\pi i \int_c \tilde{a}_0(z)e^{zt}dz, \quad Re \ z > 0 \tag{12}$$

where by c, we mean that the contour is to be on the right of all singularities contained in the rhs of Equation (11). If $a_0(t)$ is thus determined, the rest of the basis functions can be obtained one by one by

RR II. Hence, $A(t)$ (see Equation (3)) is completed solved if formally. This recurrence relation analysis can be implemented for a harmonic chain, described in Section 3.

3. Local Dynamics in a Harmonic Chain

Consider a classical harmonic chain of N equal masses in periodic boundary conditions (N even number, m mass and κ the coupling constant) defined by the Hamiltonian:

$$H = \sum_{-N/2}^{N/2-1} p_i^2/2m \; + \; 1/2\kappa \; (q_i - q_{i+1})^2 \tag{13}$$

where p_i and q_i are the momentum and the coordinate of mass m at site i, and sites $-N/2$ and $N/2-1$ are nns. Let $A = p_0$ the momentum of mass m at Site 0. The time evolution of p_0 follows from the method of recurrence relations: in units $m = \kappa = 1$,

$$p_0(t) = a_0(t) \; p_0 \; + \; a_1(t) \; ((q_{-1} + q_1)/2 \; - \; q_0) \; + \; a_2(t) \; (p_{-1} + p_1) + ... \tag{14}$$

Let HC denote a harmonic chain of N masses defined by Equation (13). It has been shown that for HC, $d = N + 1$ and that there are N recurrants in the hypersurface [25]. If the recurrants are expressed in our dimensionless units, the hypersurface has a symmetric structure in the form: $\sigma(N = 2) = (2, 2)$, $\sigma(N = 4) = (2, 1, 1, 2)$, $\sigma(N = 6) = (2, 1, 1, 1, 1, 2)$, *etc*. We can conclude that for N oscillators (N even number), Δ_1 and $\Delta_N = 2$ and $\Delta_k = 1$, $k = 2, 3, ..N - 1$, giving a general form:

$$\sigma(N) = (2, 1, 1, , ..1, 1, 2) \tag{15}$$

If these recurrants are substituted in Equation (11), they will realize Equation (11). If $N \to \infty (d \to \infty)$,

$$\sigma = (2, 1, 1, ...) \tag{16}$$

Taking this limit breaks the front-end symmetry. Equation (11) is summable:

$$\tilde{a}_0(z) = \frac{1}{\sqrt{4 + z^2}} \tag{17}$$

By taking the inverse transform, see Equation (12), we obtain:

$$a_0(t) = J_0(2t) \tag{18}$$

where J is the Bessel function. This is a known result [26,27]. By RR II, we obtain:

$$a_k(t) = J_k(2t), \; k = 1, 2, .. \tag{19}$$

Therewith, we have obtained the complete time evolution of p_0 in an infinite HC.

Observe that $a_0(t \to \infty) = 0$. The vanishing of the autocorrelation function at $t = \infty$ is an indication of irreversibility. It is possible in a Hermitian system only by the thermodynamic limit being taken. This property is an important consideration for the ergodicity of the dynamical variable $A = p_0$, to be considered later [28].

Langevin Dynamics

The equation of motion for A may also be expressed by the generalized Langevin equation [29]:

$$d/dt \; A(t) + \int_0^t M(t - t')A(t')dt' = F(t) \tag{20}$$

where M and F are the memory function and the random force, resp. They are important quantities in many dynamical issues, most often given phenomenologically or approximately [23]. For an infinite HC, we can provide exact expressions for them.

In obtaining a continued fraction for $\tilde{a}_0(z)$, we have introduced $\tilde{b}_k = \tilde{a}_k/\tilde{a}_{k-1}$, $k = 1, 2, ..d-1$. By convolution, we can determine b_k. They are the basis functions for $\bar{S}_1$, a subspace of $\bar{S}$, spanned by f_k, $k = 1, 2, ..d-1$. They satisfy RR II with the boundary condition that $b_1(t = 0) = 1$ and $b_k(t = 0) = 0$ if $k \neq 1$, with $b_0 = 0$. The hypersurface for this subspace is the same as Equation (7) with Δ_1 removed. One can also express $\tilde{b}_1(z)$ in a continued fraction:

$$\tilde{b}_1(z) = 1/(z + \Delta_2/(z + \Delta_3/(z + .. + \Delta_{d-1}/z))..) \tag{21}$$

The random force is a vector in $\bar{S}_1$; thus,

$$F(t) = \sum b_k(t) f_k \tag{22}$$

and:

$$M(t) = \Delta_1 b_1(t) \tag{23}$$

For the infinite HC, $\sigma_1 = (1, 1, 1, ..)$, summable to:

$$\tilde{b}_1(z) = 1/2 \; (\sqrt{z^2 + 4} - z) \tag{24}$$

By the inverse Laplace transform, we obtain:

$$b_1(t) = J_1(2t)/t \tag{25}$$

and the rest by RR II. Therewith, we have obtained exact expressions for the two Langevin quantities.

4. Dispersion Relation for Harmonic Chain

Equation (11) for $\tilde{a}_0$ shows that if d the dimensionality of $\bar{S}$ is finite, the continued fraction may be expressed as a ratio of two polynomials in z. For HC, let us denote the lhs of Equation (11) by $\ddot{\Psi}_N(z)$ and the rhs of Equation (11) the continued fraction by two polynomials as:

$$\ddot{\Psi}_N(z) = P_N(z)/Q_N(z) \tag{26}$$

Since every Q_N is found to contain $z(z^2 + 4)$ as a common factor, we express it as:

$$Q_N = z(z^2 + 4)q_N, \; N = 2, 4, 6, .. \tag{27}$$

Below, we list $P's$ and $q's$ for several values of N, sufficient to draw a general conclusion therefrom:

(a) $N = 2$, $\sigma = (2, 2)$
$P_2 = z^2 + 2$,
$q_2 = 1$

(b) $N = 4$, $\sigma = (2, 1, 1, 2)$
$P_4 = z^4 + 4z^2 + 2$
$q_4 = z^2 + 2$

(c) $N = 6$; $\sigma = (2, 1, 1, 1, 1, 2)$
$P_6 = z^6 + 6z^4 + 9z^2 + 2$
$q_6 = z^4 + 4z^2 + 3$

(d) $N = 8$; $\sigma = (2, 1, 1, 1, 1, 1, 1, 2)$
$P_8 = z^8 + 8z^6 + 20z^4 + 16z^2 + 2$

$q_8 = z^6 + 6z^4 + 10z^2 + 4$

(e) $N = 10;\ \sigma = (2, 1, 1, 1, 1, 1, 1, 1, 1, 2)$
$P_{10} = z^{10} + 10z^8 + 35z^6 + 50z^4 + 25z^4 + 2$
$q_{10} = z^8 + 8z^6 + 21z^4 + 20z^2 + 5$

(f) $N = 12;\ \sigma = (2, 1, 1, 1, 1, 1, 1, 1, 1, 1, 1, 2)$
$P_{12} = z^{12} + 12z^{10} + 54z^8 + 112z^6 + 105z^4 + 36z^2 + 2$
$q_{12} = z^{10} + 10z^8 + 36z^6 + 56z^4 + 35z^2 + 6$

If $z = 2i\sin\alpha$, $\alpha \neq 0$, the above polynomials have simple expressions for all orders of N:

$$P_N = 2\cos N\alpha \tag{28}$$

$$q_N = \sin N\alpha / \sin 2\alpha,\ \sin 2\alpha \neq 0 \tag{29}$$

4.1. Zeros of q_N

The dispersion relation can be deduced from z_k the zeros of q_N:

$$q_N(z) = \Pi(z - z_k) \tag{30}$$

From Equation (29),

$$\sin N\alpha_k = 0 \tag{31}$$

with $\sin 2\alpha_k \neq 0$ and $\alpha_k \neq 0$. Hence,

$$\alpha_k = (\pi / N)k,\ k = \pm 1, \pm 2, .. \pm (N/2 - 1) \tag{32}$$

Hence, with k given above,

$$z_k = 2i\sin\alpha_k \tag{33}$$

One may also write:

$$\Pi(z - z_k)|_{z=2i\sin\alpha} = \sin N\alpha / \sin 2\alpha \tag{34}$$

Since $Q_N = z(z^2 + 4)q_N$ (see Equation (26)), the prefactor contributes to the zeros of Q_N. They may be included in Equation (32) if the range of k is made to includes zero and $N/2$.

4.2. $a_0(t)$ for Finite N

Given the zeros of Q_N, it is now straightforward to obtain $a_0(t)$ by Equation (12). For example, if $N = 6$,

$$a_0(t) = 1/6[1 + 2\cos t + 2\cos\sqrt{3}t + \cos 2t] \tag{35}$$

A general expression would be:

$$a_0(t) = 1/N \sum_k \cos\omega_k t \tag{36}$$

where:

$$\omega_k = 2|\sin(\pi\nu_k)|, \nu_k = k/N \tag{37}$$

$k = -N/2, ., -1, 0, 1, ..N/2$. Since Equation (36) is a dispersion relation, $\nu's$ will be termed "wave vectors".

4.3. $a_0(t)$ When $N \to \infty$

If $N \to \infty$, the sum in Equation (36) may be converted to an integral:

$$rhs \ of \ Equation(36) = 1/2\pi \int_{-\pi}^{\pi} e^{2itsin\theta} d\theta \tag{38}$$

The rhs of Equation (38) is an integral representation of $J_0(2t)$. Hence, $a_0(t) = J_0(2t)$, the same as Equation (18).

It is worth noting here that the zeros of $J_0(2t)$ can thus be obtained from Equation (36) by taking $N \to \infty$ by the condition:

$$\omega_k t = \pi/2(2n+1), \quad n = 0, 1, 2, .. \tag{39}$$

If we write $J_0(2t) = \Pi(2t - 2t_k)$, by Equation (37):

$$2t_k = \pi(2n+1)/|2sin\pi k/N|, \ k/N = (-1/2, 1/2) \tag{40}$$

Evidently, there are infinitely many zeros in J_0 [30]. This result will be significant in Section 6.

4.4. $\tilde{a}_0(z) = \Psi_N(z)$ When $N \to \infty$

By Equations (26)–(29),

$$\check{\Psi}_N(z) = VcosN\alpha/sinN\alpha \tag{41}$$

where $V = 2sin2\alpha/(z(z^2+4)) = d\alpha/dz$ *(by $z = 2isin\alpha$)*. Furthermore:

$$cosN\alpha/sinN\alpha = 1/N \ d/d\alpha(logsinN\alpha)$$
$$= 1/N \ d/d\alpha[log(sinN\alpha/sin2\alpha) + logsin2\alpha] \tag{42}$$

The second term on the rhs of Equation (42) may be dropped if $N \to \infty$. For the first term, by Equations (28) and (29),

$$rhs \ of \ Equation(42) = dz/d\alpha \ d/dz \ log\Pi(z - z_k) = dz/d\alpha \ \sum 1/(z - z_k) \tag{43}$$

The prefactor $dz/d\alpha = 1/V$. Since $N \to \infty$, we can convert the above sum into an integral: writing $\check{\Psi} = \check{\Psi}_N, N \to \infty$,

$$\Psi(z) = \frac{1}{\pi} \int_{-\pi/2}^{\pi/2} \frac{d\theta}{z - 2isin\theta} = \frac{1}{\sqrt{4+z^2}} \tag{44}$$

The above result is the same as Equation (17).

The asymptotic results Equations (16) and (17) were obtained by taking the $N \to \infty$ limit first on the hypersurface. What is shown in Section 4 is that the same results are also obtained from finite N solutions for $a_0(t)$.

5. Ergodicity of Dynamical Variable $A = p_0$

If A is a variable of a Hermitian system of N particles, $N \to \infty$, it is possible to determine whether it is ergodic. According to the ergometric theory of the ergodic hypothesis [31], A is ergodic if $W_A \neq 0$ or ∞, where:

$$W_A = \int_0^{\infty} r_A(t) \ dt \tag{45}$$

where $r_A(t) = (A(t), A)/(A, A) = a_0(t)$, the autocorrelation function of A. By Equation (12),

$$W_A = \tilde{r}_A(z = 0) \tag{46}$$

If $d \to \infty$ as $N \to \infty$, which is the case of HC, $z \to 0$ on Equation (11) yields an infinite product of the following form:

$$W_A = \frac{\Delta_2 \times \Delta_4 \times \ldots \Delta_{2n}}{\Delta_1 \times \Delta_3 \times \ldots \Delta_{2n+1}}, \quad n \to \infty \tag{47}$$

Ordinarily, infinite products are difficult to evaluate, as they seem to require product rules that differ from those for finite products. However, they can be determined by Equation (45) or Equation (46) as illustrated below.

5.1. Infinite Harmonic Chain

If $A = p_0$ of HC, we can determine whether A is ergodic by evaluating Equations (45)–(47). If $N \to \infty$, $\sigma = (2, 1, 1, \ldots)$ (see (16)), and $\Psi(t) = J_0(2t)$ (see Equation (18)). Hence, by Equation (45), $W_A = 1/2$.

It was shown that $\tilde{\Psi}(z) = 1/\sqrt{z^2 + 4}$; see Equation (17). Hence, by Equation (46), $W_A = 1/2$. Finally, by σ, we can write down the infinite product:

$$W_A = \frac{1 \times 1 \times 1 \times \ldots}{2 \times 1 \times 1 \times \ldots} = \frac{1}{2} \tag{48}$$

in agreement with the previous results. As noted above, computing infinite products is a delicate matter. The order of terms in an infinite product may not be altered, nor the terms themselves. In Equation (48), such a nicety did not enter since all elements are one but one. Compare with another example in Section 5.2 below.

5.2. Infinite Harmonic Chain with One End Attached to a Wall

We shall now change HC defined by Equation (13) slightly. Let the coupling between the oscillators at q_{-2} and q_{-1} be cut. Furthermore, let the mass of the oscillator at q_{-1} be infinitely heavy, so that the oscillator at q_0 is attached as if to a wall. The rest of the chain is unchanged. The oscillators in this new configuration are labeled $0, 1, 2 .. N - 1$, with one end attached to a wall and the other end free. Finally, let $N \to \infty$.

If $A = p_0$, the recurrants are found to have the following form [27,32]:
$\Delta_1 = 2/1$, $\Delta_3 = 3/2$, $\Delta_5 = 4/3$, .. $\Delta_2 = 1/2$, $\Delta_4 = 2/3$, $\Delta_6 = 3/4$, ...

Evidently, they may be put in the form: $\Delta_{2n-1} = (n + 1)/n$ and $\Delta_{2n} = n/(n + 1)$, $n = 1, 2, 3, \ldots$ These recurrants imply that for $A = p_0$ [27,32],

$$a_0(t) = J_0(2t) - J_4(2t) \tag{49}$$

$$\tilde{a}_0(z) = 1/\sqrt{(z^2 + 4)} \, [1 - 1/16 \, (\sqrt{z^2 + 4} - z)^4] \tag{50}$$

By Equation (47),

$$W_A = \frac{1/2 \times 2/3 \times 3/4 \times \ldots \times n/(n + 1)}{2/1 \times 3/2 \times 4/3 \times \ldots \times (n + 1)/n}, \quad n \to \infty \tag{51}$$

Each term in the numerator is less than one, while each term in the denominator greater than one. If the terms and the order are preserved, $W_A \to 0$. By Equations (45) and (46), it may be tested using Equations (49) and (50). In both cases, we obtain $W_A = 0$ verifying the infinite product.

Since $W_A = 0$, $A = p_0$ is not ergodic in this chain. For this variable, the phase space is not transitive. If mass at Site 0 is slightly perturbed, the perturbed energy is not delocalized everywhere [33].

6. Harmonic Chain and Logistic Map

The logistic map (LM) is sometimes called the Ising model of chaos for being possibly the simplest model exhibiting chaos [34]. If x is a real number in an interval (0,1), the map is defined by:

$$f(x) = ax(1 - x), \quad x = (0, 1) \tag{52}$$

where a is a control parameter, a real number limited to $1 < a \leq 4$. Thus, the map is real and bounded as x. If there exists $x = x^*$, such that $f(x^*) = x^*$, it is termed a fixed point of $f(x)$. If f^n is an n-fold nested function of f, i.e., $f^n(x) = f(f^{n-1}) = f(f(...f(x)...))$, with $f^1 \equiv f$, there may be fixed points for $f^n : f^n(x^*) = x^*$. The values of the fixed points and the number of the fixed points will depend on the size of the control parameter a.

If $a < 3$, there is only one fixed point for any n. There is a remarkable theorem due to Sharkovskii [35] on 1d continuous maps on the interval, such as LM. As applied to this map, this theorem says that if $a \geq 1 + \sqrt{8}$, there are infinitely many fixed points as $n \to \infty$. This implies that a trajectory starting from almost any point in (0,1) is chaotic. At $a = 4$ (the largest possible value), the fixed points fill the interval $x = (0, 1)$ densely with a unique distribution ρ_x, $\int \rho_x dx = 1$. This distribution is known as the invariant density of fixed points, first deduced by Ulam [36,37]:

$$\rho_x = \frac{1}{\pi\sqrt{x(1 - x)}}, \quad 0 < x < 1 \tag{53}$$

The invariant density refers to the spectrum of fixed points in (0,1). The square-root singularity in Equation (53), a branch cut from 0–1, indicates that the spectrum is dense. If μ is a Lebesgue measure, $d\mu(x) = \rho_x dx$. Hence, $\mu = 1$.

We wish to see whether ρ_x, a distribution of fixed points, bears a relationship to ρ_ω, the power spectrum of frequencies in HC. For this purpose, consider the following transformations of variables:

$$x = 1/2 + 1/4\, \omega \tag{54}$$

and:

$$\rho_x dx = \rho_\omega d\omega \tag{55}$$

By substituting Equation (54) in (53), we obtain by Equation (55):

$$\rho_\omega = \frac{1}{\pi\sqrt{4 - \omega^2}}, \quad -2 < \omega < 2 \tag{56}$$

$$= 0 \text{ if otherwise.}$$

For an infinite HC, $\tilde{a}_0(z = i\omega) = \pi\rho_\omega$. By Equation (17), or Equation (44), the rhs of Equation (56) is precisely the power spectrum for $A = p_0$. Equation (56) shows that the fixed points of LM at $a = 4$ (LM_4) correspond to the frequencies of HC.

Since the frequencies in the power spectrum are positive quantities, let us express Equation (54) as:

$$\omega = 2|1 - 2x|, \quad 0 < x < 1 \tag{57}$$

For LM_4,

$$x = \sin^2 \pi y/2 \tag{58}$$

$$y/2 = l/(2N + 1), \quad l = 1, 2.., N \tag{59}$$

y being the pre-fixed points of x the fixed points. If Equation (59) is substituted in Equation (57) and y replaced by $v + 1/2$:

$$\omega = 2|sin\pi v| \tag{60}$$

The above is identical to Equation (37), the dispersion relation for HC. In the limit $N \to \infty$, both v and y lie in the same interval $(-1/2, 1/2)$. This property shows that the pre-fixed points of LM_4 also correspond to the wave vectors of HC.

The correspondence between x and ω and also between y and v indicate that the iteration dynamics of LM_4 and the time evolution in HC are isomorphic in their local variables. This implies that if a variable in HC is ergodic, a corresponding variable in LM_4 is also ergodic. If the trajectory of an initial value in LM_4 is chaotic, we must also conclude that the trajectory of a local variable in HC must also be chaotic.

Chaos in HC? Let us first examine chaos in LM_4. According to Sharkovskii, chaos is implied where there are infinitely many periods. By our work, they form a set of uncountable pre-fixed points of Lebesgue measure 1. This results in an aleph cycle, which can never return to the initial point [34]. In an infinite HC, there are also infinitely many periods. See Equation (40). Thus, the HC has the necessary and possibly sufficient property for chaos.

In an infinite HC attached to a wall (see Section 5.2), there is chaos also, as there are infinitely many periods. However, as was already shown, its variables are not ergodic. This indicates that ergodicity is a subtler property than chaos. In a continuous map, there may be chaos, but not ergodicity.

7. Concluding Remarks

In this work, we have dwelt with the dynamics of a monatomic chain with which to illustrate some of the finer points of the dynamics contained in it. This simplest of harmonic chains can be made richer in a variety of ways. One can make one oscillator to have a different mass than its neighbors [25]. It would be a model for an impurity or a defect. One could make it a periodic diatomic chain [8] or even an aperiodic diatomic chain [8]. We are providing a list of recent advances made by the method of recurrence relations on others [38–44]. For related studies on HC by Fokker–Planck dynamics and non-exponential decay, see [7,45,46].

Acknowledgments: I thank Joao Florencio for having kindled my interest in the dynamics of harmonic chains through our collaboration in the 1980s. I thank the University of Georgia Franklin College for supporting my research through the regents professorship. This work is dedicated to the memory of Bambi Hu.

Conflicts of Interest: The author declares no conflict of interest.

References

1. Mazur, P.; Montroll, E. Poincaré cycles, ergodicity, and irreversibility in assemblies of coupled harmonic oscillators. *J. Math. Phys.* **1960**, *1*, 70–84.
2. Lee, M.H. Solutions of the generalized Langevin equation by a method of recurrence relations. *Phys. Rev. B* **1982**, *26*, 2547–2551.
3. Pires, A.S.T. The memory function formalism in the study of the dynamics of a many body system. *Helv. Phys. Acta* **1988**, *61*, 988.
4. Viswanath, V.S.; Mueller, G. *Recursion Method*; Springer-Verlag: Berlin, Germany, 1994.
5. Balucani, U.; Lee, M.H.; Tognetti, V. Dynamical correlations. *Phys. Rep.* **2003**, *373*, 409–492.
6. Mokshin, A.V. Self-consistent approach to the description of relaxation processes in classical multiparticle systems. *Theory Math. Phys.* **2015**, *183*, 449–477.
7. Sen, S. Solving the Liouville equation for conservative systems: Continued fraction formalism and a simple application. *Phys. A* **2006**, *360*, 304–324.
8. Kim, J.; Sawada, I. Dynamics of a harmonic oscillator on the Bethe lattice. *Phys. Rev. E* **2000**, *61*, R2172–R2175.
9. Sawada, I. Dynamics of the S = 1/2 alternating chains at T = ∞. *Phys. Rev. Lett.* **1999**, *83*, 1668–1671.

10. Sen, S. Exact solution of the Heisenberg equation of motion for the surface spin in a semi-infinite S=1/2 XY chain at infinite temperatures. *Phys. Rev. B* **1991**, *44*, 7444–7450.

11. Florencio, J.; Sá Barreto, F.C.S. Dynamics of the random one-dimensional transverse Ising model. *Phys. Rev. B* **1999**, *60*, 9555–9560.

12. Silva Nunez, M.E.; Florencio, J. Effects of disorder on the dynamics of the XY chain. *Phys. Rev. B* **2003**, *68*, 144061–114065.

13. Daligault, J.; Murillo, M.S. Continued fraction matrix representation of response functions in multicomponent systems. *Phys. Rev. E* **2003**, *68*, 154011–154014.

14. Mokshin, A.V.; Yulmatyev, R.M.; Hanggi, P. Simple measure of memory for dynamical processes described by a generalized langevin equation. *Phys. Rev. Lett.* **2005**, *95*, 200601.

15. Hong, J.; Kee, H.Y. Analytic treatment of Mott-Hubbard transition in the half-filled Hubbard model and its thermodynamics. *Phys. Rev. B* **1995**, *52*, 2415–2421.

16. Liu, Z.-Q.; Kong, X.-M.; Chen, X.-S. Effects of Gaussian disorder on the dynamics of the random transverse Ising model. *Phys. Rev. B* **2006**, *73*, 224412.

17. Chen, X.-S.; Shen, Y.-Y.; Kong, X.-M. Crossover of the dynamical behavior in two-dimensional random transverse Ising model. *Phys. Rev. B* **2010**, *82*, 174404.

18. De Mello Silva, E. Time evolution in a two-dimensional ultrarelativistic-like electron gas by recurrence relations method. *Acta Phys. Pol. B* **2015**, *46*, 1135–1141.

19. De Mello Silva, E. Dynamical class of a two-dimensional plasmonic Dirac system. *Phys. Rev. E* **2015**, *92*, 042146.

20. Guimaraes, P.R.C.; Plascak, J.A.; de Alcantara Bonfim, O.F.; Florencio, J. Dynamics of the transverse Ising model with next-nearest-neighbor interactions. *Phys. Rev. E* **2015**, *92*, 042115.

21. Sharma, N.L. Response and relaxation of a dense electron gas in D dimensions at long wavelengths. *Phys. Rev. B* **1992**, *45*, 3552–3556.

22. Lee, M.H. Can the velocity autocorrelation function decay exponentially? *Phys. Rev. Lett.* **1983**, *51*, 1227–1230.

23. Kubo, R. The fluctuation-dissipation theorem. *Rep. Prog. Phys.* **1966**, *29*, 255–284.

24. Lee, M.H. Orthogonalization process by recurrence relations. *Phys. Rev. Lett.* **1982**, *49*, 1072–1075.

25. Lee, M.H.; Florencio, J., Jr.; Hong, J. Dynamic equivalence of a two-dimensional quantum electron gas and a classical harmonic oscillator chain with an impurity mass. *J. Phys. A* **1989**, *22*, L331–L335.

26. Fox, R.F. Long-time tails and diffusion. *Phys. Rev. A* **1983**, *27*, 3216–3233.

27. Florencio, J., Jr.; Lee, M.H. Exact time evolution of a classical harmonic-oscillator chain. *Phys. Rev. A* **1985**, *31*, 3231–3236.

28. Lee, M.H. Why Irreversibility is not a sufficient condition for ergodicity. *Phys. Rev. Lett.* **2007**, *98*, 190601.

29. Lee, M.H. Derivation of the generalized Langevin equation by a method of recurrence relations. *J. Math. Phys.* **1983**, *24*, 2512–2514.

30. Watson, G.N. *A Treatise on the Theory of Bessel Functions*; Cambridge U.P.: London, UK, 1980; Chapter 15.

31. Lee, M.H. Ergodic theory, infinite products, and long time behavior in Hermitian models. *Phys. Rev. Lett.* **2001**, *87*, 250601/1–250601/4.

32. Pestana Marino, E. Ph.D. Thesis, University of Georgia, Athens, GA, USA, 2011, unpublished.

33. Lee, M.H. Birkhoff's theorem, many-body response functions, and the ergodic condition. *Phys. Rev. Lett.* **2007**, *98*, 110403.

34. Lee, M.H. Solving for the fixed points of 3-cycle in the logistic map and toward realizing chaos by the theorems of Sharkovskii and Li-Yorke. *Commu. Theor. Phys.* **2014**, *62*, 485–496.

35. Sharkovskii, A.N. Coexistence of cycles of a continuous transformation of a line into itself. *Ukrainian Math. J.* **1964**, *16*, 61–71 (in Russian); *English transt.*: *Int. J. Bifurc. Chaos* **1995**, *5*, 1363–1273.

36. Ulam, S.M. *A Collection of Mathematical Problems*; Interscience: New York, NY, USA, 1960; pp. 73–74.

37. Lee, M.H. Cyclic solutions in chaos and the Sharkowskii theorem. *Acta Phys. Pol. B* **2012**, *43*, 1053–1063.

38. Yu, M.B. Momentum autocorrelation function of Fibonacci chains with finite number oscillators. *Eur. J. Phys. B* **2012**, *85*, 379.

39. Yu, M.B. Momentum autocorrelation function of a classical oscillator chain with alternating masses. *Eur. J. Phys. B* **2013**, *86*, 57.

40. Yu, M.B. Momentum autocorrelation function of an impurity in a classical oscillator chain with alternating masses - I. General theory. *Phys. A* **2014**, *398*, 252–263.

41. Yu, M.B. Momentum autocorrelation function of an impurity in a classical oscillator chain with alternating masses II. Illustrations. *Phys. A* **2015**, *438*, 469–486.

42. Yu, M.B. Momentum autocorrelation function of an impurity in a classical oscillator chain with alternating masses III. Some limiting cases. *Phys. A* **2016**, *447*, 411–421.

43. Wierling, A.; Sawada, I. Wave-number dependent current correlation for a harmonic oscillator. *Phys. Rev. E* **2010**, *82*, 051107.

44. Wierling, A. Dynamic structure factor of linear harmonic chain - A recurrence relation approach. *Eur. J. Phys. B* **2012**, *85*, 20.

45. Vitali, D.; Grigolini, P. Subdynamics, Fokker-Planck equation, and exponential decay of relaxation processes. *Phys. Rev. A* **1989**, *39*, 1486–1499.

46. Grigolini, P. *Quantum Mechanical Irreversibility and Measurement*; World Scientific: Singapore, Singapore, 1993.

© 2016 by the author. Licensee MDPI, Basel, Switzerland. This article is an open access article distributed under the terms and conditions of the Creative Commons Attribution (CC BY) license (http://creativecommons.org/licenses/by/4.0/).

symmetry

MDPI

Essay

Old Game, New Rules: Rethinking the Form of Physics

Christian Baumgarten

5244 Birrhard, Switzerland; christian-baumgarten@gmx.net

Academic Editor: Young Suh Kim
Received: 26 February 2016; Accepted: 28 April 2016; Published: 6 May 2016

Abstract: We investigate the modeling capabilities of sets of coupled *classical harmonic oscillators* (CHO) in the form of a modeling game. The application of the simple but restrictive rules of the game lead to conditions for an isomorphism between Lie-algebras and real Clifford algebras. We show that the correlations between two coupled classical oscillators find their natural description in the Dirac algebra and allow to model aspects of special relativity, inertial motion, electromagnetism and quantum phenomena including spin in one go. The algebraic properties of Hamiltonian motion of low-dimensional systems can generally be related to certain types of interactions and hence to the dimensionality of emergent space-times. We describe the intrinsic connection between phase space volumes of a 2-dimensional oscillator and the Dirac algebra. In this version of a phase space interpretation of quantum mechanics the (components of the) spinor wavefunction in momentum space are abstract canonical coordinates, and the integrals over the squared wave function represents second moments in phase space. The wave function in ordinary space-time can be obtained via Fourier transformation. Within this modeling game, 3+1-dimensional space-time is interpreted as a structural property of electromagnetic interaction. A generalization selects a series of Clifford algebras of specific dimensions with similar properties, specifically also 10- and 26-dimensional real Clifford algebras.

Keywords: Hamiltonian mechanics; coupled oscillators; Lorentz transformation; Dirac equation

PACS: 45.20.Jj, 47.10.Df, 41.75, 41.85, 03.65.Pm, 05.45.Xt, 03.30.+p, 03.65.-w, 29.27.-a

1. Introduction

D. Hestenes had the joyful idea to describe physics as a modeling game [1]. We intend to play a modeling game with (ensembles of) classical harmonic oscillators (CHO). The CHO is certainly one of the most discussed and analyzed systems in physics and one of the few exactly solveable problems. One would not expect any substantially new discoveries related to this subject. Nevertheless there are aspects that are less well-known than others. One of these aspects concerns the transformation group of the symplectic transformations of n coupled oscillators, $Sp(2n)$. We invite the reader to join us playing "a modeling game" and to discover some fascinating features related to possible reinterpretations of systems of two (or more) coupled oscillators. We will show that special relativity can be reinterpreted as a transformation theory of the second moments of the abstract canonical variables of coupled oscillator systems (The connection of the Dirac matrices to the symplectic group has been mentioned by Dirac in Reference [2]. For the connection of oscillators and Lorentz transformations (LTs) see also the papers of Kim and Noz [3–5] and references therein. The use of CHOs to model quantum systems has been recently described-for instance-by Briggs and Eisfeld [6–8]). We extend the application beyond pure LTs and show that the Lorentz force can be reinterpreted by the second moments of two coupled oscillators in proper time. Lorentz transformations can be modeled as symplectic transformations [4]. We shall show how Maxwell's equations find their place within the game.

The motivation for this game is to show that many aspects of modern physics can be understood on the basis of the classical notions of harmonic oscillation if these notions are appropriately reinterpreted.

In Section 2 we introduce the rules of our game, in Section 3 we introduce the algebraic notions of the Hamilton formalism. In Section 4 we describe how geometry emerges from coupled oscillator systems, in Section 5 we describe the use of symplectic transformations and introduce the Pauli- and Dirac algebra. In Section 6 we introduce a physical interpretation of oscillator moments and in Section 7 we relate the phase space of coupled oscillators to the real Dirac algebra. Section 8 contains a short summary.

2. The Rules Of The Game

The first rule of our game is the principle of reason (POR): *No distinction without reason*-we should not add or remove something *specific* (an asymmetry, a concept, a distinction) from our model without having a clear and explicit reason. If there is no reason for a specific asymmetry or choice, then all possibilities are considered equivalently.

The second rule is the principle of variation (POV): We postulate that change is immanent to all fundamental quantities in our game. From these two rules, we take that the mathematical object of our theory is a list (n-tuple) of quantities (variables) ψ, each of which varies at all times.

The third rule is the principle of *objectivity* (POO): Any law within this game refers to measurements, defined as comparison of quantities (object properties) in relation to other object properties of the same type (*i.e.*, unit). Measurements require reference standards (rulers). A measurement is objective if it is based on (properties of) the objects of the game. This apparent self-reference is unavoidable, as it models the *real* situation of physics as experimental science. Since all fundamental objects (quantities) in our model *vary at all times*, the only option to construct a constant quantity that might serve as a ruler, is given by *constants of motion* (COM). Hence the principle of objectivity requires that measurement standards are derived from constants of motion.

This third rule implies that the fundamental variables can not be directly measured, but only functions of the fundamental variables of the same dimension (unit) of a COM. Thus the model has two levels: The level of the fundamental variable list ψ, which is experimentally not directly accessible and a level of *observables* which are (as we shall argue) even moments of the fundamental variables ψ.

2.1. Discussion of the Rules

E.T. Jaynes wrote that "Because of their empirical origins, QM and QED are not physical theories at all. In contrast, Newtonian celestial mechanics, Relativity, and Mendelian genetics are physical theories, because their mathematics was developed by reasoning out the consequences of clearly stated physical principles from which constraint the possibilities". And he continues "To this day we have no constraining principle from which one can deduce the mathematics of QM and QED; [...] In other words, the mathematical system of the present quantum theory is [...] unconstrained by any physical principle" [9]. This remarkably harsh criticism of quantum mechanics raises the question of what we consider to be a physical principle. Are the rules of our game physical principles? We believe that they are no substantial physical principles but *formal* first principles, they are *preconditions* of a sensible theory. They contain no immediate physical content, but they define the *form* or the *idea* of physics.

It is to a large degree immanent to science and specifically to physics to presuppose the existence of *reason*: Apples do not fall down by chance—there is a reason for this tendency. Usually this believe in reason implies the believe in causality, *i.e.*, that we can also (at least in principle) explain why a specific apple falls at a specific time, but practically this latter believe can rarely be confirmed experimentally and therefore remains to some degree metaphysical. Thus, if, as scientists, we postulate that things have reason, then this is not a *physical* principle but a precondition, a first principle.

The second rules (POV), is specific to the form (or idea) of physics, e.g., that it is the *sense* of physics to *recognize the pattern* of motion and to *predict future*. Therefore the notion of time in the form of change is indeed immanent to the physical description of reality.

The principle of objectivity (POO) is immanent to the very idea of physics: A measurement is the comparison of properties of objects with compatible properties of reference objects, e.g., requires "constant" rulers. Hence the rules of the game are to a large degree unavoidable: They follow from the very form of physics and therefore certain laws of physics are not substantial results of a physical theory. For instance a consistent "explanation" of the stability of matter is impossible *as we presumed it already within the idea of measurement*. More precisely: if this presumption does *not* follow within the framework of a physical theory, then the theory is flawed, since it can not reproduce it's own presumptions.

Einstein wrote with respect to relativity that "It is striking that the theory (except for the four-dimensional space) introduces two kinds of things, *i.e.*, (1) measuring rods and clocks; (2) all other things, e.g., the electromagnetic field, the material point, *etc*. This, in a certain sense, is inconsistent; strictly speaking, measuring rods and clocks should emerge as solutions of the basic equations [...], not, as it were, as theoretically self-sufficient entities". [10]. The more it may surprise that the stability of matter can not be obtained from classical physics as remarked by Elliott H. Lieb: "A fundamental paradox of classical physics is why matter, which is held together by Coulomb forces, does not collapse" [11]. This single sentence seems to rule out the possibility of a fundamental classical theory and uncovers the uncomfortable situation of theoretical physics today: Despite the overwhelming experimental and technological success, there is a deep-seated confusion concerning the theoretical foundations. Our game is therefore a meta-experiment. It is not the primary goal to find "new" laws of nature or new experimental predictions, but it is a conceptional "experiment" that aims to further develop our understanding of the consequences of principles: which ones are *really* required to derive central "results" of contemporary physics. In this short essay final answers can not be given, but maybe some new insights are possible.

2.2. What about Space-Time?

A theory has to make the choice between postulate and proof. If a 3+1 dimensional space-time is presumed, then it cannot be proven within the same theoretical framework. More precisely, the value of such a proof remains questionable. This is a sufficient reason to avoid postulates concerning the dimensionality of space-time. Another, even stronger, reason to avoid a direct postulate of space-time and its geometry has been given above: The fundamental variables that we postulated, can not be directly measured. This excludes space-time coordinates as primary variables (which can be directly measured), but with it almost all other apriori assumed concepts like velocity, acceleration, momentum, energy and so on. At some point these concepts certainly have to be introduced, but we suggest an approach to the formation of concepts that differs from the Newtonian axiomatic method. The POR does not allow to introduce distinctions between the fundamental variables into coordinates and momenta without reason. Therefore we are forced to use an interpretational method, which one might summarize as *function follows form*. We shall first derive equations and then we shall interpret the equations according to some formal criteria. This implies that we have to refer to already existing notions if we want to identify quantities according to their appearance within a certain formalism. The consequence for the game is, that we have to show how to give rise to *geometrical* notions: If we do not postulate space-time then we have to suggest a method to construct it.

A consequence of our conception is that both, objects and fields have to be identified with dynamical structures, as there is simply nothing else available. This fits to the framework of structure preserving (symplectic) dynamics that we shall derive from the described principles.

3. Theory of Small Oscillations

In this section we shall derive the theory of coupled oscillators from the rules of our game. According to the POO there exists a function (COM) $\mathcal{H}(\psi)$ such that (Let us first (for simplicity) assume that $\frac{\partial \mathcal{H}}{\partial t} = 0$.):

$$\frac{d\mathcal{H}}{dt} = \sum_k \frac{\partial \mathcal{H}}{\partial \psi_k} \dot{\psi}_k = 0 \tag{1}$$

or in vector notation

$$\frac{d\mathcal{H}}{dt} = (\nabla_\psi \mathcal{H}) \cdot \dot{\psi} = 0 \tag{2}$$

The simplest solution is given by an arbitrary skew-symmetric matrix $\mathcal{X}$:

$$\dot{\psi} = \mathcal{X} \nabla_\psi \mathcal{H} \tag{3}$$

Note that it is only the *skew-symmetry* of $\mathcal{X}$, which ensures that it is always a solution to Equation (2) and which ensures that $\mathcal{H}$ is constant. If we now consider a state vector ψ of dimension k, then there is a theorem in linear algebra, which states that for *any* skew-symmetric matrix $\mathcal{X}$ there exists a non-singular matrix $\mathbf{Q}$ such that we can write [12]:

$$\mathbf{Q}^T \mathcal{X} \mathbf{Q} = \text{diag}(\eta_0, \eta_0, \eta_0, \dots, 0, 0, 0) \tag{4}$$

where η_0 is the matrix

$$\eta_0 = \begin{pmatrix} 0 & 1 \\ -1 & 0 \end{pmatrix} \tag{5}$$

If we restrict us to orthogonal matrices $\mathbf{Q}$, then we may still write

$$\mathbf{Q}^T \mathcal{X} \mathbf{Q} = \text{diag}(\lambda_0 \eta_0, \lambda_1 \eta_0, \lambda_2 \eta_0, \dots, 0, 0, 0) \tag{6}$$

In both cases we may leave away the zeros, since they correspond to non-varying variables, which would be in conflict with the second rule of our modeling game. Hence $k = 2n$ must be even and the square matrix $\mathcal{X}$ has the dimension $2n \times 2n$. As we have no specific reason to assume asymmetries between the different degrees of freedom (DOF), we have to choose all $\lambda_k = 1$ in Equation (6) and return to Equation (4) without zeros and define the block-diagonal so-called *symplectic unit matrix* (SUM) γ_0:

$$\mathbf{Q}^T \mathcal{X} \mathbf{Q} = \text{diag}(\eta_0, \eta_0, \dots, \eta_0) \equiv \gamma_0 \tag{7}$$

These few basic rules thus lead us directly to Hamiltonian mechanics: Since the state vector has even dimension and due to the form of γ_0, we can interpret ψ as an ensemble of n classical DOF-each DOF represented by a canonical pair of coordinate and momentum: $\psi = (q_1, p_1, q_2, p_2, \dots, q_n, p_n)^T$. In this notation and after the application of the transformation $\mathbf{Q}$, Equation (3) can be written in form of the Hamiltonian equations of motion (HEQOM):

$$\begin{aligned} \dot{q}_i &= \frac{\partial \mathcal{H}}{\partial p_i} \\ \dot{p}_i &= -\frac{\partial \mathcal{H}}{\partial q_i} \end{aligned} \tag{8}$$

The validity of the HEQOM is of fundamental importance as it allows for the use of the results of Hamiltonian mechanics, of statistical mechanics and thermodynamics-but without the intrinsic presupposition that the q_i have to be understood as positions in real space and the p_i as the corresponding canonical momenta. This is legitimate as the theory of canonical transformations is *independent from any specific physical interpretation of what the coordinates and momenta represent physically.* As no other interpretation is at hand, we say that these canonical pairs are coordinates q_i, p_i in an abstract phase space and they are canonical coordinates and momenta only due to the *form* of the HEQOM. The choice of the specific form of γ_0 is for $n > 1$ DOF not unique. It could for instance be written as

$$\gamma_0 \equiv \eta_0 \otimes \mathbf{1}_{n \times n} \tag{9}$$

which corresponds a state vector of the form

$$\psi = (q_1, \ldots, q_n, p_1, \ldots, p_n,)^T$$

or by

$$\gamma_0 \equiv \mathbf{1}_{n \times n} \otimes \eta_0 \tag{10}$$

as in Equation (7). Therefore we are forced to make an arbitrary choice (But we should keep in mind, that other "systems" with a different choice are possible. If we can not exclude their existence, then they should exist as well. With respect to the form of the SUM, we suggest that different "particle" types (different types of fermions for instance) have a different SUM). But in all cases the SUM γ_0 must be skew-symmetric and have the following properties:

$$\begin{aligned}
\gamma_0^T &= -\gamma_0 \\
\gamma_0^2 &= -1
\end{aligned} \tag{11}$$

which also implies that γ_0 is orthogonal and has unit determinant. Note also that all eigenvalues of γ_0 are purely imaginary. However, once we have chosen a specific form of γ_0, we have specified a set of canonical pairs (q_i, p_i) within the state vector. This choice fixes the set of possible canonical (structure preserving) transformations.

Now we write the Hamiltonian $\mathcal{H}(\psi)$ as a Taylor series, we remove the rule-violating constant term and cut it after the second term. We do not claim that higher terms may not appear, but we delay the discussion of higher orders to a later stage. All this is well-known in the theory of small oscillations. There is only one difference to the conventional treatment: We have no direct macroscopic interpretation for ψ and following our first rule we have to write the second-order Hamiltonian $\mathcal{H}(\psi)$ in the most general form:

$$\mathcal{H}(\psi) = \frac{1}{2} \psi^T \mathcal{A} \psi \tag{12}$$

where $\mathcal{A}$ is only restricted to be *symmetric* as all non-symmetric terms *do not contribute* to $\mathcal{H}$. Since it is not unlikely to find more than a single constant of motion in systems with multiple DOFs, we distinguish systems with singular matrix $\mathcal{A}$ from those with a positive or negative definite matrix $\mathcal{A}$. Positive definite matrices are favoured in the sense that they allow to identify $\mathcal{H}$ with the amount of a substance or an amount of energy (It is immanent to the concept of substance that it is understood as something positive semidefinite).

Before we try to interprete the elements in $\mathcal{A}$, we will explore some general algebraic properties of the Hamiltonian formalism. If we plug Equations (12) into (3), then the equations of motion can be written in the general form

$$\dot{\psi} = \gamma_0 \mathcal{A} \psi = \mathbf{F} \psi \tag{13}$$

The matrix $\mathbf{F} = \gamma_0 \mathcal{A}$ is the product of the symmetric (positive semi-definite) matrix $\mathcal{A}$ and the skew-symmetric matrix γ_0. As known from linear algebra, the trace of such products is zero:

$$\mathrm{Tr}(\mathbf{F}) = 0 \tag{14}$$

Pure harmonic oscillation of ψ is described by matrices $\mathbf{F}$ with purely imaginary eigenvalues and those are the only stable solutions [12]. Note that Equation (13) may represent a tremendous amount of different types of systems-all linearly coupled systems in any dimension, chains or d-dimensional lattices of linear coupled oscillators and wave propagation (However the linear approximation does not allow for the description of the transport of heat).

One quickly derives from the properties of γ_0 and $\mathcal{A}$ that

$$\mathbf{F}^T = \mathcal{A}^T \gamma_0^T = -\mathcal{A}\,\gamma_0 = \gamma_0^2 \mathcal{A}\,\gamma_0 = \gamma_0\,\mathbf{F}\,\gamma_0 \tag{15}$$

Since any square matrix can be written as the sum of a symmetric and a skew-symmetric matrix, it is nearby to also consider the properties of products of γ_0 with a skew-symmetric real square matrices $\mathcal{B}$. If $\mathbf{C} = \gamma_0\,\mathcal{B}$, then

$$\mathbf{C}^T = \mathcal{B}^T \gamma_0^T = \mathcal{B}\,\gamma_0 = -\gamma_0^2 \mathcal{B}\,\gamma_0 = -\gamma_0\,\mathbf{C}\,\gamma_0 \tag{16}$$

Symmetric $2n \times 2n$-matrices contain $2n\,(2n+1)/2$ different matrix elements and skew-symmetric ones $2n\,(2n-1)/2$ elements, so that there are ν_s linear independent matrix elements in $\mathcal{A}$

$$\nu_s = n\,(2n+1) \tag{17}$$

and ν_c matrix elements in $\mathcal{B}$ with

$$\nu_c = n\,(2n-1) \tag{18}$$

In the theory of linear Hamiltonian dynamics, matrices of the form of $\mathbf{F}$ are known as "Hamiltonian" or "infinitesimal symplectic" and those of the form of $\mathbf{C}$ as "skew-Hamiltonian" matrices. This convention is a bit odd as $\mathbf{F}$ does not appear in the Hamiltonian and it is in general not symplectic. Furthermore the term "Hamiltonian matrix" has a different meaning in quantum mechanics - in close analogy to $\mathcal{A}$. But it is known that this type of matrix is closely connected to symplectic matrices as every symplectic matrix is a matrix exponential of a matrix $\mathbf{F}$ [12]. We consider the matrices as defined by Equations (15) and (16) as too important and fundamental to have no meaningful and unique names: Therefore we speak of a **symplex** (plural *symplices*), if a matrix holds Equation (15) and of a **cosymplex** if it holds Equation (16).

Symplectic Motion and Second Moments

So what is a symplectic matrix anyway? The concept of symplectic transformations is a specific formulation of the theory of canonical transformations. Consider we define a new state vector (or new coordinates) $\phi(\psi)$-with the additional requirement, that the transformation is reversible. Then the Jacobian matrix of the transformation is given by

$$\mathbf{J}_{ij} = \left(\frac{\partial \phi_i}{\partial \psi_j} \right) \tag{19}$$

and the transformation is said to be symplectic, if the Jacobian matrix holds [12]

$$\mathbf{J}\,\gamma_0\,\mathbf{J}^T = \gamma_0 \tag{20}$$

Let us see what this implies in the linear case:

$$\begin{aligned}
\mathbf{J}\dot{\psi} &= \mathbf{J}\mathbf{F}\mathbf{J}^{-1}\mathbf{J}\,\psi \\
\tilde{\psi} &= \mathbf{J}\,\psi \\
\dot{\tilde{\psi}} &= \mathbf{J}\mathbf{F}\mathbf{J}^{-1}\,\tilde{\psi} \\
\dot{\tilde{\psi}} &= \tilde{\mathbf{F}}\,\tilde{\psi}
\end{aligned} \tag{21}$$

and-by the use of Equation (20) one finds that $\tilde{\mathbf{F}}$ is still a symplex:

$$
\begin{aligned}
\tilde{\mathbf{F}}^T &= (\mathbf{J}^{-1})^T \, \mathbf{F}^T \, \mathbf{J}^T \\
\tilde{\mathbf{F}}^T &= (\mathbf{J}^{-1})^T \, \gamma_0 \, \mathbf{F} \, \gamma_0 \, \mathbf{J}^T \\
\tilde{\mathbf{F}}^T &= -\gamma_0^2 \, (\mathbf{J}^{-1})^T \, \gamma_0 \, \mathbf{F} \, \mathbf{J}^{-1} \, \gamma_0 \\
\tilde{\mathbf{F}}^T &= -\gamma_0 \, \mathbf{J} \, \gamma_0^2 \, \mathbf{F} \, \mathbf{J}^{-1} \, \gamma_0 \\
\tilde{\mathbf{F}}^T &= \gamma_0 \, \mathbf{J} \, \mathbf{F} \, \mathbf{J}^{-1} \, \gamma_0 \\
\tilde{\mathbf{F}}^T &= \gamma_0 \, \tilde{\mathbf{F}} \, \gamma_0
\end{aligned}
\tag{22}
$$

Hence a symplectic transformation is first of all a similarity transformation, but secondly, it preserves the structure of all involved equations. Therefore the transformation is said to be *canonical* or *structure preserving*. The distinction between canonical and non-canonical transformations can therefore be traced back to the skew-symmetry of γ_0 and the symmetry of $\mathcal{A}$- both of them consequences of the rules of our physics modeling game.

Recall that we argued that the matrix $\mathcal{A}$ should be symmetric *because* skew-symmetric terms do not contribute to the Hamiltonian. Let us have a closer look what this means. Consider the matrix of second moments Σ that can be build from the variables ψ:

$$
\Sigma \equiv \langle \psi \, \psi^T \rangle
\tag{23}
$$

in which the angles indicate some (yet unspecified) sort of average. The equation of motion of this matrix is given by

$$
\begin{aligned}
\dot{\Sigma} &= \langle \dot{\psi} \, \psi^T \rangle + \langle \psi \, \dot{\psi}^T \rangle \\
\dot{\Sigma} &= \langle \mathbf{F} \, \psi \, \psi^T \rangle + \langle \psi \, \psi^T \, \mathbf{F}^T \rangle
\end{aligned}
\tag{24}
$$

Now, as long as $\mathbf{F}$ does not depend on ψ, we obtain

$$
\begin{aligned}
\dot{\Sigma} &= \mathbf{F} \Sigma + \Sigma \, \mathbf{F}^T \\
\dot{\Sigma} &= \mathbf{F} \Sigma + \Sigma \, \gamma_0 \, \mathbf{F} \, \gamma_0 \\
(\dot{\Sigma} \, \gamma_0) &= \mathbf{F} \, (\Sigma \, \gamma_0) - (\Sigma \, \gamma_0) \, \mathbf{F} \\
\dot{\mathbf{S}} &= \mathbf{F} \, \mathbf{S} - \mathbf{S} \, \mathbf{F}
\end{aligned}
\tag{25}
$$

where we defined the new matrix $\mathbf{S} \equiv \Sigma \, \gamma_0$. For completeness we introduce the "adjunct" spinor $\bar{\psi} = \psi^T \, \gamma_0$ so that we may write

$$
\mathbf{S} = \langle \psi \, \bar{\psi} \rangle
\tag{26}
$$

Note that $\mathbf{S}$ is also a symplex. The matrix $\mathbf{S}$ (*i.e.*, all second moments) is constant, iff $\mathbf{S}$ and $\mathbf{F}$ commute.

Now we define an *observable* to be an operator $\mathbf{O}$ with a (potentially) non-vanishing expectation value, defined by:

$$
\langle \mathbf{O} \rangle \equiv \langle \bar{\psi} \mathbf{O} \psi \rangle = \langle \psi^T \, \gamma_0 \, \mathbf{O} \psi \rangle
\tag{27}
$$

Thus, if the product $\gamma_0 \, \mathbf{O}$ is *not* skew-symmetric, *i.e.*, contains a product of γ_0 with a symmetric matrix $\mathcal{B}$, then the expectation value is potentially non-zero:

$$
\langle \mathbf{O} \rangle \equiv \langle \psi^T \, \gamma_0 \, (\gamma_0 \, \mathcal{B}) \psi \rangle = -\langle \psi^T \, \mathcal{B} \, \psi \rangle
\tag{28}
$$

This means that only the symplex-part of an operator is "observable", while cosymplices yield a vanishing expectation value. Hence Equation (25) delivers the blueprint for the general definition of observables. Furthermore we find in the last line the constituting equation for Lax pairs [13]. Peter Lax has shown that for such pairs of operators $\mathbf{S}$ and $\mathbf{F}$ that obey Equation (25) there are the following constants of motion

$$
\mathrm{Tr}(\mathbf{S}^k) = \mathrm{const}
\tag{29}
$$

for arbitrary integer $k > 0$. Since $\mathbf{S}$ is a symplex and therefore by definition the product of a symmetric matrix and the skew-symmetric γ_0, Equation (29) is always zero and hence trivially true for $k = 1$. The same is true for any odd power of $\mathbf{S}$, as it can be easily shown that any odd power of a symplex is again a symplex (see Equation (35)), so that the only non-trivial general constants of motion correspond to even powers of $\mathbf{S}$, which implies that all observables are functions of even powers of the fundamental variables.

To see the validity for $k > 1$ we have to consider the general algebraic properties of the trace operator. Let λ be an arbitrary real constant and τ be a real parameter, then

$$
\begin{aligned}
\mathrm{Tr}(\mathbf{A}) &= \mathrm{Tr}(\mathbf{A}^T) \\
\mathrm{Tr}(\lambda\,\mathbf{A}) &= \lambda\,\mathrm{Tr}(\mathbf{A}) \\
\frac{d}{d\tau}\mathrm{Tr}(\mathbf{A}(\tau)) &= \mathrm{Tr}(\frac{d\mathbf{A}}{d\tau}) \\
\mathrm{Tr}(\mathbf{A}+\mathbf{B}) &= \mathrm{Tr}(\mathbf{A})+\mathrm{Tr}(\mathbf{B}) \\
\mathrm{Tr}(\mathbf{A}\,\mathbf{B}) &= \mathrm{Tr}(\mathbf{B}\,\mathbf{A})
\end{aligned}
\tag{30}
$$

It follows that

$$
\begin{aligned}
0 &= \mathrm{Tr}(\mathbf{A}\,\mathbf{B}-\mathbf{B}\,\mathbf{A}) \\
0 &= \mathrm{Tr}(\mathbf{A}^n\,\mathbf{B}-\mathbf{A}^{n-1}\,\mathbf{B}\,\mathbf{A}) \\
0 &= \mathrm{Tr}\left[\mathbf{A}^{n-1}\left(\mathbf{A}\,\mathbf{B}-\mathbf{B}\,\mathbf{A}\right)\right]
\end{aligned}
\tag{31}
$$

From the last line of Equation (31) it follows with $\frac{d\mathbf{A}}{d\tau} = \lambda\left(\mathbf{A}\,\mathbf{B}-\mathbf{B}\,\mathbf{A}\right)$

$$
\frac{d}{d\tau}\mathrm{Tr}(\mathbf{A}^n) = 0
\tag{32}
$$

Remark: This conclusion is not limited to symplices.

However for single spinors ψ and the corresponding second moments $\mathbf{S} = \Sigma\,\gamma_0 = \psi\psi^T\,\gamma_0$ we find:

$$
\begin{aligned}
\mathrm{Tr}(\mathbf{S}^k) &= \mathrm{Tr}[\psi\,\psi^T\,\gamma_0\cdots\psi\,\psi^T\,\gamma_0] \\
&= \mathrm{Tr}[\psi\,(\psi^T\,\gamma_0\cdots\psi\,\psi^T\,\gamma_0)] \\
&= \mathrm{Tr}[(\psi^T\,\gamma_0\cdots\psi\,\psi^T\,\gamma_0)\,\psi] \\
&= \mathrm{Tr}\left[(\psi^T\,\gamma_0\,\psi)\cdots(\psi^T\gamma_0\psi)\right] = 0
\end{aligned}
\tag{33}
$$

since each single factor $(\psi^T\,\gamma_0\,\psi)$ vanishes due to the skew-symmetry of γ_0. Therefore the constants of motion as derived from Equation (29) are non-zero only for even k and *after averaging over some kind of distribution* such that $\mathbf{S} = \langle\psi\psi^T\,\gamma_0\rangle$ has non-zero eigenvalues as in Equation (34) below.

The symmetric matrix $2n \times 2n$-matrix Σ (and also $\mathcal{A}$) is positive definite, if it can be written as a product $\Sigma = \Psi\Psi^T$, where Ψ is a non-singular matrix of size $2n \times m$ with $m \geq 2n$.

For $n = m/2 = 1$, the form of Ψ may be chosen as

$$
\begin{aligned}
\Psi &= \frac{1}{\sqrt{q^2+p^2}}\begin{pmatrix} q & -p \\ p & q \end{pmatrix} = \frac{1}{\sqrt{q^2+p^2}}\left(\mathbf{1}\psi, \eta_0\,\psi\right) \\
\Rightarrow \qquad \Sigma &= \Psi\Psi^T = \Psi^T\,\Psi = 1 \\
\mathbf{S} &= \gamma_0
\end{aligned}
\tag{34}
$$

so that for $k = 2$ the average of two "orthogonal" column-vectors ψ and $\eta_0\,\psi$ gives a non-zero constant of motion via Lax pairs as $\gamma_0^2 = -\mathbf{1}$.

These findings have some consequences for the modeling game. The first is that we have found constants of motion-though some of them are physically meaningful only for a non-vanishing volume in phase space, *i.e.*, by the combination of several spinors ψ. Secondly, a stable state $\dot{\mathbf{S}} = 0$ implies that the matrix operators forming the Lax pair have the same eigenvectors: a density distribution in phase space (as described by the matrix of second moments) is stable if it is adapted or *matched* to the simplex **F**. The phase space distribution as represented by **S** and the driving terms (the components of **F**) must fit to each other in order to obtain a stable "eigenstate". But we also found a clear reason, why generators (of symplectic transformations) are always observables and vice versa: Both, the generators as well as the observables are symplices of the same type. There is a one-to-one correspondence between them, not only as *generators of infinitesimal transformations*, but also algebraically.

Furthermore, we may conclude that (anti-) commutators are an essential part of "classical" Hamiltonian mechanics and secondly that the matrix **S** has the desired properties of observables: Though **S** is based on continuously varying fundamental variables, it is constant, if it commutes with **F**, and it varies otherwise (In accelerator physics, Equation (25) describes the envelope of a beam in linear optics. The matrix of second moments Σ is a covariance matrix-and therefore our modeling game is connected to probability theory exactly when observables are introduced).

Hence it appears sensible to take a closer look on the (anti-) commutation relations of (co-) symplices and though the definitions of (co-) symplices are quite plain, the (anti-) commutator algebra that emerges from them has a surprisingly rich structure. If we denote symplices by $\mathbf{S}_k$ and cosymplices by $\mathbf{C}_k$, then the following rules can quickly be derived:

$$
\left.
\begin{array}{c}
\mathbf{S}_1\,\mathbf{S}_2 - \mathbf{S}_2\,\mathbf{S}_1 \\
\mathbf{C}_1\,\mathbf{C}_2 - \mathbf{C}_2\,\mathbf{C}_1 \\
\mathbf{C}\,\mathbf{S} + \mathbf{S}\,\mathbf{C} \\
\mathbf{S}^{2n+1}
\end{array}
\right\} \;\Rightarrow\; \text{symplex}
$$

$$
\left.
\begin{array}{c}
\mathbf{S}_1\,\mathbf{S}_2 + \mathbf{S}_2\,\mathbf{S}_1 \\
\mathbf{C}_1\,\mathbf{C}_2 + \mathbf{C}_2\,\mathbf{C}_1 \\
\mathbf{C}\,\mathbf{S} - \mathbf{S}\,\mathbf{C} \\
\mathbf{S}^{2n} \\
\mathbf{C}^{n}
\end{array}
\right\} \;\Rightarrow\; \text{cosymplex} \tag{35}
$$

This *Hamiltonian* algebra of (anti-)commutators is of fundamental importance insofar as we derived it in a few steps from first principles (*i.e.*, the rules of the game) and it defines the structure of Hamiltonian dynamics in phase space. The distinction between symplices and cosymplices is also the distinction between observables and non-observables. It is the basis of essential parts of the following considerations.

4. Geometry from Hamiltonian Motion

In the following we will demonstrate the *geometrical content* of the algebra of (co-)symplices (Equation (35)) which emerges for *specific numbers of DOF n*. As shown above, pairs of canonical variables (DOFs) are a direct consequence of the abstract rules of our game. Though single DOFs are poor "objects", it is remarkable to find physical structures emerging from our abstract rules *at all*. This suggests that there might be more structure to discover when n DOF are combined, for instance geometrical structures. The following considerations obey the rules of our game, since they are based purely on symmetry considerations like those that guided us towards Hamiltonian dynamics. The objects of interest in our algebraic interpretation of Hamiltonian dynamics are matrices. The first matrix (besides $\mathcal{A}$) with a specific form that we found, is γ_0. It is a symplex:

$$
\gamma_0^T = -\gamma_0 = \gamma_0\,\gamma_0\,\gamma_0 \tag{36}
$$

According to Equation (17) there are $\nu_s = n\,(2\,n+1)$ (*i.e.*, $\nu_s \geq 3$) symplices. Hence it is nearby to ask if other symplices with similar properties like γ_0 exist-and if so, what the relations between these matrices are. According to Equation (35) the commutator of two symplices is again a symplex, while the anti-commutator is a cosymplex. As we are primarily interested in *observables* and components of the

Hamiltonian (*i.e.*, symplices), respectively, we look for further symplices that anti-commute with γ_0 and with each other. In this case, the product of two such matrices is also a symplex, *i.e.*, another potential contribution to the general Hamiltonian matrix **F**.

Assumed we had a set of N mutually anti-commuting orthogonal symplices γ_0 and γ_k with $k \in [1 \ldots N-1]$, then a Hamiltonian matrix **F** might look like

$$\mathbf{F} = \sum_{k=0}^{N-1} f_k \, \gamma_k + \ldots \tag{37}$$

The γ_k are symplices *and* anti-commute with γ_0:

$$\gamma_0 \, \gamma_k + \gamma_k \, \gamma_0 = 0 \tag{38}$$

Multiplication from the left with γ_0 gives:

$$-\gamma_k + \gamma_0 \, \gamma_k \, \gamma_0 = -\gamma_k + \gamma_k^T = 0 \tag{39}$$

so that all other possible symplices γ_k, which anticommute with γ_0, are symmetric and square to **1**. This is an important finding for what follows, as it can (within our game) be interpreted as a classical proof of the uniqueness of (observable) time-dimension: Time is one-dimensional as there is no other skew-symmetric symplex that anti-commutes with γ_0. We can choose different forms for γ_0, but the emerging algebra allows for no second "direction of time".

The second order derivative of ψ is (for constant **F**) given by $\ddot{\psi} = \mathbf{F}^2 \, \psi$ which yields:

$$\mathbf{F}^2 = \sum_{i=0}^{N-1} f_i^2 \, \gamma_i^2 + \sum_{i \neq j} f_i f_j \, (\gamma_i \, \gamma_j + \gamma_j \, \gamma_i) \tag{40}$$

Since the anti-commutator on the right vanishes by definition, we are left with:

$$\mathbf{F}^2 = \left(\sum_{k=1}^{N-1} f_k^2 - f_0^2 \right) \mathbf{1} \tag{41}$$

Thus-we find a set of (coupled) oscillators, if

$$f_0^2 > \sum_{k=1}^{N-1} f_k^2 \tag{42}$$

such that

$$\ddot{\psi} = -\omega^2 \, \psi \tag{43}$$

Given such matrix systems exist-then they generate a Minkowski type "metric" as in Equation (41) (Indeed it appears that Dirac derived his system of matrices from the this requirement [14]). The appearance of this metric shows how a Minkowski type geometry emerges from the driving terms of oscillatory motion. This is indeed possible- at least for symplices of certain dimensions as we will show below. The first thing needed is some kind of measure to define the length of a "vector". Since the length is a measure that is invariant under certain transformations, specifically under rotations, we prefer to use a quantity with certain invariance properties to define a length. The only one we have at hand is given by Equation (29). Accordingly we define the (squared) length of a matrix representing a "vector" by

$$\|\mathbf{A}\|^2 \equiv \frac{1}{2\,n} \, \mathrm{Tr}(\mathbf{A}^2) \tag{44}$$

The division by $2\,n$ is required to make the unit matrix have unit norm. Besides the norm we need a scalar product, *i.e.*, a definition of orthogonality. Consider the Pythagorean theorem which says that two vectors $\vec{a}$ and $\vec{b}$ are orthogonal iff

$$(\vec{a} + \vec{b})^2 = \vec{a}^2 + \vec{b}^2 \tag{45}$$

The general expression is

$$(\vec{a} + \vec{b})^2 = \vec{a}^2 + \vec{b}^2 + 2\,\vec{a} \cdot \vec{b} \tag{46}$$

The equations are equal, iff $\vec{a} \cdot \vec{b} = 0$. Hence the Pythagorean theorem yields a reasonable definition of orthogonality. However, we had no method yet to define vectors within our game. Using matrices **A** and **B** we may then write

$$\begin{aligned}
\|\mathbf{A} + \mathbf{B}\|^2 &= \tfrac{1}{2n}\,\mathrm{Tr}\left[(\mathbf{A} + \mathbf{B})^2\right] \\
&= \|\mathbf{A}\|^2 + \|\mathbf{B}\|^2 + \tfrac{1}{2n}\,\mathrm{Tr}(\mathbf{A}\mathbf{B} + \mathbf{B}\mathbf{A})
\end{aligned} \tag{47}$$

If we compare this to Equations (45) and (46), respectively, then the obvious definition of the inner product is given by:

$$\mathbf{A} \cdot \mathbf{B} \equiv \frac{\mathbf{A}\,\mathbf{B} + \mathbf{B}\,\mathbf{A}}{2} \tag{48}$$

Since the anticommutator does in general not yield a scalar, we have to distinguish between inner product and scalar product:

$$(\mathbf{A} \cdot \mathbf{B})_S \equiv \frac{1}{4\,n}\,\mathrm{Tr}(\mathbf{A}\,\mathbf{B} + \mathbf{B}\,\mathbf{A}) \tag{49}$$

where we indicate the scalar part by the subscript "S". Accordingly we define the exterior product by the commutator

$$\mathbf{A} \wedge \mathbf{B} \equiv \frac{\mathbf{A}\,\mathbf{B} - \mathbf{B}\,\mathbf{A}}{2} \tag{50}$$

Now that we defined the products, we should come back to the unit vectors. The only "unit vector" that we explicitly defined so far is the symplectic unit matrix γ_0. If it represents anything at all then it must be "the direction" of change, the direction of evolution in time as it was derived in this context and is the only "dimension" found so far. As we have already shown, all other unit vectors γ_k must be symmetric, if they are symplices. And vice versa: If γ_k is symmetric and anti-commutes with γ_0, then it is a symplex. As only symplices represent observables and are generators of symplectic transformations, we can have only a single "time" direction γ_0 and a yet unknown number of *symmetric* unit vectors (Thus we found a simple answer to the question, why only a single time direction is possible, a question also debated in Reference [15]). However, for $n > 1$, there might be different equivalent choices of γ_0. Whatever the specific form of γ_0 is, we will show that in combination with some general requirements like completeness, normalizability and observability it determines the structure of the complete algebra. Though we don't yet know how many symmetric and pairwise anti-commuting unit vectors γ_k exist-we have to interpret them as unit vectors in "spatial directions" (The meaning of what a spatial direction is, especially in contrast to the direction of time γ_0, has to be derived from the form of the emerging equations, of course. As meaning follows form, we do not define space-time, but we identify structures that fit to the known concept of space-time). Of course unit vectors must have unit length, so that we have to demand that

$$\|\gamma_k\|^2 = \tfrac{1}{2n}\,\mathrm{Tr}(\gamma_k^2) = \pm 1 \tag{51}$$

Note that (since our norm is not positive definite), we explicitly allow for unit vectors with negative "length" as we find it for γ_0. Note furthermore that all skew-symmetric unit vectors square to $-\mathbf{1}$ while the symmetric ones square to $\mathbf{1}$ [16].

Indeed systems of $N = p + q$ anti-commuting real matrices are known as real representations of Clifford algebras $Cl_{p,q}$. The index p is the number of unit elements ("vectors") that square to $+1$

and q is the number of unit vectors that square to -1. Clifford algebras are not necessarily connected to Hamiltonian motion, rather they can be regarded as purely mathematical "objects". They can be defined without reference to matrices whatsoever. Hence in mathematics, sets of matrices are merely "representations" of Clifford algebras. But our game is about physics and due to the proven one-dimensionality of time we concentrate on Clifford algebras $Cl_{N-1,1}$ which link CHOs in the described way with the generators of a Minkowski type metric. Further below it will turn out that the representation by matrices is-within the game-indeed helpful, since it leads to an overlap of certain symmetry structures. The unit elements (or unit "vectors") of a Clifford algebra, $\mathbf{e}_k$, are called the *generators* of the Clifford algebra. They pairwise anticommute and they square to ± 1 (The role as *generator* of the Clifford algebra should not be confused with the role as generators of symplectic transformations (*i.e.*, symplices). Though we are especially interested in Clifford algebras in which all generators are symplices, not all symplices are generators of the Clifford algebra. Bi-vectors for instance are symplices, but not generators of the Clifford algebra). Since the inverse of the unit elements $\mathbf{e}_k$ of a Clifford algebra must be unique, the products of different unit vectors form new elements and all possible products including the unit matrix form a group. There are $\binom{N}{k}$ possible combinations (products without repetition) of k elements from a set of N generators. We therefore find $\binom{N}{2}$ *bi-vectors*, which are products of two generators, $\binom{N}{3}$ *trivectors*) and so on. The product of all N basic matrices is called *pseudoscalar*. The total number of all k-vectors then is (We identify $k = 0$ with the unit matrix **1**.):

$$\sum_{k=0}^{N} \binom{N}{k} = 2^N \tag{52}$$

If we desire to construct a *complete* system, then the number of variables of the Clifford algebra has to match the number of variables of the used matrix system:

$$2^N = (2n)^2 \tag{53}$$

Note that the root of this equation gives an even integer $2^{N/2} = 2n$ so that N must be even. Hence all Hamiltonian Clifford algebras have an even dimension. Of course not all elements of the Clifford algebra may be symplices. The unit matrix (for instance) is a cosymplex. Consider the Clifford algebra $Cl_{1,1}$ with $N = 2$, which has two generators, say γ_0 with $\gamma_0^2 = -1$ and γ_1 with $\gamma_1^2 = 1$. Since these two anticommute (by definition of the Clifford algebra), so that we find (besides the unit matrix) a fourth matrix formed by the product $\gamma_0 \gamma_1$:

$$\begin{aligned} \gamma_0 \gamma_1 &= -\gamma_1 \gamma_0 \\ (\gamma_0 \gamma_1)^2 &= \gamma_0 \gamma_1 \gamma_0 \gamma_1 \\ &= -\gamma_0 \gamma_0 \gamma_1 \gamma_1 = \mathbf{1} \end{aligned} \tag{54}$$

The completeness of the Clifford algebras as we use them here implies that any $2n \times 2n$-matrix $\mathbf{M}$ with $(2n)^2 = 2^N$ can be written as a linear combination of all elements of the Clifford algebra:

$$\mathbf{M} = \sum_{k=0}^{4n^2-1} m_k \gamma_k \tag{55}$$

The coefficients can be computed from the scalar product of the unit vectors with the matrix $\mathbf{M}$:

$$m_k = (\gamma_k \cdot \mathbf{M})_S = \frac{s_k}{4n} \mathrm{Tr}(\gamma_k \mathbf{M} + \mathbf{M} \gamma_k) \tag{56}$$

Recall that skew-symmetric γ_k have a negative length and therefore we included a factor s_k which represents the "signature" of γ_k, in order to get the correct sign of the coefficients m_k.

Can we derive more properties of the constructable space-times? One restriction results from representation theory: A theorem from the theory of Clifford algebras states that $Cl_{p,q}$ has a representation by real matrices if (and only if) [17]

$$p - q = 0 \text{ or } 2 \text{ mod } 8 \tag{57}$$

The additional requirement that all generators must be symplices so that $p = N - 1$ and $q = 1$ then restricts N to

$$N - 2 = 0 \text{ or } 2 \text{ mod } 8 \tag{58}$$

Hence the only matrix systems that have the required symmetry properties within our modeling game are those that represent Clifford algebras with the dimensions $1 + 1, 3 + 1, 9 + 1, 11 + 1, 17 + 1, 19 + 1, 25 + 1, 27 + 1$ and so on. These correspond to matrix representations of size $2 \times 2, 4 \times 4, 32 \times 32, 64 \times 64, 512 \times 512$ and so on. The first of them is called *Pauli algebra*, the second one is the *Dirac algebra*. Do these two have special properties that the higher-dimensional algebras do not have? Yes, indeed.

Firstly, since dynamics is based on canonical pairs, the real Pauli algebra describes the motion of a single DOF and the Dirac algebra decribes the *simplest system with interaction* between two DOF. This suggests the interpretation that within our game, objects (Dirac-particles) are not located "within space-time", since we did not define space *at all* up to this point, but that space-time can be modeled as an emergent phenomenon. Space-time is in between particles.

Secondly, if we equate the number of fundamental variables ($2n$) of the oscillator phase space with the dimension of the Clifford space N, then Equation (53) leads to

$$2^N = N^2 \tag{59}$$

which allows for $N = 2$ and $N = 4$ only. But why should it be meaningful to assume $N = 2n$? The reason is quite simple: If $2n > N$ as for all higher-dimensional state vectors, there are less generators of the algebra than independent variables. This discrepancy increases with n. Hence the described objects can not be pure *vectors* anymore, but must contain tensor-type components (k-vectors) (For a deeper discussion of the dimensionality of space-time, see Reference [16] and references therein).

But before we describe a formal way to interprete Equation (59), let us first investigate the physical and geometrical implications of the game as described so far.

Matrix Exponentials

We said that the unit vectors γ_0 and γ_k are symplices and therefore generators of symplectic transformations. All symplectic matrices are matrix exponentials of symplices [12]. The computation of matrix exponentials is in the general case non-trivial. However, in the special case of matrices that square to ± 1 (e.g., along the "axis" γ_k of the coordinate system), the exponentials are readily evaluated:

$$\begin{aligned} \exp\left(\gamma_a \tau\right) &= \sum_{k=0}^{\infty} \frac{(\gamma_a \tau)^k}{k!} \\ \exp\left(\gamma_a \tau\right) &= \sum_{k=0}^{\infty} s^k \frac{\tau^{2k}}{(2k)!} + \gamma_a \sum_{k=0}^{\infty} s^k \frac{\tau^{2k+1}}{(2k+1)!} \end{aligned} \tag{60}$$

where $s = \pm 1$ is the sign of the matrix square of γ_a. For $s = -1$ ($\gamma_a^2 = -1$), it follows that

$$\mathbf{R}_a(\tau) = \exp\left(\gamma_a \tau\right) = \cos\left(\tau\right) + \gamma_a \sin\left(\tau\right) \tag{61}$$

and for $s = 1$ ($\gamma_a^2 = 1$):

$$\mathbf{B}_a(\tau) = \exp\left(\gamma_a \tau\right) = \cosh\left(\tau\right) + \gamma_a \sinh\left(\tau\right) \tag{62}$$

We can indentify skew-symmetric generators with rotations and (as we will show in more detail below) symmetric generators with boosts.

The (hyperbolic) sine/cosine structure of symplectic matrices are not limited to the generators but are a general property of the matrix exponentials of the symplex $\mathbf{F}$ (These properties are the main motivation to choose the nomenclature of "symplex" and "cosymplex".):

$$\mathbf{M}(t) = \exp\left(\mathbf{F}\,t\right) = \mathbf{C} + \mathbf{S} \tag{63}$$

where the (co-) symplex $\mathbf{S}$ ($\mathbf{C}$) is given by:

$$\begin{aligned}
\mathbf{S} &= \sinh\left(\mathbf{F}\,t\right) \\
\mathbf{C} &= \cosh\left(\mathbf{F}\,t\right)
\end{aligned} \tag{64}$$

since (the linear combination of) all odd powers of a symplex is again a symplex and the sum of all even powers is a cosymplex. The inverse transfer matrix $\mathbf{M}^{-1}(t)$ is given by:

$$\mathbf{M}^{-1}(t) = \mathbf{M}(-t) = \mathbf{C} - \mathbf{S} \tag{65}$$

The physical meaning of the matrix exponential results from Equation (13), which states that (for constant symplices $\mathbf{F}$) the solutions are given by the matrix exponential of $\mathbf{F}$:

$$\psi(t) = \mathbf{M}(t)\,\psi(0) \tag{66}$$

A symplectic transformation can be regarded as the result of a *possible* evolution in time. There is no proof that non-symplectic processes are forbidden by nature, but that *only* symplectic transformations are *structure preserving*. Non-symplectic transformations are then *structure defining*. Both play a fundamental role in the physics of our model reality, *because* fundamental particles are-according to our model-represented by dynamical structures. Therefore symplectic transformations describe those processes and interactions, in which structure is preserved, *i.e.*, in which the type of the particle is not changed. The fundamental variables are just "carriers" of the dynamical structures. Non-symplectic transformations can be used to transform the structure. This could also be described by a rotation of the direction of time. Another interpretation is that of a gauge-transformation [18].

5. The Significance of (De-)Coupling

In physics it is a standard technique to reduce complexity of problems by a suitable change of variables. In case of linear systems, the change of variables is a linear canonical transformation. The goal of such transformations is usually to substitute the solution of a complicated problem by the solution of multiple simpler problems. This technique is known under various names, one of these names is *decoupling*, but it is also known as *principal component analysis* or (as we will later show) transformation into the "rest frame". In other branches of science one might refer to it as *pattern recognition*.

In the following we investigate, how to transform a general oscillatory $2n \times 2n$-dimensional symplex to normal form. Certainly it would be preferable to find a "physical method", *i.e.*, a method that matches to the concepts that we introcuded so far and that has inherently physical significance. Or at least significance and explanatory power with respect to our modeling game. Let us start from the simplest systems, *i.e.*, with the Pauli and Dirac algebras which correspond to matrices of size 2×2 and 4×4, respectively.

5.1. The Pauli Algebra

The fundamental significance of the Pauli algebra is based on the even dimensionality of (classical) phase space. The algebra of 2×2 matrices describes the motion of a single (isolated) DOF. Besides η_0, the real Pauli algebra includes the following three matrices:

$$
\begin{aligned}
\eta_1 &= \begin{pmatrix} 0 & 1 \\ 1 & 0 \end{pmatrix} \\
\eta_2 &= \eta_0\,\eta_1 = \begin{pmatrix} 1 & 0 \\ 0 & -1 \end{pmatrix} \\
\eta_3 &= \mathbf{1} = \begin{pmatrix} 1 & 0 \\ 0 & 1 \end{pmatrix}
\end{aligned}
\tag{67}
$$

All except the unit matrix η_3 are symplices. If η_0 and η_1 are chosen to represent the generators of the corresponding Clifford algebra $Cl_{1,1}$, then η_2 is the only possible bi-vector. A general symplex has the form:

$$
\begin{aligned}
\mathbf{F} &= a\,\eta_0 + b\,\eta_1 + c\,\eta_2 \\
&= \begin{pmatrix} c & a+b \\ -a+b & -c \end{pmatrix}
\end{aligned}
\tag{68}
$$

The characteristic equation is given by $\mathrm{Det}(\mathbf{F} - \lambda\,\mathbf{1}) = 0$

$$
\begin{aligned}
0 &= (c - \lambda)(-c - \lambda) - (a+b)(-a+b) \\
\lambda &= \pm\sqrt{c^2 + b^2 - a^2}
\end{aligned}
\tag{69}
$$

The eigenvalues $\lambda_\pm$ are both either real for $a^2 < c^2 + b^2$ or both imaginary $a^2 > c^2 + b^2$ (or both zero). Systems in stable oscillation have purely imaginary eigenvalues. This case is most interesting for our modeling game.

Decoupling is usually understood in the more general sense to treat the interplay of several (at least two) DOF, but here we ask, whether all possible oscillating systems of $n = 1$ are isomorphic to normal form oscillators. Since there are 3 parameters in $\mathbf{F}$ and only one COM, namely the frequency ω, we need at least two parameters in the transformation matrix. Let us see, if we can choose these two transformations along the axis of the Clifford algebra. In this case we apply subsequentially two symplectic transformations along the axis η_0 and η_2. Applying the symplectic transformation matrix $\exp\left(\eta_0\,\tau/2\right)$ we obtain:

$$
\begin{aligned}
\mathbf{F}_1 &= \exp\left(\eta_0\,\tau/2\right)\mathbf{F}\exp\left(-\eta_0\,\tau/2\right) \\
&= a'\,\eta_0 + b'\,\eta_1 + c'\,\eta_2
\end{aligned}
\tag{70}
$$

(The "half-angle" argument is for convenience). The transformed coefficients a', b' and c' are given by

$$
\begin{aligned}
a' &= a \\
b' &= b\,\cos\tau - c\,\sin\tau \\
c' &= c\,\cos\tau + b\,\sin\tau
\end{aligned}
\tag{71}
$$

so that-depending on the "duration of the pulse", we can chose to transform into a coordinate system in which either $b' = 0$ or $c' = 0$. If we choose $t = \arctan\left(-c/b\right)$, then $c' = 0$, so that

$$
\mathbf{F}' = a\,\eta_0 + \sqrt{b^2 + c^2}\,\eta_1 = a'\,\eta_0 + b'\,\eta_1
\tag{72}
$$

If we chose the next generator to be η_2, then:

$$
\begin{aligned}
a'' &= a'\,\cosh\tau - b'\,\sinh\tau \\
b'' &= b'\,\cosh\tau - a'\,\sinh\tau
\end{aligned}
\tag{73}
$$

In this case we have to dinstinguish between the case, where $a' > b'$ and $a' < b'$. The former is the oscillatory system and in this case the transformation with $\tau = \text{artanh}(b'/a')$ leads to the normal form of a 1-dim. oscillator:

$$
\begin{aligned}
a'' &= \sqrt{a^2 - b^2 - c^2} \\
b'' &= 0 \\
c'' &= 0
\end{aligned}
\tag{74}
$$

and the matrix $\mathbf{F}''$ has the form

$$
\mathbf{F}'' = \sqrt{a^2 - b^2 - c^2}\, \eta_0
\tag{75}
$$

If the eigenvalues are imaginary, then $\lambda = \pm i\,\omega$ and hence

$$
\mathbf{F}'' = \omega\,\eta_0
\tag{76}
$$

so that the solution is-for constant frequency-given by the matrix exponential:

$$
\begin{aligned}
\psi(t) &= \exp\left(\omega\,\eta_0\,t\right)\psi(0) \\
&= \left(\mathbf{1}\cos\left(\omega\,t\right) + \eta_0\sin\left(\omega\,t\right)\right)\psi(0)
\end{aligned}
\tag{77}
$$

This shows that in the context of stable oscillator algebras-the real Pauli algebra can be reduced to the complex number system: This becomes evident, if we consider possible representations of the complex numbers. Clearly we need two basic elements- the unit matrix and η_0, *i.e.*, a matrix that commutes with the unit matrix and squares to $-\mathbf{1}$. If we write "i" instead of η_0, then it is easily verified that (See also References [17,19] and Equation (34) in combination with Reference [20].):

$$
\begin{aligned}
z &= x + iy = \mathbf{Z} = \begin{pmatrix} x & y \\ -y & x \end{pmatrix} \\
\bar{z} &= x - iy = \mathbf{Z}^T = x\,\mathbf{1} + \eta_0^T\,y \\
\exp(i\phi) &= \cos(\phi) + i\sin(\phi) \\
\|z\|^2 &= \mathbf{Z}\,\mathbf{Z}^T = z\,\bar{z} = x^2 + y^2
\end{aligned}
\tag{78}
$$

The theory of holomorphic functions is based on series expansions and can be equally well formulated with matrices. Viewed from our perspective the complex numbers are a special case of the real Pauli algebra- since we have shown above that any one-dimensional oscillator can be canonically transformed into a system of the form of Equation (76). Nevertheless we emphasize that the complex numbers interpreted this way can only represent the *normal form* of an oscillator. The normal form excludes a different scaling of coordinates and momenta as used in classical mechanics, *i.e.*, it avoids intrinsically the appearance of different "spring constants" and masses (There have been several attempts to explain the appearance of the complex numbers in quantum mechanics [21–27]. A general discussion of the use of complex numbers in physics is beyond the scope of this essay, therefore we add just a remark. Gary W. Gibbons wrote that "In particular there can be no evolution if ψ is real" [24]. We agree with Gibbons that the unit imaginary can be related to evolution in time as it implies oscillation, but we do not agree with his conclusion. Physics was able to describe evolution in time without imaginaries before quantum mechanics and it still is. The unconscious use of the unit imaginary did not prevent quantum mechanics from being experimetally successful. But it prevents physicists from understanding its structure).

5.2. The Dirac Algebra

In this subsection we consider the oscillator algebra for two coupled DOF, the algebra of 4×4 matrices. In contrast to the real Pauli algebra, where the parameters a, b and c did not suggest a specific physical meaning, the structure of the Dirac algebra bears geometrical significance as has been pointed out by David Hestenes and others [28–30]. The (real) Dirac algebra is the simplest real algebra that enables for a description of two DOF and the interaction between them. Furthermore the eigenfrequencies of a Dirac symplex **F** may be complex, while the spectrum of the Pauli matrices does not include complex numbers off the real and imaginary axis. The spectrum of general $2n \times 2n$-symplices has a certain structure - since the coefficients of the characteristic polynomial are real: If λ is an eigenvalue of **F**, then its complex conjugate $\bar{\lambda}$ as well as λ and $-\bar{\lambda}$ are also eigenvalues. As we will show, this is the spectrum of the Dirac algebra and therefore any $2n \times 2n$-system can, at least in principle, be block-diagonalized using 4×4-blocks. The Dirac algebra is therefore the simplest algebra that covers the general case.

The structure of Clifford algebras follows Pascal's triangle. The Pauli algebra has the structure $1 - 2 - 1$ (scalar-vector-bivector), the Dirac algebra has the structure $1 - 4 - 6 - 4 - 1$, standing for unit element (scalar), vectors, bi-vectors, tri-vectors and pseudoscalar. The vector elements are by convention indexed with γ_μ with $\mu = 0 \ldots 3$, *i.e.*, the generators of the algebra (According to Pauli's fundamental theorem of the Dirac algebra, all possible choices of the Dirac matrices are, as long as the "metric tensor" $g_{\mu\nu}$ remains unchanged, equivalent [31].):

$$
\gamma_0 = \begin{pmatrix} 0 & 1 & 0 & 0 \\ -1 & 0 & 0 & 0 \\ 0 & 0 & 0 & 1 \\ 0 & 0 & -1 & 0 \end{pmatrix} \quad
\gamma_1 = \begin{pmatrix} 0 & -1 & 0 & 0 \\ -1 & 0 & 0 & 0 \\ 0 & 0 & 0 & 1 \\ 0 & 0 & 1 & 0 \end{pmatrix}
$$

$$
\gamma_2 = \begin{pmatrix} 0 & 0 & 0 & 1 \\ 0 & 0 & 1 & 0 \\ 0 & 1 & 0 & 0 \\ 1 & 0 & 0 & 0 \end{pmatrix} \quad
\gamma_3 = \begin{pmatrix} -1 & 0 & 0 & 0 \\ 0 & 1 & 0 & 0 \\ 0 & 0 & -1 & 0 \\ 0 & 0 & 0 & 1 \end{pmatrix}
\tag{79}
$$

We define the following numbering scheme for the remaining matrices (The specific choice of the matrices is not unique. A table of the different systems can be found in Reference ([32]).):

$$
\begin{aligned}
\gamma_{14} &= \gamma_0\,\gamma_1\,\gamma_2\,\gamma_3; & \gamma_{15} &= \mathbf{1} \\
\gamma_4 &= \gamma_0\,\gamma_1; & \gamma_7 &= \gamma_{14}\,\gamma_0\,\gamma_1 = \gamma_2\,\gamma_3 \\
\gamma_5 &= \gamma_0\,\gamma_2; & \gamma_8 &= \gamma_{14}\,\gamma_0\,\gamma_2 = \gamma_3\,\gamma_1 \\
\gamma_6 &= \gamma_0\,\gamma_3; & \gamma_9 &= \gamma_{14}\,\gamma_0\,\gamma_3 = \gamma_1\,\gamma_2 \\
\gamma_{10} &= \gamma_{14}\,\gamma_0 & = \gamma_1\,\gamma_2\,\gamma_3 \\
\gamma_{11} &= \gamma_{14}\,\gamma_1 & = \gamma_0\,\gamma_2\,\gamma_3 \\
\gamma_{12} &= \gamma_{14}\,\gamma_2 & = \gamma_0\,\gamma_3\,\gamma_1 \\
\gamma_{13} &= \gamma_{14}\,\gamma_3 & = \gamma_0\,\gamma_1\,\gamma_2
\end{aligned}
\tag{80}
$$

According to Equation (17) we expect 10 symplices and since the 4 vectors and 6 bi-vectors are symplices, all other elements are cosymplices. With this ordering, the general 4×4-symplex **F** can be written as (instead of Equation (55)):

$$
\mathbf{F} = \sum_{k=0}^{9} f_k\,\gamma_k
\tag{81}
$$

In Reference [32] we presented a detailed survey of the Dirac algebra with respect to symplectic Hamiltonian motion. The essence of this survey is the insight that the real Dirac algebra describes Hamiltonian motion of an ensembles of two-dimensional oscillators, but as well the motion of a "point particle" in 3-dimensional space, *i.e.*, that Equation (25) is, when expressed by the real Dirac algebra, *isomorphic to the Lorentz force equation* as we are going to show in Section 6.3. Or, in other words, the Dirac algebra allows to model a point particle and its interaction with the electromagnetic field in terms of the classical statistical ensemble of abstract oscillators.

6. Electromechanical Equivalence (EMEQ)

The number and type of symplices within the Dirac algebra (80) suggests to use the following vector notation for the coefficients [32,33] of the observables:

$$
\begin{aligned}
\mathcal{E} &\equiv f_0 \\
\vec{P} &\equiv (f_1, f_2, f_3)^T \\
\vec{E} &\equiv (f_4, f_5, f_6)^T \\
\vec{B} &\equiv (f_7, f_8, f_9)^T
\end{aligned}
\tag{82}
$$

where the "clustering" of the coefficients into 3-dimensional vectors will be explained in the following. The first four elements $\mathcal{E}$ and $\vec{P}$ are the coefficients of the generators of the Clifford algebra and the remaining symplices are 3 symmetric bi-vectors $\vec{E}$ and skew-symmetric bi-vectors $\vec{B}$. As explained above, the matrix exponentials of pure Clifford elements are readily evaluated (Equations (61) and (62)). The effect of a symplectic similarity transformation on a symplex

$$
\begin{aligned}
\check{\psi} &= \mathbf{R}(\tau/2)\, \psi \\
\check{\mathbf{F}} &= \mathbf{R}(\tau/2)\, \mathbf{F}\, \mathbf{R}^{-1}(\tau/2) \\
&= \mathbf{R}(\tau/2)\, \mathbf{F}\, \mathbf{R}(-\tau/2)
\end{aligned}
\tag{83}
$$

can then be computed component-wise as in the following case of a rotation (using Equation (81)):

$$
\begin{aligned}
\tilde{\mathbf{F}} &= \sum_{k=0}^{9} f_k\, \mathbf{R}_a\, \gamma_k\, \mathbf{R}_a^{-1} \\
\mathbf{R}_a\, \gamma_k\, \mathbf{R}_a^{-1} &= (\cos(\tau/2) + \gamma_a \sin(\tau/2))\, \gamma_k\, (\cos(\tau/2) - \gamma_a \sin(\tau/2)) \\
&= \gamma_k \cos^2(\tau/2) - \gamma_a \gamma_k \gamma_a \sin^2(\tau/2) + (\gamma_a \gamma_k - \gamma_k \gamma_a) \cos(\tau/2) \sin(\tau/2)
\end{aligned}
\tag{84}
$$

Since all Clifford elements either commute or anti-commute with each other, we have two possible solutions. The first (γ_k and γ_a commute) yields with $\gamma_a^2 = -1$:

$$
\mathbf{R}_a\, \gamma_k\, \mathbf{R}_a^{-1} = \gamma_k \cos^2(\tau/2) - \gamma_a^2 \gamma_k \sin^2(\tau/2) = \gamma_k
\tag{85}
$$

but if (γ_k and γ_a anti-commute) we obtain a rotation:

$$
\begin{aligned}
\mathbf{R}_a\, \gamma_k\, \mathbf{R}_a^{-1} &= \gamma_k \left(\cos^2(\tau/2) - \sin^2(\tau/2)\right) + \gamma_a \gamma_k\, 2 \cos(\tau/2) \sin(\tau/2) \\
&= \gamma_k \cos(\tau) + \gamma_a \gamma_k \sin(\tau)
\end{aligned}
\tag{86}
$$

For $a = 9$ ($\gamma_a = \gamma_1\, \gamma_2$) for instance we find:

$$
\begin{aligned}
\tilde{\gamma}_1 &= \gamma_1 \cos(\tau) + \gamma_1 \gamma_2 \gamma_1 \sin(\tau) = \gamma_1 \cos(\tau) - \gamma_2 \sin(\tau) \\
\tilde{\gamma}_2 &= \gamma_2 \cos(\tau) + \gamma_1 \gamma_2 \gamma_2 \sin(\tau) = \gamma_2 \cos(\tau) + \gamma_1 \sin(\tau) \\
\tilde{\gamma}_3 &= \gamma_3 ,
\end{aligned}
\tag{87}
$$

which is formally equivalent to a rotation of $\vec{P}$ about the "z-axis". If the generator γ_a of the transformation is symmetric, we obtain:

$$
\begin{aligned}
\mathbf{R}_a\, \gamma_k\, \mathbf{R}_a^{-1} &= (\cosh(\tau/2) + \gamma_a \sinh(\tau/2))\, \gamma_k\, (\cosh(\tau/2) - \gamma_a \sinh(\tau/2)) \\
&= \gamma_k \cosh^2(\tau/2) - \gamma_a \gamma_k \gamma_a \sinh^2(\tau/2) + (\gamma_a \gamma_k - \gamma_k \gamma_a) \cosh(\tau/2) \sinh(\tau/2)
\end{aligned}
\tag{88}
$$

so that (if γ_a and γ_k commute):

$$
\begin{aligned}
\tilde{\gamma}_k &= \gamma_k \cosh^2(\tau/2) - \gamma_a^2 \gamma_k \sinh^2(\tau/2) \\
\tilde{\gamma}_k &= \gamma_k \left(\cosh^2(\tau/2) - \sinh^2(\tau/2)\right) = \gamma_k
\end{aligned}
\tag{89}
$$

and if γ_a and γ_k anticommute:

$$
\begin{aligned}
\tilde{\gamma}_k &= \gamma_k \left(\cosh^2(\tau/2) + \sinh^2(\tau/2)\right) + 2\,\gamma_a\,\gamma_k\,\cosh(\tau/2)\,\sinh(\tau/2)) \\
&= \gamma_k\,\cosh(\tau) + \gamma_a\,\gamma_k\,\sinh(\tau),
\end{aligned}
\tag{90}
$$

which is equivalent to a boost when the following parametrization of "rapidity" τ is used:

$$
\begin{aligned}
\tanh(\tau) &= \beta \\
\sinh(\tau) &= \beta\,\gamma \\
\cosh(\tau) &= \gamma \\
\gamma &= \frac{1}{\sqrt{1-\beta^2}}
\end{aligned}
\tag{91}
$$

A complete survey of these transformations and the (anti-) commutator tables can be found in Reference [32] (This formalism corresponds exactly to the relativistic invariance of a Dirac spinor in QED as described for instance in Reference [34], although the Dirac theory uses complex numbers and a different sign-convention for the metric tensor). The "spatial" rotations are generated by the bi-vectors associated with $\vec{B}$ and Lorentz boosts by the components associated with $\vec{E}$. The remaining 4 generators of symplectic transformations correspond to $\mathcal{E}$ and $\vec{P}$. They where named *phase-rotation* (generated by γ_0) and *phase-boosts* (generated by $\vec{\gamma} = (\gamma_1, \gamma_2, \gamma_3)$) and have been used for instance for symplectic decoupling as described in Reference [33].

It is nearby (and already suggested by our notation) to consider the possibility that the EMEQ (Equation (82)) allows to model a relativistic particle as represented by energy $\mathcal{E}$ and momentum $\mathbf{P}$ either in an external electromagnetic field given by $\vec{E}$ and $\vec{B}$ or-alternatively-in an accelerating and/or rotating reference frame, where the elements $\vec{E}$ and $\vec{B}$ correspond to the axis of acceleration and rotation, respectively. We assumed, that all components of the state vector ψ are equivalent in meaning and unit. Though we found that the state vector is formally composed of canonical pairs, the units are unchanged and identical for all elements of ψ. From Equation (13) we take, that the simplex $\mathbf{F}$ (and also $\mathcal{A}$) have the unit of a frequency. If the Hamiltonian $\mathcal{H}$ is supposed to represent energy, then the components of ψ have the unit of the square root of action.

If the coefficients are supposed to represent the electromagnetic field, then we need to express these fields in the unit of frequency. This can be done, but it requires to involve natural conversion factors like $\hbar$, charge e, velocity c and a mass, for instance the electron mass m_e. The magnetic field (for instance) is related to a "cyclotron frequency" ω_c by $\omega_c \propto \frac{e}{m_e}\,B$.

However, according to the rules of the game, the distinction between particle properties and "external" fields requires a reason, an explanation. Especially as it is physically meaningless for macroscopic coupled oscillators. In References [32,33] this nomenclature was used in a merely *formal* way, namely to find a descriptive scheme to order the symplectic generators, so to speak an *equivalent circuit* to describe the general possible coupling terms for two-dimensional coupled linear optics as required for the description of charged particles beams.

Here we play the reversed modeling game: Instead of using the EMEQ as an equivalent circuit to describe ensembles of oscillators, we now use ensembles of oscillators as an equivalent circuit to describe point particles. The motivation for Equation (82) is nevertheless similar, *i.e.*, it follows from the formal structure of the Dirac Clifford algebra. The grouping of the coefficients comes along with the number of vector- and bi-vector-elements, 4 and 6, respectively. The second criterium is to distinguish between generators of rotations and boost, *i.e.*, between symmetric and skew-symmetric symplices, which separates energy from momentum and electric from magnetic elements. Third of all, we note that even (*Even k-vectors are those with even $k = 2\,m$, where m is a natural number*) elements (scalar, bi-vectors, 4-vectors *etc.*) of even-dimensional Clifford algebras form a sub-algebra. This means that we can generate the complete Clifford algebra from the vector-elements by matrix multiplication (this is why we call them generators), but we can not generate vectors from bi-vectors by multiplication. And therefore the vectors are the particles (which are understood as the sources of fields) and the bi-vectors are the fields, which are generated by the objects and influence their motion. The full Dirac symplex-algebra includes the description of a particle (vector) in a field (bi-vector). But why would the field be *external*? Simply, because it is impossible to generate bi-vectors from a single vector-type object, since any single vector-type object written as $\mathcal{E}\,\gamma_0 + \vec{P}\cdot\vec{\gamma}$ squares to a scalar. Therefore, the fields must be the result of interaction with other particles and hence we call them "external". This is in some way a "first-order" approach, since there might be higher order processes that we did not consider yet. But in the linear approach (*i.e.*, for second-order Hamiltonians), this distinction is reasonable and hence a legitimate move in the game.

Besides the Hamiltonian structure (symplices *vs.* co-symplices) and the Clifford algebraic structure (distinguishing vectors, bi-vectors, tri-vectors *etc.*) there is a third essential symmetry, which is connected to the real matrix representation of the Dirac algebra and to the fact that it describes the general Hamiltonian motion of coupled oscillators: To distinguish the even from the odd elements with respect to the block-diagonal matrix structure. We used this property in Reference [33] to develop a general geometrical decoupling algorithm (see also Section 6.2).

Now it may appear that we are cheating somehow, as relativity is usually "derived" from the constancy of the speed of light, while in our modeling game, we did neither introduce spatial notions nor light at all. Instead we directly arrive at notions of quantum electrodynamics (QED). How can this be? The definition of "velocity" within wave mechanics usually involves the dispersion relation of waves, *i.e.*, the velocity of a wave packet is given by the group velocity $\vec{v}_{gr}$ defined by

$$\vec{v}_{gr} \equiv \vec{\nabla}_{\vec{k}}\,\omega(\vec{k}) \tag{92}$$

and the so-called phase velocity v_{ph} defined by

$$v_{ph} = \frac{\omega}{k} \tag{93}$$

It is then typically mentioned that the product of these two velocities is a constant $v_{gr}\,v_{ph} = c^2$. By the use of the EMEQ and Equation (29), the eigenvalues of $\mathbf{F}$ can be written as:

$$
\begin{aligned}
K_1 &= -\text{Tr}(\mathbf{F}^2)/4 \\
K_2 &= \text{Tr}(\mathbf{F}^4)/16 - K_1^2/4 \\
\omega_1 &= \sqrt{K_1 + 2\sqrt{K_2}} \\
\omega_2 &= \sqrt{K_1 - 2\sqrt{K_2}} \\
\omega_1^2\,\omega_2^2 &= K_1^2 - 4\,K_2 = \text{Det}(\mathbf{F}) \\
K_1 &= \mathcal{E}^2 + \vec{B}^2 - \vec{E}^2 - \vec{P}^2 \\
K_2 &= (\mathcal{E}\,\vec{B} + \vec{E}\times\vec{P})^2 - (\vec{E}\cdot\vec{B})^2 - (\vec{P}\cdot\vec{B})^2
\end{aligned}
\tag{94}
$$

Since symplectic transformations are similarity transformations, they do not alter the eigenvalues of the matrix $\mathbf{F}$ and since all possible evolutions in time (which can be described by the Hamiltonian) are symplectic transformations, the eigenvalues (of closed systems) are conserved. If we consider a "free particle", the we obtain from Equation (94):

$$\omega_{1,2} = \pm\sqrt{\mathcal{E}^2 - \vec{P}^2} \tag{95}$$

As we mentioned before both, energy and momentum, have (within this game) the unit of frequencies. If we take into account that $\omega_{1,2} \equiv m$ is fixed, then the dispersion relation for "the energy" $\mathcal{E} = \omega$ is

$$\mathcal{E} = \omega = \sqrt{m^2 + \vec{P}^2} \tag{96}$$

which is indeed the correct relativistic dispersion. But how do we make the step from pure oscillations to *waves*? (The question if Quantum theory requires Planck's constant $\hbar$, has been answered negative by John P. Ralston [35]).

6.1. Moments and The Fourier Transform

In case of "classical" probability distribution functions (PDFs) $\phi(x)$ we may use the Taylor terms of the *characteristic function* $\tilde{\phi}_x(t) = \langle \exp i\, t\, x \rangle_x$, which is the Fourier transform of $\phi(x)$, at the origin. The k-th moment is then given by

$$\langle x^k \rangle = i^k \, \tilde{\phi}^{(k)}(0) \tag{97}$$

where $\phi^{(k)}$ is the k-th derivative of $\tilde{\phi}_x(t)$.

A similar method would be of interest for our modeling game. Since a (phase space-) density is positive definite, we can always take the square root of the density instead of the density itself: $\phi = \sqrt{\rho}$. The square root can also defined to be a complex function, so that the density is $\rho = \phi\phi^\star = \|\phi\|^2$ and, if mathematically well-defined (convergent), we can also define the Fourier transform of the complex root, *i.e.*,

$$\tilde{\phi}(\omega, \vec{k}) = N \int \phi(t, \vec{x}) \, \exp\left(i\,\omega\,t - i\,\vec{k}\vec{x}\right) dt\, d^3x \tag{98}$$

and vice versa:

$$\tilde{\phi}(t, \vec{x}) = \tilde{N} \int \phi(\omega, \vec{k}) \, \exp\left(-i\,\omega\,t + i\,\vec{k}\vec{x}\right) d\omega\, d^3k \tag{99}$$

In principle, we may *define* the density no only by real and imaginary part, but by an arbitrary number of components. Thus, if we consider a four-component spinor, we may of course mathematically define its Fourier transform. But in order to see, why this might be more than a mathematical "trick", but *physically meaningful*, we need to go back to the notions of classical statistical mechanics. Consider that we replace the single state vector by an "ensemble", where we leave the question open, if the ensemble should be understood as a single phase space trajectory, averaged over time, or as some (presumably large) number of different trajectories. It is well-known, that the phase space density $\rho(\psi)$ is stationary, if it depends only on constants of motion, for instance if it depends only on the Hamiltonian itself. With the Hamiltonian of Equation (12), the density could for example have the form

$$\rho(\mathcal{H}) \propto \exp\left(-\beta\,\mathcal{H}\right) = \exp\left(-\beta\,\psi\,\mathcal{A}\,\psi/2\right) \tag{100}$$

which corresponds to a multivariate Gaussian. But more important is the insight, that the density exclusively depends on the second moments of the phase space variables as given by the Hamiltonian, *i.e.*, in case of a "free particle" it depends on $\mathcal{E}$ and $\vec{P}$. And therefore we should be able to use energy and momentum as frequency ω and wave-vector $\vec{k}$.

But there are more indications in our modeling game that suggest the use of a Fourier transform as we will show in the next section.

6.2. The Geometry of (De-)Coupling

In the following we give a (very) brief summary of Reference [33]. As already mentioned, decoupling is meant-despite the use of the EMEQ-first of all purely technical-mathematical. Let us delay the question, if the notions that we define in the following have any physical relevance. Here we refer first of all to block-diagonalization, *i.e.*, we treat the symplex **F** just as a "Hamiltonian" matrix. From the definition of the real Dirac matrices we obtain **F** in explicit 4×4 matrix form:

$$
\mathbf{F} = \begin{pmatrix} -E_x & E_z + B_y & E_y - B_z & B_x \\ E_z - B_y & E_x & -B_x & -E_y - B_z \\ E_y + B_z & B_x & E_x & E_z - B_y \\ -B_x & -E_y + B_z & E_z + B_y & -E_x \end{pmatrix}
$$

$$
+ \begin{pmatrix} -P_z & \mathcal{E} - P_x & 0 & P_y \\ -\mathcal{E} - P_x & P_z & P_y & 0 \\ 0 & P_y & -P_z & \mathcal{E} + P_x \\ P_y & 0 & -\mathcal{E} + P_x & P_z \end{pmatrix} \tag{101}
$$

If we find a (sequence of) symplectic similarity transformations that would allow to reduce the 4×4-form to a block-diagonal form, then we would obtain two separate systems of size 2×2 and we could continue with the transformations of Section 5.1.

Inspection of Equation (101) unveils that **F** is block-diagonal, if the coefficents E_y, P_y, B_x and B_z vanish. Obviously this implies that $\vec{E} \cdot \vec{B} = 0$ and $\vec{P} \cdot \vec{B} = 0$. Or vice versa, if we find a symplectic method that transforms into a system in which $\vec{E} \cdot \vec{B} = 0$ and $\vec{P} \cdot \vec{B} = 0$, then we only need to apply appropriate rotations to achieve block-diagonal form. As shown in Reference [33] this can be done in different ways, but in general it requires the use of the "phase rotation" γ_0 and "phase boosts" $\vec{\gamma}$. Within the conceptional framework of our game, the application of these transformations equals the use of "matter fields". But furthermore, this shows that block-diagonalization has also geometric significance within the Dirac algebra and, with respect to the Fourier transformation, the requirement $\vec{P} \cdot \vec{B} = 0$ indicates a divergence free magnetic field, as the replacement of $\vec{P}$ by $\vec{\nabla}$ yields $\vec{\nabla} \cdot \vec{B} = 0$. The additional requirement $\vec{E} \cdot \vec{B} = 0$ also fits well to our physical picture of e.m. waves. Note furthermore, that there is no analogous requirement to make $\vec{P} \cdot \vec{E}$ equal to zero. Thus (within this analogy) we *can accept* $\vec{\nabla} \cdot \vec{E} \neq 0$.

But this is not everything to be taken from this method. If we analyze in more detail, which expressions are *required* to vanish and which may remain, then it appears that $\vec{P} \cdot \vec{B}$ is explicitely given by

$$
\begin{aligned}
P_x B_x \gamma_1 \gamma_2 \gamma_3 + P_y B_y \gamma_2 \gamma_3 \gamma_1 + P_z B_z \gamma_3 \gamma_1 \gamma_2 &= (\vec{P} \cdot \vec{B}) \gamma_{10} \\
E_x B_x \gamma_4 \gamma_2 \gamma_3 + E_y B_y \gamma_5 \gamma_3 \gamma_1 + E_z B_z \gamma_6 \gamma_1 \gamma_2 &= (\vec{E} \cdot \vec{B}) \gamma_{14} \\
P_x E_x \gamma_1 \gamma_4 \gamma_3 + P_y E_y \gamma_2 \gamma_5 \gamma_1 + P_z E_z \gamma_3 \gamma_6 \gamma_2 &= -(\vec{P} \cdot \vec{E}) \gamma_0
\end{aligned} \tag{102}
$$

That means that exactly those products have to vanish which yield *cosymplices*. This can be interpreted via the structure preserving properties of symplectic motion. Since within our game, the particle *type* can only be represented by the structure of the dynamics, and since electromagnetic processes do not change the type of a particle, then they are quite obviously *structure preserving* which then implies the non-appearance of co-symplices. Or-in other words-electromagnetism is of Hamiltonian nature. We will come back to this point in Section 6.4.

6.3. The Lorentz Force

In the previous section we constituted the distinction between the "mechanical" elements $\mathbf{P} = \mathcal{E} \gamma_0 + \vec{\gamma} \cdot \vec{P}$ of the general matrix **F** and the electrodynamical elements $\mathbf{F} = \gamma_0 \vec{\gamma} \cdot E + \gamma_{14} \gamma_0 \vec{\gamma} \cdot \vec{B}$. Since the matrix $\mathbf{S} = \Sigma \gamma_0$ is a symplex, let us assume to be equal to **P** and apply Equation (25). We then find (with the appropriate relative scaling between **P** and **F** as explained above):

$$
\frac{d\mathbf{P}}{d\tau} = \dot{\mathbf{P}} = \frac{q}{2m} (\mathbf{F} \mathbf{P} - \mathbf{P} \mathbf{F}) \tag{103}
$$

which yields written with the coefficients of the real Dirac matrices:

$$\frac{d\mathcal{E}}{d\tau} = \frac{q}{m} \vec{P} \cdot \vec{E}$$
$$\frac{d\vec{P}}{d\tau} = \frac{q}{m} \left(\mathcal{E} \vec{E} + \vec{P} \times \vec{B} \right) \tag{104}$$

where τ is the proper time. If we convert to the lab frame time t using $dt = \frac{d\tau}{\gamma}$ Equation (103) yields (setting $c = 1$):

$$\gamma \frac{d\mathcal{E}}{dt} = q \gamma \vec{v} \cdot \vec{E}$$
$$\gamma \frac{d\vec{P}}{dt} = \frac{q}{m} \left(m \gamma \vec{E} + m \gamma \vec{v} \times \vec{B} \right)$$
$$\frac{dE}{dt} = q \vec{v} \cdot \vec{E} \tag{105}$$
$$\frac{d\vec{P}}{dt} = q \left(\vec{E} + \vec{v} \times \vec{B} \right)$$

which is the Lorentz force. Therefore the Lorentz force acting on a charged particle in 3 spatial dimensions can be modeled by an ensemble of 2-dimensional CHOs. The isomorphism between the observables of the perceived 3-dimensional world and the second moments of density distributions in the phase space of 2-dimensional oscillators is remarkable.

In any case, Equation (103) clarifies two things within the game. Firstly, that both, energy $\mathcal{E}$ and momentum $\vec{p}$, have to be interpreted as mechanical energy and momentum (and not canonical), secondly the relative normalization between fields and mechanical momentum is fixed and last, but not least, it clarifies the relation between the time related to mass (proper time) and the time related to γ_0 and energy, which appears to be the laboratory time.

6.4. The Maxwell Equations

As we already pointed out, waves are (within this game) the result of a Fourier transformation (FT). But there are different ways to argue this. In Reference [16] we argued that Maxwell's equations can be derived within our framework by (a) the postulate that space-time emerges from interaction, i.e., that the fields $\vec{E}$ and $\vec{B}$ have to be constructed from the 4-vectors. $\mathbf{X} = t \gamma_0 + \vec{x} \cdot \vec{\gamma}$, $\mathbf{J} = \rho \gamma_0 + \vec{j} \cdot \vec{\gamma}$ and $\mathbf{A} = \Phi \gamma_0 + \vec{A} \cdot \vec{\gamma}$ with (b) the requirement that no co-symplices emerge. But we can also argue with the FT of the density (see Section 6.1).

If we introduce the 4-derivative

$$\partial \equiv -\partial_t \gamma_0 + \partial_x \gamma_1 + \partial_y \gamma_2 + \partial_z \gamma_3 \tag{106}$$

The non-abelian nature of matrix multiplication requires to distinguish differential operators acting to the right and to the left, i.e., we have ∂ as defined in Equation (106), $\overrightarrow{\partial}$ and $\overleftarrow{\partial}$ which is written to the right of the operand (thus indicating the order of the matrix multiplication) so that

$$\mathbf{H} \overleftarrow{\partial} \equiv -\partial_t \mathbf{H} \gamma_0 + \partial_x \mathbf{H} \gamma_1 + \partial_y \mathbf{H} \gamma_2 + \partial_z \mathbf{H} \gamma_3$$
$$\overrightarrow{\partial} \mathbf{H} \equiv -\gamma_0 \partial_t \mathbf{H} + \gamma_1 \partial_x \mathbf{H} + \gamma_2 \partial_y \mathbf{H} + \gamma_3 \partial_z \mathbf{H} \tag{107}$$

The we find the following general rules (see Equation (35)) that prevent from non-zero cosymplices:

$$\frac{1}{2} \left(\overrightarrow{\partial} \text{ vector} - \text{vector} \overleftarrow{\partial} \right) \Rightarrow \text{bi-vector}$$
$$\frac{1}{2} \left(\overrightarrow{\partial} \text{ bi-vector} - \text{bi-vector} \overleftarrow{\partial} \right) \Rightarrow \text{vector}$$
$$\frac{1}{2} \left(\overrightarrow{\partial} \text{ bi-vector} + \text{bi-vector} \overleftarrow{\partial} \right) \Rightarrow \text{axial vector} = 0 \tag{108}$$
$$\frac{1}{2} \left(\overrightarrow{\partial} \text{ vector} + \text{vector} \overleftarrow{\partial} \right) \Rightarrow \text{scalar} = 0$$

Application of these derivatives yields:

$$
\begin{aligned}
\mathbf{F} &= \tfrac{1}{2}\left(\overrightarrow{\partial}\mathbf{A} - \mathbf{A}\overleftarrow{\partial}\right) \\
4\pi\mathbf{J} &= \tfrac{1}{2}\left(\overrightarrow{\partial}\mathbf{F} - \mathbf{F}\overleftarrow{\partial}\right) \\
0 &= \overrightarrow{\partial}\mathbf{F} + \mathbf{F}\overleftarrow{\partial} \\
0 &= \tfrac{1}{2}\left(\overrightarrow{\partial}\mathbf{A} + \mathbf{A}\overleftarrow{\partial}\right) \\
0 &= \tfrac{1}{2}\left(\overrightarrow{\partial}\mathbf{J} + \mathbf{J}\overleftarrow{\partial}\right)
\end{aligned}
\tag{109}
$$

The first row of Equation (109) corresponds to the usual definition of the bi-vector fields from a vector potential $\mathbf{A}$ and is (written by components) given by

$$
\begin{aligned}
\vec{E} &= -\vec{\nabla}\phi - \partial_t\vec{A} \\
\vec{B} &= \vec{\nabla}\times\vec{A}
\end{aligned}
\tag{110}
$$

The second row of Equation (109) corresponds to the usual definition of the 4-current $\mathbf{J}$ as sources of the fields and the last three rows just express the impossibility of the appearance of cosymplices. They explicitely represent the homogenuous Maxwell equations

$$
\begin{aligned}
\vec{\nabla}\cdot\vec{B} &= 0 \\
\vec{\nabla}\times\vec{E} + \partial_t\vec{B} &= 0
\end{aligned}
\tag{111}
$$

the continuity equation

$$
\partial_t\rho + \vec{\nabla}\cdot\vec{j} = 0
\tag{112}
$$

and the so-called "Lorentz gauge"

$$
\partial_t\Phi + \vec{\nabla}\cdot\vec{A} = 0
\tag{113}
$$

The simplest idea about the 4-current within QED is to assume that it is proportional to the "probability current", which is within our game given by the vector components of $\mathbf{S} = \Sigma\,\gamma_0$.

7. The Phase Space

Up to now, our modeling game referred to the second moments and the elements of $\mathbf{S}$ are second moments such that the observables are given by (averages over) the following quadratic forms:

$$
\begin{aligned}
\mathcal{E} &\propto \psi^T\psi = q_1^2 + p_1^2 + q_2^2 + p_2^2 \\
p_x &\propto -q_1^2 + p_1^2 + q_2^2 - p_2^2 \\
p_y &\propto 2\,(q_1 q_2 - p_1 p_2) \\
p_z &\propto 2\,(q_1 p_1 + q_2 p_2) \\
E_x &\propto 2\,(q_1 p_1 - q_2 p_2) \\
E_y &\propto -2\,(q_1 p_2 + q_2 p_1) \\
E_z &\propto q_1^2 - p_1^2 + q_2^2 - p_2^2 \\
B_x &\propto 2\,(q_1 q_2 + p_1 p_2) \\
B_y &\propto q_1^2 + p_1^2 - q_2^2 - p_2^2 \\
B_z &\propto 2\,(q_1 p_2 - p_1 q_2)
\end{aligned}
\tag{114}
$$

If we analyze the real Dirac matrix coefficents of $\mathbf{S} = \psi\,\psi^T\,\gamma_0$ in terms of the EMEQ and evaluate the quadratic relations between those coefficients, then we obtain:

$$
\begin{aligned}
\vec{P}^2 &= \vec{E}^2 = \vec{B}^2 = \mathcal{E}^2 \\
0 &= \vec{E}^2 - \vec{B}^2 \\
\mathcal{E}^2 &= \tfrac{1}{2}\left(\vec{E}^2 + \vec{B}^2\right) \\
\mathcal{E}\,\vec{P} &= \vec{E} \times \vec{B} \\
\mathcal{E}^3 &= \vec{P} \cdot (\vec{E} \times \vec{B}) \\
m^2 &\propto \mathcal{E}^2 - \vec{P}^2 = 0 \\
\vec{P} \cdot \vec{E} &= \vec{E} \cdot \vec{B} = \vec{P} \cdot \vec{B} = 0
\end{aligned}
\tag{115}
$$

Besides a missing renormalization these equations describe an object without mass but with the geometric properties of light as decribed by electrodynamics, e.g., by the electrodynamic description of electromagnetic waves, which are $\vec{E} \cdot \vec{B} = 0$, $\vec{P} \propto \vec{E} \times \vec{B}$, $\vec{E}^2 = \vec{B}^2$ and so on. Hence single spinors are *light-like* and can not represent massive particles.

Consider the spinor as a vector in a four-dimensional Euclidean space. We write the symmetric matrix $\mathcal{A}$ (or Σ, respectively) as a product in the form of a Gramian:

$$
\mathcal{A} = \mathcal{B}^T \mathcal{B}
\tag{116}
$$

or-componentwise:

$$
\begin{aligned}
\mathcal{A}_{ij} &= \textstyle\sum_k (\mathcal{B}^T)_{ik}\,\mathcal{B}_{kj} \\
&= \textstyle\sum_k \mathcal{B}_{ki}\,\mathcal{B}_{kj}
\end{aligned}
\tag{117}
$$

The last line can be read such that matrix element $\mathcal{A}_{ij}$ is the conventional 4-dimensional scalar product of column vector $\mathcal{B}_i$ with column vector $\mathcal{B}_j$.

From linear algebra we know that Equation (116) yields a non-singular matrix $\mathcal{A}$, iff the column-vectors of the matrix $\mathcal{B}$ are linearly independent. In the orthonormal case, the matrix $\mathcal{A}$ simply is the pure form of a non-singular matrix, *i.e.*, the unit matrix. Hence, if we want to construct a massive object from spinors, we need several spinors to fill the columns of $\mathcal{B}$. The simplest case is the orthogonal case: the combination of four mutual orthogonal vectors. Given a general 4-component Hamiltonian spinor $\psi = (q_1, p_1, q_2, p_2)$, how do we find a spinor that is orthogonal to this one? In 3 (*i.e.*, odd) space dimensions, we know that there are two vectors that are perpendicular to any vector $(x, y, z)^T$, but without fixing the first vector, we can't define the others. In even dimensions this is different: it suffices to find a non-singular skew-symmetric matrix like γ_0 to generate a vector that is orthogonal to ψ, namely $\gamma_0\,\psi$. As in Equation (3), it is the skew-symmetry of the matrix that ensures the orthogonality. A third vector $\gamma_k\,\psi$ must then be orthogonal to ψ *and* to $\gamma_0\,\psi$. It must be skew-symmetric and it must hold $\psi^T \gamma_k^T \gamma_0\,\psi = 0$. This means that the product $\gamma_k^T \gamma_0$ must also be skew-symmetric and hence that γ_k must anti-commute with γ_0:

$$
\begin{aligned}
(\gamma_k^T \gamma_0)^T &= \gamma_0^T \gamma_k = -\gamma_k^T \gamma_0 \\
\Rightarrow\quad & \gamma_0^T \gamma_k + \gamma_k^T \gamma_0 = 0 \\
0 &= \gamma_0 \gamma_k + \gamma_k \gamma_0
\end{aligned}
\tag{118}
$$

Now let us for a moment return to the question of dimensionality. There are in general $2n\,(2n-1)/2$ non-zero independent elements in a skew-symmetric square $2n \times 2n$ matrix. But how many matrices are there in the considered phase space dimensions, *i.e.*, in $1+1$, $3+1$ and $9+1$ (*etc.*) dimensions which anti-commute with γ_0? We need at least $2n-1$ skew-symmetric anti-commuting elements to obtain a diagonal $\mathcal{A}$. However, this implies at least $N-1$ anticommuting elements of the Clifford algebra that square to -1. Hence the ideal case is $2n = N$, which is only true for the Pauli and Dirac algebra. For the Pauli algebra, there is one skew-symmetric element, namely η_0. In the Dirac algebra there are 6 skew-symmetric generators that contain two sets of mutually anti-commuting skew-symmetric matrices: γ_0, γ_{10} and γ_{14} on the one hand and γ_7, γ_8 and γ_9 on the other hand. The next considered Clifford algebra with $N = 9 + 1$ dimensions requires a representation by $2n = 32 = \sqrt{2^{10}}$ -dimensional real matrices. Hence this algebra may not represent a Clifford algebra with more than 10 unit elements-certainly not $2n$. Hence, we can not use the algebra to generate purely massive objects (e.g., diagonal matrices) without further restrictions (i.e., projections) of the spinor ψ.

But what exactly does this mean? Of course we can easily find 32 linearly independent spinors to generate an orthogonal matrix $\mathcal{B}$. So what exactly is special in the Pauli- and Dirac algebra? To see this, we need to understand, what it means that we can use the matrix $\mathcal{B}$ of mutually orthogonal column-spinors

$$\mathcal{B} = (\psi, \gamma_0 \, \psi, \gamma_{10} \, \psi, \gamma_{14} \, \psi) \tag{119}$$

This form implies that we can define the *mass* of the "particle" *algebraically*, and since we have $N - 1 = 3$ anticommuting skew-symmetric matrices in the Dirac algebra, we can find a multispinor $\mathcal{B}$ for *any* arbitrary point in phase space. This does not seem to be sensational at first sight, since this appears to be a property of any Euclidean space. The importance comes from the fact that ψ is a "point" in a very special space-a point in phase space. In fact, we will argue in the following that this possibility to factorize ψ and the density ρ is everything but self-evident.

If we want to simulate a phase space distribution, we can either define a phase space density $\rho(\psi)$ or we use the technique of Monte-Carlo simulations and represent the phase space by (a huge number of random) samples. If we generate a random sample and we like to implement a certain exact symmetry of the density in phase space, then we would (for instance) form a symmetric sample by appending not only a column-vector to $\mathcal{B}$, but also its negative $-\psi$. In this way we obtain a sample with an exact symmetry. In a more general sense: If a phase space symmetry can be represented by a matrix γ_s that allows to associate to an arbitrary phase space point ψ a second point $\gamma_s \, \psi$ where γ_s is skew-symmetric, then we have a certain continuous linear rotational symmetry in this phase space. As we have shown, phase-spaces are intrinsically structured by γ_0 and insofar much more restricted than Euclidean spaces. This is due to the distinction of symplectic from non-symplectic transformations and due to the intrinsic relation to Clifford algebras: *Phase spaces are spaces structured by time*. Within our game, the phase space is the only possible fundamental space.

We may imprint the mentioned symmetry to an arbitrary phase space density ρ by taking all phase space samples that we have so far and adding the same number of samples, each column multiplied by γ_s. Thus, we have a single rotation in the Pauli algebra and two of them in the Dirac algebra:

$$
\begin{aligned}
& \mathcal{B}_0 = \psi \\
\gamma_0 \quad \to \quad & \mathcal{B}_1 = (\psi, \gamma_0 \, \psi) \\
\gamma_{14} \quad \to \quad & \mathcal{B}_2 = (\psi, \gamma_0 \, \psi, \gamma_{14} \, \psi, \gamma_{14} \, \gamma_0 \, \psi) \\
& \quad = (\psi, \gamma_0 \, \psi, \gamma_{14} \, \psi, \gamma_{10} \, \psi)
\end{aligned}
\tag{120}
$$

or:

$$
\begin{aligned}
& \mathcal{B}_0 = \psi \\
\gamma_7 \quad \to \quad & \mathcal{B}_1 = (\psi, \gamma_7 \, \psi) \\
\gamma_8 \quad \to \quad & \mathcal{B}_2 = (\psi, \gamma_7 \, \psi, \gamma_8 \, \psi, \gamma_8 \, \gamma_7 \, \psi) \\
& \quad = (\psi, \gamma_7 \, \psi, \gamma_8 \, \psi, -\gamma_9 \, \psi)
\end{aligned}
\tag{121}
$$

Note that order and sign of the column-vectors in $\mathcal{B}$ are irrelevant—at least with respect to the autocorrelation matrix $\mathcal{B} \, \mathcal{B}^T$. Thus we find that there are two fundamental ways to represent a positive mass in the Dirac algebra and one in the Pauli-algebra. The 4-dimensional phase space of the Dirac algebra is in two independent ways self-matched.

Our starting point was the statement that $2\,n$ linear independent vectors are needed to generate mass. If we can't find $2\,n$ vectors in the way described above for the Pauli and Dirac algebra, then this does (of course) not automatically imply that there are not $2\,n$ linear independent vectors.

But what does it mean that the dimension of the Clifford algebra of observables (N) does not match the dimension of the phase space ($2\,n$) in higher dimensions? There are different physical descriptions given. Classically we would say that a positive definite $2\,n$-component spinor describes a system of n (potentially) coupled oscillators with n frequencies. If $\mathcal{B}$ is orthogonal, then all oscillators have the same frequency, *i.e.*, the system is degenerate. But for $n > 2$ we find that not all eigenmodes can involve the complete $2\,n$-dimensional phase space. This phenomenon is already known in 3 dimensions: The trajectory of the isotropic three-dimensional oscillator always happens in a 2-dimensional plane, *i.e.*, in a subspace. If it did not, then the angular momentum would not be conserved. In this case the isotropy of space would be broken. Hence one may say in some sense that the *isotropy of space* is the reason for a 4-dimensional phase-space and hence the reason for the $3 + 1$-dimensional observable space-time of objects. Or in other words: higher-dimensional spaces are incompatible with isotropy, *i.e.*, with the conservation of angular momentum. There is an intimate connection of these findings to the impossibility of Clifford algebras $Cl_{p,1}$ with $p > 3$ to create a homogeneous "Euclidean" space: Let γ_0 represent time and γ_k with $k \in [1, \dots, N-1]$ the spatial coordinates. The spatial rotators are products of two spatial basis vectors. The generator of rotations in the $(1,2)$-plane is $\gamma_1 \gamma_2$. Then we have 6 rotators in 4 "spatial" dimensions:

$$\gamma_1 \gamma_2, \quad \gamma_1 \gamma_3, \quad \gamma_1 \gamma_4, \quad \gamma_2 \gamma_3, \quad \gamma_2 \gamma_4, \quad \gamma_3 \gamma_4 \tag{122}$$

However, we find that some generators commute and while others anticommute and it can be taken from combinatorics that only sets of 3 mutual anti-commuting rotators can be formed from a set of symmetric anti-commuting γ_k. The 3 rotators

$$\gamma_1 \gamma_2, \quad \gamma_2 \gamma_3, \quad \gamma_1 \gamma_3 \tag{123}$$

mutually anticommute, but $\gamma_1 \gamma_2$ and $\gamma_3 \gamma_4$ commute. Furthermore, in $9 + 1$ dimensions, the spinors are either projections into 4-dimensional subspaces or there are non-zero off-diagonal terms in $\mathcal{A}$, *i.e.*, there is "internal interaction".

Another way to express the above considerations is the following: Only in 4 phase space dimensions we may construct a massive object from a matrix $\mathcal{B}$ that represents a multispinor Ψ of exactly $N = 2n$ single spinors and construct a wave-function according to

$$\Psi = \phi\, \mathcal{B} \tag{124}$$

where $\rho = \phi^2$ is the phase space density.

It is easy to prove and has been shown in Reference [16] that the elements γ_0, γ_{10} and γ_{14} represent parity, time reversal and charge conjugation. The combination of these operators to form a multispinor, may lead (with normalization) to the construction of symplectic matrices $\mathbf{M}$. Some examples are:

$$\begin{aligned}
\mathbf{M} &= (\mathbf{1}\,\psi, \gamma_0\,\psi, -\gamma_{14}\,\psi, -\gamma_{10}\,\psi)/\sqrt{\psi^T\psi} \\
\mathbf{M}\,\gamma_0\,\mathbf{M}^T &= \gamma_0
\end{aligned}$$

$$\begin{aligned}
\mathbf{M} &= (\mathbf{1}\,\psi, -\gamma_{14}\,\psi, -\gamma_{10}\,\psi, \gamma_0\,\psi)/\sqrt{\psi^T\psi} \\
\mathbf{M}\,\gamma_{10}\,\mathbf{M}^T &= \gamma_{10}
\end{aligned} \tag{125}$$

$$\begin{aligned}
\mathbf{M} &= (\gamma_{10}\,\psi, -\mathbf{1}\,\psi, -\gamma_{14}\,\psi, \gamma_0\,\psi)/\sqrt{\psi^T\psi} \\
\mathbf{M}\,\gamma_{14}\,\mathbf{M}^T &= \gamma_{14}
\end{aligned}$$

Hence the combination of the identity and CPT-operators can be arranged such that the multispinor **M** is symplectic with respect to the directions of time γ_0, γ_{10} and γ_{14}, but not with respect to γ_7, γ_8 or γ_9. As we tried to explain, the specific choice of the skew-symmetric matrix γ_0 is determined by a structure defining transformation. Since particles are nothing but dynamical structures in this game, the 6 possible SUMs should stand for 6 different particle types. However, for each direction of time, there are also two choices of the spatial axes. For γ_0 we have chosen γ_1, γ_2 and γ_3, but we could have used $\gamma_4 = \gamma_0 \gamma_1$, $\gamma_5 = \gamma_0 \gamma_2$ and $\gamma_6 = \gamma_0 \gamma_3$ as well.

Thus, there should be either 6 or 12 different types of structures (types of fermions) that can be constructed within the Dirac algebra. The above construction allows for three different types corresponding to three different forms of the symplectic unit matrix, further three types are expected to be related to γ_7, γ_8 and γ_9:

$$
\begin{aligned}
\mathbf{M} &= (\mathbf{1}\,\psi, -\gamma_9\,\psi, -\gamma_8\,\psi, -\gamma_7\,\psi)/\sqrt{\psi^T \psi} \\
\mathbf{M}\,\gamma_7\,\mathbf{M}^T &= \gamma_7
\end{aligned}
$$

$$
\begin{aligned}
\mathbf{M} &= (\mathbf{1}\,\psi, -\gamma_8\,\psi, -\gamma_7\,\psi, -\gamma_9\,\psi)/\sqrt{\psi^T \psi} \\
\mathbf{M}\,\gamma_8\,\mathbf{M}^T &= \gamma_8
\end{aligned}
\tag{126}
$$

$$
\begin{aligned}
\mathbf{M} &= (\gamma_7\,\psi, -\mathbf{1}\,\psi, -\gamma_8\,\psi, -\gamma_9\,\psi)/\sqrt{\psi^T \psi} \\
\mathbf{M}\,\gamma_9\,\mathbf{M}^T &= \gamma_9
\end{aligned}
$$

These matrices describe specific symmetries of the 4-dimensional phase space, *i.e.*, geometric objects in phase space. Therefore massive multispinors can be described as volumes in phase space. If we deform the figure by stretching parameters a, b, c, d such that

$$
\tilde{\mathbf{M}} = (a\,\mathbf{1}\,\psi, -b\,\gamma_0\,\psi, -c\,\gamma_{14}\,\psi, -d\,\gamma_{10}\,\psi)/\sqrt{\psi^T \psi}
\tag{127}
$$

then one obtains with f_k taken from Equation (114):

$$
\begin{aligned}
\tilde{\mathbf{M}}\,\tilde{\mathbf{M}}^T\,\gamma_0 &= \sum_{k=0}^{9} g_k\, f_k\, \gamma_k / \sqrt{\psi^T \psi} \\
g_0 &= a^2 + b^2 + c^2 + d^2 \\
g_1 &= -g_2 = g_3 = a^2 - b^2 + c^2 - d^2 \\
g_4 &= -g_5 = g_6 = a^2 - b^2 - c^2 + d^2 \\
g_7 &= g_8 = g_9 = a^2 + b^2 - c^2 - d^2
\end{aligned}
\tag{128}
$$

This result reproduces the quadratic forms f_k of Equation (114), but furthermore the phase space radii a, b, c and d reproduce the structure of the Clifford algebra, *i.e.*, the classification into the 4 types of observables $\mathcal{E}$, $\vec{P}$, $\vec{E}$ and $\vec{B}$. This means that a deformation of the phase space "unit cell" represents momenta and fields, *i.e.*, the dimensions of the phase space unit cell are related to the appearance of certain symplices:

$$
\begin{aligned}
(a = b)\ \text{AND}\ (c = d) &\Rightarrow \vec{P} = \vec{E} = 0 \\
(a = c)\ \text{AND}\ (b = d) &\Rightarrow \vec{E} = \vec{B} = 0 \\
(a = d)\ \text{AND}\ (b = c) &\Rightarrow \vec{P} = \vec{B} = 0
\end{aligned}
\tag{129}
$$

while for $a = b = c = d$ all vectors but $\mathcal{E}$ vanish. Only in this latter case, the matrix **M** is symplectic for $a = b = c = d = 1$. These relations confirm the intrinsic connection between a classical 4-dimensional Hamiltonian phase space and Clifford algebras in dimension 3+1.

8. Summary and Discussion

Based on three fundamental principles, which describe the form of physics, we have shown that the algebraic structure of coupled classical degrees of freedom is (depending on the number of the DOFs) isomorph to certain Clifford algebras that allow to explain the dimensionality of space-time, to model Lorentz-transformations, the relativistic energy-momentum relation and even Maxwell's equations.

It is usually assumed that we have to define the properties of space-time in the first place: "In Einstein's theory of gravitation matter and its dynamical interaction are based on the notion of an intrinsic geometric structure of the space-time continuum" [36]. However, as we have shown within this "game", it has far more explanatory power to derive and explain space-time from the principles of interaction. Hence we propose to reverse the above statement: The intrinsic geometric structure of the space-time continuum is based on the dynamical interaction of matter. A rigorous consequence of this reversal of perspective is that "space-time" does not need to have a fixed and unique dimensionality at all. It appears that the dimensionality is a property of the type of interaction. However, supposed higher-dimensional space-times (see Reference [16]) would emerge in analogy to the method presented here, for instance in nuclear interaction, then these space-times would not simply be Euclidean spaces of higher dimension. Clifford algebras, especially if they are restricted by symplectic conditions by a Hamiltonian function, have a surprisingly complicated intrinsic structure. As we pointed out, if all generators of a Clifford algebra are symplices, then in $9 + 1$ dimensions, we find k-vectors with $k \in [0..10]$ but k-vectors generated from symplices are themselves symplices only for $k \in [1, 2, 5, 6, 9, 10, \dots]$. However, if space-time is constraint by Hamiltonian motion, then ensembles of oscillators may also clump together to form "objects" with $9 + 1$ or $25 + 1$-dimensional interactions, despite the fact that we gave strong arguments for the fundamentality of the $3 + 1$-dimensional Hamiltonian algebra.

There is no a priori reason to exclude higher order terms-whenever they include constants of motion. However, as the Hamiltonian then involves terms of higher order, we might then need to consider higher order moments of the phase space distribution. In this case we would have to invent an action constant in order to scale ψ.

Our game is based a few general rules and symmetry considerations. The math used in our derivation-taken the results of representation theory for granted-is simple and can be understood on an undergraduate level. And though we never intended to find a connection to string theory, we found-besides the $3 + 1$-dimensional interactions a list of possible higher-dimensional candidates, two of which are also in the focus of string theories, namely $9 + 1 = 10$-dimensional and $25 + 1 = 26$-dimensional theories [37].

We understand this modeling game as a contribution to the demystification (and unification) of our understanding of space-time, relativity, electrodynamics and quantum mechanics. Despite the fact that it has become tradition to write all equations of motion of QED and QM in a way that requires the use of the unit imaginary, our model seems to indicate that it does not have to be that way. Though it is frequently postulated that evolution in time has to be unitary within QM, it appears that symplectic motion does not only suffice, but is superior as it yields the correct number of relevant operators. While in the unitary case, one should expect 16 (15) unitary (traceless) operators for a 4-component spinor, but the natural number of generators in the corresponding symplectic treatment is 10 as found by Dirac himself in QED [2,38]. If a theory contains things which are *not required*, then we have added something arbitrary and artificial. The theory as we described it indicates that in momentum space, which is used here, there is no immediate need for the use of the unit imaginary and no need for more than 10 fundamental generators. The use of the unit imaginary however appears unavoidable when we switch via Fourier transform to the "real space".

There is a dichotomy in physics. On the one hand all *causes* are considered to inhabit space-time (*local causality*), but on the other hand the *physical reasoning* mostly happens in energy-momentum space: There are no Feyman-graphs, no scattering amplitudes, no fundamental physical relations, that do not refer in some way to energy or momentum (-conservation). We treat problems in solid state physics as well as in high energy physics mostly in Fourier space (reciprocal lattice).

We are aware that the rules of the game are, due to their rigour, difficult to accept. However, maybe it does not suffice to speculate that the world might be a hologram (As t'Hooft suggested [39] and Leonard Susskind sketched in his celebrated paper, Reference [40])-we really should play modeling games that might help to decide, if and *how* it could be like that.

Conflicts of Interest: "The author declares no conflict of interest."

Appendix Microcanonical Ensemble

Einstein once wrote that "A theory is the more impressive the greater the simplicity of its premises, the more different kinds of things it relates, and the more extended its area of applicability. Hence the deep impression that classical thermodynamics made upon me. It is the only physical theory of universal content concerning which I am convinced that, within the framework of the applicability of its basic concepts, it will never be overthrown [...]" [10]. We agree with him and we will try to show in the following that this holds also for the branch of thermodynamics that is called statistical mechanics.

By the use of the EMEQ it has been shown, that the expectation values

$$f_k = \frac{\text{Tr}(\gamma_k^2)}{16} \, \bar{\psi} \, \gamma_k \, \psi \tag{A1}$$

can be associated with energy $\mathcal{E}$ and momentum $\vec{p}$ of and with the electric (magnetic) field $\vec{E}$ and $\vec{B}$ as *seen by* a relativistic charged particle. It has also been shown that stable systems can always be transformed in such a way as to bring $\mathcal{H}$ into a diagonal form:

$$\mathbf{F} = \begin{pmatrix} 0 & \omega_1 & 0 & 0 \\ -\omega_1 & 0 & 0 & 0 \\ 0 & 0 & 0 & \omega_2 \\ 0 & 0 & -\omega_2 & 0 \end{pmatrix} \tag{A2}$$

In the following we will use the classical model of the microcanonical ensemble to compute some phase space averages. Let the constant value of the Hamiltonian be $\mathcal{H} = U$ where U is some energy, the volume in phase space $\Phi^\star$ that is limited by the surface of constant energy U is given by [41]:

$$\Phi^\star = \int\limits_{\mathcal{H}<U} dq_1 \, dp_1 \, dq_2 \, dp_2 \tag{A3}$$

and the partition function $\omega^\star$ is the derivative

$$\omega^\star = \frac{d\Phi^\star}{dU} \tag{A4}$$

which is the phase space integral over all states of constant energy U. The average value of any phase space function $\overline{f(p,q)}$ is then given by

$$\overline{f(p,q)} = \frac{1}{\omega^\star} \frac{d}{dU} \int\limits_{\mathcal{H}<U} f(p,q) \, dq_1 \, dp_1 \, dq_2 \, dp_2 \tag{A5}$$

In case of a 2-dimensional harmonic oscillator, for instance, we may take the following parametrization of the phase space:

$$
\begin{aligned}
q_1 &= r \cos{(\alpha)} \cos{(\beta)} \\
p_1 &= r \cos{(\alpha)} \sin{(\beta)} \\
q_2 &= r \sin{(\alpha)} \cos{(\gamma)} \\
p_2 &= r \sin{(\alpha)} \sin{(\gamma)}
\end{aligned}
\tag{A6}
$$

Note that Equation (A6) describes a solution of the equations of motion

$$
\dot{\psi} = \mathbf{F}\,\psi
\tag{A7}
$$

when we replace

$$
\begin{aligned}
\beta &\rightarrow -\omega_1 t \\
\gamma &\rightarrow -\omega_2 t
\end{aligned}
\tag{A8}
$$

This means that the (normalized) integration over β and γ is mathematically identical to an integration over all times (time average). From Equation (A5) one would directly conclude

$$
\overline{f(p,q)} = \frac{1}{\omega^\star} \frac{d}{dU} \int\limits_{\mathcal{H}<U} f(p,q)\,\sqrt{g}\,dr\,d\alpha\,d\beta\,d\gamma
\tag{A9}
$$

where g is the Gramian determinant. However the relative amplitude controlled by the parameter α can not be changed by symplectic transformations and hence remains constant in a closed system. Therefore the phase space trajectory of the oscillator can not cover the complete 3-dim. energy surface, but only a 2-dim. subset thereof. This is known very well in accelerator physics as the emittance preservation of decoupled DOF. And we have shown in Reference [33] that all stable harmonic oscillators are symplectically similar to a decoupled system. Consequently α has to be excluded from the integration of a "single particle" average and has to be treated instead as an additional parameter or "boundary condition":

$$
\overline{f(p,q)} = \frac{1}{\omega^\star} \frac{d}{dU} \int\limits_{\mathcal{H}<U} f(p,q)\,\sqrt{g}\,dr\,d\beta\,d\gamma
\tag{A10}
$$

The Gramian determinant hence is given by:

$$
\begin{aligned}
g &= \mathrm{Det}(\mathbf{G}^T\,\mathbf{G}) \\[4pt]
\mathbf{G} &= \begin{pmatrix}
\frac{\partial q_1}{\partial r} & \frac{\partial q_1}{\partial \beta} & \frac{\partial q_1}{\partial \gamma} \\
\frac{\partial p_1}{\partial r} & \frac{\partial p_1}{\partial \beta} & \frac{\partial p_1}{\partial \gamma} \\
\frac{\partial q_2}{\partial r} & \frac{\partial q_2}{\partial \beta} & \frac{\partial q_2}{\partial \gamma} \\
\frac{\partial p_2}{\partial r} & \frac{\partial p_2}{\partial \beta} & \frac{\partial p_2}{\partial \gamma}
\end{pmatrix}
\end{aligned}
\tag{A11}
$$

so that one finds

$$
\sqrt{g} = r^2 \cos{(\alpha)} \sin{(\alpha)}
\tag{A12}
$$

Accordingly, Equation (A3) has to be written as

$$
\Phi^\star = \int\limits_{\mathcal{H}<U} \sqrt{g}\,dr\,d\beta\,d\gamma
\tag{A13}
$$

We use the abbreviations

$$
\begin{aligned}
\bar{\omega} &= \tfrac{\omega_1 + \omega_2}{2} \\
\Delta\omega &= \tfrac{\omega_1 - \omega_2}{2} \\
\Omega &= \bar{\omega} + \Delta\omega \, \cos(2\alpha)
\end{aligned}
\tag{A14}
$$

The Hamilton function is given in the new coordinates by

$$
\mathcal{H} = \frac{r^2 \, \Omega}{2}
\tag{A15}
$$

so that the condition $\mathcal{H} < U$ translates into

$$
r \leq \sqrt{2\,\varepsilon}
\tag{A16}
$$

where $\varepsilon = U/\Omega$. The integration over β and γ is taken from 0 to 2π. The complete integration results in

$$
\begin{aligned}
\Phi^\star &= \tfrac{2\pi^2}{3} \sin(2\alpha) \left(\tfrac{2U}{\Omega}\right)^{3/2} \\
\omega^\star &= \tfrac{3}{2U} \, \Phi^\star
\end{aligned}
\tag{A17}
$$

The following average values are computed from Equation (A10):

$$
\begin{aligned}
\overline{\mathcal{H}} &= U \\
\overline{\mathcal{H}^2} &= U^2 \\
\overline{f_k} &= \begin{cases} \varepsilon & \text{for} \quad k = 0 \\ \varepsilon \cos(2\alpha) & \text{for} \quad k = 8 \\ 0 & \text{for} \quad k \in \{1-7, 9\} \end{cases} \\
\overline{f_k^2} &= \begin{cases} \overline{f_k}^2 & \text{for} \quad k = 0, 8 \\ \tfrac{1}{4}\varepsilon^2 \left(1 + \cos^2(2\alpha)\right) & \text{for} \quad k \in \{1, 3, 4, 6\} \\ \tfrac{1}{2}\varepsilon^2 \sin^2(2\alpha) & \text{for} \quad k \in \{2, 5, 7, 9\} \end{cases}
\end{aligned}
\tag{A18}
$$

Hence we find that f_0 (energy), f_8 (one spin component) and $\mathcal{H}$ (mass) are "sharp" (*i.e.*, operators with an eigenvalue), while the other "expectation values" have a non-vanishing variance. The fact that spin always has a "direction of quantization", *i.e.*, that only *one single* "sharp" component, can therefore be modelled within our game. It is a consequence of symplectic motion. Note also that the squared expectation values of all even (γ_1, γ_3, γ_4 and γ_6, except γ_0 and γ_8) and all odd (γ_2, γ_5, γ_7 and γ_9) operators are equal (The *even* Dirac matrices are block-diagonal, the *odd* ones not. There are six even symplices and four odd (γ_2, γ_5, γ_7 and γ_9) ones [32]. Obviously this pattern is the reason for the grouping in Equation (A18)).

Consider the coordinates are given by the fields ($\vec{E} \propto \vec{x}$) $\vec{Q} = (f_4, f_5, f_6)^T$ and the momenta as usual by $\vec{P} = (f_1, f_2, f_3)^T$, then the angular momentum $\vec{L}$ should be given by $\vec{L} = \vec{Q} \times \vec{P}$. We obtain the following expectation values from the microcanonical ensemble:

$$
\begin{aligned}
\overline{L_x} &= \overline{L_z} = 0 \\
\overline{L_x^2} &= \overline{L_z^2} = \tfrac{1}{2}\varepsilon^4 \sin^2(2\alpha) \\
\overline{L_y} &= \varepsilon^2 \cos(2\alpha) \\
\overline{L_y^2} &= \varepsilon^4 \cos^2(2\alpha)
\end{aligned}
\tag{A19}
$$

That is-up to a common scale factor of ε (or ε^2, respectively)-we have the same results as in Equation (A18). Consider now the quantum mechanical postulate that the spin component of a fermion is $s_z = \pm s = \pm\tfrac{1}{2}$ and $|\vec{s}|^2 = s_x^2 + s_y^2 + s_z^2 = s(s+1) = \tfrac{3}{4}$. We can "derive" this result (up to a factor) from an isotropy requirement for the 4th order moments, *i.e.*, from the condition that $\langle P_x^2 \rangle = \langle P_y^2 \rangle = \langle P_z^2 \rangle$:

$$
\sin^2(2\alpha) = 2\cos^2(2\alpha)
\tag{A20}
$$

so that

$$\alpha = \frac{1}{2} \arctan \sqrt{2} = 27.3678° \tag{A21}$$

or equivalently with

$$1 + 3 \cos^2 (2\alpha) = 0 \tag{A22}$$

we obtain

$$\cos (2\alpha) = \pm \frac{1}{\sqrt{3}} \tag{A23}$$

With respect to the symplex **F** as defined in Equation (A2), we have

$$\mathbf{F} = \bar{\omega}\,\gamma_0 + \Delta\omega\,\gamma_8 \tag{A24}$$

so that with Equation (A18) one finds

$$f_8 / f_0 = \overline{f_8} / \overline{f_0} = \cos (2\alpha) \tag{A25}$$

The total spin would then be given by the 4th order moments as $\vec{S}^2 = \varepsilon^2$, so that for a spin-$\frac{1}{2}$-particle we would have to normalize to $\varepsilon^2 = (s(s+1))\hbar^2 = \frac{3}{4}\hbar^2$ and hence $f_8 = \sqrt{\frac{3}{4}}\hbar\frac{1}{\sqrt{3}} = \frac{\hbar}{2}$. However, the mass formula (Equation (94)), refers to the second moments, so that in linear theory we would have

$$\mathcal{H} = m = \omega_0 \sqrt{f_0^2 + f_8^2} = \omega_0 \sqrt{\varepsilon^2 + \frac{1}{3}\varepsilon^2} = \omega_0 \sqrt{\frac{4}{3}}\varepsilon = \omega_0\,\hbar \tag{A26}$$

In order to relate this to a frequency difference, we use Equation (A25):

$$
\begin{aligned}
\frac{f_8}{f_0} &= \frac{\Delta\omega}{\bar{\omega}} = \frac{\omega_1 - \omega_2}{\omega_1 + \omega_2} = \cos(2\alpha) = \frac{1}{\sqrt{3}} \\
\Rightarrow \qquad \frac{\omega_1}{\omega_2} &= 2 + \sqrt{3} \\
\Omega &= 4\,\Delta\omega = \frac{4}{\sqrt{3}}\bar{\omega}
\end{aligned}
\tag{A27}
$$

Then from $r = \sqrt{2\varepsilon}$ and Equation (A15) we find

$$
\begin{aligned}
\mathcal{H} &= \frac{r^2 \Omega}{2} = \varepsilon \Omega = \sqrt{\frac{3}{4}}\hbar\,\Omega = \sqrt{\frac{3}{4}}\hbar\frac{4}{\sqrt{3}}\bar{\omega} \\
&= 2\hbar\bar{\omega} = \hbar(\omega_1 + \omega_2)
\end{aligned}
\tag{A28}
$$

To conclude, classical statistical mechanics allows for a description of spin, if the rules of symplectic motion are taken into account. This alone is remarkable. Secondly, assumed that the microcanonical ensemble is the right approach, then the isotropy of the emergent 3+1-dimensional space-time (with respect to 4th-order moments) apparently requires a certain ratio between the frequencies and amplitudes of the two coupled oscillators, *i.e.*, an asymmetry on the fundamental level.

Appendix .1 Entropy and Heat Capacity

The entropy $\mathcal{S}$ of the microcanonical ensemble can be written as [41]:

$$\mathcal{S} = k \log \Phi^\star \tag{A29}$$

The temperature T of the system is given by

$$
\begin{aligned}
\frac{\partial \mathcal{S}}{\partial \mathcal{U}} &= \frac{1}{T} = \frac{\partial (k \log \Phi^\star)}{\partial \mathcal{U}} = k\frac{\omega^\star}{\Phi^\star} \\
&= \frac{3k}{2\mathcal{U}}
\end{aligned}
\tag{A30}
$$

so that energy as a function of temperature is

$$U = \frac{3}{2} k T \tag{A31}$$

and the heat capacity $C_V = \frac{\partial U}{\partial T}$ is (per particle)

$$C_V = \frac{3}{2} k \tag{A32}$$

This important result demonstrates—according to statistical mechanics— that the 3-dimensionality of the "particle" as the energy per DOF is $\frac{kT}{2}$: a two-dimensional harmonic oscillator of fundamental variables is equivalent to an free 3-dimensional "point particle". To our knowledge this is the first *real physical* model of a relativistic point particle.

References

1. Hestenes, D. Modeling games in the Newtonian World. *Am. J. of Phys.* **1992**, *60*, 732–748.
2. Dirac, P.A.M. A remarkable representation of the 3 + 2 de Sitter group. *J. Math. Phys.* **1963**, *4*, 901–909.
3. Kim, Y.S.; Noz, M.E. Symplectic formulation of relativistic quantum mechanics. *J. Math. Phys.* **1981**, *22*, 2289–2293.
4. Kim, Y.S.; Noz, M.E. Dirac's light—Cone coordinate system. *Am. J. Phys.* **1982**, *50*, 721–724.
5. Kim, Y.S.; Noz, M.E. Coupled oscillators, entangled oscillators, and Lorentz-covariant harmonic oscillators. *J. Opt. B Quantum Semiclass. Opt.* **2005**, *7*, S458–S467.
6. Briggs, J.; Eisfeld, A. Equivalence of quantum and classical coherence in electronic energy transfer. *Phys. Rev. E* **2011**, *83*, 051911, doi:10.1103/PhysRevE.83.051911.
7. Briggs, J.; Eisfeld, A. Coherent quantum states from classical oscillator amplitudes. *Phys. Rev. A* **2012**, *85*, 052111, doi:10.1103/PhysRevA.85.052111.
8. Briggs, J.; Eisfeld, A. Quantum dynamics simulations with classical oscillators. *Phys. Rev. A* **2013**, *88*, 062104, doi:10.1103/PhysRevA.88.062104.
9. Jaynes, E.T. *The Electron*; Hestenes, D., Weingartshofer, A., Eds; Kluwer Academic Publishers: Dordrecht, Holland, 1991.
10. Schilpp, P.A. *Albert Einstein: Autobiographical Notes*; Open Court Publishing Company: Chicago, IL, USA, 1979.
11. Lieb, E.H. The stability of matter, *Rev. Mod. Phys.* **1976**, *48*, 553–569.
12. Meyer, K.R.; Hall, G.R.; Offin, D. *Introduction to Hamiltonian Dynamical Systems and the N-Body Problem*, 2nd ed.; Springer, New York, NY, USA, 2000.
13. Lax, P.D. Integrals of Nonlinear Equations of Evolution and Solitary Waves. *Comm. Pure Appl. Math.* **1968**, *21*, 467–490.
14. Dirac, P.A.M. The Quantum Theory of the Electron. *Proc. R. Soc. A* **1928**, *117*, 610–624.
15. Borstnik, N.M.; Nielson, H.B.; Why odd-space and odd-time dimensions in even-dimensional spaces? *Phys. Lett. B* **2000**, *486*, 314–321.
16. Baumgarten, C. Minkowski Spacetime and QED from Ontology of Time. Kastner, R.E., Jeknic-Dugic, J., Jaroszkiewicz, G., Eds; **2014/2015**. arXiv:1409.5338v5; to appear in *Quantum Structural Studies*, forthcoming from World Scientific.
17. Lounesto, P. *Clifford Algebras and Spinors*, 2nd ed.; Cambridge University Press: New York, NY, USA, 2001.
18. Baumgarten, C. A new look at linear (non-?) symplectic ion beam optics in magnets. *Nucl. Instr. Meth. A* **2014**, *735*, 546–551.
19. Dyson, F.J. The Threefold Way. Algebraic Structure of Symmetry Groups and Ensembles in Quantum Mechanics. *J. Math. Phys.* **1962**, *3*, 1199–1215.
20. Ralston, J.P. Berrys phase and the symplectic character of quantum time evolution. *Phys. Rev. A* **1989**, *40*, 4872–4884.
21. Hestenes, D. Observables, Operators, and Complex Numbers in the Dirac Theory. *J. Math. Phys.* **1975**, *16*, 556–572.

22. Baylis, W.E.; Huschlit, J.; Wei, J. Why *i*? *Am. J. Phys.* **1992**, *60*, 788–796.
23. Horn, M.E. Living in a world without imaginaries. *J. Phys.: Conf. Ser.* **2012**, *380*, 012006, doi:10.1088/1742-6596/380/1/012006.
24. Gibbons, G.W. The Emergent Nature of Time and the Complex Numbers in Quantum Cosmology. In *The Arrows of Time: A debate in Cosmology*; Mersini-Houghton, L., Vass, R., Eds; Springer Berlin Heidelberg: Berlin, Germany, 2012; pp. 109–148.
25. Schiff, J.; Poirier, B. Communication: Quantum mechanics without wavefunctions. *J. Chem. Phys.* **2012**, *136*, 031102, doi:10.1063/1.3680558.
26. Strocchi, F. Complex coordinates and quantum mechanics. *Rev. Mod. Phys.* **1966**, *38*, 36–40.
27. Stückelberg, E.C.G. Quantum theory in real Hilbert space. *Helv. Phys. Acta* **1960**, *33*, 727–752.
28. Hestenes, D. *Space-Time-Algebra*; Gordon and Breach: New York, NY, USA, 1966.
29. Hestenes, D. Spacetime Physics with Geometric Algebra. *Am. J. Phys.* **2003**, *71*, 691–714.
30. Gull, S.; Lasenby, A.; Doran, C. Imaginary numbers are not real—The geometric algebra of spacetime. *Found. Phys.* **1993**, *23*, 1175–1201.
31. Pauli, W. Elementary Particles and Their Interactions: Concepts and Phenomena. *Ann. de l'Inst. Henry Poincare* **1936**, *6*, 109–136.
32. Baumgarten, C. Use of real Dirac matrices in two-dimensional coupled linear optics. *Phys. Rev. ST Accel. Beams.* **2011**, *14*, 114002, doi:10.1103/PhysRevSTAB.15.124001.
33. Baumgarten, C. Geometrical method of decoupling. *Phys. Rev. ST Accel. Beams* **2012**, *15*, 124001, doi:10.1103/PhysRevSTAB.15.124001.
34. Schmüser, P. *Feynman-Graphen und Eichtheorien für Experimentalphysiker* (Lecture Notes in Physics No. 295); Springer-Verlag: Heidelburg, Germany, 1988. (in German)
35. Ralston, J.P. Quantum Theory without Planck's Constant. **2012**, arXiv:1203.5557.
36. Schrödinger, E. *Space-Time Structure*; Cambridge University Press: New York, NY, USA, 1950, ISBN 0-521-31520-4.
37. Blumenhagen, R.; Lüst, D.; Theissen, S. *Basic Concepts of String Theory*; Springer-Verlag Berlin Heidelberg: Berlin, Germany, 2013.
38. Dirac, P.A.M. Forms of Relativistic Dynamics. *Rev. Mod. Phys.* **1949**, *21*, 392–399.
39. 't Hooft, G. Dimensional Reduction in Quantum Gravity. **2009**, arxiv:gr-qc/9310026.
40. Susskind, L. The world as a hologram. *J. Math. Phys.* **1995**, *36*, 6377-6396.
41. Becker, R. *Theory of Heat*, 2nd ed.; Springer-Verlag Berlin Heidelberg: Berlin, Germany, 1967.

© 2016 by the author. Licensee MDPI, Basel, Switzerland. This article is an open access article distributed under the terms and conditions of the Creative Commons Attribution (CC BY) license (http://creativecommons.org/licenses/by/4.0/).

symmetry

MDPI

Article

Higher Order Nonclassicality from Nonlinear Coherent States for Models with Quadratic Spectrum

Anaelle Hertz [1], Sanjib Dey [2,3,*], Véronique Hussin [2,4] and Hichem Eleuch [5,6]

[1] Centre for Quantum Information and Communication, École Polytechnique, Université Libre de Bruxelles, Bruxelles 1050, Belgium; ahertz@ulb.ac.be
[2] Centre de Recherches Mathématiques, Université de Montréal, Montréal, QC H3C 3J7, Canada; dey@crm.umontreal.ca (S.D.); veronique.hussin@umontreal.ca (V.H.)
[3] Department of Mathematics and Statistics, Concordia University, Montréal, QC H3G 1M8, Canada
[4] Department de Mathématiques et de Statistique, Université de Montréal, Montréal, QC H3C 3J7, Canada
[5] Department of Physics, McGill University, Montréal, QC H3A 2T8, Canada; heleuch@fulbrightmail.org
[6] Institute for Quantum Science and Engineering, Texas A&M University, College Station, Texas 77843, USA
* Correspondence: dey@crm.umontreal.ca; Tel.: +1-514-343-6111 (ext. 4731)

Academic Editor: Young Suh Kim
Received: 29 February 2016; Accepted: 13 May 2016; Published: 19 May 2016

Abstract: Harmonic oscillator coherent states are well known to be the analogue of classical states. On the other hand, nonlinear and generalised coherent states may possess nonclassical properties. In this article, we study the nonclassical behaviour of nonlinear coherent states for generalised classes of models corresponding to the generalised ladder operators. A comparative analysis among them indicates that the models with quadratic spectrum are more nonclassical than the others. Our central result is further underpinned by the comparison of the degree of nonclassicality of squeezed states of the corresponding models.

Keywords: nonlinear coherent states; nonclassicality; squeezed states; entanglement entropy

1. Introduction

In 1926, Erwin Schrödinger first introduced coherent states while searching for classical like states [1]. In the very first proposal, coherent states were interpreted as nonspreading wavepackets when they move in the harmonic oscillator potential. They minimise the Heisenberg's uncertainty relation, with equal uncertainties in each quadrature. Thus, they are the best quantum mechanical representation of a point in phase space, or in other words, they are the closest possible quantum mechanical states whose behaviour resemble that of classical particles. Coherent states were introduced later by Glauber and Sudarshan [2,3] as a starting point of nonclassicality in terms of the P-function for arbitrary density matrices ρ,

$$\rho = \int P(z)|z\rangle\langle z|d\mathrm{Re}z\,d\mathrm{Im}z \quad \int P(z)d\mathrm{Re}z\,d\mathrm{Im}z = 1 \tag{1}$$

For coherent states, the weight function $P(z)$ can be interpreted as a probability density, as in this case the P-function is a delta function. Glauber defined the nonclassical states as those for which the P-distribution fails to be a probability density. More specifically, if the singularities of P-functions are either of types stronger than those of the delta functions or they are negative, the corresponding states have no classical analogue. A different argument is given in [4], where the author says that a quantum state may be nonclassical even though the P-distribution is a probability density, rather, the nonclassicality is associated with the failure of the Margenau-Hill distribution to be a probability distribution. Nevertheless, we will stick to the convention of Glauber throughout our discussion. It is

170

important to know that while the coherent states play important roles in various branches of physics due to their classical like behaviour, the nonclassical states also have several interesting characteristics for the purpose of quantum information processing. Over last few decades, there have been numerous experimental attempts in search of nonclassical states [5,6]. The underlying inspiration behind all these efforts is that the nonclassical states are the prerequisites for creating entangled states [7], which are the most fundamental requirements for quantum teleportation.

There exist different types of *nonclassical states* in the literature; such as, squeezed states [8,9], photon added coherent states [10], Schrödinger cat states [11], pair coherent states [12], photon subtracted squeezed states [13] and many more. For a concrete review on the subject, one can follow; for instance, [14]. Apart from the aforementioned nonclassical states, the nonclassical properties of different type of coherent states have also been explored, for instance, for Gazeau-Klauder coherent states [15], nonlinear coherent states [16–20] and coherent states in noncommutative space [21,22]. In this article, we explore the comparative analysis of nonclassical nature of nonlinear coherent states for two different type of models associated to the nonlinear ladder operators, and realise that the models with quadratic spectrum produce higher amount of nonclassicality than the others.

Our manuscript is organised as follows: In Section 2, we discuss the general construction procedure of nonlinear coherent and squeezed states. In Section 3, we revisit a well known technique of computing quantum entanglement from nonclassical states by utilising a quantum beam splitter. In Section 4, nonclassical properties of nonlinear coherent states for two general classes of models are analysed. Our conclusions are stated in Section 5.

2. Nonlinear Coherent and Squeezed States

We start with a brief review of coherent states $|z\rangle$, which are defined as the eigenstates of the boson annihilation operator $a|z\rangle = z|z\rangle$, with z being complex eigenvalues [2]. In terms of Fock states they are represented in a compact form by

$$|z\rangle \;=\; \frac{1}{\mathcal{N}(z)} \sum_{n=0}^{\infty} \frac{z^n}{\sqrt{n!}} |n\rangle \qquad \mathcal{N}(z) = e^{\frac{|z|^2}{2}} \tag{2}$$

It was shown that the above expression is equivalent to the one obtained by applying the Glauber's unitary displacement operator, $D(z) = \exp(za^\dagger - z^*a)$, on the vacuum $|0\rangle$. The striking feature of coherent states is that they minimise the uncertainty relation, *i.e.*, $\Delta x^2 \Delta p^2 = 1/4$, with equal uncertainties in each coordinate and, therefore, exhibit reduced noise in optical communications [23]. On the other hand, it is well known that for squeezed states the uncertainty in one of the coordinates x, p becomes squeezed, so that they produce much lower noise in the corresponding quadrature in comparison to the coherent states [8,9]. Squeezed states $|z, \gamma\rangle$ are constructed in two equivalent ways as for coherent states. First, from the eigenvalue definition, $(a + \gamma a^\dagger)|z, \gamma\rangle = z|z, \gamma\rangle$ [24,25] and, second, by acting the squeezing operator, $S(\gamma) = \exp[(\gamma a^\dagger a^\dagger - \gamma^* aa)/2]$, on the coherent state, *i.e.*, $|z, \gamma\rangle = S(\gamma)D(z)|0\rangle$, with $z, \gamma \in \mathbb{C}, |\gamma| < 1$. Consequently, one can express the squeezed states in the number state basis as follows

$$|z, \gamma\rangle \;=\; \frac{1}{\mathcal{N}(z, \gamma)} \sum_{n=0}^{\infty} \frac{1}{\sqrt{n!}} \left(\frac{\gamma}{2}\right)^{n/2} \mathcal{H}_n \left(\frac{z}{\sqrt{2\gamma}}\right) |n\rangle \tag{3}$$

where, $\mathcal{H}_n(\alpha)$ are the Hermite polynomials in the complex variable α. Therefore, the coherent states Equation (2) appear to be a special solution of squeezed states Equation (3) for $\gamma = 0$. A direct generalisation of the above formalism is carried out by the replacement of boson creation and annihilation operators $a, a^\dagger$ with the generalised ladder operators $A, A^\dagger$, such that

$$A = af(a^\dagger a) = f(a^\dagger a + 1)a \tag{4}$$
$$A^\dagger = f(a^\dagger a)a^\dagger = a^\dagger f(a^\dagger a + 1) \tag{5}$$

with $f(a^\dagger a)$ being an operator valued function of the number operator [25,26]. For the choice of $f(a^\dagger a) = 1$, the generalised ladder operators A, $A^\dagger$ fall into the canonical creation and annihilation operators a, $a^\dagger$. Note that, the operator valued function $f(a^\dagger a)$ can be associated with the eigenvalues e_n of a system composed of the ladder operators A and $A^\dagger$, as follows

$$A^\dagger A = f(a^\dagger a)a^\dagger a f(a^\dagger a) = f^2(a^\dagger a)a^\dagger a \sim f^2(n)n = e_n \tag{6}$$

which holds in general for the function $f(n)$. The appearance of additional constant terms in the eigenvalues can be realised by rescaling the composite system of A and $A^\dagger$ correspondingly. In the generalised framework, the coherent states Equation (2) are modified according to the eigenvalue definition, $A|z,f\rangle = z|z,f\rangle$, to

$$|z,f\rangle = \frac{1}{\mathcal{N}(z,f)} \sum_{n=0}^{\infty} \frac{z^n}{\sqrt{n!f(n)!}}|n\rangle \qquad f(0)! = 1 \tag{7}$$

which are familiar as the nonlinear coherent states [27–29], with the nonlinearity arising from the function $f(n)$. The normalisation constant $\mathcal{N}(z,f)$ can be computed by the requirement $\langle z,f|z,f\rangle = 1$. For more informations regarding the generalisation of the coherent states, we refer the readers to [14,30]. By following Equation (6), also it is possible to express the coherent states Equation (7) in terms of the eigenvalues of the corresponding systems as given below [21,31]

$$|z,f\rangle = \frac{1}{\mathcal{N}(z,f)} \sum_{n=0}^{\infty} \frac{z^n}{\sqrt{e_n!}}|n\rangle \qquad e_0 = 0 \tag{8}$$

Construction of squeezed states in the generalised formalism is more involved. Indeed, one requires to expand the squeezed states $|z,\gamma,f\rangle$ in terms of Fock states

$$|z,\gamma,f\rangle = \frac{1}{\mathcal{N}(z,\gamma,f)} \sum_{n=0}^{\infty} \frac{\mathcal{I}(z,\gamma,n)}{\sqrt{n!f(n)!}}|n\rangle = \frac{1}{\mathcal{N}(z,\gamma,f)} \sum_{n=0}^{\infty} \frac{\mathcal{I}(z,\gamma,n)}{\sqrt{e_n!}}|n\rangle \tag{9}$$

followed by the substitution into the corresponding eigenvalue equation, $(A + \gamma A^\dagger)|z,\gamma,f\rangle = z|z,\gamma,f\rangle$, to yield a three term recurrence relation [21,26]

$$\mathcal{I}(z,\gamma,n+1) - z\,\mathcal{I}(z,\gamma,n) + \gamma n f^2(n)\mathcal{I}(z,\gamma,n-1) = 0 \tag{10}$$

with $\mathcal{I}(z,\gamma,0) = 1$ and $\mathcal{I}(z,\gamma,1) = z$. It is worthwhile to mention that due to the inadequacy of general formalism in the literature, one requires to solve the recurrence relation Equation (10) each time to obtain the squeezed states for the model corresponding to the particular value of $f(n)$. To this end, we would like to mention that the nonlinear coherent Equation (7) and squeezed states Equation (9) are very interesting areas of research in recent days. Let us mention, for example, oscillating motion of a particle in a quadratic potential is described by the coherent states. For a non-quadratic confining potential, the nonlinear coherent state could be a good model for such motion states. Also, the nonlinear coherent state could be an adapted model for studying several physical systems with a non-harmonic potential; such as, the Bose-Einstein condensate [32], Hall effects [33], nonlinear field theories [34], etc. Experimentally, the nonlinearity can be realised by the optical pumping (with a laser) of nonlinear mediums; such as, the Kerr medium inside a cavity [35]. The emitted light from such nonlinear mediums can be modelled as a nonlinear coherent light, which are nonclassical in nature [35,36]. The existence of squeezed states in nonlinear mediums have also been found experimentally; see, for example [37].

3. Nonclassicality via Entanglement

The modern quantum optics is undergoing through an extensive research of detecting various nonclassical states and analysing the qualitative behaviour among them. A well known fact is that in terms of nonclassicality the coherent states define a boundary, where the nonclassical effects are absent. There exist several approaches of examining the nonclassicality of states, for instance, by analysing quadrature and photon number squeezing [8,9] or higher order squeezing [38], by examining the negativity of the Wigner function [39], by testing the separability with the Peres-Horodecki criterion [12,40,41], *etc.* Here, we would like to implement a well established protocol, *i.e.*, the quantum beam splitter [7,21,42]. The output states of a beam splitter are realised by acting a unitary operator $\mathcal{B}$ on the input states:

$$|\text{out}\rangle = \mathcal{B}|\text{in}\rangle = e^{\frac{\theta}{2}(a^\dagger b e^{i\phi} - ab^\dagger e^{-i\phi})}|\text{in}\rangle \qquad |\text{in}\rangle = |\psi_1\rangle \otimes |\psi_2\rangle \qquad (11)$$

with $a, a^\dagger$ and $b, b^\dagger$ being the sets of canonical ladder operators operating on the input fields $|\psi_1\rangle$ and $|\psi_2\rangle$, respectively. ϕ is the phase difference between the reflected and transmitted fields and $\theta \in [0, \pi]$ denotes the angle of the beam splitter. For more details on the device we refer the readers to [42]. It is well known that when we pass the coherent state Equation (2) through one of the input ports and a vacuum state $|0\rangle$ through the other, we obtain two coherent states at the output ports with the intensity in each of the outputs being halved. However, passing nonclassical states through the inputs create inseparable/entangled states at the output. Thus it can be utilised as a simple and efficient tool to test the nonclassicality of the input states by checking whether the output states are entangled or not. Let us briefly review the detailed procedure of computing the outputs for arbitrary inputs. Consider, for instance, the generalised squeezed states Equation (9) at one of the inputs, such that the output states become

$$|\text{out}\rangle = \mathcal{B}(|z, \gamma, f\rangle \otimes |0\rangle) = \frac{1}{\mathcal{N}(z, \gamma, f)} \sum_{n=0}^{\infty} \frac{\mathcal{I}(z, \gamma, n)}{\sqrt{n! f(n)!}} \mathcal{B}(|n\rangle \otimes |0\rangle) \qquad (12)$$

where the action of the beam splitter on the Fock state is known [7]

$$\mathcal{B}(|n\rangle \otimes |0\rangle) = \sum_{q=0}^{n} \binom{n}{q}^{1/2} t^q r^{n-q} (|q\rangle \otimes |n-q\rangle) \qquad (13)$$

with $t = \cos(\theta/2)$ and $r = -e^{i\phi}\sin(\theta/2)$ being the transmission and reflection coefficients, respectively. By substituting Equation (13) into Equation (12) and following the similar steps as in [21], we obtain the linear entropy of the output states, $S = 1 - \text{Tr}(\rho_a^2)$, with ρ_a^2 being the reduced density matrix of system a obtained by performing a trace over system b of the density matrix $\rho_{ab} = |\text{out}\rangle\langle\text{out}|$, as follows

$$
\begin{aligned}
S = {} & 1 - \frac{1}{\mathcal{N}^4(z, \gamma, f)} \sum_{q=0}^{\infty} \sum_{s=0}^{\infty} \sum_{m=0}^{\infty} \sum_{n=0}^{\infty} |t|^{2(q+s)} |r|^{2(m+n)} \\
& \times \frac{\mathcal{I}(z, \gamma, m+q)\mathcal{I}^*(z, \gamma, m+s)\mathcal{I}(z, \gamma, n+s)\mathcal{I}^*(z, \gamma, n+q)}{q!s!m!n!f(m+q)!f(m+s)!f(n+s)!f(n+q)!}
\end{aligned} \qquad (14)
$$

The linear entropy S, which varies between 0 and 1 is used to measure entanglement, 0 corresponding to the case of a pure state while 1 refers to a maximally entangled state. Note that the phase ϕ between the transmitted and reflected fields is not relevant for the present case as they appear with their square of their norms in Equation (14). We always work on 50:50 beam splitter, *i.e.*, with $\theta = \pi/2$. The reason is that the symmetric beam splitter produces the highest amount of entanglement among all other cases. Moreover, we perform a comparative study between different models so that the angle of the beam splitter, θ, also becomes irrelevant to us.

4. Nonclassical Models

In this section, we construct coherent states and squeezed states of several types of models associated to the operators $A, A^\dagger$ and compare their qualitative behaviour in terms of nonclassicality. To start with, let us recall some well known facts of the harmonic oscillator along with a deformed version of it.

4.1. Linear versus Quadratic Spectrum

The general form of coherent and squeezed states associated with the model corresponding to $f(n)$ are given in Equations (8) and (9), respectively. The harmonic oscillator is the limit of the generalised expressions for $f(n) = 1$. For the case when $f(n) = \sqrt{n}$, the corresponding eigenvalues of $A^\dagger A$ are $e_n = f^2(n)n = n^2$, which we call a deformed version of the harmonic oscillator with quadratic spectrum. Let us first study the squeezing properties of the above two systems in the quadrature components

$$x = \frac{1}{2}(a + a^\dagger) \qquad p = \frac{1}{2i}(a - a^\dagger) \tag{15}$$

which are nothing but dimensionless position and momentum operators, respectively. It is well known that for the coherent states of harmonic oscillator the uncertainty relation is saturated, *i.e.*, $\Delta x \Delta p = 1/2$, with $\Delta x = \Delta p = 1/\sqrt{2}$. When we consider the quadratic case, *i.e.*, $f(n) = \sqrt{n}$, one of the quadratures Δx becomes squeezed while the the other one Δp is expanded correspondingly. Quadrature squeezing is a well known phenomenon, which indicates the nonclassical nature of a quantum state [8,9]. A comparative analysis of the two cases in Figure 1 shows that the quadratic case is nonclassical, while the harmonic oscillator resembles the classical behaviour.

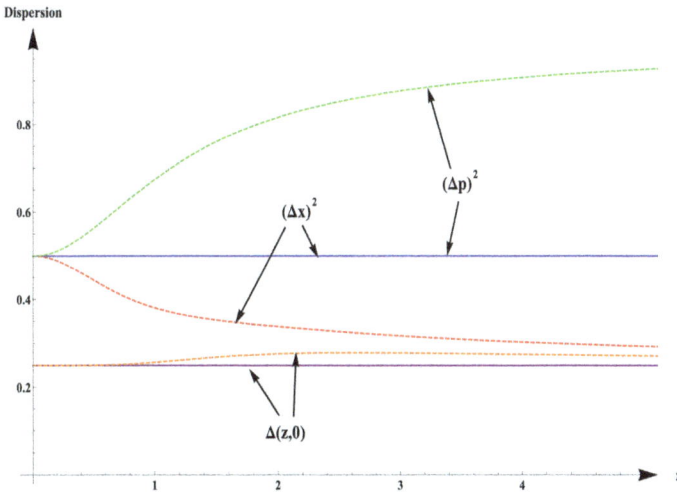

Figure 1. Dispersion in position $(\Delta x)^2$, momentum $(\Delta p)^2$ and product of dispersions $\Delta(z, 0) = (\Delta x)^2(\Delta p)^2$ for the usual (solid lines) and quadratic (dashed lines) coherent states of the harmonic oscillator.

Next we compare the density probabilities between the two cases, which are shown in Figure 2. For the harmonic oscillator case, the probability density in x remains uniform for all values of z. Whereas, in the quadratic case it becomes narrower with the increase of the value of z. This provides us an additional evidence of nonclassicality of the states.

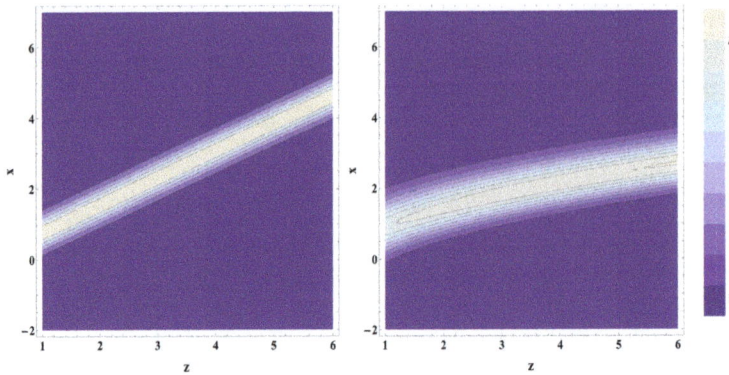

Figure 2. Density probabilities $|\psi(z,f)|^2$ for usual (**left**) *versus* the quadratic (**right**) coherent states of the harmonic oscillator.

Let us now move on to the squeezed states. We first solve the recurrence relation Equation (10) for $f(n) = \sqrt{n}$, in terms of the Gauss hypergeometric function $_2F_1$ as follows

$$\mathcal{I}(z,\gamma,n) = i^n \gamma^{n/2} n! \,_2F_1\left[-n, \frac{1}{2} + \frac{iz}{2\sqrt{\gamma}}; 1; 2\right] \tag{16}$$

which, when substituted in Equation (9), we obtain the explicit form of the squeezed states for the corresponding model. We know that the squeezed states of any model are always nonclassical [8,9]. However, from the previous analysis we can claim that the squeezed states in the quadratic case are more nonclassical than those of the harmonic oscillator. To verify, we utilise the protocol discussed in Section 3. We compute the linear entropies of the output states for the input states $|z,\gamma\rangle \otimes |0\rangle$ and $|z,\gamma,f\rangle \otimes |0\rangle$, respectively, corresponding to the two different cases. The comparison of the outcomes are demonstrated in Figure 3 for different values of the squeezing parameter γ. Although for very small values of z, we obtain lower amount of entropy in the quadratic case than in the case of harmonic oscillator. It may happen because of the fact that the lower value of z corresponds to the case when the uncertainties are almost same for both of the cases. However, we do not have a concrete argument here behind this limitation, and we leave it as an open problem. Nevertheless, our results are valid for higher values of z.

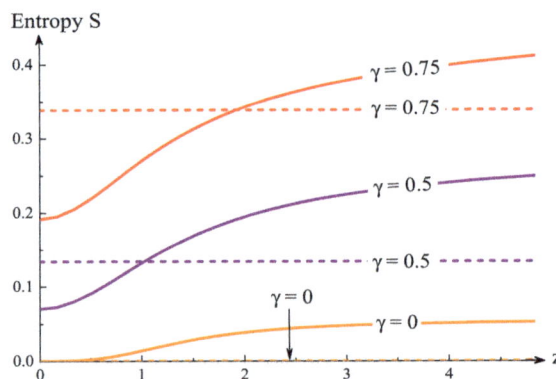

Figure 3. Linear entropy for squeezed states of harmonic oscillator (dashed lines) *versus* quadratic spectrum (solid lines).

4.2. Linear Plus Quadratic Spectrum

In this section, we would like to explore the nonclassical behaviour of the models with eigenvalues of type $e_n = An + Bn^2$, such that $f(n) = \sqrt{A + Bn}$, where A, B being some parameters, $B \neq 0$. Note that there exists various models which belong to this class of eigenvalues for different parameters, for instance, the Pöschl-Teller model [43], the harmonic oscillator in noncommutative space [44], *etc.* The squeezed states for the corresponding class of models are constructed by solving the recurrence relation Equation (10)

$$\mathcal{I}(z, \gamma, n) = i^n (\gamma B)^{n/2} \left(1 + \frac{A}{B}\right)^{(n)} {}_2F_1\left[-n, \frac{1}{2} + \frac{A}{2B} + \frac{iz}{2\sqrt{\gamma B}}; 1 + \frac{A}{B}; 2\right] \tag{17}$$

and computing

$$f^2(n)! = B^n \left(1 + \frac{A}{B}\right)^{(n)} \tag{18}$$

where $Q^{(n)} := \prod_{k=0}^{n-1}(Q + k)$ denotes the Pochhammer symbol with the raising factorial. We calculate the entanglement of the coherent and squeezed states of the corresponding models for different values of A and B, as shown in Figure 4a,b, respectively. Note that, the expression of linear entropy Equation (14) contains infinite sums and, therefore, it is not an easy task to compute the entanglement. In addition, one needs to be careful when one deals with the entropy for finite number of levels. We have carefully investigated the minimum requirement of the number of levels for the convergence of the series for the corresponding values of the parameters that we have chosen in our computation. For instance, in case of Figure 3, we have considered the number of levels equals to 40 and in case of Figure 4 it is 30.

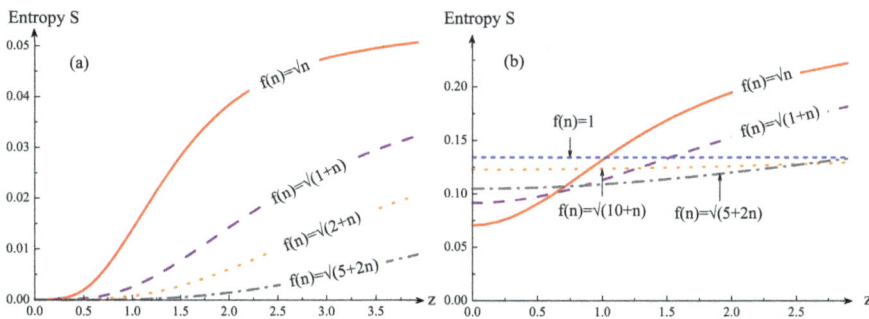

Figure 4. Linear entropy for (**a**) coherent states; (**b**) squeezed states for $\gamma = 0.5$.

As indicated in the previous section, apart from the case of the harmonic oscillator we obtain finite amount of entropies for coherent states in each case corresponding to different values of the function $f(n)$ as shown in Figure 4a. This suggests the general nonclassical nature of the nonlinear coherent states. In addition, we observe that the quadratic case, $f(n) = \sqrt{n}$, produces the highest amount of entanglement among all other cases. The similar effects are also found for the case of squeezed states as demonstrated in Figure 4b.

5. Conclusions

We have studied the nonclassical behaviour of nonlinear coherent states for several classes of models. We observed that the nonclassicality depends on the nature of the function $f(n)$. By comparing the entanglement properties of the coherent states and the squeezed states for different models corresponding to different values of $f(n)$, we showed that the model with quadratic spectrum produces the highest amount of entanglement. Thus, our analysis might help someone to find out the right models, whose degree of nonclassicality may be higher and, therefore, may produce more entanglement in comparison to other models. For further investigations, one can study few other nonclassical properties [8,9,45] as discussed in Section 3 of our models to confirm our findings. One can also perform a similar analysis as presented here, or utilise any other powerful techniques available in the literature to verify whether the quadratic spectrum of other classes of models delivers the highest nonclassicality and entanglement, which we leave as open problems.

Acknowledgments: Anaelle Hertz acknowledges the support from the Canadian NSERC research fellowship and from the Belgian FNRS research fellowship. Sanjib Dey is supported by the Postdoctoral Fellowship jointly funded by the Laboratory of Mathematical Physics of the Centre de Recherches Mathématiques and by Syed Twareque Ali, Marco Bertola and Véronique Hussin. Veronique Hussin acknowledges the support of research grants from NSERC of Canada.

Author Contributions: All of the authors contributed equally to this article.

Conflicts of Interest: The authors declare no conflict of interest.

References

1. Schrödinger, E. Der stetige übergang von der mikro-zur makromechanik. *Naturwissenschaften* **1926**, *14*, 664–666.
2. Glauber, R.J. Coherent and incoherent states of the radiation field. *Phys. Rev.* **1963**, *131*, 2766, doi:10.1103/PhysRev.131.2766.
3. Sudarshan, E.C.G. Equivalence of semiclassical and quantum mechanical descriptions of statistical light beams. *Phys. Rev. Lett.* **1963**, *10*, 277–279.
4. Johansen, L.M. Nonclassical properties of coherent states. *Phys. Lett. A* **2004**, *329*, 184–187.
5. Meekhof, D.M.; Monroe, C.; King, B.E.; Itano, W.M.; Wineland, D.J. Generation of nonclassical motional states of a trapped atom. *Phys. Rev. Lett.* **1996**, *76*, 1796, doi:10.1103/PhysRevLett.76.1796.
6. Bose, S.; Jacobs, K.; Knight, P.L. Preparation of nonclassical states in cavities with a moving mirror. *Phys. Rev. A* **1997**, *56*, 4175, doi:10.1103/PhysRevA.56.4175.
7. Kim, M.S.; Son, W.; Bužek, V.; Knight, P.L. Entanglement by a beam splitter: Nonclassicality as a prerequisite for entanglement, *Phys. Rev. A* **2002**, *65*, 032323.
8. Walls, D.F. Squeezed states of light. *Nature* **1983**, *306*, 141–146.
9. Loudon, R.; Knight, P.L. Squeezed light. *J. Mod. Opt.* **1987**, *34*, 709–759.
10. Agarwal, G.S.; Tara, K. Nonclassical properties of states generated by the excitations on a coherent state. *Phys. Rev. A* **1991**, *43*, 492, doi:10.1103/PhysRevA.43.492.
11. Xia, Y.; Guo, G. Nonclassical properties of even and odd coherent states. *Phys. Lett. A* **1989**, *136*, 281–283.
12. Agarwal, G.S.; Biswas, A. Quantitative measures of entanglement in pair-coherent states. *J. Opt. B* **2005**, *7*, 350–354.
13. Wakui, K.; Takahashi, H.; Furusawa, A.; Sasaki, M. Photon subtracted squeezed states generated with periodically poled KTiOPO$_4$. *Opt. Exp.* **2007**, *15*, 3568–3574.
14. Dodonov, V.V. Nonclassical states in quantum optics: A squeezed review of the first 75 years. *J. Opt. B* **2002**, *4*, R1.
15. Roy, B.; Roy, P. Gazeau-Klauder coherent state for the Morse potential and some of its properties. *Phys. Lett. A* **2002**, *296*, 187–191.
16. Roy, B.; Roy, P. New nonlinear coherent states and some of their nonclassical properties. *J. Phys. B* **2000**, *2*, 65.
17. Choquette, J.J.; Cordes, J.G.; Kiang, D. Nonlinear coherent states: Nonclassical properties. *J. Opt. B* **2003**, *5*, 56.
18. Obada, A.S.; Darwish, M.; Salah, H.H. Some non-classical properties of a class of new nonlinear coherent states. *J. Mod. Opt.* **2005**, *52*, 1263–1274.
19. Récamier, J.; Gorayeb, M.; Mochán, W.L.; Paz, J.L. Nonlinear coherent states and some of their properties. *Int. J. Theor. Phys.* **2008**, *47*, 673–683.

20. Tavassoly, M.K. On the non-classicality features of new classes of nonlinear coherent states. *Opt. Commun.* **2010**, *283*, 5081–5091.

21. Dey, S.; Hussin, V. Entangled squeezed states in noncommutative spaces with minimal length uncertainty relations. *Phys. Rev. D* **2015**, *91*, 124017.

22. Dey, S.; Fring, A.; Hussin, V. Nonclassicality *versus* entanglement in a noncommutative space. arXiv:1506.08901.

23. Gerry, C.; Knight, P. *Introductory Quantum Optics*; Cambridge University Press: Cambridge, UK, 2005.

24. Fu, H.-C.; Sasaki, R. Exponential and Laguerre squeezed states for $su(1,1)$ algebra and the Calogero-Sutherland model. *Phys. Rev. A* **1996**, *53*, 3836.

25. Alvarez-Moraga, N.; Hussin, V. Generalized coherent and squeezed states based on the $h(1) \oplus su(2)$ algebra. *J. Math. Phys.* **2002**, *43*, 2063–2096.

26. Angelova, M.; Hertz, A.; Hussin, V. Squeezed coherent states and the one-dimensional Morse quantum system. *J. Phys. A* **2012**, *45*, 244007.

27. Man'ko, V.I.; Marmo, G.; Solimeno, S.; Zaccaria, F. Physical nonlinear aspects of classical and quantum q-oscillators. *Int. J. Mod. Phys. A* **1993**, *8*, 3577–3597.

28. Filho, R.L.M.; Vogel, W. Nonlinear coherent states. *Phys. Rev. A* **1996**, *54*, 4560.

29. Sivakumar, S. Studies on nonlinear coherent states. *J. Opt. B* **2000**, *2*, R61.

30. Perelomov, A. *Generalized Coherent States and Their Applications*; Springer-Verlag: Berlin, Germany, 2012.

31. Gazeau, J.P.; Klauder, J.R. Coherent states for systems with discrete and continuous spectrum. *J. Phys. A* **1999**, *32*, 123.

32. Greiner, M.; Mandel, O.; Hänsch, T.W.; Bloch, I. Collapse and revival of the matter wave field of a Bose-Einstein condensate. *Nature* **2002**, *419*, 51–54.

33. Ezawa, Z.F. *Quantum Hall Effects: Field Theoretical Approach and Related Topics*; World Scientific: Singapore, 2008.

34. Ashtekar, A.; Lewandowski, J.; Marolf, D.; Mourao, J.; Thiemann, T. Coherent State Transforms for Spaces of Connections. *J. Funct. Anal.* **1996**, *135*, 519–551.

35. Ourjoumtsev, A.; Kubanek, A.; Koch, M.; Sames, C.; Pinkse, P.W.H.; Rempe, G.; Murr, K. Observation of squeezed light from one atom excited with two photons. *Nature* **2011**, *474*, 623–626.

36. Baas, A.; Karr, J.P.; Eleuch, H.; Giacobino, E. Optical bistability in semiconductor microcavities. *Phys. Rev. A* **2004**, *69*, 023809.

37. Huang, K.; Le Jeannic, H.; Ruaudel, J.; Verma, V.B.; Shaw, M.D.; Marsili, F.; Nam, S.W.; Wu, E.; Zeng, H.; Jeong, Y.-C.; *et al.* Optical synthesis of large-amplitude squeezed coherent-state superpositions with minimal resources. *Phys. Rev. Lett.* **2015**, *115*, 023602.

38. Hong, C.K.; Mandel, L. Higher-order squeezing of a quantum field. *Phys. Rev. Lett.* **1985**, *54*, 323, doi:10.1103/PhysRevLett.54.323.

39. Wigner, E. On the quantum correction for thermodynamic equilibrium. *Phys. Rev.* **1932**, *40*, 749, doi:10.1103/PhysRev.40.749.

40. Duan, L.M.; Giedke, G.; Cirac, J.I.; Zoller, P. Inseparability criterion for continuous variable systems. *Phys. Rev. Lett.* **2000**, *84*, 2722, doi:10.1103/PhysRevLett.84.2722.

41. Simon, R. Peres-Horodecki separability criterion for continuous variable systems. *Phys. Rev. Lett.* **2000**, *84*, 2726, doi:10.1103/PhysRevLett.84.2726.

42. Campos, R.A.; Saleh, B.E.A.; Teich, M.C. Quantum-mechanical lossless beam splitter: $SU(2)$ symmetry and photon statistics. *Phys. Rev. A* **1989**, *40*, 1371–1384.

43. Antoine, J.P.; Gazeau, J.P.; Monceau, P.; Klauder, J.R.; Penson, K.A. Temporally stable coherent states for infinite well and Pöschl-Teller potentials. *J. Math. Phys.* **2001**, *42*, 2349–2387.

44. Dey, S.; Fring, A.; Khantoul, B. Hermitian *versus* non-Hermitian representations for minimal length uncertainty relations. *J. Phys. A* **2013**, *46*, 335304.

45. Dey, S. q-deformed noncommutative cat states and their nonclassical properties. *Phys. Rev. D* **2015**, *91*, 044024.

© 2016 by the authors. Licensee MDPI, Basel, Switzerland. This article is an open access article distributed under the terms and conditions of the Creative Commons Attribution (CC BY) license (http://creativecommons.org/licenses/by/4.0/).

symmetry

MDPI

Article

On Solutions for Linear and Nonlinear Schrödinger Equations with Variable Coefficients: A Computational Approach

Gabriel Amador [1], Kiara Colon [1], Nathalie Luna [1], Gerardo Mercado [1], Enrique Pereira [1] and Erwin Suazo [2],*

[1] Department of Mathematical Sciences, University of Puerto Rico at Mayagüez, Mayagüez, Puerto Rico, PR 00681-9018, USA; gabriel.amador@upr.edu (G.A.); kiara.colon1@upr.edu (K.C.); nathalie.luna@upr.edu (N.L.); gerardo.mercado1@upr.edu (G.M.); enrique.pereira@upr.edu (E.P.)

[2] School of Mathematical and Statistical Sciences, University of Texas at Rio Grande Valley, Edinburg, TX 78539-2999, USA

* Correspondence: erwin.suazo@utrgv.edu; Tel.: +1-956-665-7087

Academic Editor: Young Suh Kim
Received: 2 March 2016; Accepted: 6 May 2016; Published: 27 May 2016

Abstract: In this work, after reviewing two different ways to solve Riccati systems, we are able to present an extensive list of families of integrable nonlinear Schrödinger (NLS) equations with variable coefficients. Using Riccati equations and similarity transformations, we are able to reduce them to the standard NLS models. Consequently, we can construct bright-, dark- and Peregrine-type soliton solutions for NLS with variable coefficients. As an important application of solutions for the Riccati equation with parameters, by means of computer algebra systems, it is shown that the parameters change the dynamics of the solutions. Finally, we test numerical approximations for the inhomogeneous paraxial wave equation by the Crank-Nicolson scheme with analytical solutions found using Riccati systems. These solutions include oscillating laser beams and Laguerre and Gaussian beams.

Keywords: generalized harmonic oscillator; paraxial wave equation; nonlinear schrödinger-type equations; riccati systems; solitons

PACS: J0101

---

1. Introduction

In modern nonlinear sciences, some of the most important models are the variable coefficient nonlinear Schrödinger-type ones. Applications include long distance optical communications, optical fibers and plasma physics, (see [1–25] and references therein).

In this paper, we first review a generalized pseudoconformal transformation introduced in [26] (lens transform in optics [27] see also [28]). As the first main result, we will use this generalized lens transformation to construct solutions of the general variable coefficient nonlinear Schrödinger equation (VCNLS):

$$i\psi_t = -a(t)\psi_{xx} + \left(b(t)x^2 - f(t)x + G(t)\right)\psi - ic(t)x\psi_x - id(t)\psi + ig(t)\psi_x + h(t)|\psi|^{2s}\psi, \quad (1)$$

extending the results in [1]. If we make $a(t) = \Lambda/4\pi n_0$, Λ being the wavelength of the optical source generating the beam, and choose $c(t) = g(t) = 0$, then Equation (**??**) models a beam propagation inside of a planar graded-index nonlinear waveguide amplifier with quadratic refractive index represented by $b(t)x^2 - f(t)x + G(t)$, and $h(t)$ represents a Kerr-type nonlinearity of the waveguide amplifier, while $d(t)$ represents the gain coefficient. If $b(t) > 0$ [11] (resp. $b(t) < 0$, see [13]) in the low-intensity limit, the graded-index waveguide acts as a linear defocusing (focusing) lens.

Depending on the selections of the coefficients in Equation (1), its applications vary in very specific problems (see [16] and references therein):

- Bose-Einstein condensates: $b(\cdot) \neq 0$, a, h constants and other coefficients are zero.
- Dispersion-managed optical fibers and soliton lasers [9,14,15]: $a(\cdot), h(\cdot), d(\cdot) \neq 0$ are respectively dispersion, nonlinearity and amplification, and the other coefficients are zero. $a(\cdot)$ and $h(\cdot)$ can be periodic as well, see [29].
- Pulse dynamics in the dispersion-managed fibers [10]: $h(\cdot) \neq 0$, a is a constant and other coefficients are zero.

In this paper, to obtain the main results, we use a fundamental approach consisting of the use of similarity transformations and the solutions of Riccati systems with several parameters inspired by the work in [30]. Similarity trasformations have been a very popular strategy in nonlinear optics since the lens transform presented by Talanov [27]. Extensions of this approach have been presented in [26,28]. Applications include nonlinear optics, Bose-Einstein condensates, integrability of NLS and quantum mechanics, see for example [3,31–33], and references therein. E. Marhic in 1978 introduced (probably for the first time) a one-parameter $\{\alpha(0)\}$ family of solutions for the linear Schrödinger equation of the one-dimensional harmonic oscillator, where the use of an explicit formulation (classical Melher's formula [34]) for the propagator was fundamental. The solutions presented by E. Marhic constituted a generalization of the original Schrödinger wave packet with oscillating width.

In addition, in [35], a generalized Melher's formula for a general linear Schrödinger equation of the one-dimensional generalized harmonic oscillator of the form Equation (1) with $h(t) = 0$ was presented. For the latter case, in [36–38], multiparameter solutions in the spirit of Marhic in [30] have been presented. The parameters for the Riccati system arose originally in the process of proving convergence to the initial data for the Cauchy initial value problem Equation (1) with $h(t) = 0$ and in the process of finding a general solution of a Riccati system [38,39]. In addition, Ermakov systems with solutions containing parameters [36] have been used successfully to construct solutions for the generalized harmonic oscillator with a hidden symmetry [37], and they have also been used to present Galilei transformation, pseudoconformal transformation and others in a unified manner, see [37]. More recently, they have been used in [40] to show spiral and breathing solutions and solutions with bending for the paraxial wave equation. In this paper, as the second main result, we introduce a family of Schrödinger equations presenting periodic soliton solutions by using multiparameter solutions for Riccati systems. Furthermore, as the third main result, we show that these parameters provide a control on the dynamics of solutions for equations of the form Equation (1). These results should deserve numerical and experimental studies.

This paper is organized as follows: In Section 2, by means of similarity transformations and using computer algebra systems, we show the existence of Peregrine, bright and dark solitons for the family Equation (1). Thanks to the computer algebra systems, we are able to find an extensive list of integrable VCNLS, in the sense that they can be reduced to the standard integrable NLS, see Table 1. In Section 3, we use different similarity transformations than those used in Section 3. The advantage of the presentation of this section is a multiparameter approach. These parameters provide us a control on the center axis of bright and dark soliton solutions. Again in this section, using Table 2 and by means of computer algebra systems, we show that we can produce a very extensive number of integrable VCNLS allowing soliton-type solutions. A supplementary Mathematica file is provided where it is evident how the variation of the parameters change the dynamics of the soliton solutions. In Section 4, we use a finite difference method to compare analytical solutions described in [41] (using similarity transformations) with numerical approximations for the paraxial wave equation (also known as linear Schrödinger equation with quadratic potential).

Table 1. Families of NLS with variable coefficients.

#	Variable Coefficient NLS	Solutions ($j=1,2,3$)
1	$i\psi_t = l_0\psi_{xx} - \frac{bmt^{m-1}+b^2t^{2m}}{4l_0}x^2\psi$ $-ibt^m x\psi_x - \lambda l_0 e^{\frac{-bt^{m+1}}{m+1}}\|\psi\|^2\psi$	$\psi_j(x,t) = \frac{1}{\sqrt{e^{\frac{-bt^{m+1}}{m+1}}}}e^{i\left(\frac{bt^m}{4}l_0x^2\right)}u_j(x,t)$
2	$i\psi_t = l_0\psi_{xx} - \frac{t^{-2}}{2l_0}x^2\psi$ $+i\frac{1}{t}x\psi_x - \lambda l_0 t\|\psi\|^2\psi$	$\psi_j(x,t) = \frac{1}{\sqrt{t}}e^{i\left(\frac{-1}{4t}l_0x^2\right)}u_j(x,t)$
3	$i\psi_t = l_0\psi_{xx} - \left(\frac{c^2}{4}l_0\right)x^2\psi$ $+icx\psi_x - \lambda l_0 e^{ct}\|\psi\|^2\psi$	$\psi_j(x,t) = \frac{1}{\sqrt{e^{ct}}}e^{i\left(\frac{-c}{4}l_0x^2\right)}u_j(x,t)$
4	$i\psi_t = l_0\psi_{xx} - \frac{b^2}{4l_0}t^k x^2\psi$ $+ibx\psi_x - \lambda l_0 e^{bt}\|\psi\|^2\psi$	$\psi_j(x,t) = \frac{1}{\sqrt{e^{bt}}}e^{i\left(\frac{-b}{4}l_0x^2\right)}u_j(x,t)$
5	$i\psi_t = l_0\psi_{xx} - \frac{abe^{bt}+a^2e^{2bt}}{4l_0}x^2\psi$ $-iae^{bt}x\psi_x - \lambda l_0 e^{\frac{a-ae^{bt}}{b}}\|\psi\|^2\psi$	$\psi_j(x,t) = \frac{1}{\sqrt{e^{\frac{a-ae^{bt}}{b}}}}e^{i\left(\frac{ae^{bt}}{4}l_0x^2\right)}u_j(x,t)$
6	$i\psi_t = l_0\psi_{xx} - \frac{1}{4l_0}x^2\psi$ $-icoth(t)x\psi_x - \lambda l_0 csch(t)\|\psi\|^2\psi$	$\psi_j(x,t) = \frac{1}{\sqrt{csch(t)}}e^{i\left(\frac{coth(t)}{4}l_0x^2\right)}u_j(x,t)$
7	$i\psi_t = l_0\psi_{xx} - \frac{1}{4l_0}x^2\psi$ $-itan(t)x\psi_x - \lambda l_0 cos(t)\|\psi\|^2\psi$	$\psi_j(x,t) = \frac{1}{\sqrt{cos(t)}}e^{i\left(\frac{tan(t)}{4}l_0x^2\right)}u_j(x,t)$
8	$i\psi_t = l_0\psi_{xx} - \frac{bt^{-1}+b^2ln^2(t)}{4l_0}x^2\psi$ $-ibln(t)x\psi_x - \lambda l_0 t^{-bt}e^{bt}\|\psi\|^2\psi$	$\psi_j(x,t) = \frac{1}{\sqrt{-t^{-bt}e^{bt}}}e^{i\left(\frac{bln(t)}{4}l_0x^2\right)}u_j(x,t)$
9	$i\psi_t = l_0\psi_{xx} + \frac{1}{4l_0}x^2\psi + icot(-t)x\psi_x$ $-\lambda l_0 csc(t)\|\psi\|^2\psi$	$\psi_j(x,t) = \frac{1}{\sqrt{csc(t)}}e^{i\left(\frac{-cot(-t)}{4}l_0x^2\right)}u_j(x,t)$
10	$i\psi_t = l_0\psi_{xx} + \frac{1}{4l_0}x^2\psi - itan(-t)x\psi_x$ $-\lambda l_0 sec(t)\|\psi\|^2\psi$	$\psi_j(x,t) = \frac{1}{\sqrt{sec(t)}}e^{i\left(\frac{tan(-t)}{4}l_0x^2\right)}u_j(x,t)$
11	$i\psi_t = l_0\psi_{xx} - \frac{2abte^{bt^2}+a^2e^{2bt^2}}{4l_0}x^2\psi$ $-iae^{bt^2}x\psi_x - \lambda l_0 e^{\frac{-a}{2}\sqrt{\frac{\pi}{b}}erfi(\sqrt{b}t)}\|\psi\|^2\psi$	$\psi_j(x,t) = \frac{1}{\sqrt{e^{\frac{-a}{2}\sqrt{\frac{\pi}{b}}erfi(\sqrt{b}t)}}}e^{\frac{ae^{bt^2}}{4}l_0x^2}u_j(x,t)$
12	$i\psi_t = l_0\psi_{xx} + \frac{atanh^2(bt)(b-a)-ab}{4l_0}x^2\psi$ $-iatanh(bt)x\psi_x - \lambda l_0 \|cosh(bt)\|^{\frac{a}{b}}\|\psi\|^2\psi$	$\psi_j(x,t) = \frac{1}{\sqrt{\|cosh(bt)\|^{\frac{a}{b}}}}e^{i\left(\frac{atanh(bt)}{4}l_0x^2\right)}u_j(x,t)$
13	$i\psi_t = l_0\psi_{xx} + \frac{acoth^2(bt)(b-a)-ab}{4l_0}x^2\psi$ $-iacoth(bt)x\psi_x - \lambda l_0 \|sinh(bt)\|^{\frac{a}{b}}\|\psi\|^2\psi$	$\psi_j(x,t) = \frac{1}{\sqrt{\|sinh(bt)\|^{\frac{a}{b}}}}e^{i\left(\frac{acoth(bt)}{4}l_0x^2\right)}u_j(x,t)$
14	$i\psi_t = l_0\psi_{xx} - \left(\frac{a^2+absinh(bt)+a^2sinh^2(bt)}{4l_0}\right)x^2\psi$ $-iacosh(bt)x\psi_x - \lambda l_0 e^{\frac{-asinh(bt)}{b}}\|\psi\|^2\psi$	$\psi_j(x,t) = \frac{1}{\sqrt{e^{\frac{-asinh(bt)}{b}}}}e^{i\left(\frac{acosh(bt)}{4}l_0x^2\right)}u_j(x,t)$
15	$i\psi_t = l_0\psi_{xx} - \left(\frac{a^2+absin(bt)-a^2sin^2(bt)}{4l_0}\right)x^2\psi$ $+iacos(bt)x\psi_x - \lambda l_0 e^{\frac{asin(bt)}{b}}\|\psi\|^2\psi$	$\psi_j(x,t) = \frac{1}{\sqrt{e^{\frac{asin(bt)}{b}}}}e^{i\left(\frac{-acos(bt)}{4}l_0x^2\right)}u_j(x,t)$
16	$i\psi_t = l_0\psi_{xx} - \left(\frac{a^2+abcos(bt)-a^2cos^2(bt)}{4l_0}\right)x^2\psi$ $-iasin(bt)x\psi_x + \lambda l_0 e^{\frac{acos(bt)}{b}}\|\psi\|^2\psi$	$\psi_j(x,t) = \frac{1}{\sqrt{e^{\frac{acos(bt)}{b}}}}e^{i\left(\frac{-asin(bt)}{4}l_0x^2\right)}u_j(x,t)$
17	$i\psi_t = l_0\psi_{xx} - \frac{atan^2(bt)(a+b)+ab}{4l_0}x^2\psi$ $-iatan(bt)x\psi_x - \lambda l_0 \|cos(bt)\|^{\frac{a}{b}}\|\psi\|^2\psi$	$\psi_j(x,t) = \frac{1}{\sqrt{\|cos(bt)\|^{\frac{a}{b}}}}e^{i\left(\frac{atan(bt)}{4}l_0x^2\right)}u_j(x,t)$
18	$i\psi_t = l_0\psi_{xx} - \frac{acot^2(bt)(a+b)+ab}{4l_0}x^2\psi$ $+iacot(bt)x\psi_x - \lambda l_0 \|sin(bt)\|^{\frac{a}{b}}\|\psi\|^2\psi$	$\psi_j(x,t) = \frac{1}{\sqrt{\|sin(bt)\|^{\frac{a}{b}}}}e^{i\left(\frac{acot(bt)}{4}l_0x^2\right)}u_j(x,t)$

Table 2. Riccati equations used to generate the similarity transformations.

#	Riccati Equation	Similarity Transformation from Table 1
1	$y'_x = ax^n y^2 + bmx^{m-1} - ab^2 x^{n+2m}$	1
2	$(ax^n + b)y'_x = by^2 + ax^{n-2}$	2
3	$y'_x = ax^n y^2 + bx^m y + bcx^m - ac^2 x^n$	3
4	$y'_x = ax^n y^2 + bx^m y + ckx^{k-1} - bcx^{m+k} - ac^2 x^{n+2k}$	1
5	$xy'_x = ax^n y^2 + my - ab^2 x^{n+2m}$	3
6	$(ax^n + bx^m + c)y'_x = \alpha x^k y^2 + \beta x^s y - \alpha b^2 x^k + \beta bx^s$	4
7	$y'_x = be^{\mu x} y^2 + ace^{cx} - a^2 be^{(\mu + 2c)x}$	5
8	$y'_x = ae^{\mu x} y^2 + cy - ab^2 e^{(\mu + 2c)x}$	3
9	$y'_x = ae^{cx} y^2 + bnx^{n-1} - ab^2 e^{cx} x^{2n}$	1
10	$y'_x = ax^n y^2 + bce^{cx} - ab^2 x^n e^{2cx}$	8
11	$y'_x = ax^n y^2 + cy - ab^2 x^n e^{2cx}$	3
12	$y'_x = \left[a \sinh^2(cx) - c \right] y^2 - a \sinh^2(cx) + c - a$	6
13	$2y'_x = [a - b + a \cosh(bx)] y^2 + a + b - a \cosh(bx)$	7
14	$y'_x = a(\ln x)^n y^2 + bmx^{m-1} - ab^2 x^{2m} (\ln x)^n$	1
15	$xy'_x = ax^n y^2 + b - ab^2 x^n \ln^2 x$	8
16	$y'_x = \left[b + a \sin^2(bx) \right] y^2 + b - a + a \sin^2(bx)$	9
17	$2y'_x = [b + a + a \cos(bx)] y^2 + b - a + a \cos(bx)$	10
18	$y'_x = \left[b + a \cos^2(bx) \right] y^2 + b - a + a \cos^2(bx)$	10
19	$y'_x = c(\arcsin x)^n y^2 + ay + ab - b^2 c(\arcsin x)^n$	3
20	$y'_x = a(\arcsin x)^n y^2 + \beta mx^{m-1} - a\beta^2 x^{2m} (\arcsin x)^n$	1
21	$y'_x = c(\arccos x)^n y^2 + ay + ab - b^2 c(\arccos x)^n$	3
22	$y'_x = a(\arccos x)^n y^2 + \beta mx^{m-1} - a\beta^2 x^{2m} (\arccos x)^n$	1
23	$y'_x = c(\arctan x)^n y^2 + ay + ab - b^2 c(\arctan x)^n$	3
24	$y'_x = a(\arctan x)^n y^2 + bmx^{m-1} - ab^2 x^{2m} (\arctan x)^n$	1
25	$y'_x = c(\text{arccot } x)^n y^2 + ay + ab - b^2 c(\text{arccot } x)^n$	3
26	$y'_x = a(\text{arccot } x)^n y^2 + bmx^{m-1} - ab^2 x^{2m} (\text{arccot } x)^n$	1
27	$y'_x = fy^2 + ay - ab - b^2 f$	3
28	$y'_x = fy^2 + anx^{n-1} - a^2 x^{2n} f$	1
29	$y'_x = fy^2 + gy - a^2 f - ag$	3
30	$y'_x = fy^2 + gy + anx^{n-1} - ax^n g - a^2 fx^{2n}$	1
31	$y'_x = fy^2 - ax^n gy + anx^{n-1} - a^2 x^{2n} (g - f)$	1
32	$y'_x = fy^2 + abe^{bx} - a^2 e^{2bx} f$	5
33	$y'_x = fy^2 + gy + abe^{bx} - ae^{bx} g - a^2 e^{2bx} f$	5
34	$y'_x = fy^2 - ae^{bx} gy + abe^{bx} + a^2 e^{2bx} (g - f)$	5
35	$y'_x = fy^2 + 2abxe^{bx^2} - a^2 fe^{2bx^2}$	11
36	$y'_x = fy^2 - a \tanh^2(bx)(af + b) + ab$	12
37	$y'_x = fy^2 - a \coth^2(bx)(af + b) + ab$	13
38	$y'_x = fy^2 - a^2 f + ab \sinh(bx) - a^2 f \sinh^2(bx)$	14
39	$y'_x = fy^2 - a^2 f + ab \sin(bx) + a^2 f \sin^2(bx)$	15
40	$y'_x = fy^2 - a^2 f + ab \cos(bx) + a^2 f \cos^2(bx)$	16
41	$y'_x = fy^2 - a \tan^2(bx)(af - b) + ab$	17
42	$y'_x = fy^2 - a \cot^2(bx)(af - b) + ab$	18

2. Soliton Solutions for VCNLS through Riccati Equations and Similarity Transformations

In this section, by means of a similarity transformation introduced in [42], and using computer algebra systems, we show the existence of Peregrine, bright and dark solitons for the family Equation (1). Thanks to the computer algebra systems, we are able to find an extensive list of integrable variable coefficient nonlinear Schrödinger equations (see Table 1). For similar work and applications to Bose-Einstein condensates, we refer the reader to [1]

Lemma 1. *([42]) Suppose that $h(t) = -l_0 \lambda \mu(t)$ with $\lambda \in \mathbb{R}$, $l_0 = \pm 1$ and that $c(t)$, $\alpha(t)$, $\delta(t)$, $\kappa(t)$, $\mu(t)$ and $g(t)$ satisfy the equations:*

$$\alpha(t) = l_0 \frac{c(t)}{4}, \delta(t) = -l_0 \frac{g(t)}{2}, h(t) = -l_0 \lambda \mu(t), \tag{2}$$

$$\kappa(t) = \kappa(0) - \frac{l_0}{4} \int_0^t g^2(z) dz, \tag{3}$$

$$\mu(t) = \mu(0) exp\left(\int_0^t (2d(z) - c(z)) dz \right) \mu(0) \neq 0, \tag{4}$$

$$g(t) = g(0) - 2l_0 exp\left(-\int_0^t c(z) dz \right) \int_0^t exp\left(\int_0^z c(y) dy \right) f(z) dz. \tag{5}$$

Then,

$$\psi(t,x) = \frac{1}{\sqrt{\mu(t)}} e^{i(\alpha(t)x^2 + \delta(t)x + \kappa(t))} u(t,x) \tag{6}$$

is a solution to the Cauchy problem for the nonautonomous Schrödinger equation

$$i\psi_t - l_0 \psi_{xx} - b(t)x^2 \psi + ic(t)x\psi_x + id(t)\psi + f(t)x\psi - ig(t)\psi_x - h(t)|\psi|^2 \psi = 0, \tag{7}$$

$$\psi(0,x) = \psi_0(x), \tag{8}$$

if and only if $u(t,x)$ is a solution of the Cauchy problem for the standard Schrödinger equation

$$iu_t - l_0 u_{xx} + l_0 \lambda |u|^2 u = 0, \tag{9}$$

with initial data

$$u(0,x) = \sqrt{\mu(0)} e^{-i(\alpha(0)x^2 + \delta(0)x + \kappa(0))} \psi_0(x). \tag{10}$$

Now, we proceed to use Lemma 1 to discuss how we can construct NLS with variable coefficients equations that can be reduced to the standard NLS and therefore be solved explicitly. We start recalling that

$$u_1(t,x) = A \exp\left(2iA^2 t\right) \left(\frac{3 + 16iA^2 t - 16A^4 t^2 - 4A^2 x^2}{1 + 16A^4 t^2 + 4A^2 x^2} \right), A \in \mathbb{R} \tag{11}$$

is a solution for ($l_0 = -1$ and $\lambda = -2$)

$$iu_t + u_{xx} + 2|u|^2 u = 0, t, x \in \mathbb{R}. \tag{12}$$

In addition,

$$u_2(\xi, \tau) = A \tanh(A\xi) e^{-2iA^2 \tau} \tag{13}$$

is a solution of ($l_0 = -1$ and $\lambda = 2$)

$$iu_\tau + u_{\xi\xi} - 2|u|^2 u = 0, \tag{14}$$

and

$$u_3(\tau, \xi) = \sqrt{v}\operatorname{sech}(\sqrt{v}\xi)\exp(-iv\tau), v > 0 \tag{15}$$

is a solution of ($l_0 = 1$ and $\lambda = -2$),

$$iu_\tau - u_{\xi\xi} - 2|u|^2 u = 0. \tag{16}$$

Example 1. Consider the NLS:

$$i\psi_t + \psi_{xx} - \frac{c^2}{4}x^2\psi - icx\psi_x \pm 2e^{ct}|\psi|^2\psi = 0. \tag{17}$$

Our intention is to construct a similarity transformation from Equation (17) to standard NLS Equation (9) by means of Lemma 1. Using the latter, we obtain

$$b(t) = \frac{c^2}{4}, c(t) = c, \mu(t) = e^{ct},$$

and

$$\alpha(t) = -\frac{c}{4}, h(t) = \pm 2e^{ct}.$$

Therefore,

$$\psi(x, t) = \frac{e^{-i\frac{c}{4}x^2}}{\sqrt{e^{ct}}}u_j(x, t), j = 1, 2$$

is a solution of the form Equation (6), and $u_j(x, t)$ are given by Equations (12) and (13).

Example 2. Consider the NLS:

$$i\psi_t + \psi_{xx} - \frac{1}{2t^2}x^2\psi - i\frac{1}{t}x\psi_x \pm 2t|\psi|^2\psi = 0. \tag{18}$$

By Lemma 1, a Riccati equation associated to the similarity transformation is given by

$$\frac{dc}{dt} + c(t)^2 - 2t^{-2} = 0, \tag{19}$$

and we obtain the functions

$$b(t) = \frac{1}{2t^2}, c(t) = -\frac{1}{t}, \mu(t) = t,$$

$$\alpha(t) = -\frac{1}{4t}, h_1(t) = -2t, h_2(t) = 2t.$$

Using $u_j(x, t), j = 1$ and 2, given by Equations (12) and (13), we get the solutions

$$\psi_j(x, t) = \frac{e^{-i\frac{1}{4t}x^2}}{\sqrt{t}}u_i(x, t). \tag{20}$$

Table 1 shows integrable variable coefficient NLS and the corresponding similarity transformation to constant coefficient NLS. Table 2 lists some Riccati equations that can be used to generate these transformations.

Example 3. If we consider the following family (m and B are parameters) of variable coefficient NLS,

$$i\psi_t + \psi_{xx} - \frac{Bmt^{m-1} + Bt^{2m}}{4} x^2\psi + iBt^m x\psi_x + \gamma e^{-\frac{Bt^{m+1}}{m+1}} |\psi|^2\psi = 0, \tag{21}$$

by means of the Riccati equation

$$y_t = At^n y^2 + Bmt^{m-1} - AB^2 t^{n+2m}, \tag{22}$$

and Lemma 1, we can construct soliton-like solutions for Equation (21). For this example, we restrict ourselves to taking $A = -1$ and $n = 0$. Furthermore, taking in Lemma 1 $l_0 = -1$, $\lambda = -2$, $a(t) = 1$, $b(t) = \frac{Bmt^{m-1} + Bt^{2m}}{4}$, $c(t) = Bt^m$, $\mu(t) = e^{-\frac{Bt^{m+1}}{m+1}}$, $h(t) = -2e^{-\frac{Bt^{m+1}}{m+1}}$, and $\alpha(t) = -Bt^m/4$, soliton-like solutions to the Equation (21) are given by

$$\psi_j(x,t) = e^{i\frac{-Bx^2 t^m}{4}} e^{\frac{Bt^{m+1}}{2(m+1)}} u_j(x,t), \tag{23}$$

where using $u_j(x,t)$, $j = 1$ and 2, given by Equations (12) and (15), we get the solutions. It is important to notice that if we consider $B = 0$ in Equation (21) we obtain standard NLS models.

3. Riccati Systems with Parameters and Similarity Transformations

In this section, we use different similarity trasformations than those used in Section 2, but they have been presented previously [26,35,39,42]. The advantage of the presentation of this section is a multiparameter approach. These parameters provide us with a control on the center axis of bright and dark soliton solutions. Again in this section, using Table 2, and by means of computer algebra systems, we show that we can produce a very extensive number of integrable VCNLS allowing soliton-type solutions. The transformations will require:

$$\frac{d\alpha}{dt} + b(t) + 2c(t)\alpha + 4a(t)\alpha^2 = 0, \tag{24}$$

$$\frac{d\beta}{dt} + (c(t) + 4a(t)\alpha(t))\beta = 0, \tag{25}$$

$$\frac{d\gamma}{dt} + l_0 a(t)\beta^2(t) = 0, l_0 = \pm 1, \tag{26}$$

$$\frac{d\delta}{dt} + (c(t) + 4a(t)\alpha(t))\delta = f(t) + 2\alpha(t)g(t), \tag{27}$$

$$\frac{d\varepsilon}{dt} = (g(t) - 2a(t)\delta(t))\beta(t), \tag{28}$$

$$\frac{d\kappa}{dt} = g(t)\delta(t) - a(t)\delta^2(t). \tag{29}$$

Considering the standard substitution

$$\alpha(t) = \frac{1}{4a(t)} \frac{\mu'(t)}{\mu(t)} - \frac{d(t)}{2a(t)}, \tag{30}$$

it follows that the Riccati Equation (24) becomes

$$\mu'' - \tau(t)\mu' + 4\sigma(t)\mu = 0, \tag{31}$$

with

$$\tau(t) = \frac{a'}{a} - 2c + 4d, \sigma(t) = ab - cd + d^2 + \frac{d}{2}\left(\frac{a'}{a} - \frac{d'}{d}\right). \tag{32}$$

We will refer to Equation (31) as the characteristic equation of the Riccati system. Here, $a(t)$, $b(t)$, $c(t)$, $d(t)$, $f(t)$ and $g(t)$ are real value functions depending only on the variable t. A solution of the Riccati system Equations (24)–(29) with multiparameters is given by the following expressions (with the respective inclusion of the parameter l_0) [26,35,39]:

$$\mu(t) = 2\mu(0)\mu_0(t)(\alpha(0) + \gamma_0(t)), \tag{33}$$

$$\alpha(t) = \alpha_0(t) - \frac{\beta_0^2(t)}{4(\alpha(0) + \gamma_0(t))}, \tag{34}$$

$$\beta(t) = -\frac{\beta(0)\beta_0(t)}{2(\alpha(0) + \gamma_0(t))} = \frac{\beta(0)\mu(0)}{\mu(t)}w(t), \tag{35}$$

$$\gamma(t) = l_0\gamma(0) - \frac{l_0\beta^2(0)}{4(\alpha(0) + \gamma_0(t))}, l_0 = \pm 1, \tag{36}$$

$$\delta(t) = \delta_0(t) - \frac{\beta_0(t)(\delta(0) + \varepsilon_0(t))}{2(\alpha(0) + \gamma_0(t))}, \tag{37}$$

$$\varepsilon(t) = \varepsilon(0) - \frac{\beta(0)(\delta(0) + \varepsilon_0(t))}{2(\alpha(0) + \gamma_0(t))}, \tag{38}$$

$$\kappa(t) = \kappa(0) + \kappa_0(t) - \frac{(\delta(0) + \varepsilon_0(t))^2}{4(\alpha(0) + \gamma_0(t))}, \tag{39}$$

subject to the initial arbitrary conditions $\mu(0)$, $\alpha(0)$, $\beta(0) \neq 0$, $\gamma(0)$, $\delta(0)$, $\varepsilon(0)$ and $\kappa(0)$. α_0, β_0, γ_0, δ_0, ε_0 and κ_0 are given explicitly by:

$$\alpha_0(t) = \frac{1}{4a(t)}\frac{\mu_0'(t)}{\mu_0(t)} - \frac{d(t)}{2a(t)}, \tag{40}$$

$$\beta_0(t) = -\frac{w(t)}{\mu_0(t)}, w(t) = \exp\left(-\int_0^t (c(s) - 2d(s))ds\right), \tag{41}$$

$$\gamma_0(t) = \frac{d(0)}{2a(0)} + \frac{1}{2\mu_1(0)}\frac{\mu_1(t)}{\mu_0(t)}, \tag{42}$$

$$\delta_0(t) = \frac{w(t)}{\mu_0(t)}\int_0^t \left[\left(f(s) - \frac{d(s)}{a(s)}g(s)\right)\mu_0(s) + \frac{g(s)}{2a(s)}\mu_0'(s)\right]\frac{ds}{w(s)}, \tag{43}$$

$$\varepsilon_0(t) = -\frac{2a(t)w(t)}{\mu_0'(t)}\delta_0(t) + 8\int_0^t \frac{a(s)\sigma(s)w(s)}{(\mu_0'(s))^2}(\mu_0(s)\delta_0(s))ds \\ + 2\int_0^t \frac{a(s)w(s)}{\mu_0'(s)}\left[f(s) - \frac{d(s)}{a(s)}g(s)\right]ds, \tag{44}$$

$$\kappa_0(t) = \frac{a(t)\mu_0(t)}{\mu_0'(t)}\delta_0^2(t) - 4\int_0^t \frac{a(s)\sigma(s)}{(\mu_0'(s))^2}(\mu_0(s)\delta_0(s))^2 ds \\ - 2\int_0^t \frac{a(s)}{\mu_0'(s)}(\mu_0(s)\delta_0(s))\left[f(s) - \frac{d(s)}{a(s)}g(s)\right]ds, \tag{45}$$

with $\delta_0(0) = g_0(0)/(2a(0))$, $\varepsilon_0(0) = -\delta_0(0)$, $\kappa_0(0) = 0$. Here, μ_0 and μ_1 represent the fundamental solution of the characteristic equation subject to the initial conditions $\mu_0(0) = 0$, $\mu_0'(0) = 2a(0) \neq 0$ and $\mu_1(0) \neq 0$, $\mu_1'(0) = 0$.

Using the system Equations (34)–(39), in [26], a generalized lens transformation is presented. Next, we recall this result (here we use a slight perturbation introducing the parameter $l_0 = \pm 1$ in order to use Peregrine type soliton solutions):

Lemma 2 ($l_0 = 1$, [26]). *Assume that* $h(t) = \lambda a(t)\beta^2(t)\mu(t)$ *with* $\lambda \in \mathbb{R}$. *Then, the substitution*

$$\psi(t, x) = \frac{1}{\sqrt{\mu(t)}} e^{i(\alpha(t)x^2 + \delta(t)x + \kappa(t))} u(\tau, \xi), \tag{46}$$

where $\xi = \beta(t)x + \varepsilon(t)$ *and* $\tau = \gamma(t)$, *transforms the equation*

$$i\psi_t = -a(t)\psi_{xx} + b(t)x^2\psi - ic(t)x\psi_x - id(t)\psi - f(t)x\psi + ig(t)\psi_x + h(t)|\psi|^2\psi$$

into the standard Schrödinger equation

$$iu_\tau - l_0 u_{\xi\xi} + l_0\lambda|u|^2 u = 0, l_0 = \pm 1, \tag{47}$$

as long as α, β, γ, δ, ε *and* κ *satisfy the Riccati system Equations (24)–(29) and also Equation (30).*

Example 4. Consider the NLS:

$$i\psi_t = \psi_{xx} - \frac{x^2}{4}\psi + h(0)\operatorname{sech}(t)|\psi|^2\psi. \tag{48}$$

It has the associated characteristic equation $\mu'' + a\mu = 0$, and, using this, we will obtain the functions:

$$\alpha(t) = \frac{\coth(t)}{4} - \frac{1}{2}\operatorname{csch}(t)\operatorname{sech}(t), \delta(t) = -\operatorname{sech}(t), \tag{49}$$

$$\kappa(t) = 1 - \frac{\tanh(t)}{2}, \mu(t) = \cosh(t), \tag{50}$$

$$h(t) = h(0)\operatorname{sech}(t), \beta(t) = \frac{1}{\cosh(t)}, \tag{51}$$

$$\varepsilon(t) = -1 + \tanh(t), \gamma(t) = 1 - \frac{\tanh(t)}{2}. \tag{52}$$

Then, we can construct solution of the form

$$\psi_j(t, x) = \frac{1}{\sqrt{\mu(t)}} e^{i(\alpha(t)x^2 + \delta(t)x + \kappa(t))} u_j\left(1 - \frac{\tanh(t)}{2}, \frac{x}{\cosh(t)} - 1 + \tanh(t)\right), \tag{53}$$

with u_j, $j = 1$ and 2, given by Equations (12) and (13).

Example 5. Consider the NLS:

$$i\psi_t(x, t) = \psi_{xx}(x, t) + \frac{h(0)\beta(0)^2\mu(0)}{1 + \alpha(0)2c_2t}|\psi(x, t)|^2\psi(x, t).$$

It has the characteristic equation $\mu'' + a\mu = 0$, and, using this, we will obtain the functions:

$$\alpha(t) = \frac{1}{4t} - \frac{1}{2 + \alpha(0)4c_2^2t^2}, \delta(t) = \frac{\delta(0)}{1 + \alpha(0)2c_2t}, \tag{54}$$

$$\kappa(t) = \kappa(0) - \frac{\delta(0)^2c_2t}{2 + 4\alpha(0)c_2t}, h(t) = \frac{h(0)\beta(0)^2\mu(0)}{1 + \alpha(0)2c_2t}, \tag{55}$$

$$\mu(t) = (1 + \alpha(0)2c_2t)\mu(0), \beta(t) = \frac{\beta(0)}{1 + \alpha(0)2c_2t},$$

$$\gamma(t) = \gamma(0) - \frac{\beta(0)^2 c_2 t}{2 + 4\alpha(0)c_2 t}, \epsilon(t) = \epsilon(0) - \frac{\beta(0)\delta(0)c_2 t}{1 + 2\alpha(0)c_2 t}.$$

Then, we can construct a solution of the form

$$
\begin{aligned}
\psi_j(t,x) &= \frac{1}{\sqrt{\mu(t)}} e^{i(\alpha(t)x^2 + \delta(t)x + \kappa(t))} \\
&\quad u_j\left(\gamma(0) - \frac{\beta(0)^2 c_2 t}{2 + 4\alpha(0)c_2 t}, \frac{\beta(0)x}{1+\alpha(0)2c_2 t} + \epsilon(0) - \frac{\beta(0)\delta(0)c_2 t}{1+2\alpha(0)c_2 t}\right),
\end{aligned}
\tag{56}
$$

with u_j, $j = 1$ and 2, Equations (12) and (13).

Following Table 2 of Riccati equations, we can use Equation (24) and Lemma 2 to construct an extensive list of integrable variable coefficient nonlinear Schrödinger equations.

4. Crank-Nicolson Scheme for Linear Schrödinger Equation with Variable Coefficients Depending on Space

In addition, in [35], a generalized Melher's formula for a general linear Schrödinger equation of the one-dimensional generalized harmonic oscillator of the form Equation (1) with $h(t) = 0$ was presented. As a particular case, if $b = \lambda\frac{\omega^2}{2}$; $f = b$, $\omega > 0$, $\lambda \in \{-1,0,1\}$, $c = g = 0$, then the evolution operator is given explicitly by the following formula (note—this formula is a consequence of Mehler's formula for Hermite polynomials):

$$\psi(x,t) = U_V(t)f := \frac{1}{\sqrt{2i\pi\mu_j(t)}} \int_{\mathbb{R}^n} e^{iS_V(x,y,t)} f(y)dy, \tag{57}$$

where

$$S_V(x,y,t) = \frac{1}{\mu_j(t)}\left(\frac{x_j^2 + y_j^2}{2} l_j(t) - x_j y_j\right),$$

$$
\{\mu_j(t), l_j(t)\} = \left\{
\begin{array}{l}
\\
\\
\\
\\
\end{array}
\right.
\tag{58}
$$

Using Riccati-Ermakov systems in [41], it was shown how computer algebra systems can be used to derive the multiparameter formulas (33)–(45). This multi-parameter study was used also to study solutions for the inhomogeneous paraxial wave equation in a linear and quadratic approximation including oscillating laser beams in a parabolic waveguide, spiral light beams, and more families of propagation-invariant laser modes in weakly varying media. However, the analytical method is restricted to solve Riccati equations exactly as the ones presented in Table 2. In this section, we use a finite differences method to compare analytical solutions described in [41] with numerical approximations. We aim (in future research) to extend numerical schemes to solve more general cases that the analytical method exposed cannot. Particularly, we will pursue to solve equations of the general form:

$$i\psi_t = -\Delta\psi + V(\mathbf{x},t)\psi, \tag{59}$$

using polynomial approximations in two variables for the potential function $V(\mathbf{x},t)$ ($V(\mathbf{x},t) \approx b(t)(x_1^2 + x_2^2) + f(t)x_1 + g(t)x_2 + h(t)$). For this purpose, it is necessary to analyze stability of different methods applied to this equation.

We also will be interested in extending this process to nonlinear Schrödinger-type equations with potential terms dependent on time, such as

$$i\psi_t = -\Delta\psi + V(\mathbf{x}, t)\psi + s|\psi|^2\psi. \tag{60}$$

In this section, we show that the Crank-Nicolson scheme seems to be the best method to deal with reconstructing numerically the analytical solutions presented in [41].

Numerical methods arise as an alternative when it is difficult to find analytical solutions of the Schrödinger equation. Despite numerical schemes not providing explicit solutions to the problem, they do yield approaches to the real solutions which allow us to obtain some relevant properties of the problem. Most of the simplest and often-used methods are those based on finite differences.

In this section, the Crank-Nicolson scheme is used for linear Schrödinger equation in the case of coefficients depending only on the space variable because it is absolutely stable and the matrix of the associate system does not vary for each iteration.

A rectangular mesh $(\mathbf{x}_m, t_n)$ is introduced in order to discretize a bounded domain $\Omega \times [0, T]$ in space and time. In addition, τ and $\mathbf{h}$ represent the size of the time step and the size of space step, respectively. $\mathbf{x}_m$ and $\mathbf{h}$ are in $\mathbb{R}$ if one-dimensional space is considered; otherwise, they are in $\mathbb{R}^2$.

The discretization is given by the matrix system

$$\left(I + \frac{ia\tau}{2h^2}\Delta + \frac{i\tau}{2}V(\mathbf{x})\right)\psi^{n+1} = \left(I - \frac{ia\tau}{2h^2}\Delta - \frac{i\tau}{2}V(\mathbf{x})\right)\psi^n, \tag{61}$$

where I is the identity matrix, Δ is the discrete representation of the Laplacian operator in space, and $V(\mathbf{x})$ is the diagonal matrix that represents the operator of the external potential depending on $\mathbf{x}$.

The paraxial wave equation (also known as harmonic oscillator)

$$2i\psi_t + \Delta\psi - \mathbf{r}^2\psi = 0, \tag{62}$$

where $\mathbf{r} = x$ for $\mathbf{x} \in \mathbb{R}$ or $\mathbf{r} = \sqrt{x_1^2 + x_2^2}$ for $\mathbf{x} \in \mathbb{R}^2$, describes the wave function for a laser beam [40].

One solution for this equation can be presented as Hermite-Gaussian modes on a rectangular domain:

$$
\begin{aligned}
\psi_{nm}(\mathbf{x}, t) =\ & A_{nm}\frac{exp[i(\kappa_1 + \kappa_2) + 2i(n+m+1)\gamma]}{\sqrt{2^{n+m}n!m!\pi}}\beta \\
\times\ & exp\left[i(\alpha\mathbf{r}^2 + \delta_1 x_1 + \delta_2 x_2) - (\beta x_1 + \varepsilon_1)^2/2 - (\beta x_2 + \varepsilon_2)^2/2\right] \\
\times\ & H_n(\beta x_1 + \varepsilon_1)H_m(\beta x_2 + \varepsilon_2),
\end{aligned}
\tag{63}
$$

where $H_n(x)$ is the n-th order Hermite polynomial in the variable x, see [40,41].

In addition, some solutions of the paraxial equation may be expressed by means of Laguerre–Gaussian modes in the case of cylindrical domains (see [43]):

$$
\begin{aligned}
\psi_n^m(\mathbf{x}, t) =\ & A_n^m\sqrt{\frac{n!}{\pi(n+m)!}}\beta \\
\times\ & exp\left[i(\alpha\mathbf{r}^2 + \delta_1 x_1 + \delta_2 x_2 + \kappa_1 + \kappa_2) - (\beta x_1 + \varepsilon_1)^2/2 - (\beta x_2 + \varepsilon_2)^2/2\right] \\
\times\ & exp[i(2n + m + 1)\gamma](\beta(x_1 \pm ix_2) + \varepsilon_1 \pm i\varepsilon_2)^m \\
\times\ & L_n^m\left((\beta x_1 + \varepsilon_1)^2 + (\beta x_2 + \varepsilon_2)^2\right),
\end{aligned}
\tag{64}
$$

with $L_n^m(x)$ being the n-th order Laguerre polynomial with parameter m in the variable x.

α, β, γ, δ_1, δ_2, ε_1, ε_2, κ_1 and κ_2 given by Equations (34)–(39) for both Hermite-Gaussian and Laguerre-Gaussian modes.

Figures 1 and 2 show two examples of solutions of the one-dimensional paraxial equation with $\Omega = [-10, 10]$ and $T = 12$. The step sizes are $\tau = \frac{10}{200}$ and $h = \frac{10}{200}$.

(a)

(b)

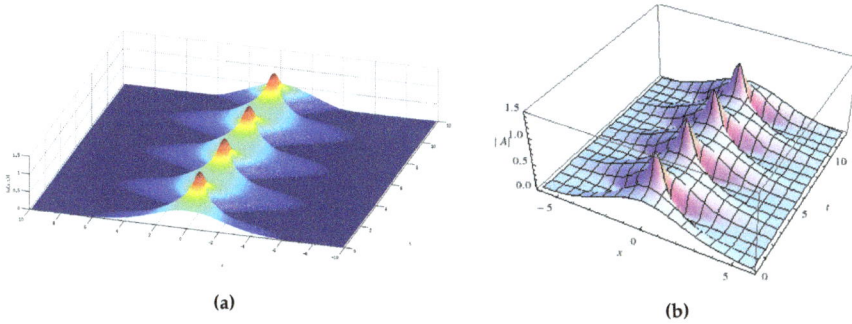

Figure 1. (**a**) corresponding approximation for the one-dimensional Hermite-Gaussian beam with $t = 10$. The initial condition is $\sqrt{\frac{2}{3\sqrt{\pi}}} e^{\left(\frac{2}{3}x\right)^2/2}$; (**b**) the exact solution for the one-dimensional Hermite-Gaussian beam with $t = 10$, $A_n = 1$, $\mu_0 = 1$, $\alpha_0 = 0$, $\beta_0 = \frac{4}{9}$, $n_0 = 0$, $\delta_0 = 0$, $\gamma_0 = 0$, $\epsilon_0 = 0$, $\kappa_0 = 0$.

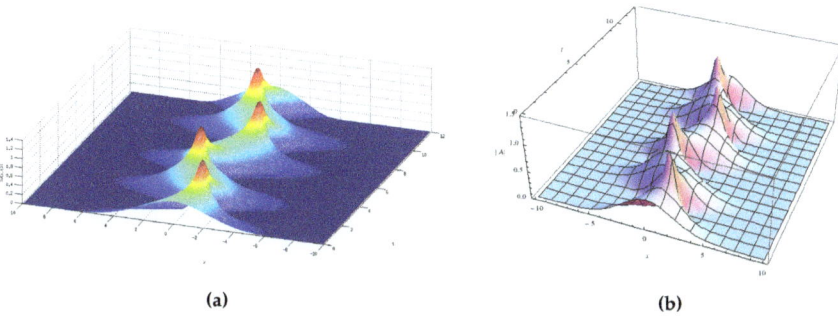

(a)

(b)

Figure 2. (**a**) corresponding approximation for the one-dimensional Hermite-Gaussian beam with $t = 10$. The initial condition is $\sqrt{\frac{2}{3\sqrt{\pi}}} e^{\left(\frac{2}{3}x\right)^2/2 + ix}$; (**b**) the exact solution for the one-dimensional Hermite-Gaussian beam with $t = 10$, $A_n = 1$, $\mu_0 = 1$, $\alpha_0 = 0$, $\beta_0 = \frac{4}{9}$, $n_0 = 0$, $\delta_0 = 1$, $\gamma_0 = 0$, $\epsilon_0 = 0$, $\kappa_0 = 0$.

Figure 3 shows four profiles of two-dimensional Hermite-Gaussian beams considering $\Omega = [-6, 6] \times [-6, 6]$ and $T = 10$. The corresponding step sizes are $\tau = \frac{10}{40}$ and $\mathbf{h} = \left(\frac{12}{48}, \frac{12}{48} \right)$.

(a)

(b)

(c)

(d)

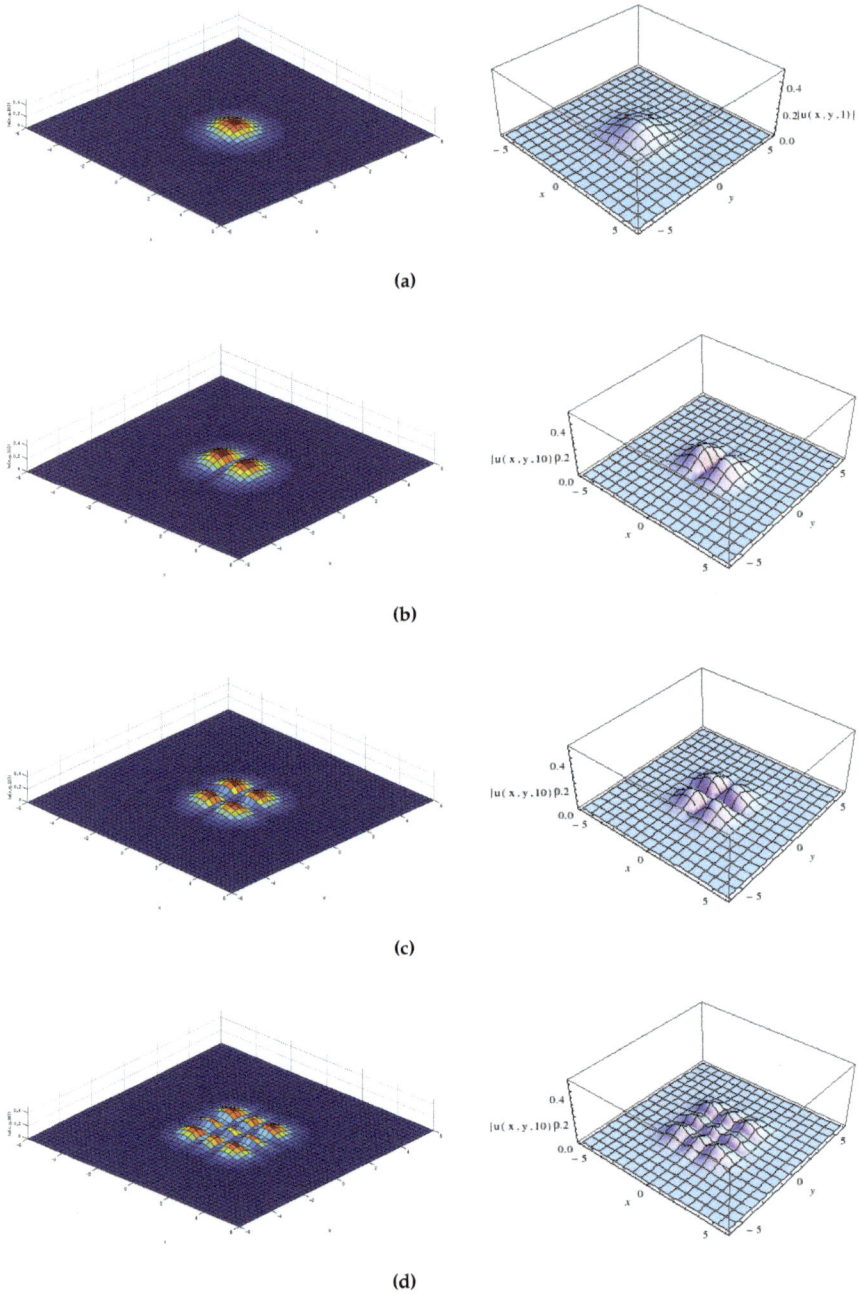

Figure 3. (**Left**): corresponding approximations for the two-dimensional Hermite-Gaussian beams with $t = 10$. The initial conditions are (**a**) $\frac{1}{\sqrt{8\pi}}e^{-(x^2+y^2)}$; (**b**) $\frac{1}{\sqrt{2\pi}}e^{-(x^2+y^2)}x$; (**c**) $\sqrt{\frac{2}{\pi}}e^{-(x^2+y^2)}xy$; (**d**) $\frac{1}{4\sqrt{32\pi}}e^{-(x^2+y^2)}\left(8x^2 - 2\right)\left(8y^2 - 2\right)$. (**Right**): the exact solutions for the two-dimensional Hermite-Gaussian beams with $t = 10$ and parameters $A_{nm} = \frac{1}{4}$, $\alpha_0 = 0$, $\beta_0 = \sqrt{2}$, $\delta_{0,1} = 1$, $\gamma_{0,1} = 0$, $\epsilon_{0,1} = 0$, $\kappa_{0,1} = 0$. For (**a**) $n = 0$ and $m = 0$, for (**b**) $n = 1$ and $m = 0$, for (**c**) $n = 1$ and $m = 1$, for (**d**) $n = 2$ and $m = 2$.

Figure 4 shows two profiles of two-dimensional Laguerre–Gaussian beams considering $\Omega = [-6,6] \times [-6,6]$ and $T = 10$. The corresponding step sizes are $\tau = \frac{10}{40}$ and $\mathbf{h} = \left(\frac{12}{48}, \frac{12}{48} \right)$.

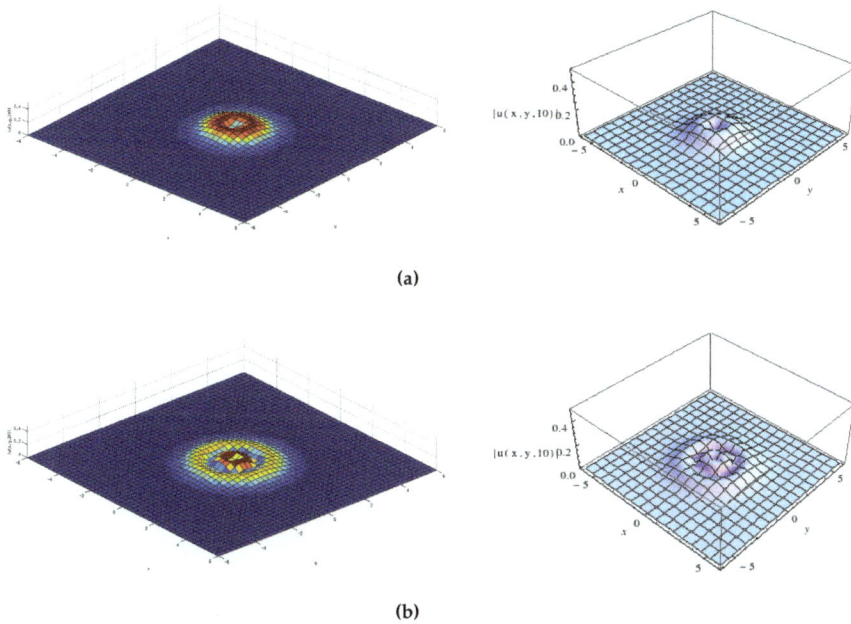

(a)

(b)

Figure 4. (**Left**): corresponding approximations for the two-dimensional Laguerre–Gaussian beams with $t = 10$. The initial conditions are (**a**) $\frac{1}{\sqrt{4\pi}} e^{-(x^2+y^2)} (x + iy)$; (**b**) $\frac{1}{\sqrt{2\pi}} e^{-(x^2+y^2)} (x + iy)(1 - x^2 - y^2)$. (**Right**): the exact solutions for the two-dimensional Laguerre–Gaussian beams with $t = 10$ and parameters $A_n^m = \frac{1}{4}$, $\alpha_0 = 0$, $\beta_0 = \sqrt{2}$, $\delta_{0,1} = 1$, $\gamma_{0,1} = 0$, $\epsilon_{0,1} = 0$, $\kappa_{0,1} = 0$.

5. Conclusions

Rajendran *et al.* in [1] used similarity transformations introduced in [28] to show a list of integrable NLS equations with variable coefficients. In this work, we have extended this list, using similarity transformations introduced by Suslov in [26], and presenting a more extensive list of families of integrable nonlinear Schrödinger (NLS) equations with variable coefficients (see Table 1 as a primary list. In both approaches, the Riccati equation plays a fundamental role. The reader can observe that, using computer algebra systems, the parameters (see Equations (33)–(39)) provide a change of the dynamics of the solutions; the Mathematica files are provided as a supplement for the readers. Finally, we have tested numerical approximations for the inhomogeneous paraxial wave equation by the Crank-Nicolson scheme with analytical solutions. These solutions include oscillating laser beams and Laguerre and Gaussian beams. The explicit solutions have been found previously thanks to explicit solutions of Riccati-Ermakov systems [41].

Supplementary Materials: The following are available online at http://www.mdpi.com/2073-8994/8/5/38/s1, Mathematica supplement file.

Acknowledgments: The authors were partially funded by the Mathematical American Association through NSF (grant DMS-1359016) and NSA (grant DMS-1359016). Also, the authors are thankful for the funding received from the Department of Mathematics and Statistical Sciences and the College of Liberal Arts and Sciences at University of Puerto Rico, Mayagüez. E. S. is funded by the Simons Foundation Grant # 316295 and by the National Science Foundation Grant DMS-1440664. E.S is also thankful for the start up funds and the "Faculty

Development Funding Program Award" received from the School of Mathematics and Statistical Sciences and the College of Sciences at University of Texas, Rio Grande Valley.

Author Contributions: The original results presented in this paper are the outcome of a research collaboration started during the Summer 2015 and continuous until Spring 2016. Similarly, the selection of the examples, tables, graphics and extended bibliography is the result of a continuous long interaction between the authors.

Conflicts of Interest: The authors declare no conflict of interest.

References

1. Rajendran, S.; Muruganandam, P.; Lakshmanan, M. Bright and dark solitons in a quasi-1D Bose–Einstein condensates modelled by 1D Gross–Pitaevskii equation with time-dependent parameters. *Phys. D Nonlinear Phenom.* **2010**, *239*, 366–386. [CrossRef]
2. Agrawal, G.-P. *Nonlinear Fiber Optics*, 4th ed.; Academic Press: New York, NY, USA, 2007.
3. Al Khawaja, U. A comparative analysis of Painlevé, Lax Pair and similarity transformation methods in obtaining the integrability conditions of nonlinear Schrödinger equations. *J. Phys. Math.* **2010**, *51*. [CrossRef]
4. Brugarino, T.; Sciacca, M. Integrability of an inhomogeneous nonlinear Schrödinger equation in Bose-Einstein condensates and fiber optics. *J. Math. Phys.* **2010**, *51*. [CrossRef]
5. Chen, H.-M.; Liu, C.S. Solitons in nonuniform media. *Phys. Rev. Lett.* **1976**, *37*, 693–697. [CrossRef]
6. He, X.G.; Zhao, D.; Li, L.; Luo, H.G. Engineering integrable nonautonomous nonlinear Schrödinger equations. *Phys. Rev. E.* **2009**, *79*. [CrossRef] [PubMed]
7. He, J.; Li, Y. Designable inegrability of the variable coefficient nonlinear Schrödinger equations. *Stud. Appl. Math.* **2010**, *126*, 1–15. [CrossRef]
8. He, J.S.; Charalampidis, E.G.; Kevrekidis, P.G.; Frantzeskakis, D.J. Rogue waves in nonlinear Schrödinger models with variable coefficients: Application to Bose-Einstein condensates. *Phys. Lett. A* **2014**, *378*, 577–583. [CrossRef]
9. Kruglov, V.I.; Peacock, A.C.; Harvey, J.D. Exact solutions of the generalized nonlinear Schrödinger equation with distributed coefficients. *Phys. Rev. E* **2005**, *71*. [CrossRef] [PubMed]
10. Marikhin, V.G.; Shabat, A.B.; Boiti, M.; Pempinelli, F. Self-similar solutions of equations of the nonlinear Schrödinger type. *J. Exp. Theor. Phys.* **2000**, *90*, 553–561. [CrossRef]
11. Ponomarenko, S.A.; Agrawal, G.P. Do Solitonlike self-similar waves exist in nonlinear optical media? *Phys. Rev. Lett.* **2006**, *97*. [CrossRef] [PubMed]
12. Ponomarenko, S.A.; Agrawal, G.P. Optical similaritons in nonlinear waveguides. *Opt. Lett.* **2007**, *32*, 1659–1661. [CrossRef] [PubMed]
13. Raghavan, S.; Agrawal, G.P. Spatiotemporal solitons in inhomogeneous nonlinear media. *Opt. Commun.* **2000**, *180*, 377–382. [CrossRef]
14. Serkin, V.N.; Hasegawa, A. Novel Soliton solutions of the nonlinear Schrödinger Equation model. *Phys. Rev. Lett.* **2000**, *85*. [CrossRef] [PubMed]
15. Serkin, V.; Matsumoto, M.; Belyaeva, T. Bright and dark solitary nonlinear Bloch waves in dispersion managed fiber systems and soliton lasers. *Opt. Commun.* **2001**, *196*, 159–171. [CrossRef]
16. Tian, B.; Shan, W.; Zhang, C.; Wei, G.; Gao, Y. Transformations for a generalized variable-coefficient nonlinear Schrödinger model from plasma physics, arterial mechanics and optical fibers with symbolic computation. *Eur. Phys. J. B* **2005**, *47*, 329–332. [CrossRef]
17. Dai, C.-Q.; Wang, Y.-Y. Infinite generation of soliton-like solutions for complex nonlinear evolution differential equations via the NLSE-based constructive method. *Appl. Math. Comput.* **2014**, *236*, 606–612. [CrossRef]
18. Wang, M.; Shan, W.-R.; Lü, X.; Xue, Y.-S.; Lin, Z.-Q.; Tian, B. Soliton collision in a general coupled nonlinear Schrödinger system via symbolic computation. *Appl. Math. Comput.* **2013**, *219*, 11258–11264. [CrossRef]
19. Yu, F.; Yan, Z. New rogue waves and dark-bright soliton solutions for a coupled nonlinear Schrödinger equation with variable coefficients. *Appl. Math. Comput.* **2014**, *233*, 351–358. [CrossRef]
20. Fibich, G. *The Nonlinear Schrödinger Equation, Singular Solutions and Optical Collapse*; Springer: Berlin/Heidelberg, Germany, 2015.
21. Kevrekidis, P.G.; Frantzeskakis, D.J.; Carretero-Gonzáles, R. *Emergent Nonlinear Phenomena in Bose-Einstein Condensates: Theory and Experiment*; Springer Series of Atomic, Optical and Plasma Physics; Springer: Berlin/Heidelberg, Germany, 2008; Volume 45.

22. Suazo, E.; Suslov, S.-K. Soliton-Like solutions for nonlinear Schrödinger equation with variable quadratic Hamiltonians. *J. Russ. Laser Res.* **2010**, *33*, 63–83. [CrossRef]
23. Sulem, C.; Sulem, P.L. *The Nonlinear Schrödinger Equation*; Springer: New York, NY, USA, 1999.
24. Tao, T. Nonlinear dispersive equations: Local and global analysis. In *CBMS Regional Conference Series in Mathematics*; American Mathematical Society: Providence, RI, USA, 2006.
25. Zakharov, V.-E.; Shabat, A.-B. Exact theory of two-dimensional self-focusing and one-dimensional self-modulation of waves in nonlinear media. *Soviet. Phys. JETP* **1972**, *34*, 62–69.
26. Suslov, S.-K. On integrability of nonautonomous nonlinear Schrödinger equations. *Proc. Am. Math. Soc.* **2012**, *140*, 3067–3082. [CrossRef]
27. Talanov, V.I. Focusing of light in cubic media. *JETP Lett.* **1970**, *11*, 199–201.
28. Perez-Garcia, V.M.; Torres, P.J.; Konotop, V.K. Similarity transformations for nonlinear Schrödinger equations with time-dependent coefficients. *Physica D* **2006**, *221*, 31–36. [CrossRef]
29. Ablowitz, M.; Hooroka, T. Resonant intrachannel pulse interactions in dispersion-managed transmission systems. *IEEE J. Sel. Top. Quantum Electron.* **2002**, *8*, 603–615. [CrossRef]
30. Marhic, M.E. Oscillating Hermite-Gaussian wave functions of the harmonic oscillator. *Lett. Nuovo Cim.* **1978**, *22*, 376–378. [CrossRef]
31. Carles, R. Nonlinear Schrödinger equation with time dependent potential. *Commun. Math. Sci.* **2010**, *9*, 937–964. [CrossRef]
32. López, R.M.; Suslov, S.K.; Vega-Guzmán, J.M. On a hidden symmetry of quantum harmonic oscillators. *J. Differ. Equ. Appl.* **2013**, *19*, 543–554. [CrossRef]
33. Aldaya, V.; Cossío, F.; Guerrero, J.; López-Ruiz, F.F. The quantum Arnold transformation. *J. Phys. A Math. Theor.* **2011**, *44*, 1–6. [CrossRef]
34. Feynman, R.P.; Hibbs, A.R. *Quantum Mechanics and Path Integrals*; McGraw-Hill: New York, NY, USA, 1965.
35. Cordero-Soto, R.; Lopez, R.M.; Suazo, E.; Suslov, S.K. Propagator of a charged particle with a spin in uniform magnetic and perpendicular electric fields. *Lett. Math. Phys.* **2008**, *84*, 159–178. [CrossRef]
36. Lanfear, N.; López, R.M.; Suslov, S.K. Exact wave functions for a generalized harmonic oscillators. *J. Russ. Laser Res.* **2011**, *32*, 352–361. [CrossRef]
37. López, R.M.; Suslov, S.K.; Vega-Guzmán, J.M. Reconstructing the Schrödinger groups. *Phys. Scr.* **2013**, *87*, 1–6. [CrossRef]
38. Suazo, E.; Suslov, S.K. Cauchy problem for Schrödinger equation with variable quadratic Hamiltonians. **2011**. to be submitted.
39. Suazo, E. Fundamental Solutions of Some Evolution Equations. Ph.D. Thesis, Arizona State University, Tempe, AZ, USA, September 2009.
40. Mahalov, A.; Suazo, E.; Suslov, S.K. Spiral laser beams in inhomogeneous media. *Opt. Lett.* **2013**, *38*, 2763–2766. [CrossRef] [PubMed]
41. Koutschan, C.; Suazo, E.; Suslov, S.K. Fundamental laser modes in paraxial optics: From computer algebra and simulations to experimental observation. *Appl. Phys. B* **2015**, *121*, 315–336. [CrossRef]
42. Escorcia, J.; Suazo, E. Blow-up results and soliton solutions for a generalized variable coefficient nonlinear Schrödinger equation. Available online: http://arxiv.org/abs/1605.07554 (accessed on 24 May 2016).
43. Andrews, L.C.; Phillips, R.L. *Laser Beam Propagation through Random Media*, 2nd ed.; SPIE Press: Bellingham, WA, USA, 2005.

© 2016 by the authors. Licensee MDPI, Basel, Switzerland. This article is an open access article distributed under the terms and conditions of the Creative Commons Attribution (CC BY) license (http://creativecommons.org/licenses/by/4.0/).

symmetry

MDPI

Article

Coherent States of Harmonic and Reversed Harmonic Oscillator

Alexander Rauh

Department of Physics, University of Oldenburg, Oldenburg D-26111, Germany;
alexander.rauh@uni-oldenburg.de; Tel.:+49-441-798-3460

Academic Editor: Young Suh Kim
Received: 16 January 2016; Accepted: 3 June 2016; Published: 13 June 2016

Abstract: A one-dimensional wave function is assumed whose logarithm is a quadratic form in the configuration variable with time-dependent coefficients. This trial function allows for general time-dependent solutions both of the harmonic oscillator (HO) and the reversed harmonic oscillator (RO). For the HO, apart from the standard coherent states, a further class of solutions is derived with a time-dependent width parameter. The width of the corresponding probability density fluctuates, or "breathes" periodically with the oscillator frequency. In the case of the RO, one also obtains normalized wave packets which, however, show diffusion through exponential broadening with time. At the initial time, the integration constants give rise to complete sets of coherent states in the three cases considered. The results are applicable to the quantum mechanics of the Kepler-Coulomb problem when transformed to the model of a four-dimensional harmonic oscillator with a constraint. In the classical limit, as was shown recently, the wave packets of the RO basis generate the hyperbolic Kepler orbits, and, by means of analytic continuation, the elliptic orbits are also obtained quantum mechanically.

Keywords: inverted harmonic oscillator; harmonic trap; Kepler-Coulomb problem; Kustaanheimo-Stiefel transformation

1. Introduction

Coherent states of the harmonic oscillator (HO) were introduced already at the beginning of wave mechanics [1]. Much later, such states were recognized as being useful as a basis to describe radiation fields [2] and optical correlations [3]. The reversed harmonic oscillator (RO) refers to a model with repulsive harmonic forces, and was discussed in [4] in the context of irreversibility. Recently, in [5], which also communicates historical remarks, the RO was applied to describe nonlinear optical phenomena. As mentioned in [5], the term "inverted harmonic oscillator" (IO) originally refers to a model with negative kinetic and potential energy, as proposed in [6]. Nevertheless, most articles under the headline IO, actually consider the RO model, see, e.g., [7–9].

The RO model formally can be obtained by assuming a purely imaginary oscillator frequency. It is then not anymore possible to construct coherent states by means of creation and annihilation operators; for a text book introduction see [10]. In [9], the RO was generalized by the assumption of a time-dependent mass and frequency. The corresponding Schrödinger equation was solved by means of an algebraic method with the aim to describe quantum tunneling.

In the present study, emphasis is laid on the derivation of complete sets of coherent states both for the HO and the RO model, together with their time evolution. In the case of the HO, in addition to the standard coherent states, a further function set is found with a time-dependent width parameter. Both in the HO and RO case, the integration constants of the time-dependent solutions induce complete function sets which, at time $t = 0$, are isomorphic to the standard coherent states of the HO.

In Section 6, an application to the quantum mechanics of the Kepler-Coulomb problem will be briefly discussed. As has first been observed by Fock [11], the underlying four-dimensional rotation symmetry of the non-relativistic Hamiltonian of the hydrogen atom permits the transformation to the problem of four isotropic harmonic oscillators with a constraint; for applications see, e.g., [12–14]. The transformation proceeds conveniently by means of the Kustaanheimo-Stiefel transformation [15]. In [14], the elliptic Kepler orbits were derived in the classical limit on the basis of coherent HO states. By means of coherent RO states, the classical limit for hyperbolic Kepler orbits was achieved in [16,17], whereby the elliptic regime could be obtained by analytic continuation from the hyperbolic side. Recently, by means of the same basis, a first order quantum correction to Kepler's equation was derived in [18], whereby the smallness parameter was defined by the reciprocal angular momentum in units of $\hbar$.

As compared to the classical elliptic Kepler orbits, the derivation of hyperbolic orbits from quantum mechanics was accomplished quite recently [16,17]. For this achievement, it was crucial to devise a suitable time-dependent ansatz for the wave function, see (1) below, in order to construct coherent RO states. As it turns out, the wave function (1) contains also the usual coherent HO states, and, unexpectedly, a further set of coherent states, which we call type-II states. The latter are characterized by a time-dependent width parameter and are solutions of the time-dependent Schrödinger equation of the HO. Section 4 contains the derivation. Essentially, the type-II states offer a disposable width parameter which allows us, for instance, to describe arbitrarily narrowly peaked initial states together with their time evolution in a harmonic potential. In this paper, a unified derivation is presented of coherent states of the HO, RO, and type-II HO states. Furthermore, the connection of HO and RO with the quantum mechanics of the Kepler-Coulomb problem is briefly discussed in the context of the derivation of the classical Kepler orbits from quantum mechanics.

2. Introducing a Trial Wave Function

In order to solve the Schrödinger equation for the harmonic oscillator (HO) and the reversed oscillator (RO), a trial wave function of Gaussian type is assumed as follows

$$\psi(x,t) = C_0 \exp\left[C(t) + B(t)x - \Gamma(t)x^2\right], \quad x \in \mathbf{R}, \quad \text{Real}(\Gamma) > 0, \tag{1}$$

where C, B, Γ are complex functions of time t and C_0 the time-independent normalization constant. When the Schrödinger operator $[i\,\hbar\partial_t - H]$ is applied to ψ for a Hamiltonian with harmonic potential, then the wave function ψ is reproduced up to a factor which is a quadratic polynomial and must vanish identically in the configuration variable x:

$$0 = p_0(t) + p_1(t)x + p_2(t)x^2. \tag{2}$$

The conditions $p_0 = 0$, $p_1 = 0$, and $p_2 = 0$, give rise to three first-order differential equations for the functions $C(t)$, $B(t)$, and $\Gamma(t)$. In the following we examine two cases for the HO: type-I and type-II are characterized by a constant and time-dependent function Γ, respectively. In the case of the RO, only a time-dependent Γ leads to a solution. By a suitable choice of the parameters, the ansatz (1) solves the time-dependent Schrödinger equation both for the HO and the RO Hamiltonian

$$H = p^2/(2m) + (m\omega^2/2)x^2 \quad \text{and} \quad H_\Omega = p^2/(2m) - (m\Omega^2/2)x^2, \quad \omega, \ \Omega > 0,$$

respectively.

3. Standard (Type-I) Coherent States of the HO

In the following, the time-dependent solutions are derived, within the trial function scheme, for the Hamiltonian

$$H = p^2/(2m) + (m\omega^2/2)x^2 = (\hbar\omega/2)\left[-\partial_\zeta^2 + \zeta^2\right], \tag{3}$$

where $\zeta = \alpha x$ is dimensionless with $\alpha^2 = m\omega/\hbar$. For later comparison, we list the standard definition of coherent states from the textbook [10], see Equations (4.72) and (4.75):

$$|z\rangle = \exp\left[-\frac{1}{2}zz^*\right]\sum_{n=0}^{\infty}\frac{z^n}{\sqrt{n!}}|n\rangle, \tag{4}$$

$$\psi_z(\zeta) = \pi^{-1/4}\exp\left[-\frac{1}{2}(zz^*+z^2)\right]\exp\left[-\frac{1}{2}\zeta^2+\sqrt{2}\zeta z\right], \ \zeta = \alpha x, \ \alpha^2 = \frac{m\omega}{\hbar}, \tag{5}$$

where $\psi_z(\zeta) = \langle\zeta|z\rangle$, $|n\rangle$ denotes the n-th energy eigenvector, and the star superscript means complex conjugation. The time evolution gives rise to, see [10],

$$|z,t\rangle = \exp[-i\omega t/2]\,|z\exp[-i\omega t]\rangle, \tag{6}$$

$$\psi_z(\zeta,t) = \exp[-i\omega t/2]\psi_{(z\exp[-i\omega t])}(\zeta). \tag{7}$$

The state $|z\rangle$ is minimal with respect to the position-momentum uncertainty product $\Delta x\,\Delta p$, and there exists the following completeness property, see [3],

$$\frac{1}{\pi}\int_0^{\infty}u\mathrm{d}u\int_0^{2\pi}\mathrm{d}\varphi\,|z\rangle\langle z| = \sum_n|n\rangle\langle n|, \quad z = u\exp[i\,\varphi]. \tag{8}$$

The relation (8) follows immediately from the definition (4). An equivalent statement is

$$\frac{1}{\pi}\int_0^{\infty}u\mathrm{d}u\int_0^{2\pi}\mathrm{d}\varphi\,\langle\zeta_2|z\rangle\langle z|\zeta_1\rangle = \delta(\zeta_2-\zeta_1), \tag{9}$$

which corresponds to the completeness of the energy eigenfunctions of the harmonic oscillator. In Appendix B, we reproduce a proof of (9), which is appropriate, since the proof has to be extended to the modified coherent states in the type-II HO and the RO cases.

In terms of the scaled variables ζ and $\tau = t\omega$, the trial ansatz reads

$$\psi(\zeta,\tau) = C_0\exp\left[c(\tau)+\beta(\tau)\zeta-\gamma(\tau)\zeta^2/2\right], \tag{10}$$

where c,β,γ are dimensionless functions of τ, and the re-scaling factor of the probability density, $1/\sqrt{\alpha}$, is taken into the normalization constant C_0.

We assume that $\gamma = \gamma_0 = const$. Then, the polynomial (2) gives rise to the equations

$$\gamma_0^2 = 1, \quad i\,\beta'(\tau) = \beta(\tau), \quad 2i\,c'(t) = 1-\beta^2(t), \tag{11}$$

which implies that $\gamma_0 = 1$ is fixed. The further solutions emerge easily as

$$\beta(\tau) = C_2\exp[-i\,\tau], \quad c(\tau) = -i\,\tau/2 - (C_2^2/4)\exp[-2i\,\tau] + C_3, \tag{12}$$

where C_2 and C_3 are complex integration constants. A comparison with (5), at $t = 0$, suggests to set

$$C_2 = \sqrt{2}\,z, \quad C_3 = -(1/2)zz^*, \tag{13}$$

which specifies the functions β and c as follows

$$\beta(\tau) = \sqrt{2}\,(z\exp[-i\,\tau]), \quad c(t) = -i\,\tau/2 - (1/2)\left[zz^* + (z\exp[-i\,\tau])^2\right]. \tag{14}$$

The normalization integral with respect to ζ amounts to the condition

$$C_0^2 \sqrt{\pi} \exp[zz^*] = 1;$$ (15)

hence (7) with (5) is reproduced.

4. Type-II Solutions of the Harmonic Oscillator

With γ being a function of time, one obtains the following differential equations with prime denoting the derivative with respect to the scaled time τ:

$$i\gamma' = \gamma^2 - 1, \quad i\beta' = \gamma\beta; \quad 2ic' = \gamma - \beta^2.$$ (16)

The solution for γ is

$$\gamma(\tau) = \frac{\exp(2i\tau) - C_1}{\exp(2i\tau) + C_1} \quad C_1 = \frac{1 - \gamma_0}{1 + \gamma_0}. \quad \gamma_0 = \gamma(0).$$ (17)

Splitting γ into its real and imaginary parts, one can write

$$\begin{aligned} \gamma(\tau) &= \gamma_R + i\gamma_I; \quad \gamma_R = (1 - C_1^2)N_1^{-1}, \quad \gamma_I = 2C_1N_1^{-1}\sin(2\tau), \\ N_1(\tau) &= 1 + C_1^2 + 2C_1\cos(2\tau) = 4(1 + \gamma_0)^{-2}\left[1 + (\gamma_0^2 - 1)\sin^2(\tau)\right]. \end{aligned}$$ (18)

In order that the wave function is square integrable, γ_R has to be positive, which implies that

$$C_1^2 < 1 \quad \text{or} \quad \gamma_0 > 0.$$ (19)

The initial value $\gamma(t = 0) \equiv \gamma_0 > 0$ emerges as a disposable parameter.

The probability density, $P = |\psi(\zeta, \tau)|^2$, is characterized by a width of order of magnitude $d = 1/\sqrt{\gamma_R}$:

$$d(\tau) = \sqrt{\left[1 + (\gamma_0^2 - 1)\sin^2(\tau)\right]/\gamma_0}.$$ (20)

Obviously, the width fluctuates, or "breathes", periodically with time. Of course, this is not a breathing mode as observed in systems of confined interacting particles, see [19,20], e.g.,

Integration of the β equation leads to

$$\beta = C_2 \exp(i\tau)\left[\exp(2i\tau) + C_1\right]^{-1} = C_2 N_1^{-1}\left[\exp(-i\tau) + \exp(i\tau)C_1\right].$$ (21)

Later on, the complex integration constant $C_2 \equiv A_2 + iB_2$ will serve as a state label. The third differential equation of (16) amounts to

$$c(\tau) = i\tau/2 - C_2^2\left[4\left(\exp(2i\tau) + C_1\right)\right]^{-1} - (1/2)\ln\left(\sqrt{\exp(2i\tau) + C_1}\right) + C_3.$$ (22)

By reasons explained in Appendix A, we dispose of the integration constant C_3 as follows

$$C_3 = -(1 + \gamma_0)(8\gamma_0)^{-1}(A_2^2 + \gamma_0 B_2^2), \quad C_2 = A_2 + iB_2.$$ (23)

In Appendix A, the probability density P is derived in the following form

$$P(\xi, \tau) = \frac{C_0^2}{\sqrt{N_1}}\exp\left[-\gamma_R\left(\xi - \beta_R/\gamma_R\right)^2\right],$$ (24)

where the time-dependent functions γ_R and N_1 are defined through (17) and (18), and β_R comes out as

$$\beta_R(\tau) = (1/8)(1+\gamma_0)^{-1}N_1^{-1}\left[A_2\cos(\tau) + B_2\sin(\tau)\right]. \tag{25}$$

The complex integration constant C_2 corresponds to the familiar complex quantum number z in the case of the standard coherent states; hence, the real numbers A_2, B_2 characterize different states. The normalization constant C_0 obeys the following condition, see Appendix A,

$$1 = (1/2)C_0^2\sqrt{\pi/\gamma_0}(1+\gamma_0). \tag{26}$$

4.1. Completeness of Type-II States

Combining the above results, we write the time-dependent wave function as follows

$$\psi(\zeta,\tau) = \frac{C_0}{\sqrt{\exp(2i\,\tau)+C_1}}\exp\left[C_3 - \frac{C_2^2\,(\exp(-2i\,\tau)+C_1)}{4N_1} + \beta(\tau)\zeta - \gamma(\tau)\zeta^2/2\right], \tag{27}$$

where γ, β, and C_3 are defined in (18), (21), and (23), respectively. Let us consider ψ at zero time:

$$\psi(\zeta,0) = \frac{C_0}{\sqrt{1+C_1}}\exp\left[C_3 - \frac{C_2^2}{4(1+C_1)} + C_2(1+\gamma_0)\,\zeta/2 - \gamma_0\zeta^2/2\right]. \tag{28}$$

In (28), we set $\zeta = \tilde{\zeta}/\sqrt{\gamma_0}$ to write

$$\psi(\tilde{\zeta},0) = \frac{C_0\gamma_0^{-1/4}}{\sqrt{1+C_1}}\exp\left[C_3 - \frac{C_2^2}{4(1+C_1)} + C_2(1+\gamma_0)/\sqrt{\gamma_0}\,\tilde{\zeta}/2 - \tilde{\zeta}^2/2\right]. \tag{29}$$

Now we substitute the complex variable z for the integration constant C_2 as follows

$$C_2\frac{1+\gamma_0}{2\sqrt{\gamma_0}} = \sqrt{2}\,z \tag{30}$$

and obtain

$$\psi(\tilde{\zeta},0) = \frac{C_0}{\sqrt{1+C_1}}\exp\left[C_3 - z^2\frac{\gamma_0}{1+\gamma_0} + \sqrt{2}\,z\tilde{\zeta} - \tilde{\zeta}^2/2\right]. \tag{31}$$

In C_3, given in (23), we make the following replacements which are induced by (30):

$$A_2 \to \kappa(z+z^*), \quad B_2 \to -i\kappa(z-z^*), \quad \kappa = \sqrt{2\gamma_0}/(1+\gamma_0). \tag{32}$$

There occur some nice cancelations, and one obtains

$$\psi_z(\tilde{\zeta}) = \frac{C_0\gamma_0^{-1/4}}{\sqrt{1+C_1}}\exp\left[-\frac{1}{2}\left(zz^*+z^2\right) + iD + \sqrt{2}\,z\tilde{\zeta} - \tilde{\zeta}^2/2\right], \quad D = \frac{1-\gamma_0}{2(1+\gamma_0)}\mathrm{Im}(z^2). \tag{33}$$

Comparison with (5) shows that the wave function (33) has the same structure apart from the purely imaginary phase iD. The latter drops out in the completeness proof, see (A15) in Appendix B. As a consequence, the states (33) form a complete set of states with respect to the state label z.

At $\tau = 0$, the states (33) differ from the standard coherent states (5) by the state dependent phase D, through the variables ζ and $\tilde{\zeta}$ which denote the differently scaled space variable x, and also through the different definition of the quantum number z, which for simplicity was denoted by the same symbol in (30). Essentially, type-I and type-II states differ by their time evolution and width parameter γ_0 which is equal to $\alpha^2 = m\omega/\hbar$ and to an arbitrary positive number, respectively.

4.2. Mean Values and Uncertainty Product

In the following, we list mean values for the time-dependent states (27) including the position momentum uncertainty product Δ_{xp}. They are periodic in time with the oscillator angular frequency $\omega \equiv 2\pi/T$. The uncertainty product is minimal at the discrete times $t_n = (1/4)nT$, $n = 0, 1, \ldots$. For comparison, the traditional coherent states are always minimal [10]. We use the abbreviations $(\Delta_x)^2 = \langle x^2 \rangle - \langle x \rangle^2$ and $(\Delta_v)^2 = \langle v^2 \rangle - \langle v \rangle^2$ for the mean square deviations of position and velocity, respectively.

$$\langle x(\tau) \rangle = (1/\alpha)(1+\gamma_0)(2\gamma_0)^{-1}[A_2 \cos(\tau) + B_2\gamma_0 \sin(\tau)]; \tag{34}$$

$$\langle v(\tau) \rangle = \hbar\alpha(2m\gamma_0)^{-1}[-A_2 \sin(\tau) + \gamma_0 B_2 \cos(\tau)]; \tag{35}$$

$$(\Delta_x)^2 = (4\alpha^2\gamma_0)^{-1}\left[1 + \gamma_0^2 + (1-\gamma_0^2)\cos(2\tau)\right]; \tag{36}$$

$$(\Delta_v)^2 = \hbar^2\alpha^2(4m^2\gamma_0)^{-1}\left[1 + \gamma_0^2 + (\gamma_0^2 - 1)\cos(2\tau)\right]; \tag{37}$$

$$\langle H \rangle = \hbar\omega(8\gamma_0^2)^{-1}\left[(1+\gamma_0)^2\left(A_2^2 + \gamma_0^2 B_2^2\right) + 2\gamma_0(1+\gamma_0^2)\right]. \tag{38}$$

It is noticed that the mean square deviations do not depend on the state label (A_2, B_2). The uncertainty product follows immediately from (36) and (37) as

$$\Delta_{xp} := (\Delta_x)^2(\Delta_p)^2 \equiv m^2(\Delta_x)^2(\Delta_v)^2 = \frac{\hbar^2}{16\gamma_0^2}\left[(1+\gamma_0^2)^2 - (1-\gamma_0^2)^2\cos^2(2\tau)\right]. \tag{39}$$

In the special case $\gamma_0 = 1$, the product is always minimal. As a matter of fact, $\gamma_0 = 1$ is the type-I case of Section 3.

By (38), the mean energy does not depend on time and is positive definite, as it must be. The limit to the standard case with $\gamma_0 = 1$, gives the known result

$$\langle H \rangle_{\gamma_0=1} = \hbar\omega(zz^* + 1/2). \tag{40}$$

and the state with $z = 0$ is the ground state of the HO with zero point energy $\hbar\omega/2$.

5. Wave Packet Solutions for the RO

For convenience, we will keep the same symbols for the trial functions $\gamma(\tau)$, $\beta(\tau)$, and $c(\tau)$. Setting $\omega = i\Omega$ with $\Omega > 0$, implies that $\alpha^2 = -m\Omega/\hbar$. In the coherent state (5), the exponential part, $-\zeta^2/2 \equiv -(m\omega/\hbar)x^2/2$, is then replaced by $+(m\Omega/\hbar)x^2/2$, which precludes normalization.

We introduce $1/\alpha_\Omega$ as the new length parameter and define the dimensionless magnitudes

$$\zeta = \alpha_\Omega x, \quad \tau = t\Omega, \quad \text{with} \quad \alpha_\Omega^2 = m\Omega/\hbar. \tag{41}$$

The Schrödinger equation, with the ansatz (10), has to be solved for the RO Hamiltonian

$$H_\Omega = p^2/(2m) - m\Omega^2/2\, x^2 = -\hbar\Omega/2\left[\partial_\zeta^2 + \zeta^2\right]. \tag{42}$$

From (2), the following differential equations result:

$$i\,\gamma'(\tau) = 1 + \gamma^2(\tau), \quad i\,\beta'(\tau) = \gamma(\tau)\beta(\tau), \quad 2i\,c'(\tau) = \gamma(\tau) - \beta^2(\tau), \tag{43}$$

where, as compared with the HO case in (16), only the equation for γ differs. Beginning with γ, one successively obtains the following solutions

$$\gamma(\tau) = -i \tanh(\tau + i C_1), \tag{44}$$

$$\beta(\tau) = C_2/\cosh(\tau + i C_1), \tag{45}$$

$$c(\tau) = C_3 - (1/2)\ln\left(\cosh(\tau + i C_1)\right) + (i/2)C_2^2 \tanh(\tau + i C_1), \tag{46}$$

where C_1, C_2, C_3 are integration constants. We assume that

$$\gamma_0 \equiv \gamma(0) = \tan(C_1) > 0, \quad 0 < C_1 < \pi/2, \tag{47}$$

which implies that

$$\cos(C_1) = (1 + \gamma_0^2)^{-1/2}, \quad \sin(C_1) = \gamma_0(1 + \gamma_0^2)^{-1/2}. \tag{48}$$

In order to decompose the functions $c(\tau), \beta(\tau), \gamma(\tau)$ into their real and imaginary parts, we take over the following abbreviations from [16]

$$f(\tau) = \cosh(\tau) - i \gamma_0 \sinh(\tau), \quad h(\tau) = [ff^*]^{-1}. \tag{49}$$

After the decompositions $\beta = \beta_R + i \beta_I$, $\gamma = \gamma_R + i \gamma_I$, $C_2 = A_2 + i B_2$, we infer from (44) to (46):

$$\gamma_R = h(\tau)\gamma_0, \quad \gamma_I = -(h(\tau)/2)(1 + \gamma_0^2)\sinh(2\tau); \tag{50}$$

$$\beta_R = h(\tau)\sqrt{1 + \gamma_0^2}\left[A_2\cosh(\tau) + \gamma_0 B_2 \sinh(\tau)\right],$$

$$\beta_I = h(\tau)\sqrt{1 + \gamma_0^2}\left[B_2\cosh(\tau) - \gamma_0 A_2 \sinh(\tau)\right]; \tag{51}$$

$$\exp[c(\tau)] = [\cosh(\tau + i C_1)]^{-1/2}\exp\left[C_3 - C_2^2\gamma(\tau)/2\right]. \tag{52}$$

According to (50), γ_R is larger zero, which makes the wave function (10) a normalizable wave packet. The probability density reads:

$$P(\zeta, \tau) = C_0^2 \exp\left[c + c^* + 2\beta_R\zeta - \gamma_R\zeta^2\right]. \tag{53}$$

Integration with respect to ζ leads to the normalization condition

$$1 = C_0^2\sqrt{\pi/\gamma_R}\exp\left[c(\tau) + c^*(\tau) + \beta_R^2/\gamma_R\right]. \tag{54}$$

The normalization constant C_0 was determined in [16] for real constants C_2. With $C_2 = A_2 + i B_2$, we dispose of the integration constant C_3 as

$$C_3 = -(1/2)(A_2^2/\gamma_0 + B_2^2\gamma_0) \tag{55}$$

to obtain in a straightforward manner

$$C_0^2 = \sqrt{\pi(\gamma_0^{-1} + \gamma_0)}, \tag{56}$$

which is a time independent condition as it must be.

With the aid of elementary trigonometric manipulations and the normalization constant C_0 given in (56), the wave function can be written as follows

$$\psi(\zeta, \tau) = (\gamma_0/\pi)^{1/4}\sqrt{h(\tau)f(\tau)}\exp\left[C_3 - (1/2)C_2^2\gamma(\tau) + \beta(\tau)\zeta - \gamma(\tau)\zeta^2/2\right]. \tag{57}$$

5.1. Coherent States of the RO

As before, let us consider the wave function at time $t = 0$, where in particular $h = f = 1$:

$$\psi(\zeta, 0) \equiv \psi(\zeta, \tau = 0) = (\gamma_0/\pi)^{1/4} \exp\left[C_3 - 1/2C_2^2\gamma_0 + C_2\sqrt{1 + \gamma_0^2}\,\zeta - \gamma_0\zeta^2/2\right]. \tag{58}$$

After the re-scaling $\zeta \to \tilde{\zeta}$ with $\tilde{\zeta} = \sqrt{\gamma_0}\,\zeta$, one obtains

$$\Psi(\tilde{\zeta}, 0) = \pi^{-1/4} \exp\left[C_3 - 1/2C_2^2\gamma_0 + C_2\sqrt{(1 + \gamma_0^2)/\gamma_0}\,\tilde{\zeta} - \tilde{\zeta}^2/2\right]. \tag{59}$$

In view of the standard HO wave function (5), we replace the integration constant C_2 by z:

$$C_2\sqrt{(1 + \gamma_0^2)/\gamma_0} = \sqrt{2}\,z \tag{60}$$

and obtain

$$\Psi_z(\tilde{\zeta}) = \pi^{-1/4} \exp\left[C_3 - \gamma_0^2 z^2/(1 + \gamma_0^2) + \sqrt{2}z\tilde{\zeta} - \tilde{\zeta}^2/2\right]. \tag{61}$$

In C_3, given in (55), the relation (60) gives rise to the substitutions

$$A_2 \to \kappa_1(z + z^*), \quad B_2 \to -i\kappa_1(z - z^*), \quad \kappa_1 = (1/2)\sqrt{2\gamma_0/(1 + \gamma_0^2)}, \tag{62}$$

and hence to

$$C_3 = \left[4(1 + \gamma_0^2)\right]^{-1}\left[(\gamma_0^2 - 1)(z^2 + z^*z^*) - 2(1 + \gamma_0^2)zz^*\right]. \tag{63}$$

After some elementary re-arrangements, one finds

$$\Psi_z(\tilde{\zeta}) = \frac{1}{\pi^{1/4}} \exp\left[-\frac{1}{2}(zz^* + z^2) + i D_1 + \sqrt{2}z\tilde{\zeta} - \frac{\tilde{\zeta}^2}{2}\right], \quad D_1 = \frac{1 - \gamma_0^2}{2(1 + \gamma_0^2)}\mathrm{Im}(z^2). \tag{64}$$

Apart from the purely imaginary phase $i D_1$, the wave functions Ψ_z are the same as the standard coherent states (5). Since in the completeness proof the D_1 phase drops out, see (A15) in Appendix B, the states Ψ_z form a complete function set.

5.2. Mean Values

With the aid of Mathematica [21], we get the following mean values for position x, velocity v, their mean square deviations $(\Delta x)^2$, $(\Delta v)^2$, and the mean energy $\langle H_\Omega \rangle$:

$$\langle x \rangle = (\alpha_\Omega)^{-1}\sqrt{1 + \gamma_0^{-2}}\left[A_2\cosh(\tau) + \gamma_0 B_2\sinh(\tau)\right]; \tag{65}$$

$$(\Delta x)^2 = \left(2\alpha_\Omega^2\gamma_0\right)^{-1}\left[\cosh^2(\tau) + \gamma_0^2\sinh^2(\tau)\right]; \tag{66}$$

$$\langle v \rangle = (\hbar\alpha_\Omega/m)\sqrt{1 + \gamma_0^{-2}}\left[A_2\sinh(\tau) + \gamma_0 B_2\cosh(\tau)\right]; \tag{67}$$

$$(\Delta v)^2 = (\hbar\alpha_\Omega/(2m))^2\,\gamma_0^{-1}\left[\gamma_0^2 - 1 + (1 + \gamma_0^2)\cosh(2\tau)\right]; \tag{68}$$

$$\langle H_\Omega \rangle = \hbar\Omega(4\gamma_0)^{-1}\left[\gamma_0^2 - 1 + 2(\gamma_0 + \gamma_0^{-1})\left(\gamma_0^2 B_2^2 - A_2^2\right)\right]. \tag{69}$$

The mean energy does not depend on time, as it must be. With the aid of (62), the mean energy could also be expressed in terms of the complex state label z. Since A_2 and B_2 are arbitrary real numbers, the mean energy can have any positive or negative value. From (66) and (68) one infers the position-momentum uncertainty product Δ_{xp} as

$$\Delta_{xp}^2(\tau) = \hbar^2/(8\gamma_0^2)\left[\cosh^2(\tau) + \gamma_0^2\sinh(\tau)\right]\left[\gamma_0^2 - 1 + (1 + \gamma_0^2)\cosh(2\tau)\right]. \tag{70}$$

This product obeys the inequality

$$\Delta_{xp}^2(\tau) > \Delta_{xp}^2(0) = \hbar^2/4, \quad \tau > 0. \tag{71}$$

Obviously, the uncertainty product is minimal at $\tau = 0$, which means for the coherent states (64). By (66), the wave packets broaden exponentially with time.

6. Application to the Kepler-Coulomb Problem

The connection of the non-relativistic Hamiltonian for the hydrogen atom with the model of a four-dimensional oscillator is conveniently achieved by means of the Kustaanheimo-Stiefel transformation [15], which we write as follows [16,22]

$$\begin{aligned} u_1 &= \sqrt{r}\cos(\theta/2)\cos(\varphi - \Phi); \quad u_2 = \sqrt{r}\cos(\theta/2)\sin(\varphi - \Phi); \\ u_3 &= \sqrt{r}\sin(\theta/2)\cos(\Phi); \qquad u_4 = \sqrt{r}\sin(\theta/2)\sin(\Phi), \end{aligned} \tag{72}$$

where r, θ, φ are three-dimensional polar coordinates with $r > 0$, $0 < \theta < \pi$, $0 \le \varphi < 2\pi$, and $0 \le \Phi < 2\pi$ generates the extension to the fourth dimension. The vector $\mathbf{u} = \{u_1, u_2, u_3, u_3\}$ covers the $\mathbf{R}^4$ and the volume elements are related as [16]

$$du_1 du_2 du_3 du_4 = (1/8)r\sin(\theta)drd\theta d\varphi d\Phi. \tag{73}$$

The stationary Schrödinger equation $H\psi = E\psi$ for the Hamiltonian $H = p^2/(2m) - \lambda/r$ is transformed into the following form of a four-dimensional harmonic oscillator [14]:

$$H_u\Psi(\mathbf{u}) = \lambda\Psi(\mathbf{u}), \quad H_u = -\hbar^2/(8m)\Delta_\mathbf{u} - E\,\mathbf{u}\cdot\mathbf{u}, \quad \Delta_\mathbf{u} = \partial_{u_1}^2 + ...\partial_{u_4}^2 \tag{74}$$

with the constraint

$$\partial_\Phi\Psi(\mathbf{u}) = 0. \tag{75}$$

It should be noticed that, by (72), the components u_i^2 have the dimension of a length rather than length square. As a consequence, in the evolution equation $i\hbar\partial_\sigma\Psi = H_u\Psi$, the parameter σ, which has the dimension time/length, is not the time parameter of the original problem. For negative energies with $E < 0$, four-dimensional coherent oscillator states (of type-I) were used in [14] to show that elliptic orbits emerge in the classical limit whereby σ turns out being proportional to the eccentric anomaly.

In the spectrum of positive energies (ionized states of the hydrogen atom) with $E > 0$, coherent states of the RO were constructed in [16] and gave rise to hyperbolic orbits in the classical limit; by analytic continuation, also the elliptic orbits were derived from the RO states in the classical limit [17]. In addition, Kepler's equation was obtained by the assumption that time-dependence enters through the curve parameter σ only. Recently [18], based on the coherent RO states, the first order quantum correction to Kepler's equation could be established for the smallness parameter $\epsilon = \hbar/L$ where L denotes the orbital angular momentum.

7. Conclusions

Besides the standard coherent states of the harmonic oscillator (H0), a further solution family of the time-dependent Schrödinger equation was derived with the following properties: (i) The functions are normalizable of Gaussian type and contain a disposable width parameter. The latter allows us, for instance, to use arbitrarily concentrated one-particle states independently of the parameters of a harmonic trap; (ii) The functions are complete and isomorphic to the standard coherent states at time $t = 0$; (iii) The states minimize the position-momentum uncertainty product at the discrete times $T_n = n\,\pi/(2\omega)$, $n = 0, 1, \ldots$; (iv) The width of the wave packets "breathes" periodically with period $T/2 = \pi/\omega$. (v) There is no diffusion, $T = 2\pi/\omega$ is the recurrence time of the states.

In the case of the reversed harmonic oscillator (RO), there exists only one family of time-dependent solutions. They share the properties (i) and (ii) of the type-II HO states, and (iii) is fulfilled at time $t = 0$, only. There is no recurrence, instead there is diffusion with a broadening which increases exponentially with time. The application to the Kepler-Coulomb problem was briefly discussed. The HO coherent states of type-I and the RO coherent states served as basis to derive, in the classical limit, the elliptic Kepler orbits [14] and the hyperbolic ones [16,17], respectively.

Acknowledgments: The author expresses his gratitude to Jürgen Parisi for his constant encouragement and support. He also profited from his critical reading of the manuscript.

Conflicts of Interest: The author declares no conflict of interest.

Appendix. Probability Density for Type-II States

We have to decompose the functions $\beta(\tau)$ and $c(\tau)$, as given by (21) and (22), into their real and imaginary parts. To this end, we set $C_2 = A_2 + i\,B_2$ with real constants A_2 and B_2 and $\beta = \beta_R + i\,\beta_I$. Using the definitions of N_1 and C_1 in terms of γ_0, we obtain

$$
\begin{aligned}
\beta_R &= \frac{1 + \gamma_0}{2}\, \frac{A_2 \cos(\tau) + B_2 \gamma_0 \sin(\tau)}{1 + (\gamma_0^2 - 1)\sin^2(\tau)}, \\
\beta_I &= \frac{1 + \gamma_0}{2}\, \frac{B_2 \cos(\tau) - A_2 \gamma_0 \sin(\tau)}{1 + (\gamma_0^2 - 1)\sin^2(\tau)}.
\end{aligned} \tag{A1}
$$

In view of the function $c(\tau)$, we make use of the following auxiliary relations

$$
\begin{aligned}
F_c &\equiv -C_2^2 \left[4\left(\exp(2i\,\tau) + C_1\right)\right]^{-1} = F_R + i\,F_I, \\
F_R &= (1/(4N_1)) \left[(B_2^2 - A_2^2)\cos(2\tau) - 2A_2 B_2 \sin(2\tau) + (B_2^2 - A_2^2)C_1\right], \\
F_I &= (1/(4N_1)) \left[(A_2^2 - B_2^2)\sin(2\tau) - 2A_2 B_2 \cos(2\tau) - 2A_2 B_2 C_1\right], \tag{A2}
\end{aligned}
$$

$$
\exp\left[c(\tau) + c^*(\tau)\right] = (1/\sqrt{N_1}) \exp\left[2C_3 + 2F_R\right], \tag{A3}
$$

where the integration constant C_3 is assumed being real and the star suffix means complex conjugation.

The probability density P results from the wave function (10) in the form

$$
P(\xi, \tau) = \frac{C_0^2}{\sqrt{N_1}}\, \exp\left[2C_3 + 2F_R + 2\beta_R \xi - \gamma_R \xi^2\right], \tag{A4}
$$

where C_0 is defined through the normalization integral

$$
1 = \int_{-\infty}^{\infty} d\xi\, P(\xi, \tau) = \frac{C_0^2 \sqrt{\pi}}{\sqrt{N_1 \gamma_R}}\, \exp(G), \quad G = 2C_3 + 2F_R + \beta_R^2/\gamma_R. \tag{A5}
$$

From the expression of G, it is not obvious that C_0 is independent of τ which was assumed in (10). Clearly, since $\Phi := \psi/C_0$ obeys the Schrödinger equation and H is hermitian, one has the property

$$
\partial_\tau \langle \Phi | \Phi \rangle = 0. \tag{A6}
$$

As a matter of fact, it is straightforward to show that

$$
2F_R + \beta_R^2/\gamma_R = \left[B_2^2(C_1 - 1) - A_2^2(1 + C_1)\right]\left[2(C_1^2 - 1)\right]^{-1} \tag{A7}
$$

does not depend on τ. We now dispose of the integration constant C_3 such that the exponent G vanishes:

$$C_3 = -\left[B_2^2(C_1 - 1) - A_2^2(1 + C_1) \right] \left[4(C_1^2 - 1) \right]^{-1}. \tag{A8}$$

In view of $G = 0$, we replace $2C_3 + 2F_R$ by $-\beta_R^2/\gamma_R$, so that

$$P(\xi, \tau) = \frac{C_0^2}{\sqrt{N_1}} \exp\left[-\gamma_R \, (\xi - \beta_R/\gamma_R)^2 \right], \tag{A9}$$

which is the result (24). The normalization condition comes out immediately in the form

$$1 = \frac{C_0^2\sqrt{\pi}}{\sqrt{N_1 \gamma_R}} = \frac{C_0^2\sqrt{\pi}}{\sqrt{1 - C_1^2}} = \frac{C_0^2\sqrt{\pi}(1 + \gamma_0)}{2\sqrt{\gamma_0}}. \tag{A10}$$

Appendix. Proof of Completeness

In order to prove the completeness of the functions (5), *i.e.*, for the type-I HO case, we take advantage of the following generating function of the Hermite polynomials [23]:

$$\exp\left[2XZ - Z^2 \right] = \sum_{n=0}^{\infty} \frac{Z^n}{n!} H_n(X). \tag{A11}$$

In the function (5), we replace z by $\sqrt{2}\, Z$ to obtain

$$\psi_z(\zeta) = \pi^{-1/4} \exp\left[-ZZ^* - (1/2)\zeta^2 \right] \exp\left[-Z^2 + 2\zeta Z \right]. \tag{A12}$$

With the aid of (A11), one can write

$$\psi_z(\zeta) = \exp\left[-(1/2)zz^* \right] \sum_{n=0}^{\infty} \frac{z^n}{\sqrt{n!}} \varphi_n(\zeta), \tag{A13}$$

where

$$\varphi_n(\zeta) = \frac{1}{\sqrt{n!2^n \sqrt{\pi}}} H_n(\zeta) \exp\left[-(1/2)\zeta^2 \right]. \tag{A14}$$

By means of (A13) and setting $z = u \exp[i\,\varphi]$, we obtain

$$\langle \zeta_2|z\rangle\langle z|\zeta_1\rangle = \exp\left[-u^2 \right] \sum_{m,n=0}^{\infty} \frac{u^{n+m} \exp\left[i\,(m-n)\varphi \right]}{\sqrt{m!n!}} \varphi_m(\zeta_2)\varphi_n(\zeta_1). \tag{A15}$$

In (A15), the φ integration projects out the terms $n = m$ with the result

$$\frac{1}{\pi} \int_0^\infty u\,du \int_0^{2\pi} d\varphi \, \langle \zeta_2|z\rangle\langle z|\zeta_1\rangle = 2\int_0^\infty u\,du \, \exp[-u^2] \sum_{n=0}^{\infty} \frac{u^{2n}}{n!} \varphi_n(\zeta_2)\varphi_n(\zeta_1). \tag{A16}$$

After changing the integration variable $u \to v$ with $v = u^2$ with $u\,du = dv/2$, one uses

$$\int_0^\infty dv\, \frac{v^n}{n!} \exp[-v] = 1, \quad n = 0, 1, \ldots \tag{A17}$$

and, in view of the completeness of the Hermite polynomials, arrives at

$$\frac{1}{\pi} \int_0^\infty u\,du \int_0^{2\pi} d\varphi \, \langle \zeta_2|z\rangle\langle z|\zeta_1\rangle = \sum_{n=0}^{\infty} \varphi_n(\zeta_2)\varphi_n(\zeta_1) = \delta\,(\zeta_2 - \zeta_1). \tag{A18}$$

In the type-II HO and the RO cases, there appear additional purely imaginary phases in the wave function, which do not depend on ζ_1, ζ_2, and drop out at the step (A15) of the completeness proof above.

References

1. Schrödinger, E. Der stetige Übergang von der Mikro-zur Makromechanik. *Naturwissenschaften* **1926**, *14*, 664–666.
2. Glauber, R.J. Coherent and incoherent states of the radiation field. *Phys. Rev.* **1963**, *131*, 2766.
3. Glauber, R.J. Photon Correlations. *Phys. Rev. Lett.* **1963**, *10*, 84.
4. Antoniou, I.E.; Progogine, I. Intrinsic irreversibility and integrability of dynamics. *Phys. A Stat. Mech. Appl.* **1993**, *192*, 443–464.
5. Gentilini, S.; Braidotti, M.C.; Marcucci, G.; DelRe, E.; Conti, C. Physical realization of the Glauber quantum oscillator. *Sci. Rep.* **2015**, *5*, 15816.
6. Glauber, R.J. Amplifiers, attenuators, and schrödinger's cat. *Ann. N. Y. Acad. Sci.* **1986**, *480*, 336–372.
7. Barton, G. Quantum mechanis of the inverted oscillator potential. *Ann. Phys.* **1986**, *166*, 322–363.
8. Bhaduri, R.K.; Khare, A.; Reimann, S.M.; Tomisiek, E.L. The riemann zeta function and the inverted harmonic oscillator. *Ann. Phys.* **1997**, *264*, 25–40.
9. Guo, G.-J.; Ren, Z.-Z.; Ju, G.-X.; Guo, X.-Y. Quantum tunneling effect of a time-dependent inverted harmonic oscillator. *J. Phys. A Math. Theor.* **2011**, *44*, 185301.
10. Galindo, A.; Pascual, P. *Quantum Mechanics I*; Springer: Berlin, Germany, 1990.
11. Fock, V.A. Zur Theorie des Wassenstoffatoms. *Z. Phys.* **1935**, *98*, 145–154.
12. Chen, A.C. Hydrogen atom as a four-dimensional oscillator. *Phys. Rev. A* **1980**, *22*, 333–335.
13. Gracia-Bondia, J.M. Hydrogen atom in the phase-space formulation of quantum mechanics. *Phys. Rev. A* **1984**, *30*, 691–697.
14. Gerry, C.C. Coherent states and the Kepler-Coulomb problem. *Phys. Rev. A* **1986**, *33*, 6–11.
15. Kustaanheimo, P.; Stiefel, E. Perturbation theory of Kepler motion based on spinor regularization. *J. Reine Angew. Math.* **1965**, *218*, 204–219.
16. Rauh, A.; Parisi, J. Quantum mechanics of hyperbolic orbits in the Kepler problem. *Phys. Rev. A* **2011**, *83*, 042101.
17. Rauh, A.; Parisi, J. Quantum mechanics of Kepler orbits. *Adv. Stud. Theor. Phys.* **2014**, *8*, 889–938.
18. Rauh, A.; Parisi, J. Quantum mechanical correction to Kepler's equation. *Adv. Stud. Theor. Phys.* **2016**, *10*, 1–22.
19. Baletto, F.; Riccardo, F. Structural properties of nanoclusters: Energetic, thermodynamic, and kinetic effects. *Rev. Mod. Phys.* **2015**, *77*, 371–423.
20. Bauch, S.; Balzer, K.; Bonitz, M. Quantum breathing mode of trapped bosons and fermions at arbitrary coupling. *Phys. Rev. B* **2009**, *80*, 054515.
21. Wolfram Research, Inc. *Mathematica*; Version 10.1.0.0; Wolfram Research, Inc.: Champaign, IL, USA, 2015.
22. Chen, C.; Kibler, M. Connection between the hydrogen atom and the four-dimensional oscillator. *Phys. Rev. A* **1985**, *31*, 3960–3963.
23. Gradshteyn, I.S.; Ryzhik, I.M. *Table of Integrals, Series, and Products*; Academic Press: New York, NY, USA, 1965.

© 2016 by the author. Licensee MDPI, Basel, Switzerland. This article is an open access article distributed under the terms and conditions of the Creative Commons Attribution (CC BY) license (http://creativecommons.org/licenses/by/4.0/).

symmetry

MDPI

Article

Entangled Harmonic Oscillators and Space-Time Entanglement

Sibel Başkal [1], Young S. Kim [2,*] and Marilyn E. Noz [3]

[1] Department of Physics, Middle East Technical University, 06800 Ankara, Turkey;
baskal@newton.physics.metu.edu.tr
[2] Center for Fundamental Physics, University of Maryland College Park, College Park, MD 20742, USA
[3] Department of Radiology, New York University School of Medicine, New York, NY 10016, USA;
marilyn.noz@med.nyu.edu
* Correspondence: yskim@umd.edu; Tel.: +1-301-937-6306

Academic Editor: Sergei D. Odintsov
Received: 26 February 2016; Accepted: 20 June 2016; Published: 28 June 2016

Abstract: The mathematical basis for the Gaussian entanglement is discussed in detail, as well as its implications in the internal space-time structure of relativistic extended particles. It is shown that the Gaussian entanglement shares the same set of mathematical formulas with the harmonic oscillator in the Lorentz-covariant world. It is thus possible to transfer the concept of entanglement to the Lorentz-covariant picture of the bound state, which requires both space and time separations between two constituent particles. These space and time variables become entangled as the bound state moves with a relativistic speed. It is shown also that our inability to measure the time-separation variable leads to an entanglement entropy together with a rise in the temperature of the bound state. As was noted by Paul A. M. Dirac in 1963, the system of two oscillators contains the symmetries of the $O(3,2)$ de Sitter group containing two $O(3,1)$ Lorentz groups as its subgroups. Dirac noted also that the system contains the symmetry of the $Sp(4)$ group, which serves as the basic language for two-mode squeezed states. Since the $Sp(4)$ symmetry contains both rotations and squeezes, one interesting case is the combination of rotation and squeeze, resulting in a shear. While the current literature is mostly on the entanglement based on squeeze along the normal coordinates, the shear transformation is an interesting future possibility. The mathematical issues on this problem are clarified.

Keywords: Gaussian entanglement; two coupled harmonic oscillators; coupled Lorentz groups; space-time separation; Wigner's little groups; $O(3,2)$ group; Dirac's generators for two coupled oscillators

PACS: 03.65.Fd, 03.65.Pm, 03.67.-a, 05.30.-d

1. Introduction

Entanglement problems deal with fundamental issues in physics. Among them, the Gaussian entanglement is of current interest not only in quantum optics [1–4], but also in other dynamical systems [3,5–8]. The underlying mathematical language for this form of entanglement is that of harmonic oscillators. In this paper, we present first the mathematical tools that are and may be useful in this branch of physics.

The entangled Gaussian state is based on the formula:

$$\frac{1}{\cosh \eta} \sum_k (\tanh \eta)^k \chi_k(x) \chi_k(y) \tag{1}$$

where $\chi_n(x)$ is the nth excited-state oscillator wave function.

In Chapter 16 of their book [9], Walls and Milburn discussed in detail the role of this formula in the theory of quantum information. Earlier, this formula played the pivotal role for Yuen to formulate his two-photon coherent states or two-mode squeezed states [10]. The same formula was used by Yurke and Patasek in 1987 [11] and by Ekert and Knight [12] for the two-mode squeezed state where one of the photons is not observed. The effect of entanglement is to be seen from the beam splitter experiments [13,14].

In this paper, we point out first that the series of Equation (1) can also be written as a squeezed Gaussian form:

$$\frac{1}{\sqrt{\pi}} \exp\left\{ -\frac{1}{4} \left[e^{-2\eta}(x+y)^2 + e^{2\eta}(x-y)^2 \right] \right\} \tag{2}$$

which becomes:

$$\frac{1}{\sqrt{\pi}} \exp\left\{ -\frac{1}{2} \left(x^2 + y^2 \right) \right\} \tag{3}$$

when $\eta = 0$.

We can obtain the squeezed form of Equation (2) by replacing x and y by x' and y', respectively, where:

$$\begin{pmatrix} x' \\ y' \end{pmatrix} = \begin{pmatrix} \cosh\eta & -\sinh\eta \\ -\sinh\eta & \cosh\eta \end{pmatrix} \begin{pmatrix} x \\ y \end{pmatrix} \tag{4}$$

If x and y are replaced by z and t, Equation (4) becomes the formula for the Lorentz boost along the z direction. Indeed, the Lorentz boost is a squeeze transformation [3,15].

The squeezed Gaussian form of Equation (2) plays the key role in studying boosted bound states in the Lorentz-covariant world [16–20], where z and t are the space and time separations between two constituent particles. Since the mathematics of this physical system is the same as the series given in Equation (1), the physical concept of entanglement can be transferred to the Lorentz-covariant bound state, as illustrated in Figure 1.

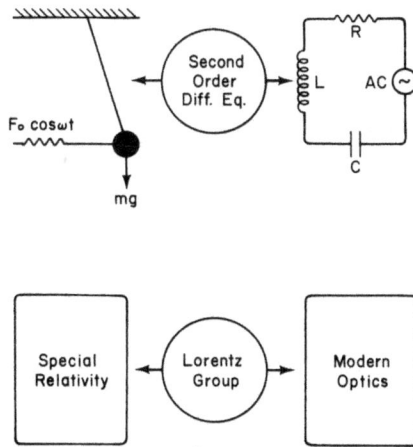

Figure 1. One mathematics for two branches of physics. Let us look at Equations (1) and (2) applicable to quantum optics and special relativity, respectively. They are the same formula from the Lorentz group with different variables as in the case of the Inductor-Capacitor-Resistor (LCR) circuit and the mechanical oscillator sharing the same second-order differential equation.

We can approach this problem from the system of two harmonic oscillators. In 1963, Paul A. M. Dirac studied the symmetry of this two-oscillator system and discussed all possible transformations applicable to this oscillator [21]. He concluded that there are ten possible generators of transformations satisfying a closed set of commutation relations. He then noted that this closed set corresponds to the Lie algebra of the $O(3,2)$ de Sitter group, which is the Lorentz group applicable to three space-like and two time-like dimensions. This $O(3,2)$ group has two $O(3,1)$ Lorentz groups as its subgroups.

We note that the Lorentz group is the language of special relativity, while the harmonic oscillator is one of the major tools for interpreting bound states. Therefore, Dirac's two-oscillator system can serve as a mathematical framework for understanding quantum bound systems in the Lorentz-covariant world.

Within this formalism, the series given in Equation (1) can be produced from the ten-generator Dirac system. In discussing the oscillator system, the standard procedure is to use the normal coordinates defined as:

$$u = \frac{x+y}{\sqrt{2}}, \quad \text{and} \quad v = \frac{x-y}{\sqrt{2}} \tag{5}$$

In terms of these variables, the transformation given in Equation (4) takes the form:

$$\begin{pmatrix} u' \\ v' \end{pmatrix} = \begin{pmatrix} e^{-\eta} & 0 \\ 0 & e^{\eta} \end{pmatrix} \begin{pmatrix} u \\ v \end{pmatrix} \tag{6}$$

where this is a squeeze transformation along the normal coordinates. While the normal-coordinate transformation is a standard procedure, it is interesting to note that it also serves as a Lorentz boost [18].

With these preparations, we shall study in Section 2 the system of two oscillators and coordinate transformations of current interest. It is pointed out in Section 3 that there are ten different generators for transformations, including those discussed in Section 2. It is noted that Dirac derived ten generators of transformations applicable to these oscillators, and they satisfy the closed set of commutation relations, which is the same as the Lie algebra of the $O(3,2)$ de Sitter group containing two Lorentz groups among its subgroups. In Section 4, Dirac's ten-generator symmetry is studied in the Wigner phase-space picture, and it is shown that Dirac's symmetry contains both canonical and Lorentz transformations.

While the Gaussian entanglement starts from the oscillator wave function in its ground state, we study in Section 5 the entanglements of excited oscillator states. We give a detailed explanation of how the series of Equation (1) can be derived from the squeezed Gaussian function of Equation (2).

In Section 6, we study in detail how the sheared state can be derived from a squeezed state. It appears to be a rotated squeezed state, but this is not the case. In Section 7, we study what happens when one of the two entangled variables is not observed within the framework of Feynman's rest of the universe [22,23].

In Section 8, we note that most of the mathematical formulas in this paper have been used earlier for understanding relativistic extended particles in the Lorentz-covariant harmonic oscillator formalism [20,24–28]. These formulas allow us to transport the concept of entanglement from the current problem of physics to quantum bound states in the Lorentz-covariant world. The time separation between the constituent particles is not observable and is not known in the present form of quantum mechanics. However, this variable effects the real world by entangling itself with the longitudinal variable.

2. Two-Dimensional Harmonic Oscillators

The Gaussian form:

$$\left[\frac{1}{\sqrt{\pi}}\right]^{1/4} \exp\left(-\frac{x^2}{2}\right) \tag{7}$$

is used for many branches of science. For instance, we can construct this function by throwing dice.

In physics, this is the wave function for the one-dimensional harmonic oscillator in the ground state. This function is also used for the vacuum state in quantum field theory, as well as the zero-photon state in quantum optics. For excited oscillator states, the wave function takes the form:

$$\chi_n(x) = \left[\frac{1}{\sqrt{\pi}2^n n!}\right]^{1/2} H_n(x) \exp\left(\frac{-x^2}{2}\right) \tag{8}$$

where $H_n(x)$ is the Hermite polynomial of the nth degree. The properties of this wave function are well known, and it becomes the Gaussian form of Equation (7) when $n = 0$.

We can now consider the two-dimensional space with the orthogonal coordinate variables x and y and the same wave function with the y variable:

$$\chi_m(y) = \left[\frac{1}{\sqrt{\pi}2^m m!}\right]^{1/2} H_m(y) \exp\left(\frac{-y^2}{2}\right) \tag{9}$$

and construct the function:

$$\psi^{n,m}(x,y) = [\chi_n(x)]\,[\chi_m(y)] \tag{10}$$

This form is clearly separable in the x and y variables. If n and m are zero, the wave function becomes:

$$\psi^{0,0}(x,y) = \frac{1}{\sqrt{\pi}} \exp\left\{-\frac{1}{2}\left(x^2 + y^2\right)\right\} \tag{11}$$

Under the coordinate rotation:

$$\begin{pmatrix} x' \\ y' \end{pmatrix} = \begin{pmatrix} \cos\theta & -\sin\theta \\ \sin\theta & \cos\theta \end{pmatrix} \begin{pmatrix} x \\ y \end{pmatrix} \tag{12}$$

this function remains separable. This rotation is illustrated in Figure 2. This is a transformation very familiar to us.

We can next consider the scale transformation of the form:

$$\begin{pmatrix} x' \\ y' \end{pmatrix} = \begin{pmatrix} e^\eta & 0 \\ 0 & e^{-\eta} \end{pmatrix} \begin{pmatrix} x \\ y \end{pmatrix} \tag{13}$$

This scale transformation is also illustrated in Figure 2. This area-preserving transformation is known as the squeeze. Under this transformation, the Gaussian function is still separable.

If the direction of the squeeze is rotated by $45°$, the transformation becomes the diagonal transformation of Equation (6). Indeed, this is a squeeze in the normal coordinate system. This form of squeeze is most commonly used for squeezed states of light, as well as the subject of entanglements. It is important to note that, in terms of the x and y variables, this transformation can be written as Equation (4) [18]. In 1905, Einstein used this form of squeeze transformation for the longitudinal and time-like variables. This is known as the Lorentz boost.

In addition, we can consider the transformation of the form:

$$\begin{pmatrix} x' \\ y' \end{pmatrix} = \begin{pmatrix} 1 & 2\alpha \\ 0 & 1 \end{pmatrix} \begin{pmatrix} x \\ y \end{pmatrix} \tag{14}$$

This transformation shears the system as is shown in Figure 2.

After the squeeze or shear transformation, the wave function of Equation (10) becomes non-separable, but it can still be written as a series expansion in terms of the oscillator wave functions. It can take the form:

$$\psi(x,y) = \sum_{n,m} A_{n,m}\chi_n(x)\chi_m(y) \tag{15}$$

with:

$$\sum_{n,m} |A_{n,m}|^2 = 1$$

if $\psi(x,y)$ is normalized, as was the case for the Gaussian function of Equation (11).

2.1. Squeezed Gaussian Function

Under the squeeze along the normal coordinate, the Gaussian form of Equation (11) becomes:

$$\psi_\eta(x,y) = \frac{1}{\sqrt{\pi}} \exp\left\{ -\frac{1}{4} \left[e^{-2\eta}(x+y)^2 + e^{2\eta}(x-y)^2 \right] \right\} \tag{16}$$

which was given in Equation (2). This function is not separable in the x and y variables. These variables are now entangled. We obtain this form by replacing, in the Gaussian function of Equation (11), the x and y variables by x' and y', respectively, where:

$$x' = (\cosh \eta)x - (\sinh \eta)y, \quad \text{and} \quad y' = (\cosh \eta)y - (\sinh \eta)x \tag{17}$$

This form of squeeze is illustrated in Figure 3, and the expansion of this squeezed Gaussian function becomes the series given in Equation (1) [20,26]. This aspect will be discussed in detail in Section 5.

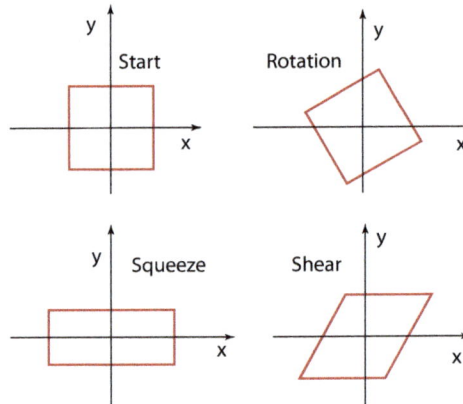

Figure 2. Transformations in the two-dimensional space. The object can be rotated, squeezed or sheared. In all three cases, the area remains invariant.

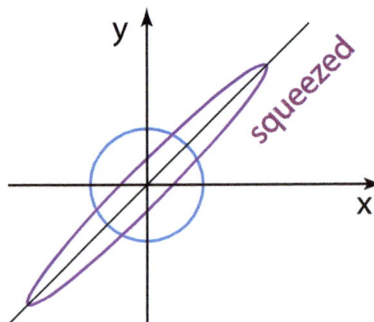

Figure 3. Squeeze along the 45 °C direction, discussed most frequently in the literature.

In 1976 [10], Yuen discussed two-photon coherent states, often called squeezed states of light. This series expansion served as the starting point for two-mode squeezed states. More recently, in 2003, Giedke et al. [1] used this formula to formulate the concept of the Gaussian entanglement.

There is another way to derive the series. For the harmonic oscillator wave functions, there are step-down and step-up operators [17]. These are defined as:

$$a = \frac{1}{\sqrt{2}} \left(x + \frac{\partial}{\partial x} \right), \quad \text{and} \quad a^\dagger = \frac{1}{\sqrt{2}} \left(x - \frac{\partial}{\partial x} \right) \tag{18}$$

If they are applied to the oscillator wave function, we have:

$$a\, \chi_n(x) = \sqrt{n}\, \chi_{n-1}(x), \quad \text{and} \quad a^\dagger \chi_n(x) = \sqrt{n+1}\, \chi_{n+1}(x) \tag{19}$$

Likewise, we can introduce b and $b^\dagger$ operators applicable to $\chi_n(y)$:

$$b = \frac{1}{\sqrt{2}} \left(y + \frac{\partial}{\partial y} \right), \quad \text{and} \quad b^\dagger = \frac{1}{\sqrt{2}} \left(y - \frac{\partial}{\partial y} \right) \tag{20}$$

Thus

$$\left(a^\dagger \right)^n \psi^0(x) = \sqrt{n!}\, \chi_n(x)$$

$$\left(b^\dagger \right)^n \psi^0(y) = \sqrt{n!}\chi_n(y) \tag{21}$$

and:

$$a\, \chi_0(x) = b\, \chi_0(y) = 0 \tag{22}$$

In terms of these variables, the transformation leading the Gaussian function of Equation (11) to its squeezed form of Equation (16) can be written as:

$$\exp\left\{ \frac{\eta}{2} \left(a^\dagger b^\dagger - a\, b \right) \right\} \tag{23}$$

which can also be written as:

$$\exp\left\{ -\eta \left(x\frac{\partial}{\partial y} + y\frac{\partial}{\partial x} \right) \right\} \tag{24}$$

Next, we can consider the exponential form:

$$\exp\left\{ (\tanh \eta) a^\dagger b^\dagger \right\} \tag{25}$$

which can be expanded as:

$$\sum_n \frac{1}{n!} (\tanh \eta)^n \left(a^\dagger b^\dagger \right)^n \tag{26}$$

If this operator is applied to the ground state of Equation (11), the result is:

$$\sum_n (\tanh \eta)^n \chi_n(x) \chi_n(y) \tag{27}$$

This form is not normalized, while the series of Equation (1) is. What is the origin of this difference?

There is a similar problem with the one-photon coherent state [29,30]. There, the series comes from the expansion of the exponential form:

$$\exp\left\{ \alpha a^\dagger \right\} \tag{28}$$

which can be expanded to:

$$\sum_n \frac{1}{n!} \alpha^n \left(a^\dagger\right)^n \tag{29}$$

However, this operator is not unitary. In order to make this series unitary, we consider the exponential form:

$$\exp\left(\alpha a^\dagger - \alpha^* a\right) \tag{30}$$

which is unitary. This expression can then be written as:

$$e^{-\alpha\alpha^*/2} \left[\exp\left(\alpha a^\dagger\right)\right] \left[\exp\left(\alpha^* a\right)\right] \tag{31}$$

according to the Baker–Campbell–Hausdorff (BCH) relation [31,32]. If this is applied to the ground state, the last bracket can be dropped, and the result is:

$$e^{-\alpha\alpha^*/2} \exp\left[\alpha a^\dagger\right] \tag{32}$$

which is the unitary operator with the normalization constant:

$$e^{-\alpha\alpha^*/2}$$

Likewise, we can conclude that the series of Equation (27) is different from that of Equation (1) due to the difference between the unitary operator of Equation (23) and the non-unitary operator of Equation (25). It may be possible to derive the normalization factor using the BCH formula, but it seems to be intractable at this time. The best way to resolve this problem is to present the exact calculation of the unitary operator leading to the normalized series of Equation (11). We shall return to this problem in Section 5, where squeezed excited states are studied.

2.2. Sheared Gaussian Function

In addition, there is a transformation called "shear," where only one of the two coordinates is translated, as shown in Figure 2. This transformation takes the form:

$$\begin{pmatrix} x' \\ y' \end{pmatrix} = \begin{pmatrix} 1 & 2\alpha \\ 0 & 1 \end{pmatrix} \begin{pmatrix} x \\ y \end{pmatrix} \tag{33}$$

which leads to:

$$\begin{pmatrix} x' \\ y' \end{pmatrix} = \begin{pmatrix} x + 2\alpha y \\ y \end{pmatrix} \tag{34}$$

This shear is one of the basic transformations in engineering sciences. In physics, this transformation plays the key role in understanding the internal space-time symmetry of massless particles [33–35]. This matrix plays the pivotal role during the transition from the oscillator mode to the damping mode in classical damped harmonic oscillators [36,37].

Under this transformation, the Gaussian form becomes:

$$\psi_{shr}(x, y) = \frac{1}{\sqrt{\pi}} \exp\left\{-\frac{1}{2}\left[(x - 2\alpha y)^2 + y^2\right]\right\} \tag{35}$$

It is possible to expand this into a series of the form of Equation (15) [38].

The transformation applicable to the Gaussian form of Equation (11) is:

$$\exp\left(-2\alpha y \frac{\partial}{\partial x}\right) \tag{36}$$

and the generator is:

$$-iy\frac{\partial}{\partial x} \tag{37}$$

It is of interest to see where this generator stands among the ten generators of Dirac.

However, the most pressing problem is whether the sheared Gaussian form can be regarded as a rotated squeezed state. The basic mathematical issue is that the shear matrix of Equation (33) is triangular and cannot be diagonalized. Therefore, it cannot be a squeezed state. Yet, the Gaussian form of Equation (35) appears to be a rotated squeezed state, while not along the normal coordinates. We shall look at this problem in detail in Section 6.

3. Dirac's Entangled Oscillators

Paul A. M. Dirac devoted much of his life-long efforts to the task of making quantum mechanics compatible with special relativity. Harmonic oscillators serve as an instrument for illustrating quantum mechanics, while special relativity is the physics of the Lorentz group. Thus, Dirac attempted to construct a representation of the Lorentz group using harmonic oscillator wave functions [17,21].

In his 1963 paper [21], Dirac started from the two-dimensional oscillator whose wave function takes the Gaussian form given in Equation (11). He then considered unitary transformations applicable to this ground-state wave function. He noted that they can be generated by the following ten Hermitian operators:

$$L_1 = \frac{1}{2}\left(a^\dagger b + b^\dagger a\right), \qquad L_2 = \frac{1}{2i}\left(a^\dagger b - b^\dagger a\right)$$

$$L_3 = \frac{1}{2}\left(a^\dagger a - b^\dagger b\right), \qquad S_3 = \frac{1}{2}\left(a^\dagger a + bb^\dagger\right)$$

$$K_1 = -\frac{1}{4}\left(a^\dagger a^\dagger + aa - b^\dagger b^\dagger - bb\right)$$

$$K_2 = \frac{i}{4}\left(a^\dagger a^\dagger - aa + b^\dagger b^\dagger - bb\right)$$

$$K_3 = \frac{1}{2}\left(a^\dagger b^\dagger + ab\right)$$

$$Q_1 = -\frac{i}{4}\left(a^\dagger a^\dagger - aa - b^\dagger b^\dagger + bb\right)$$

$$Q_2 = -\frac{1}{4}\left(a^\dagger a^\dagger + aa + b^\dagger b^\dagger + bb\right)$$

$$Q_3 = \frac{i}{2}\left(a^\dagger b^\dagger - ab\right) \tag{38}$$

He then noted that these operators satisfy the following set of commutation relations.

$$[L_i, L_j] = i\epsilon_{ijk}L_k, \qquad [L_i, K_j] = i\epsilon_{ijk}K_k, \qquad [L_i, Q_j] = i\epsilon_{ijk}Q_k$$

$$[K_i, K_j] = [Q_i, Q_j] = -i\epsilon_{ijk}L_k, \qquad [L_i, S_3] = 0$$

$$[K_i, Q_j] = -i\delta_{ij}S_3, \qquad [K_i, S_3] = -iQ_i, \qquad [Q_i, S_3] = iK_i \tag{39}$$

Dirac then determined that these commutation relations constitute the Lie algebra for the $O(3,2)$ de Sitter group with ten generators. This de Sitter group is the Lorentz group applicable to three space coordinates and two time coordinates. Let us use the notation (x, y, z, t, s), with (x, y, z) as the space coordinates and (t, s) as two time coordinates. Then, the rotation around the z axis is generated by:

$$L_3 = \begin{pmatrix} 0 & -i & 0 & 0 & 0 \\ i & 0 & 0 & 0 & 0 \\ 0 & 0 & 0 & 0 & 0 \\ 0 & 0 & 0 & 0 & 0 \\ 0 & 0 & 0 & 0 & 0 \end{pmatrix} \tag{40}$$

The generators L_1 and L_2 can also be constructed. The K_3 and Q_3 generators will take the form:

$$K_3 = \begin{pmatrix} 0 & 0 & 0 & 0 & 0 \\ 0 & 0 & 0 & 0 & 0 \\ 0 & 0 & 0 & i & 0 \\ 0 & 0 & i & 0 & 0 \\ 0 & 0 & 0 & 0 & 0 \end{pmatrix}, \quad Q_3 = \begin{pmatrix} 0 & 0 & 0 & 0 & 0 \\ 0 & 0 & 0 & 0 & 0 \\ 0 & 0 & 0 & 0 & i \\ 0 & 0 & 0 & 0 & 0 \\ 0 & 0 & i & 0 & 0 \end{pmatrix} \tag{41}$$

From these two matrices, the generators K_1, K_2, Q_1, Q_2 can be constructed. The generator S_3 can be written as:

$$S_3 = \begin{pmatrix} 0 & 0 & 0 & 0 & 0 \\ 0 & 0 & 0 & 0 & 0 \\ 0 & 0 & 0 & 0 & 0 \\ 0 & 0 & 0 & 0 & -i \\ 0 & 0 & 0 & i & 0 \end{pmatrix} \tag{42}$$

The last five-by-five matrix generates rotations in the two-dimensional space of (t, s). If we introduce these two time variables, the $O(3, 2)$ group leads to two coupled Lorentz groups. The particle mass is invariant under Lorentz transformations. Thus, one Lorentz group cannot change the particle mass. However, with two coupled Lorentz groups, we can describe the world with variable masses, such as the neutrino oscillations.

In Section 2, we used the operators Q_3 and K_3 as the generators for the squeezed Gaussian function. For the unitary transformation of Equation (23), we used:

$$\exp\left(-i\eta Q_3\right) \tag{43}$$

However, the exponential form of Equation (25) can be written as:

$$\exp\left\{-i(\tanh \eta)\left(Q_3 + iK_3\right)\right\} \tag{44}$$

which is not unitary, as was seen before.

From the space-time point of view, both K_3 and Q_3 generate Lorentz boosts along the z direction, with the time variables t and s, respectively. The fact that the squeeze and Lorentz transformations share the same mathematical formula is well known. However, the non-unitary operator iK_3 does not seem to have a space-time interpretation.

As for the sheared state, the generator can be written as:

$$Q_3 - L_2 \tag{45}$$

leading to the expression given in Equation (37). This is a Hermitian operator leading to the unitary transformation of Equation (36).

4. Entangled Oscillators in the Phase-Space Picture

Also in his 1963 paper, Dirac states that the Lie algebra of Equation (39) can serve as the four-dimensional symplectic group $Sp(4)$. This group allows us to study squeezed or entangled states in terms of the four-dimensional phase space consisting of two position and two momentum variables [15,39,40].

In order to study the $Sp(4)$ contents of the coupled oscillator system, let us introduce the Wigner function defined as [41]:

$$W(x,y;p,q) = \left(\frac{1}{\pi}\right)^2 \int \exp\left\{-2i(px' + qy')\right\}$$

$$\times \psi^*(x + x', y + y')\psi(x - x', y - y')dx'dy' \tag{46}$$

If the wave function $\psi(x,y)$ is the Gaussian form of Equation (11), the Wigner function becomes:

$$W(x,y:p,q) = \left(\frac{1}{\pi}\right)^2 \exp\left\{-\left(x^2 + p^2 + y^2 + q^2\right)\right\} \tag{47}$$

The Wigner function is defined over the four-dimensional phase space of (x,p,y,q) just as in the case of classical mechanics. The unitary transformations generated by the operators of Equation (38) are translated into Wigner transformations [39,40,42]. As in the case of Dirac's oscillators, there are ten corresponding generators applicable to the Wigner function. They are:

$$L_1 = +\frac{i}{2}\left\{\left(x\frac{\partial}{\partial q} - q\frac{\partial}{\partial x}\right) + \left(y\frac{\partial}{\partial p} - p\frac{\partial}{\partial y}\right)\right\}$$

$$L_2 = -\frac{i}{2}\left\{\left(x\frac{\partial}{\partial y} - y\frac{\partial}{\partial x}\right) + \left(p\frac{\partial}{\partial q} - q\frac{\partial}{\partial p}\right)\right\}$$

$$L_3 = +\frac{i}{2}\left\{\left(x\frac{\partial}{\partial p} - p\frac{\partial}{\partial x}\right) - \left(y\frac{\partial}{\partial q} - q\frac{\partial}{\partial y}\right)\right\}$$

$$S_3 = -\frac{i}{2}\left\{\left(x\frac{\partial}{\partial p} - p\frac{\partial}{\partial x}\right) + \left(y\frac{\partial}{\partial q} - q\frac{\partial}{\partial y}\right)\right\} \tag{48}$$

and:

$$K_1 = -\frac{i}{2}\left\{\left(x\frac{\partial}{\partial p} + p\frac{\partial}{\partial x}\right) - \left(y\frac{\partial}{\partial q} + q\frac{\partial}{\partial y}\right)\right\}$$

$$K_2 = -\frac{i}{2}\left\{\left(x\frac{\partial}{\partial x} + y\frac{\partial}{\partial y}\right) - \left(p\frac{\partial}{\partial p} + q\frac{\partial}{\partial q}\right)\right\}$$

$$K_3 = +\frac{i}{2}\left\{\left(x\frac{\partial}{\partial q} + q\frac{\partial}{\partial x}\right) + \left(y\frac{\partial}{\partial p} + p\frac{\partial}{\partial y}\right)\right\}$$

$$Q_1 = +\frac{i}{2}\left\{\left(x\frac{\partial}{\partial x} + q\frac{\partial}{\partial q}\right) - \left(y\frac{\partial}{\partial y} + p\frac{\partial}{\partial p}\right)\right\}$$

$$Q_2 = -\frac{i}{2}\left\{\left(x\frac{\partial}{\partial p} + p\frac{\partial}{\partial x}\right) + \left(y\frac{\partial}{\partial q} + q\frac{\partial}{\partial y}\right)\right\}$$

$$Q_3 = -\frac{i}{2}\left\{\left(y\frac{\partial}{\partial x} + x\frac{\partial}{\partial y}\right) - \left(q\frac{\partial}{\partial p} + p\frac{\partial}{\partial q}\right)\right\} \tag{49}$$

These generators also satisfy the Lie algebra given in Equations (38) and (39). Transformations generated by these generators have been discussed in the literature [15,40,42].

As in the case of Section 3, we are interested in the generators Q_3 and K_3. The transformation generated by Q_3 takes the form:

$$\left[\exp\left\{\eta\left(x\frac{\partial}{\partial y}+y\frac{\partial}{\partial x}\right)\right\}\right]\left[\exp\left\{-\eta\left(p\frac{\partial}{\partial q}+q\frac{\partial}{\partial p}\right)\right\}\right] \qquad (50)$$

This exponential form squeezes the Wigner function of Equation (47) in the $x\,y$ space, as well as in their corresponding momentum space. However, in the momentum space, the squeeze is in the opposite direction, as illustrated in Figure 4. This is what we expect from canonical transformation in classical mechanics. Indeed, this corresponds to the unitary transformation, which played the major role in Section 2.

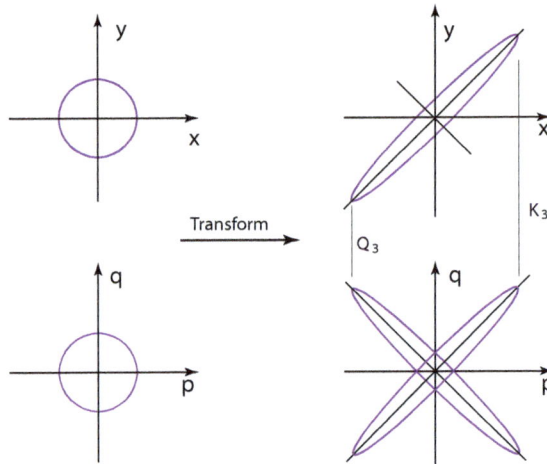

Figure 4. Transformations generated by Q_3 and K_3. As the parameter η becomes larger, both the space and momentum distribution becomes larger.

Even though shown insignificant in Section 2, K_3 had a definite physical interpretation in Section 3. The transformation generated by K_3 takes the form:

$$\left[\exp\left\{\eta\left(x\frac{\partial}{\partial q}+q\frac{\partial}{\partial x}\right)\right\}\right]\left[\exp\left\{\eta\left(y\frac{\partial}{\partial p}+p\frac{\partial}{\partial y}\right)\right\}\right] \qquad (51)$$

This performs the squeeze in the $x\,q$ and $y\,p$ spaces. In this case, the squeezes have the same sign, and the rate of increase is the same in all directions. We can thus have the same picture of squeeze for both $x\,y$ and $p\,q$ spaces, as illustrated in Figure 4. This parallel transformation corresponds to the Lorentz squeeze [20,25].

As for the sheared state, the combination:

$$Q_3 - L_2 = -i\left(y\frac{\partial}{\partial x}+q\frac{\partial}{\partial p}\right) \qquad (52)$$

generates the same shear in the $p\,q$ space.

5. Entangled Excited States

In Section 2, we discussed the entangled ground state and noted that the entangled state of Equation (1) is a series expansion of the squeezed Gaussian function. In this section, we are interested in what happens when we squeeze an excited oscillator state starting from:

$$\chi_n(x)\chi_m(y) \tag{53}$$

In order to entangle this state, we should replace x and y, respectively, by x' and y' given in Equation (17).

The question is how the oscillator wave function is squeezed after this operation. Let us note first that the wave function of Equation (53) satisfies the equation:

$$\frac{1}{2}\left\{\left(x^2 - \frac{\partial^2}{\partial x^2}\right) - \left(y^2 - \frac{\partial^2}{\partial y^2}\right)\right\}\chi_n(x)\chi_m(y) = (n-m)\chi_n(x)\chi_m(y) \tag{54}$$

This equation is invariant under the squeeze transformation of Equation (17), and thus, the eigenvalue $(n-m)$ remains invariant. Unlike the usual two-oscillator system, the x component and the y component have opposite signs. This is the reason why the overall equation is squeeze-invariant [3,25,43].

We then have to write this squeezed oscillator in the series form of Equation (15). The most interesting case is of course for $m = n = 0$, which leads to the Gaussian entangled state given in Equation (16). Another interesting case is for $m = 0$, while n is allowed to take all integer values. This single-excitation system has applications in the covariant oscillator formalism where no time-like excitations are allowed. The Gaussian entangled state is a special case of this single-excited oscillator system.

The most general case is for nonzero integers for both n and m. The calculation for this case is available in the literature [20,44]. Seeing no immediate physical applications of this case, we shall not reproduce this calculation in this section.

For the single-excitation system, we write the starting wave function as:

$$\chi_n(x)\chi_0(y) = \left[\frac{1}{\pi\, 2^n n!}\right]^{1/2} H_n(x) \exp\left\{-\left(\frac{x^2+y^2}{2}\right)\right\} \tag{55}$$

There are no excitations along the y coordinate. In order to squeeze this function, our plan is to replace x and y by x' and y', respectively, and write $\chi_n(x')\chi_0(y')$ as a series in the form:

$$\chi_n(x')\chi_0(y') = \sum_{k',k} A_{k',k}(n)\chi_{k'}(x)\chi_k(y) \tag{56}$$

Since $k' - k = n$ or $k' = n + k$, according to the eigenvalue of the differential equation given in Equation (54), we write this series as:

$$\chi_n(x')\chi_0(y') = \sum_{k',k} A_k(n)\chi_{(k+n)}(x)\chi_k(y) \tag{57}$$

with:

$$\sum_k |A_k(n)|^2 = 1 \tag{58}$$

This coefficient is:

$$A_k(n) = \int \chi_{k+n}(x)\chi_k(y)\chi_n(x')\chi_0(y')\, dx\, dy \tag{59}$$

This calculation was given in the literature in a fragmentary way in connection with a Lorentz-covariant description of extended particles starting from Ruiz's 1974 paper [45], subsequently by Kim et al. in

1979 [26] and by Rotbart in 1981 [44]. In view of the recent developments of physics, it seems necessary to give one coherent calculation of the coefficient of Equation (59).

We are now interested in the squeezed oscillator function:

$$A_k(n) = \left[\frac{1}{\pi^2 \, 2^n n! (k+n)^2 (n+k)! k^2 k!} \right]^{1/2}$$

$$\times \int H_{n+k}(x) H_k(y) H_n(x') \exp \left\{ - \left(\frac{x^2 + y^2 + x'^2 + y'^2}{2} \right) \right\} dx dy \tag{60}$$

As was noted by Ruiz [45], the key to the evaluation of this integral is to introduce the generating function for the Hermite polynomials [46,47]:

$$G(r, z) = \exp \left(-r^2 + 2rz \right) = \sum_m \frac{r^m}{m!} H_m(z) \tag{61}$$

and evaluate the integral:

$$I = \int G(r, x) G(s, y) G(r', x') \exp \left\{ - \left(\frac{x^2 + y^2 + x'^2 + y'^2}{2} \right) \right\} dx dy \tag{62}$$

The integrand becomes one exponential function, and its exponent is quadratic in x and y. This quadratic form can be diagonalized, and the integral can be evaluated [20,26]. The result is:

$$I = \left[\frac{\pi}{\cosh \eta} \right] \exp \left(2rs \tanh \eta \right) \exp \left(\frac{2rr'}{\cosh \eta} \right) \tag{63}$$

We can now expand this expression and choose the coefficients of r^{n+k}, s^k, r'^n for $H_{(n+k)}(x), H_n(y)$ and $H_n(z')$, respectively. The result is:

$$A_{n;k} = \left(\frac{1}{\cosh \eta} \right)^{(n+1)} \left[\frac{(n+k)!}{n! k!} \right]^{1/2} (\tanh \eta)^k \tag{64}$$

Thus, the series becomes:

$$\chi_n(x') \chi_0(y') = \left(\frac{1}{\cosh \eta} \right)^{(n+1)} \sum_k \left[\frac{(n+k)!}{n! k!} \right]^{1/2} (\tanh \eta)^k \chi_{k+n}(x) \chi_k(y) \tag{65}$$

If $n = 0$, it is the squeezed ground state, and this expression becomes the entangled state of Equation (16).

6. E(2)-Sheared States

Let us next consider the effect of shear on the Gaussian form. From Figures 3 and 5, it is clear that the sheared state is a rotated squeezed state.

In order to understand this transformation, let us note that the squeeze and rotation are generated by the two-by-two matrices:

$$K = \begin{pmatrix} 0 & i \\ i & 0 \end{pmatrix}, \qquad J = \begin{pmatrix} 0 & -i \\ i & 0 \end{pmatrix} \tag{66}$$

which generate the squeeze and rotation matrices of the form:

$$\exp\left(-i\eta K\right) = \begin{pmatrix} \cosh\eta & \sinh\eta \\ \sinh\eta & \cosh\eta \end{pmatrix}$$

$$\exp\left(-i\theta J\right) = \begin{pmatrix} \cos\theta & -\sin\theta \\ \sin\theta & \cos\theta \end{pmatrix} \tag{67}$$

respectively. We can then consider:

$$S = K - J = \begin{pmatrix} 0 & 2i \\ 0 & 0 \end{pmatrix} \tag{68}$$

This matrix has the property that $S^2 = 0$. Thus, the transformation matrix becomes:

$$\exp\left(-i\alpha S\right) = \begin{pmatrix} 1 & 2\alpha \\ 0 & 1 \end{pmatrix} \tag{69}$$

Since $S^2 = 0$, the Taylor expansion truncates, and the transformation matrix becomes the triangular matrix of Equation (34), leading to the transformation:

$$\begin{pmatrix} x \\ y \end{pmatrix} \rightarrow \begin{pmatrix} x + 2\alpha y \\ y \end{pmatrix} \tag{70}$$

The shear generator S of Equation (68) indicates that the infinitesimal transformation is a rotation followed by a squeeze. Since both rotation and squeeze are area-preserving transformations, the shear should also be an area-preserving transformations.

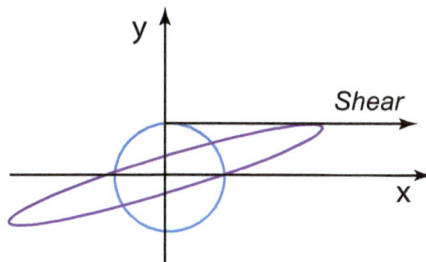

Figure 5. Sear transformation of the Gaussian form given in Equation (11).

In view of Figure 5, we should ask whether the triangular matrix of Equation (69) can be obtained from one squeeze matrix followed by one rotation matrix. This is not possible mathematically. It can however, be written as a squeezed rotation matrix of the form:

$$\begin{pmatrix} e^{\lambda/2} & 0 \\ 0 & e^{-\lambda/2} \end{pmatrix} \begin{pmatrix} \cos\omega & \sin\omega \\ -\sin\omega & \cos\omega \end{pmatrix} \begin{pmatrix} e^{-\lambda/2} & 0 \\ 0 & e^{\lambda/2} \end{pmatrix} \tag{71}$$

resulting in:

$$\begin{pmatrix} \cos\omega & e^{\lambda}\sin\omega \\ -e^{-\lambda}\sin\omega & \cos\omega \end{pmatrix} \tag{72}$$

If we let:

$$(\sin\omega) = 2\alpha e^{-\lambda} \tag{73}$$

Then:

$$\begin{pmatrix} \cos \omega & 2\alpha \\ -2\alpha e^{-2\lambda} & \cos \omega \end{pmatrix} \tag{74}$$

If λ becomes infinite, the angle ω becomes zero, and this matrix becomes the triangular matrix of Equation (69). This is a singular process where the parameter λ goes to infinity.

If this transformation is applied to the Gaussian form of Equation (11), it becomes:

$$\psi(x,y) = \frac{1}{\sqrt{\pi}} \exp\left\{ -\frac{1}{2} \left[(x - 2\alpha y)^2 + y^2 \right] \right\} \tag{75}$$

The question is whether the exponential portion of this expression can be written as:

$$\exp\left\{ -\frac{1}{2} \left[e^{-2\eta} (x \, \cos\theta + y \, \sin\theta)^2 + e^{2\eta} (x \, \sin\theta - y \, \cos\theta)^2 \right] \right\} \tag{76}$$

The answer is yes. This is possible if:

$$\tan(2\theta) = \frac{1}{\alpha}$$

$$e^{2\eta} = 1 + 2\alpha^2 + 2\alpha\sqrt{\alpha^2 + 1}$$

$$e^{-2\eta} = 1 + 2\alpha^2 - 2\alpha\sqrt{\alpha^2 + 1} \tag{77}$$

In Equation (74), we needed a limiting case of λ becoming infinite. This is necessarily a singular transformation. On the other hand, the derivation of the Gaussian form of Equation (75) appears to be analytic. How is this possible? In order to achieve the transformation from the Gaussian form of Equations (11) to (75), we need the linear transformation:

$$\begin{pmatrix} \cos\theta & -\sin\theta \\ \sin\theta & \cos\theta \end{pmatrix} \begin{pmatrix} e^{\eta} & 0 \\ 0 & e^{-\eta} \end{pmatrix} \tag{78}$$

If the initial form is invariant under rotations as in the case of the Gaussian function of Equation (11), we can add another rotation matrix on the right-hand side. We choose that rotation matrix to be:

$$\begin{pmatrix} \cos(\theta - \pi/2) & -\sin(\theta - \pi/2) \\ \sin(\theta - \pi/2) & \cos(\theta - \pi/2) \end{pmatrix} \tag{79}$$

write the three matrices as:

$$\begin{pmatrix} \cos\theta' & -\sin\theta' \\ \sin\theta' & \cos\theta' \end{pmatrix} \begin{pmatrix} \cosh\eta & \sinh\eta \\ \sinh\eta & \cosh\eta \end{pmatrix} \begin{pmatrix} \cos\theta' & -\sin\theta' \\ \sin\theta' & \cos\theta' \end{pmatrix} \tag{80}$$

with:

$$\theta' = \theta - \frac{\pi}{4}$$

The multiplication of these three matrices leads to:

$$\begin{pmatrix} (\cosh\eta)\sin(2\theta) & \sinh\eta + (\cosh\eta)\cos(2\theta) \\ \sinh\eta - (\cosh\eta)\cos(2\theta) & (\cosh\eta)\sin(2\theta) \end{pmatrix} \tag{81}$$

The lower-left element can become zero when $\sinh\eta = \cosh(\eta)\cos(2\theta)$, and consequently, this matrix becomes:

$$\begin{pmatrix} 1 & 2\sinh\eta \\ 0 & 1 \end{pmatrix} \tag{82}$$

Furthermore, this matrix can be written in the form of a squeezed rotation matrix given in Equation (72), with:

$$\cos\omega = (\cosh\eta)\sin(2\theta)$$

$$e^{-2\lambda} = \frac{\cos(2\theta) - \tanh\eta}{\cos(2\theta) + \tanh\eta} \tag{83}$$

The matrices of the form of Equations (72) and (81) are known as the Wigner and Bargmann decompositions, respectively [33,36,48–50].

7. Feynman's Rest of the Universe

We need the concept of entanglement in quantum systems of two variables. The issue is how the measurement of one variable affects the other variable. The simplest case is what happens to the first variable while no measurements are taken on the second variable. This problem has a long history since von Neumann introduced the concept of the density matrix in 1932 [51]. While there are many books and review articles on this subject, Feynman stated this problem in his own colorful way. In his book on statistical mechanics [22], Feynman makes the following statement about the density matrix.

When we solve a quantum-mechanical problem, what we really do is divide the universe into two parts—the system in which we are interested and the rest of the universe. We then usually act as if the system in which we are interested comprised the entire universe. To motivate the use of density matrices, let us see what happens when we include the part of the universe outside the system.

Indeed, Yurke and Potasek [11] and also Ekert and Knight [12] studied this problem in the two-mode squeezed state using the entanglement formula given in Equation (16). Later in 1999, Han et al. studied this problem with two coupled oscillators where one oscillator is observed while the other is not and, thus, is in the rest of the universe as defined by Feynman [23].

Somewhat earlier in 1990 [27], Kim and Wigner observed that there is a time separation wherever there is a space separation in the Lorentz-covariant world. The Bohr radius is a space separation. If the system is Lorentz-boosted, the time-separation becomes entangled with the space separation. However, in the present form of quantum mechanics, this time-separation variable is not measured and not understood.

This variable was mentioned in the paper of Feynman et al. in 1971 [43], but the authors say they would drop this variable because they do not know what to do with it. While what Feynman et al. did was not quite respectable from the scientific point of view, they made a contribution by pointing out the existence of the problem. In 1990, Kim and Wigner [27] noted that the time-separation variable belongs to Feynman's rest of the universe and studied its consequences in the observable world.

In this section, we first reproduce the work of Kim and Wigner using the x and y variables and then study the consequences. Let us introduce the notation $\psi_\eta^n(x,y)$ for the squeezed oscillator wave function given in Equation (65):

$$\psi_\eta^n(x,y) = \chi_n(x')\chi_0(y') \tag{84}$$

with no excitations along the y direction. For $\eta = 0$, this expression becomes $\chi_n(x)\chi_0(y)$.

From this wave function, we can construct the pure-state density matrix as:

$$\rho_\eta^n(x,y;r,s) = \psi_\eta^n(x,y)\psi_\eta^n(r,s) \tag{85}$$

which satisfies the condition $\rho^2 = \rho$, which means:

$$\rho_\eta^n(x, y; r, s) = \int \rho_\eta^n(x, y; u, v) \rho_\eta^n(u, v; r, s) du dv \tag{86}$$

As illustrated in Figure 6, it is not possible to make measurements on the variable y. We thus have to take the trace of this density matrix along the y axis, resulting in:

$$\rho_\eta^n(x, r) = \int \psi_\eta^n(x, y) \psi_\eta^n(r, y) dy$$

$$= \left(\frac{1}{\cosh \eta} \right)^{2(n+1)} \sum_k \frac{(n+k)!}{n!k!} (\tanh \eta)^{2k} \chi_{n+k}(x) \chi_{k+n}(r) \tag{87}$$

The trace of this density matrix is one, but the trace of ρ^2 is:

$$Tr\left(\rho^2\right) = \int \rho_\eta^n(x, r) \rho_\eta^n(r, x) dr dx$$

$$= \left(\frac{1}{\cosh \eta} \right)^{4(n+1)} \sum_k \left[\frac{(n+k)!}{n!k!} \right]^2 (\tanh \eta)^{4k} \tag{88}$$

which is less than one. This is due to the fact that we are not observing the y variable. Our knowledge is less than complete.

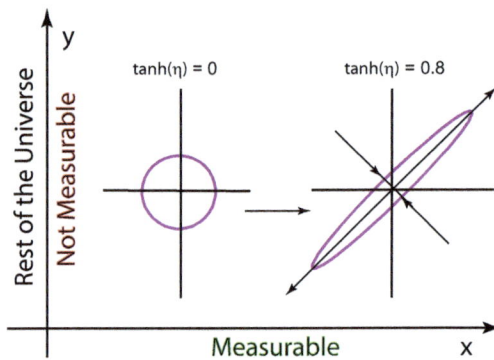

Figure 6. Feynman's rest of the universe. As the Gaussian function is squeezed, the x and y variables become entangled. If the y variable is not measured, it affects the quantum mechanics of the x variable.

The standard way to measure this incompleteness is to calculate the entropy defined as [51–53]:

$$S = -Tr\left(\rho(x, r) \ln[\rho(x, r)]\right) \tag{89}$$

which leads to:

$$S = 2(n+1)[(\cosh \eta)^2 \ln(\cosh \eta) - (\sinh \eta)^2 \ln(\sinh \eta)]$$

$$- \left(\frac{1}{\cosh \eta} \right)^{2(n+1)} \sum_k \frac{(n+k)!}{n!k!} \ln \left[\frac{(n+k)!}{n!k!} \right] (\tanh \eta)^{2k} \tag{90}$$

Let us go back to the wave function given in Equation (84). As is illustrated in Figure 6, its localization property is dictated by its Gaussian factor, which corresponds to the ground-state wave function. For this reason, we expect that much of the behavior of the density matrix or the entropy for the nth excited state will be the same as that for the ground state with n = 0. For this state, the density matrix is:

$$\rho_\eta(x,r) = \left(\frac{1}{\pi\cosh(2\eta)}\right)^{1/2}\exp\left\{-\frac{1}{4}\left[\frac{(x+r)^2}{\cosh(2\eta)} + (x-r)^2\cosh(2\eta)\right]\right\}\qquad(91)$$

and the entropy is:

$$S_\eta = 2\left[(\cosh\eta)^2\ln(\cosh\eta) - (\sinh\eta)^2\ln(\sinh\eta)\right]\qquad(92)$$

The density distribution $\rho_\eta(x,x)$ becomes:

$$\rho_\eta(x,x) = \left(\frac{1}{\pi\cosh(2\eta)}\right)^{1/2}\exp\left(\frac{-x^2}{\cosh(2\eta)}\right)\qquad(93)$$

The width of the distribution becomes $\sqrt{\cosh(2\eta)}$, and the distribution becomes wide-spread as η becomes larger. Likewise, the momentum distribution becomes wide-spread as can be seen in Figure 4. This simultaneous increase in the momentum and position distribution widths is due to our inability to measure the y variable hidden in Feynman's rest of the universe [22].

In their paper of 1990 [27], Kim and Wigner used the x and y variables as the longitudinal and time-like variables respectively in the Lorentz-covariant world. In the quantum world, it is a widely-accepted view that there are no time-like excitations. Thus, it is fully justified to restrict the y component to its ground state, as we did in Section 5.

8. Space-Time Entanglement

The series given in Equation (1) plays the central role in the concept of the Gaussian or continuous-variable entanglement, where the measurement on one variable affects the quantum mechanics of the other variable. If one of the variables is not observed, it belongs to Feynman's rest of the universe.

The series of the form of Equation (1) was developed earlier for studying harmonic oscillators in moving frames [20,24–28]. Here, z and t are the space-like and time-like separations between the two constituent particles bound together by a harmonic oscillator potential. There are excitations along the longitudinal direction. However, no excitations are allowed along the time-like direction. Dirac described this as the "c-number" time-energy uncertainty relation [16]. Dirac in 1927 was talking about the system without special relativity. In 1945 [17], Dirac attempted to construct space-time wave functions using harmonic oscillators. In 1949 [18], Dirac introduced his light-cone coordinate system for Lorentz boosts, demonstrating that the boost is a squeeze transformation. It is now possible to combine Dirac's three observations to construct the Lorentz covariant picture of quantum bound states, as illustrated in Figure 7.

If the system is at rest, we use the wave function:

$$\psi_0^n(z,t) = \chi_n(z)\chi_0(t)\qquad(94)$$

which allows excitations along the z axis, but no excitations along the t axis, according to Dirac's c-number time-energy uncertainty relation.

If the system is boosted, the z and t variables are replaced by z' and t' where:

$$z' = (\cosh\eta)z - (\sinh\eta)t, \quad\text{and}\quad t' = -(\sinh\eta)z + (\cosh\eta)t\qquad(95)$$

This is a squeeze transformation as in the case of Equation (17). In terms of these space-time variables, the wave function of Equation (84), can be written as:

$$\psi_\eta^n(z,t) = \chi_n(z')\chi_0(t')\qquad(96)$$

and the series of Equation (65) then becomes:

$$\psi_\eta^n(z,t) = \left(\frac{1}{\cosh\eta}\right)^{(n+1)} \sum_k \left[\frac{(n+k)!}{n!k!}\right]^{1/2} (\tanh\eta)^k \chi_{k+n}(z)\chi_k(t) \tag{97}$$

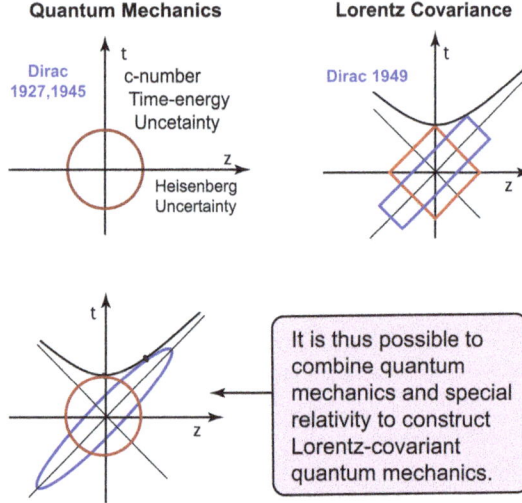

Figure 7. Dirac's form of Lorentz-covariant quantum mechanics. In addition to Heisenberg's uncertainty relation, which allows excitations along the spatial direction, there is the "c-number" time-energy uncertainty without excitations. This form of quantum mechanics can be combined with Dirac's light-cone picture of Lorentz boost, resulting in the Lorentz-covariant picture of quantum mechanics. The elliptic squeeze shown in this figure can be called the space-time entanglement.

Since the Lorentz-covariant oscillator formalism shares the same set of formulas with the Gaussian entangled states, it is possible to explain some aspects of space-time physics using the concepts and terminologies developed in quantum optics, as illustrated in Figure 1.

The time-separation variable is a case in point. The Bohr radius is a well-defined spatial separation between the proton and electron in the hydrogen atom. However, if the atom is boosted, this radius picks up a time-like separation. This time-separation variable does not exist in the Schrödinger picture of quantum mechanics. However, this variable plays the pivotal role in the covariant harmonic oscillator formalism. It is gratifying to note that this "hidden or forgotten" variable plays a role in the real world while being entangled with the observable longitudinal variable. With this point in mind, let us study some of the consequences of this space-time entanglement.

First of all, does the wave function of Equation (96) carry a probability interpretation in the Lorentz-covariant world? Since $dzdt = dz'dt'$, the normalization:

$$\int |\psi_\eta^n(z,t)|^2 dtdz = 1 \tag{98}$$

This is a Lorentz-invariant normalization. If the system is at rest, the z and t variables are completely dis-entangled, and the spatial component of the wave function satisfies the Schrödinger equation without the time-separation variable.

However, in the Lorentz-covariant world, we have to consider the inner product:

$$\left(\psi_\eta^n(z,t), \psi_{\eta'}^m(z,t)\right) = \int \left[\psi_\eta^n(z,t)\right]^* \psi_{\eta'}^m(z,t)\right) dzdt \tag{99}$$

The evaluation of this integral was carried out by Michael Ruiz in 1974 [45], and the result was:

$$\left(\frac{1}{|\cosh(\eta - \eta')|} \right)^{n+1} \delta_{nm} \tag{100}$$

In order to see the physical implications of this result, let us assume that one of the oscillators is at rest with $\eta' = 0$ and the other is moving with the velocity $\beta = \tanh(\eta)$. Then, the result is:

$$\left(\psi_\eta^n(z,t), \psi_0^m(z,t) \right) = \left(\sqrt{1-\beta^2} \right)^{n+1} \delta_{nm} \tag{101}$$

Indeed, the wave functions are orthonormal if they are in the same Lorentz frame. If one of them is boosted, the inner product shows the effect of Lorentz contraction. We are familiar with the contraction $\sqrt{1-\beta^2}$ for the rigid rod. The ground state of the oscillator wave function is contracted like a rigid rod.

The probability density $|\psi_\eta^0(z)|^2$ is for the oscillator in the ground sate, and it has one hump. For the nth excited state, there are $(n+1)$ humps. If each hump is contracted like $\sqrt{1-\beta^2}$, the net contraction factor is $\left(\sqrt{1-\beta^2} \right)^{n+1}$ for the nth excited state. This result is illustrated in Figure 8.

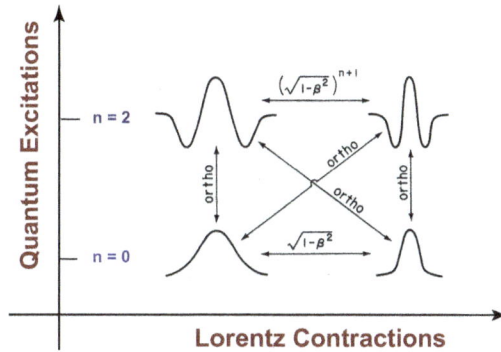

Figure 8. Orthogonality relations for two covariant oscillator wave functions. The orthogonality relation is preserved for different frames. However, they show the Lorentz contraction effect for two different frames.

With this understanding, let us go back to the entanglement problem. The ground state wave function takes the Gaussian form given in Equation (11):

$$\psi_0(z,t) = \frac{1}{\sqrt{\pi}} \exp\left\{ -\frac{1}{2} \left(z^2 + t^2 \right) \right\} \tag{102}$$

where the x and y variables are replaced by z and t, respectively. If Lorentz-boosted, this Gaussian function becomes squeezed to [20,24,25]:

$$\psi_\eta^0(z,t) = \frac{1}{\sqrt{\pi}} \exp\left\{ -\frac{1}{4} \left[e^{-2\eta}(z+t)^2 + e^{2\eta}(z-t)^2 \right] \right\} \tag{103}$$

leading to the series:

$$\frac{1}{\cosh \eta} \sum_k (\tanh \eta)^k \chi_k(z) \chi_k(t) \tag{104}$$

According to this formula, the z and t variables are entangled in the same way as the x and y variables are entangled.

Here, the z and t variables are space and time separations between two particles bound together by the oscillator force. The concept of the space separation is well defined, as in the case of the Bohr radius. On the other hand, the time separation is still hidden or forgotten in the present form of quantum mechanics. In the Lorentz-covariant world, this variable affects what we observe in the real world by entangling itself with the longitudinal spatial separation.

In Chapter 16 of their book [9], Walls and Milburn wrote down the series of Equation (1) and discussed what would happen when the η parameter becomes infinitely large. We note that the series given in Equation (104) shares the same expression as the form given by Walls and Milburn, as well as other papers dealing with the Gaussian entanglement. As in the case of Wall and Milburn, we are interested in what happens when η becomes very large.

As we emphasized throughout the present paper, it is possible to study the entanglement series using the squeezed Gaussian function given in Equation (103). It is then possible to study this problem using the ellipse. Indeed, we can carry out the mathematics of entanglement using the ellipse shown Figure 9. This figure is the same as that of Figure 6, but it illustrates the entanglement of the space and time separations, instead of the x and y variables. If the particle is at rest with $\eta = 0$, the Gaussian form corresponds to the circle in Figure 9. When the particle gains speed, this Gaussian function becomes squeezed into an ellipse. This ellipse becomes concentrated along the light cone with $t = z$, as η becomes very large.

The point is that we are able to observe this effect in the real world. These days, the velocity of protons from high-energy accelerators is very close to that of light. According to Gell-Mann [54], the proton is a bound state of three quarks. Since quarks are confined in the proton, they have never been observed, and the binding force must be like that of the harmonic oscillator. Furthermore, the observed mass spectra of the hadrons exhibit the degeneracy of the three-dimensional harmonic oscillator [43]. We use the word "hadron" for the bound state of the quarks. The simplest hadron is thus the bound state of two quarks.

In 1969 [55], Feynman observed that the same proton, when moving with a velocity close to that of light, can be regarded as a collection of partons, with the following peculiar properties.

1. The parton picture is valid only for protons moving with velocity close to that of light.
2. The interaction time between the quarks becomes dilated, and partons are like free particles.
3. The momentum distribution becomes wide-spread as the proton moves faster. Its width is proportional to the proton momentum.
4. The number of partons is not conserved, while the proton starts with a finite number of quarks.

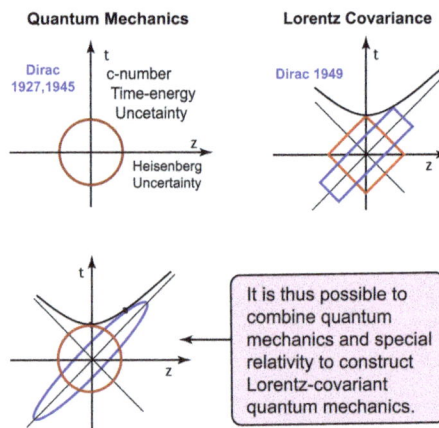

Figure 9. Feynman's rest of the universe. This figure is the same as Figure 6. Here, the space variable z and the time variable t are entangled.

Indeed, Figure 10 tells why the quark and parton models are two limiting cases of one Lorentz-covariant entity. In the oscillator regime, the three-particle system can be reduced to two independent two-particle systems [43]. Also in the oscillator regime, the momentum-energy wave function takes the same form as the space-time wave function, thus with the same squeeze or entanglement property as illustrated in this figure. This leads to the wide-spread momentum distribution [20,56,57].

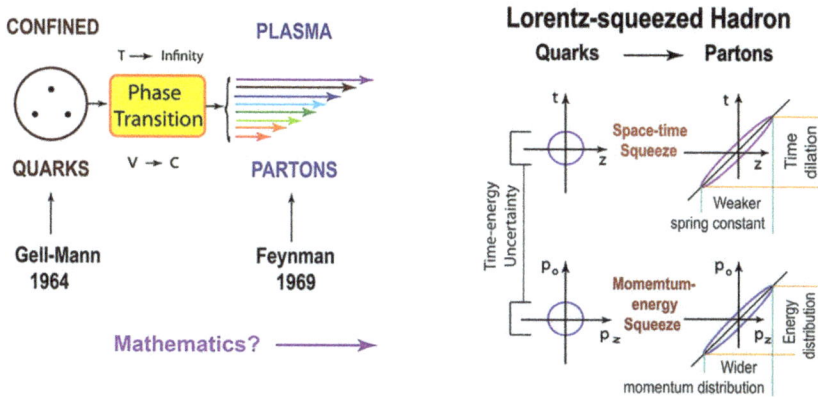

Figure 10. The transition from the quark to the parton model through space-time entanglement. When $\eta = 0$, the system is called the quark model where the space separation and the time separation are dis-entangled. Their entanglement becomes maximum when $\eta = \infty$. The quark model is transformed continuously to the parton model as the η parameter increases from zero to ∞. The mathematics of this transformation is given in terms of circles and ellipses.

Also in Figure 10, the time-separation between the quarks becomes large as η becomes large, leading to a weaker spring constant. This is why the partons behave like free particles [20,56,57].

As η becomes very large, all of the particles are confined into a narrow strip around the light cone. The number of particles is not constant for massless particles as in the case of black-body radiation [20,56,57].

Indeed, the oscillator model explains the basic features of the hadronic spectra [43]. Does the oscillator model tell the basic feature of the parton distribution observed in high-energy laboratories? The answer is yes. In his 1982 paper [58], Paul Hussar compared the parton distribution observed in a high-energy laboratory with the Lorentz-boosted Gaussian distribution. They are close enough to justify that the quark and parton models are two limiting cases of one Lorentz-covariant entity.

To summarize, the proton makes a phase transition from the bound state into a plasma state as it moves faster, as illustrated in Figure 10. The unobserved time-separation variable becomes more prominent as η becomes larger. We can now go back to the form of this entropy given in Equation (92) and calculate it numerically. It is plotted against $(\tanh \eta)^2 = \beta^2$ in Figure 11. The entropy is zero when the hadron is at rest, and it becomes infinite as the hadronic speed reaches the speed of light.

Entropy

[tanh(η)]^2

Temperature

[tanh(η)]^2

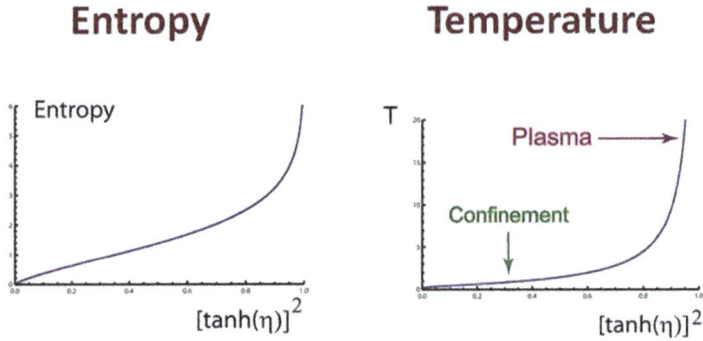

Figure 11. Entropy and temperature as functions of $[\tanh(\eta)]^2 = \beta^2$. They are both zero when the hadron is at rest, but they become infinitely large when the hadronic speed becomes close to that of light. The curvature for the temperature plot changes suddenly around $[\tanh(\eta)]^2 = 0.8$, indicating a phase transition.

Let us go back to the expression given in Equation (87). For this ground state, the density matrix becomes:

$$\rho_\eta(z, z') = \left(\frac{1}{\cosh \eta}\right)^2 \sum_k (\tanh \eta)^{2k} \chi_k(z) \chi_k(z') \tag{105}$$

We can now compare this expression with the density matrix for the thermally-excited oscillator state [22]:

$$\rho_\eta(z, z') = \left(1 - e^{-1/T}\right) \sum_k \left[e^{-1/T}\right]^k \chi_k(z) \chi_k(z') \tag{106}$$

By comparing these two expressions, we arrive at:

$$[\tanh(\eta)]^2 = e^{-1/T} \tag{107}$$

and thus:

$$T = \frac{-1}{\ln\left[(\tanh \eta)^2\right]} \tag{108}$$

This temperature is also plotted against $(\tanh \eta)^2$ in Figure 11. The temperature is zero if the hadron is at rest, but it becomes infinite when the hadronic speed becomes close to that of light. The slope of the curvature changes suddenly around $(\tanh \eta)^2 = 0.8$, indicating a phase transition from the bound state to the plasma state.

In this section, we have shown how useful the concept of entanglement is in understanding the role of the time-separation in high energy hadronic physics including Gell–Mann's quark model and Feynman's parton model as two limiting cases of one Lorentz-covariant entity.

9. Concluding Remarks

The main point of this paper is the mathematical identity:

$$\frac{1}{\sqrt{\pi}} \exp\left\{-\frac{1}{4}\left[e^{-2\eta}(x+y)^2 + e^{2\eta}(x-y)^2\right]\right\} = \frac{1}{\cosh \eta} \sum_k (\tanh \eta)^k \chi_k(x) \chi_k(y) \tag{109}$$

which says that the series of Equation (1) is an expansion of the Gaussian form given in Equation (2).

The first derivation of this series was published in 1979 [26] as a formula from the Lorentz group. Since this identity is not well known, we explained in Section 5 how this formula can be derived from the generating function of the Hermite polynomials.

While the series serves useful purposes in understanding the physics of entanglement, the Gaussian form can be used to transfer this idea to high-energy hadronic physics. The hadron, such as the proton, is a quantum bound state. As was pointed out in Section 8, the squeezed Gaussian function of Equation (109) plays the pivotal role for hadrons moving with relativistic speeds.

The Bohr radius is a very important quantity in physics. It is a spatial separation between the proton and electron in the the the hydrogen atom. Likewise, there is a space-like separation between constituent particles in a bound state at rest. When the bound state moves, it picks up a time-like component. However, in the present form of quantum mechanics, this time-like separation is not recognized. Indeed, this variable is hidden in Feynman's rest of the universe. When the system is Lorentz-boosted, this variable entangles itself with the measurable longitudinal variable. Our failure to measure this entangled variable appears in the form of entropy and temperature in the real world.

While harmonic oscillators are applicable to many aspects of quantum mechanics, Paul A. M. Dirac observed in 1963 [21] that the system of two oscillators contains also the symmetries of the Lorentz group. We discussed in this paper one concrete case of Dirac's symmetry. There are different languages for harmonic oscillators, such as the Schrödinger wave function, step-up and step-down operators and the Wigner phase-space distribution function. In this paper, we used extensively a pictorial language with circles and ellipses.

Let us go back to Equation (109); this mathematical identity was published in 1979 as textbook material in the American Journal of Physics [26], and the same formula was later included in a textbook on the Lorentz group [20]. It is gratifying to note that the same formula serves as a useful tool for the current literature in quantum information theory [59,60].

Author Contributions: Each of the authors participated in developing the material presented in this paper and in writing the manuscript.

Conflicts of Interest: The authors declare that no conflict of interest exists.

References

1. Giedke, G.; Wolf, M.M.; Krueger, O.; Werner, R.F.; Cirac, J.J. Entanglement of formation for symmetric Gaussian states. *Phys. Rev. Lett.* **2003**, *91*, 10790.1–10790.4.
2. Braunstein, S.L.; van Loock, P. Quantum information with continuous variables. *Rev. Mod. Phys.* **2005**, *28*, 513–676.
3. Kim, Y.S.; Noz, M.E. Coupled oscillators, entangled oscillators, and Lorentz-covariant Oscillators. *J. Opt. B Quantum Semiclass.* **2003**, *7*, s459–s467.
4. Ge, W.; Tasgin, M.E.; Suhail Zubairy, S. Conservation relation of nonclassicality and entanglement for Gaussian states in a beam splitter. *Phys. Rev. A* **2015**, *92*, 052328.
5. Gingrigh, R.M.; Adami, C. Quantum Engtanglement of Moving Bodies. *Phys. Rev. Lett.* **2002**, *89*, 270402.
6. Dodd, P.J.; Halliwell, J.J. Disentanglement and decoherence by open system dynamics. *Phys. Rev. A* **2004**, *69*, 052105.
7. Ferraro, A.; Olivares, S.; Paris, M.G.A. Gaussian States in Continuous Variable Quantum Information. EDIZIONI DI FILOSOFIA E SCIENZE (2005). Available online: http://arxiv.org/abs/quant-ph/0503237 (accessed on 24 June 2016).
8. Adesso, G.; Illuminati, F. Entanglement in continuous-variable systems: Recent advances and current perspectives. *J. Phys. A* **2007**, *40*, 7821–7880.
9. Walls, D.F.; Milburn, G.J. *Quantum Optics*, 2nd ed.; Springer: Berlin, Germany, 2008.
10. Yuen, H.P. Two-photon coherent states of the radiation field. *Phys. Rev. A* **1976**, *13*, 2226–2243.
11. Yurke, B.; Potasek, M. Obtainment of Thermal Noise from a Pure State. *Phys. Rev. A* **1987**, *36*, 3464–3466.
12. Ekert, A.K.; Knight, P.L. Correlations and squeezing of two-mode oscillations. *Am. J. Phys.* **1989**, *57*, 692–697.

13. Paris, M.G.A. Entanglement and visibility at the output of a Mach-Zehnder interferometer. *Phys. Rev. A* **1999**, *59*, 1615.
14. Kim, M.S.; Son, W.; Buzek, V.; Knight, P.L. Entanglement by a beam splitter: Nonclassicality as a prerequisite for entanglement. *Phys. Rev. A* **2002**, *65*, 02323.
15. Han, D.; Kim, Y.S.; Noz, M.E. Linear Canonical Transformations of Coherent and Squeezed States in the Wigner phase Space III. Two-mode States. *Phys. Rev. A* **1990**, *41*, 6233–6244.
16. Dirac, P.A.M. The Quantum Theory of the Emission and Absorption of Radiation. *Proc. Roy. Soc. (Lond.)* **1927**, *A114*, 243–265.
17. Dirac, P.A.M. Unitary Representations of the Lorentz Group. *Proc. Roy. Soc. (Lond.)* **1945**, *A183*, 284–295.
18. Dirac, P.A.M. Forms of relativistic dynamics. *Rev. Mod. Phys.* **1949**, *21*, 392–399.
19. Yukawa, H. Structure and Mass Spectrum of Elementary Particles. I. General Considerations. *Phys. Rev.* **1953**, *91*, 415–416.
20. Kim, Y.S.; Noz, M.E. *Theory and Applications of the Poincaré Group*; Reidel: Dordrecht, The Netherlands, 1986.
21. Dirac, P.A.M. A Remarkable Representation of the 3 + 2 de Sitter Group. *J. Math. Phys.* **1963**, *4*, 901–909.
22. Feynman, R.P. *Statistical Mechanics*; Benjamin Cummings: Reading, MA, USA, 1972.
23. Han, D.; Kim, Y.S.; Noz, M.E. Illustrative Example of Feynman's Rest of the Universe. *Am. J. Phys.* **1999**, *67*, 61–66.
24. Kim, Y.S.; Noz, M.E. Covariant harmonic oscillators and the quark model. *Phys. Rev. D* **1973**, *8*, 3521–3627.
25. Kim, Y.S.; Noz, M.E.; Oh, S.H. Representations of the Poincaré group for relativistic extended hadrons. *J. Math. Phys.* **1979**, *20*, 1341–1344.
26. Kim, Y.S.; Noz, M.E.; Oh, S.H. A simple method for illustrating the difference between the homogeneous and inhomogeneous Lorentz groups. *Am. J. Phys.* **1979**, *47*, 892–897.
27. Kim, Y.S.; Wigner, E.P. Entropy and Lorentz Transformations. *Phys. Lett. A* **1990**, *147*, 343–347.
28. Kim, Y.S.; Noz, M.E. Lorentz Harmonics, Squeeze Harmonics and Their Physical Applications. *Symmerty* **2011**, *3*, 16–36.
29. Klauder, J.R.; Sudarshan, E.C.G. *Fundamentals of Quantum Optics*; Benjamin: New York, NY, USA, 1968.
30. Saleh, B.E.A.; Teich, M.C. *Fundamentals of Photonics*, 2nd ed.; John Wiley and Sons: Hoboken, NJ, USA, 2007.
31. Miller, W. *Symmetry Groups and Their Applications*; Academic Press: New York, NY, USA, 1972.
32. Hall, B.C. *Lie Groups, Lie Algebras, and Representations: An Elementary Introduction*, 2nd ed.; Springer International: Cham, Switzerland, 2015.
33. Wigner, E. On Unitary Representations of the Inhomogeneous Lorentz Group. *Ann. Math.* **1939**, *40*, 149–204.
34. Weinberg, S. Photons and gravitons in S-Matrix theory: Derivation of charge conservation and equality of gravitational and inertial mass. *Phys. Rev.* **1964**, *135*, B1049–B1056.
35. Kim, Y.S.; Wigner, E.P. Space-time geometry of relativistic-particles. *J. Math. Phys.* **1990**, *31*, 55–60.
36. Başkal, S.; Kim, Y.S.; Noz, M.E. Wigner's Space-Time Symmetries Based on the Two-by-Two Matrices of the Damped Harmonic Oscillators and the Poincaré Sphere. *Symmetry* **2014**, *6*, 473–515.
37. Başkal, S.; Kim, Y.S.; Noz, M.E. *Physics of the Lorentz Group*; IOP Science; Morgan & Claypool Publishers: San Rafael, CA, USA, 2015.
38. Kim, Y.S.; Yeh, Y. E(2)-symmetric two-mode sheared states. *J. Math. Phys.* **1992**, *33*, 1237–1246
39. Kim, Y.S.; Noz, M.E. *Phase Space Picture of Quantum Mechanics*; World Scientific Publishing Company: Singapore, Singapore, 1991.
40. Kim, Y.S.; Noz, M.E. Dirac Matrices and Feynman's Rest of the Universe. *Symmetry* **2012**, *4*, 626–643.
41. Wigner, E. On the Quantum Corrections for Thermodynamic Equilibrium. *Phys. Rev.* **1932**, *40*, 749–759.
42. Han, D.; Kim, Y.S.; Noz, M.E. $O(3,3)$-like Symmetries of Coupled Harmonic Oscillators. *J. Math. Phys.* **1995**, *36*, 3940–3954.
43. Feynman, R.P.; Kislinger, M.; Ravndal, F. Current Matrix Elements from a Relativistic Quark Model. *Phys. Rev. D* **1971**, *3*, 2706–2732.
44. Rotbart, F.C. Complete orthogonality relations for the covariant harmonic oscillator. *Phys. Rev. D* **1981**, *12*, 3078–3090.
45. Ruiz, M.J. Orthogonality relations for covariant harmonic oscillator wave functions. *Phys. Rev. D* **1974**, *10*, 4306–4307.
46. Magnus, W.; Oberhettinger, F.; Soni, R.P. *Formulas and Theorems for the Special Functions of Mathematical Physics*; Springer-Verlag: Heidelberg, Germany, 1966.

47. Doman, B.G.S. *The Classical Orthogonal Polynomials*; World Scientific: Singapore, Singapore, 2016.
48. Bargmann, V. Irreducible unitary representations of the Lorentz group. *Ann. Math.* **1947**, *48*, 568–640.
49. Han, D.; Kim, Y.S. Special relativity and interferometers. *Phys. Rev. A* **1988**, *37*, 4494–4496.
50. Han, D.; Kim, Y.S.; Noz, M.E. Wigner rotations and Iwasawa decompositions in polarization optics. *Phys. Rev. E* **1999**, *1*, 1036–1041.
51. Von Neumann, J. *Die mathematische Grundlagen der Quanten-Mechanik*; Springer: Berlin, Germany, 1932. (von Neumann, I. *Mathematical Foundation of Quantum Mechanics*; Princeton University: Princeton, NJ, USA, 1955.)
52. Fano, U. Description of States in Quantum Mechanics by Density Matrix and Operator Techniques. *Rev. Mod. Phys.* **1957**, *29*, 74–93.
53. Wigner E.P.; Yanase, M.M. Information Contents of Distributions. *Proc. Natl. Acad. Sci. USA* **1963**, *49*, 910–918.
54. Gell-Mann, M. A Schematic Model of Baryons and Mesons. *Phys. Lett.* **1964**, *8*, 214–215.
55. Feynman, R.P. Very High-Energy Collisions of Hadrons. *Phys. Rev. Lett.* **1969**, *23*, 1415–1417.
56. Kim, Y.S.; Noz, M.E. Covariant harmonic oscillators and the parton picture. *Phys. Rev. D* **1977**, *15*, 335–338.
57. Kim, Y.S. Observable gauge transformations in the parton picture. *Phys. Rev. Lett.* **1989**, *63*, 348–351.
58. Hussar, P.E. Valons and harmonic oscillators. *Phys. Rev. D* **1981**, *23*, 2781–2783.
59. Leonhardt, U. *Essential Quantum Optics*; Cambridge University Press: London, UK, 2010.
60. Furusawa, A.; Loock, P.V. *Quantum Teleportation and Entanglement: A Hybrid Approach to Optical Quantum Information Processing*; Wiley-VCH: Weinheim, Germany, 2010.

© 2016 by the authors. Licensee MDPI, Basel, Switzerland. This article is an open access article distributed under the terms and conditions of the Creative Commons Attribution (CC BY) license (http://creativecommons.org/licenses/by/4.0/).

symmetry

MDPI

Article

Massless Majorana-Like Charged Carriers in Two-Dimensional Semimetals

Halina Grushevskaya [†] and George Krylov [†,*]

Physics Department, Belarusian State University, 4 Nezaleznasti Ave., 220030 Minsk, Belarus; grushevskaja@bsu.by
* Correspondence: krylov@bsu.by; Tel.: +375-296-62-44-97
† These authors contributed equally to this work.

Academic Editor: Young Suh Kim
Received: 29 February 2016; Accepted: 1 July 2016; Published: 8 July 2016

Abstract: The band structure of strongly correlated two-dimensional (2D) semimetal systems is found to be significantly affected by the spin-orbit coupling (SOC), resulting in SOC-induced Fermi surfaces. Dirac, Weyl and Majorana representations are used for the description of different semimetals, though the band structures of all these systems are very similar. We develop a theoretical approach to the band theory of two-dimensional semimetals within the Dirac–Hartree–Fock self-consistent field approximation. It reveals partially breaking symmetry of the Dirac cone affected by quasi-relativistic exchange interactions for 2D crystals with hexagonal symmetry. Fermi velocity becomes an operator within this approach, and elementary excitations have been calculated in the tight-binding approximation when taking into account the exchange interaction of $\pi(p_z)$-electron with its three nearest $\pi(p_z)$-electrons. These excitations are described by the massless Majorana equation instead of the Dirac one. The squared equation for this field is of the Klein–Gordon–Fock type. Such a feature of the band structure of 2D semimetals as the appearance of four pairs of nodes is shown to be described naturally within the developed formalism. Numerical simulation of band structure has been performed for the proposed 2D-model of graphene and a monolayer of Pb atoms.

Keywords: 2D semimetals; Dirac–Hartree–Fock self-consistent field approximation; Majorana-like field; Weyl-like nodes; Fermi velocity operator

PACS: 73.22.-f, 81.05.Bx

1. Introduction

Strongly correlated materials, such as two-dimensional (2D) complex oxides of transition metals, graphene, oxides with a perovskite structure, and IV–VI semiconductors being three-dimensional (3D) analogues of graphene, can demonstrate unusual electronic and magnetic properties, such as e.g., half-metallicity. The linear dispersion law for such materials is stipulated by the simultaneous existence of positively and negatively charged carriers [1]. Conical singularities are generic in the quantum crystals having honeycomb lattice symmetry [2]. Bipolarity of the material suggests that the state of an excitonic insulator is possible for it. Since an electron-hole pair is at the same time its own antiparticle, the Majorana representation has been used [3,4] to describe the interaction of pseudospins with the valley currents in a monolayer graphene.

The electron is a complex fermion, so if one decomposes it into its real and imaginary parts, which would be Majorana fermions, they are rapidly re-mixed by electromagnetic interactions. However, such a decomposition could be reasonable for a superconductor where, because of effective electrostatic screening, the Bogoliubov quasi-fermions behave as if they are neutral excitations [5].

A helical magnetic ordering (commensurate magnetism) occurs due to strong spin-orbit coupling (SOC) between Fe and Pb atoms in the system where a chain of ferromagnetic Fe atoms is placed on the surface of conventional superconductor composed of Pb atoms [6]. In this case, the imposition of SOC results in the appearance of Majorana-like excitations at the ends of the Fe atom chain.

The discovered p-wave function pairing in this Fe-chain is allowed to assume that there exists a new mechanism of superconductivity in high-temperature superconductors through the exchange of Majorana particles rather than phonons in the Bardeen–Cooper–Schrieffer theory. Such a novel superconducting state emerges, for example, in compound CeCoIn$_5$ in strong magnetic fields in addition to ordinary superconducting state, [7]. It has been shown [8–10] that the coupling of electrons into Cooper pairs in pnictides (LiFeAs with slabs FeAs) is mediated by the mixing of d-electron orbitals surrounding the atomic cores of transition metal. The new state is mediated by an anti-ferromagnetic order, and its fluctuations appear due to strong spin-orbit coupling [8,9,11]. It has been experimentally confirmed in [10] for LiFeAs. For antiferromagnetic itinerant-electron system LaFe$_{12}$B$_6$, ultrasharp magnetization steps have been observed [12]. The last can be only explained by the existence of anti-ferromagnetic order, and its fluctuations appear due to strong spin-orbit coupling.

Thus, there is a strong evidence that SOC may control the spin ordering in the absence of external magnetic fields. However, the mechanism that leads to such, commensurate magnetism has not been yet established.

The phenomenon of the contraction of electron density distribution in one direction is called nematicity. It is observed in pnictides BaFe$_2$(As$_{1-x}$P$_x$)$_2$ placed in a magnetic field, and such a phenomenon remains in the superconducting state [13]. The nematicity is coupled with considerable stripe spin fluctuations in FeSe [14]. The very strong spin orbit coupling leads to contraction in a factor of about 10% and rotation on 30° of the hexagonal Brillouin zone of delafossite oxide PtCoO$_2$, belonging to yet another class of topological insulators in which atoms of metal are in layers with triangular lattices [15].

Other topological insulators, namely so-called Weyl materials with a linear dispersion law, are close in properties with layered perovskite-like materials (see [16] and references therein). Currently, the first candidate for such a material has been found, namely TaAs, whose Brillouin zone has Weyl-like nodes and Fermi arcs [17–19].

Moreover, the experimental evidence of the similarities between the Fermi surfaces of insulator SmB$_6$ and metallic rare earth hexaborides (PrB$_6$ and LaB$_6$) has been presented in [20]. To explain the accompanying ordering phenomena, each associated with different symmetry breaking, it is necessary to develop a unified theory as it has been pointed out in [9].

Electrically charged carriers in the strongly correlated semimetallic systems with half-filled bands are massless fermions [15,21,22].

In a low-dimensional system, the exciton binding energy turns out to be high [23] and, respectively, the transition to the state of excitonic insulator is possible. Therefore, the Majorana rather than Weyl representation is preferable for the description of 2D semimetals. An attempt to represent the transition to the state of excitonic insulator as the appearance of Majorana zero-modes solution in graphene with trigonal warping [24] contradicts experimental data on the absence of a gap in band structure of graphene [25] and on diminishing of charged carriers mobility [26] and minimal conductivity [27]. However, at the present time, there exist experimental signatures of graphene Majorana states in graphene-superconductor junctions without the need for spin-orbit coupling [28]. However, modern Quantum Field Theory of pseudo-Dirac quasiparticles in random phase approximation predicts a strong screening that destroys the excitonic pairing instability if the fermion dynamic mass $m(p)$ dependent on momentum p is small in comparison with the chemical potential μ: $m(p) \leq \mu$ [29].

In the paper, we would like to show how the above described features of the layered materials can be formalized in 2D models, where the charged carriers are the quasiparticles of Majorana rather than of the Weyl type. We also show that, under certain conditions, these quasiparticles reveal themselves as Weyl-like states or massless Dirac pseudofermions.

However, the use of the well-known Majorana representations to describe a semimetal as a massless-quasiparticle system is encountered with such a puzzle as the absence of harmonic oscillatory solutions in ultrarelativistic limit for Majorana particles of zero mass [30]. The equations are known for massive Majorana particles only [31–33].

In the paper, we reveal different aspects of appearance of Majorana-like quasiparticle states in the band structure of semimetals. 2D Hartree–Fock approximation for graphene, however, predicts experimentally observable increase of the Fermi velocity value $v_F(\vec{p})$ at small momenta p [25] but leads to logarithmically divergent $v_F(\vec{p})$ at $p \to 0$ [34]. To take into account this effect of long range Coulomb interactions correctly, our calculation is based on the quasi-relativistic Dirac–Hartree–Fock self-consistent field approach developed earlier [35,36].

The goal is to construct a 2D-semimetal model in which a motion equation is a pseudo-relativistic massless Majorana-like one. We show that the squared equation for this field is of a Klein–Gordon–Fock type, and therefore the charged carriers in such 2D-semimetal models can be assumed massless Majorana-like quasiparticles.

We study quasiparticle excitations of the electronic subsystem of a hexagonal monoatomic layer (monolayer) of light or heavy atoms in tight-binding approximation. The simulations are performed for the atoms of C and Pb on the assumption that sp^2-hybridization for s- and p-electron orbitals is also possible for the atoms of Pb.

We demonstrate that the band-structure features for the hexagonal monolayers are similar to each other due to the similarity of external electronic shells of their atoms. Despite the similarity of the band structure, the charged carriers in such 2D-semimetal models can possess different features, e.g., the charged carriers in the monolayer of the atoms of C can be thought of as massless Dirac pseudofermions, whereas in the monolayer from the atoms of Pb, they reveal themselves as Weyl-like states.

The paper is organized as follows. In Section 2, we propose a semimetal model with coupling between pseudospin and valley currents and prove the pseudo-helicity conservation law. In Section 3, we briefly introduce the approach [3,35–37] and use it in a simple tight-binding approximation to obtain the system of equations for a Majorana secondary quantized field. In Section 4, we support the statement that the squared equation for the constructed field is of the Klein–Gordon–Fock type for different model exchange operators. We also discuss features of our model manifesting in the band structure of real semimetals. In Section 5, we discuss the proposed approximations for the exchange interactions in 2D semimetals and summarize our findings.

2. Monolayer Semimetal Model with Partial Unfolding of Dirac Bands

Semimetals are known to be bipolar materials with half-filled valence and conduction bands. A distinctive feature of the graphene band structure is the existence of Dirac cones in the Dirac points (valleys) K, K' of the Brillouin zone. In the present paper, these Dirac points are designated as K_A, K_B. We assume that pseudo-spins of hexagonally packed carbon atoms in the monoatomic layer (monolayer) graphene are anti-ordered, as it is shown schematically in Figure 1a. The fact that the pseudo-helicity (chirality) conservation law forbids massless charged carriers to be in lattice sites with the opposite signs of pseudo-spin, makes possible the existence of valley currents due to jumps through the forbidden sites. This is shown schematically in Figure 1a. Coupling between the pseudo-spin and the valley current in the Majorana representation of bispinors can be determined in the following way.

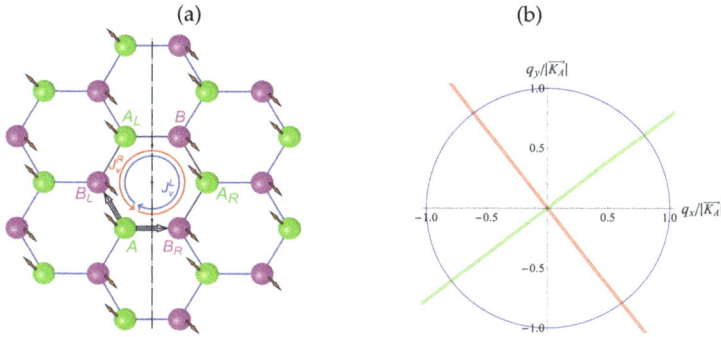

Figure 1. (a) graphene lattice, comprised of two sublattices $\{A\}$ with spin "up" and $\{B\}$ with spin "down". Right and left valley currents J_v^R and J_v^L are shown as circular curves with arrows. Double arrows from site A to site B_L and from A to B_R indicate clockwise and anti-clockwise directions. The axis of mirror reflection from A_R to B_L is marked by dash-dotted line; (b) transformations of a q-circumference into ellipses under an action of exchange operators $(\Sigma_{rel}^x)_{AB}$ and $(\Sigma_{rel}^x)_{BA}$ (in color).

According to Figure 1a, a particle can travel from a lattice site A to e.g., a lattice site A_R through right or left sites B_R or B_L, respectively. Since the particle is symmetrical, its description in the right and left reference frames has to be equivalent. Therefore, a bispinor wave function Ψ' of graphene has to be chosen in the Majorana representation, and its upper and lower spin components ψ', $\dot{\psi}'$ are transformed by left and right representations of the Lorentz group:

$$\Psi' = \left(\begin{array}{c} \psi'_\sigma \\ \dot{\psi}'_{-\sigma} \end{array} \right) = \left(\begin{array}{c} e^{\frac{\chi}{2}\vec{\sigma}\cdot\vec{n}}\psi_\sigma \\ e^{\frac{\chi}{2}(-\vec{\sigma})\cdot\vec{n}}\dot{\psi}_{-\sigma} \end{array} \right). \tag{1}$$

The wave-function $\widehat{\chi_\sigma^\dagger}(\vec{r}_A)\,|0,+\sigma\rangle$ of a particle (in our case of an electron-hole pair) located on the site A, behaves as a component ψ_σ, while the wave-function $\widehat{\chi_{-\sigma}^\dagger}(\vec{r}_B)\,|0,-\sigma\rangle$ of a particle located on the site B behaves as a component $\dot{\psi}_{-\sigma}$ of the bispinor (1).

Relativistic particles with non-zero spin possess the helicity h, which is the projection of the particle's spin to the direction of motion [32]:

$$h \equiv \vec{p}\cdot\vec{S} = \frac{1}{2}p_i \left(\begin{array}{cc} \sigma_i & 0 \\ 0 & \sigma_i \end{array} \right), \tag{2}$$

where $\vec{p}$ is the particle momentum, $\vec{S}$ is the spin operator for a particle, $\vec{\sigma}$ is the vector of the Pauli matrices σ_i, and $i = x$, y. In quantum relativistic field theory, the value of the helicity of a massless particle is preserved in the transition from one reference frame moving with the velocity v_1, to another one moving with the velocity v_2 [32,38].

Let us designate the two-dimensional spin of the quasi-particle in valleys K_A and K_B as $\vec{S}_{AB} = \hbar\vec{\sigma}_{AB}/2$ and $\vec{S}_{BA} = \hbar\vec{\sigma}_{BA}/2$, respectively.

Let us introduce two-dimensional pseudospin $\vec{S}_{AB}$ and $\vec{S}_{BA}$ of quasi-particles in valleys K_A and K_B through the transformed vector $\vec{\sigma}$ of the Pauli matrices σ_i, $i = x$, y as $\vec{S}_{AB} = \hbar\vec{\sigma}_{AB}/2$ and $\vec{S}_{BA} = \hbar\vec{\sigma}_{BA}/2$. The explicit form of this transformation is given in Section 3.

A valley current J_v^R or J_v^L, on the right or left closed contour $\{A \to B_R \to A_R \to B \to A_L \to B_L \to A\}$ or $\{A \to B_L \to A_L \to B \to A_R \to B_R \to A\}$, respectively, in Figure 1, is created by an electron (hole) with pseudo-angular momentum $\vec{l}_{AB_R}$ and momentum $\vec{p}_{AB_R}$ or by an electron (hole) with $\vec{l}_{AB_L}$ and

$\vec{p}_{AB_L}$. Pseudo-helicity of bispinors (1), describing the particles right or left the from lattice site A, is defined by the expressions, which are analogous to (2):

$$h_{B_R A} \equiv \vec{p}_{AB_R} \cdot \vec{S}_{B_R A},$$ (3)

$$h_{B_L A} \equiv \vec{p}_{AB_L} \cdot \vec{S}_{B_L A}.$$ (4)

Let us use the parity operator P, which mirrors the bispinor (1) with respect to the line passing through the points A and B. Pseudo-helicity of the mirrored bispinor is defined by the expression:

$$P h_{B_R A_R} P = h_{A_L B_L} = \vec{p}_{B_L A_L} \cdot \vec{S}_{A_L B_L}.$$ (5)

Pseudo-helicity h_{AB} does not change its value while the valley momentum and the pseudo-spin change signs: $\vec{p}_{A_L B_L} = -\vec{p}_{B_R A_R}$ and $\vec{S}_{A_L B_L} = -\vec{S}_{B_R A_R}$.

The pseudo-helicity h_{AB} is expressed through the projection $\tilde{\mathcal{M}}_{AB} = \vec{\sigma}_{BA} \cdot \left(\vec{l}_{AB} + \hbar \vec{\sigma}_{BA} / 2 \right)$ of the total angular momentum on the direction of the spin $\vec{\sigma}_{BA}$ as [39,40]:

$$\vec{\sigma}_{BA} \cdot \vec{p}_{AB} = \sigma_{BA}^r \left(p_{r,BA} + \imath \frac{\tilde{\mathcal{M}}_{AB}}{r} - \hbar/2 \right) = \sigma_{BA}^r \left(p_{r,BA} + \imath \frac{\vec{\sigma}_{BA} \cdot \vec{l}_{AB}}{r} \right),$$ (6)

where σ_{BA}^r and $p_{r,BA}$ are radial components of the spin and the momentum, respectively. According to Equation (6), the pseudo-spin-orbit scalar $\vec{\sigma}_{BA} \cdot \vec{l}_{AB}$ describes the coupling (interaction) of the spin with the valley currents flowing along a closed loop clockwise or in opposite directions, as is shown in Figure 1a. Hence, there exists a preferred direction along which the spin projection of the bispinor (1) is not changed after transition from one moving reference frame into another. At this, the spin of a particle precesses. Transformation of the electron and hole into each other in an exciton is a pseudo-precession.

As a result, the coupling of pseudo-spin and valley currents stipulates the spin precession of exciton charged carriers in graphene. In our model, the orientation of non-equilibrium spin of the states of monolayer graphene in electromagnetic fields may be retained for a long time due to prohibition of change for exciton pseudo-helicity. Pseudo-precession is possible, if spins of p_z -electrons are anti-ordered (pseudo-antiferromagnetic ordering). Therefore, the pseudo-spin precession of the exciton can be implemented through the exchange interaction. Furthermore, we determine the operators $\vec{\sigma}_{BA(AB)}, \vec{p}_{AB(BA)}$ and describe the effects of pseudo-spin and valley current coupling.

3. Effects of Coupling between Pseudo-Spin and Valley Current

In quasi-relativistic approximation (c^{-1} expansion), the eigenproblem for the equation of motion of the secondary quantized field $\hat{\chi}^\dagger_{-\sigma_A}$ in the model shown in Figure 1a has the form: [35–37]

$$\left\{ \vec{\sigma} \cdot \vec{p} \, \hat{v}_F^{qu} - \frac{1}{c} \left(i \Sigma_{rel}^x \right)_{AB} \left(i \Sigma_{rel}^x \right)_{BA} \right\} \widehat{\chi^\dagger_{-\sigma_A}} (\vec{r}) \, |0, -\sigma\rangle$$ (7)

$$= E_{qu}(p) \widehat{\chi^\dagger_{-\sigma_A}} (\vec{r}) \, |0, -\sigma\rangle,$$

where the Fermi velocity operator $\hat{v}_F^{qu}$ is defined as

$$\hat{v}_F^{qu} = \left[\left(\Sigma_{rel}^x \right)_{BA} + c\hbar \vec{\sigma} \cdot \left(\vec{K}_A + \vec{K}_B \right) \right].$$

$\left(\Sigma^x_{rel}\right)_{BA}$, $\left(\Sigma^x_{rel}\right)_{AB}$ are determined through an ordinary exchange interaction contribution, for example [39,40]:

$$\left(\Sigma^x_{rel}\right)_{AB} \widehat{\chi}^\dagger_{\sigma_B}(\vec{r})\,|0,\sigma\rangle = \sum_{i=1}^{N_v N} \int d\vec{r}_i \widehat{\chi}^\dagger_{\sigma_i{}^B}(\vec{r})\,|0,\sigma\rangle$$

$$\times \langle 0,-\sigma_i| \widehat{\chi}^\dagger_{-\sigma_i{}^A}(\vec{r}_i) V(\vec{r}_i - \vec{r}) \widehat{\chi}_{-\sigma_B}(\vec{r}_i)|0,-\sigma_{i'}\rangle.$$

$V(\vec{r}_i - \vec{r})$ is the Coulomb interaction between two valent electrons with radius-vectors $\vec{r}_i$ and $\vec{r}$; N is a total number of atoms in the system, N_v is a number of valent electrons in an atom, c is the speed of light.

After applying the non-unitary transformation to the wave function in the form

$$\widehat{\widetilde{\chi}^\dagger_{-\sigma_A}}\,|0,-\sigma\rangle = \left(\Sigma^x_{rel}\right)_{BA} \widehat{\chi^\dagger_{-\sigma_A}}\,|0,-\sigma\rangle,$$

we obtain (neglecting mixing of the states for the Dirac points) the equation that is similar to the one in 2D quantum field theory (QFT) [41–43], but it describes the motion of a particle with pseudo-spin $\vec{S}_{AB} = \hbar\vec{\sigma}_{AB}/2$:

$$\left\{\vec{\sigma}^{AB}_{2D} \cdot \vec{p}_{BA} - c^{-1}\widetilde{\Sigma_{BA}\Sigma_{AB}}\right\} \widehat{\widetilde{\chi}^\dagger_{-\sigma_A}}(\vec{r})\,|0,-\sigma\rangle = \tilde{E}_{qu}(p)\widehat{\widetilde{\chi}^\dagger_{-\sigma_A}}(\vec{r})\,|0,-\sigma\rangle \quad , \tag{8}$$

with a transformed 2D vector $\vec{\sigma}^{AB}_{2D}$ of the Pauli matrices, which are determined as $\vec{\sigma}^{AB}_{2D} = \left(\Sigma^x_{rel}\right)_{BA}\vec{\sigma}\left(\Sigma^x_{rel}\right)^{-1}_{BA}$. The following notions are introduced: $\vec{p}_{BA}\widetilde{\chi}^\dagger_{-\sigma_A} = \left(\Sigma^x_{rel}\right)_{BA}\vec{p}\left(\Sigma^x_{rel}\right)^{-1}_{BA}\widetilde{\chi}_{-\sigma_A} \equiv \left[\left(\Sigma^x_{rel}\right)_{BA}\vec{p}\right]\widetilde{\chi}^\dagger_{-\sigma_A}$, $\tilde{E}_{qu} = E_{qu}/\vartheta^{BA}_F$, $\vartheta^{BA}_F = \left(\Sigma^x_{rel}\right)_{BA}$, $\Sigma_{BA}\widetilde{\Sigma}_{AB} \equiv \left(\Sigma^x_{rel}\right)_{BA}\left(i\Sigma^x_{rel}\right)_{AB}\left(i\Sigma^x_{rel}\right)_{BA}\left(\Sigma^x_{rel}\right)^{-1}_{BA} = \left(i\Sigma^x_{rel}\right)_{BA}\left(i\Sigma^x_{rel}\right)_{AB}$; and the product of two capital sigma, as one sees from the last chain of formulas, behaves like a scalar mass term.

Further simulations are performed in nearest neighbor tight-binding approximation [44,45]. This approximation correctly predicts the graphene band structure in the energy range ± 1 eV [46]. This turns out to be sufficient for our purposes. We use the expressions for the exchange between $\pi(p_z)$-electrons only. One can find the explicit form of these expressions in [4].

The action of the matrices $\left(\Sigma^x_{rel}\right)_{BA}$, $\left(\Sigma^x_{rel}\right)_{AB}$ in the momentum space is shown in Figure 1b. As $\left(\Sigma^x_{rel}\right)_{BA} \neq \left(\Sigma^x_{rel}\right)_{AB}$, the vector $\vec{p}_{BA}$ is rotated with respect to $\vec{p}_{AB}$ and stretched. According to Figure 1b, ellipses in momentum spaces of electrons and holes are rotated 90° with respect to each other. With an account of the hexagonal symmetry of the system, the last explains the experimentally observed rotation in 30° of the hexagonal Brillouin zone of $PtCoO_2$ [15].

Thus, the sequence of exchange interactions $\left(\Sigma^x_{rel}\right)_{AB}\left(\Sigma^x_{rel}\right)_{BA}\left(\Sigma^x_{rel}\right)_{AB}$ for valley currents makes rotation initially of the electron Brillouin zone and Dirac band into the hole Brillouin zone and Dirac band, and then vice-versa. Thus, the exchange $\left(\Sigma^x_{rel}\right)_{AB(AB)} \equiv \Sigma_{AB(BA)}$ changes the sublattices wave functions:

$$|\psi_{AB}\rangle = \Sigma_{AB}\,|\psi^*_{BA}\rangle.$$

Owing to it and neglecting a very small mass term $c^{-1}\widetilde{\Sigma_{BA}\Sigma_{AB}}$, the equation in which the operator of the Fermi velocity enters, can be rewritten as follows:

$$\vec{\sigma}^{BA}_{2D} \cdot \vec{p}_{AB}\,|\psi_{AB}\rangle = E_{qu}\,|\psi^*_{BA}\rangle. \tag{9}$$

Taking into account that $E \to i\frac{\partial}{\partial t}$ and $\vec{p} = -i\vec{\nabla}$, we transform the system of equations for the Majorana bispinor $(\psi^\dagger_{AB}, (\psi^*_{BA})^\dagger)$:

$$\vec{\sigma}_{2D}^{BA} \cdot \vec{p}_{AB} \, |\psi_{AB}\rangle = i \frac{\partial}{\partial t} \, |\psi_{BA}^*\rangle \, , \tag{10}$$

$$\vec{\sigma}_{2D}^{AB} \cdot \vec{p}_{BA}^* \, |\psi_{BA}^*\rangle = -i \frac{\partial}{\partial t} \, |\psi_{AB}\rangle \, , \tag{11}$$

into the wave equation of the form:

$$(\vec{\sigma}_{2D}^{AB} \cdot \vec{p}_{BA}^*)(\vec{\sigma}_{2D}^{BA} \cdot \vec{p}_{AB}) \, |\psi_{AB}\rangle = \frac{\partial^2}{\partial t^2} \, |\psi_{AB}\rangle \, . \tag{12}$$

Equation (12) describes an oscillator with the energy operator $\hat{\omega}(\vec{p})$

$$\hat{\omega}(\vec{p}) = \frac{1}{\sqrt{2}} \left[(\vec{\sigma}_{2D}^{AB} \cdot \vec{p}_{BA})(\vec{\sigma}_{2D}^{BA} \cdot \vec{p}_{AB}) + (\vec{\sigma}_{2D}^{BA} \cdot \vec{p}_{AB})(\vec{\sigma}_{2D}^{AB} \cdot \vec{p}_{BA}) \right]^{1/2} \, . \tag{13}$$

Now, one can really see that the obtained equation is the equation of motion for a Majorana bispinor wave function of the semimetal charged carriers.

Thus, the Fermi velocity becomes an operator within this approach, and elementary excitations are fermionic excitations described by the massless Majorana-like equation rather than Dirac-like one.

4. Harmonic Analysis of the Problem

Equation (13) can be rewritten in the following form:

$$\hat{\omega}^2(\vec{p}) = \frac{1}{2} \left(\hat{H}_{AB}\hat{H}_{BA} + \hat{H}_{BA}\hat{H}_{AB} \right) . \tag{14}$$

In order to describe the proposed secondary quantized field by a set of harmonic oscillators, it is necessary to show that the squared Equation (14), obtained by the symmetrization of the product of the Hamiltonians $\hat{H}_{AB}$ and $\hat{H}_{BA}$, is the Klein–Gordon–Fock operator. This will be the case if the non-diagonal matrix elements of the operator vanish identically, and therefore the components of the equation are independent. Then, $\hat{\omega}^2(\vec{p})$ can be considered as a "square of energy operator".

Unfortunately, because of the complex form of the exchange operator, the statement is difficult to prove in the general case. Therefore, we do this for several approximations of the exchange interaction and demonstrate that the Equation (14) is a Klein–Gordon–Fock one.

As a first particular case, when the proposed Majorana-like field is proven to be a harmonic oscillators set, we consider ϵ-neighborhood ($\epsilon \to 0$) of the Dirac point $K_A(K_B)$.

Let us designate the momentum of a particle in a valley as $\vec{q}$. The momentum $\vec{q}$ is determined as $\vec{q} = \vec{p} - \hbar\vec{K}_A$. In the case of very small values of $\vec{q}$, $q \to 0$ the exchange operator $\Sigma_{AB(BA)}$ is approximated by a power series expansion up to the fourth order in q. Then, an analytical calculation of non-diagonal elements of the operator $\hat{\omega}^2(\vec{p})$ performed in the Mathematica system proves that they are identically zero.

Band structures for monolayer graphene and monolayer of atoms of Pb are shown in Figure 2a,b. One can see that the Weyl nodes in graphene are located far enough from the Dirac point. The Weyl nodes are shifted to the Dirac point for the Pb-monolayer. Therefore, Weyl-like character in the behavior of charged carriers may be exhibited for the Pb-monolayer under the condition that the contributions up to 4-th order in q are prevailing in the exchange. In accordance with Figure 1b, the exchange operator matrices transform a circumference in the momentum space into a highly stretched ellipse that allows us to assume the presence of nematicity in the model.

For a given $\vec{q}$, where the eigenfunction of Equation (9) represents 2D spinor Ψ, we choose its normalization in the form $\Psi(\vec{q}) = (\psi(\vec{q}), 1)^\dagger$ with lower component equal to unity. Then, as it can be easily shown for the massless Dirac pseudo-fermion model [47], the absolute value of the upper component $|\psi(\vec{q})|$ does not depend upon the wave vector $\vec{q}$, demonstrating the equivalence of all directions in $\vec{q}$ space. We construct $|\psi(\vec{q})|^2$ for Equation (9) in q^4-approximation for the exchange. The results are shown in Figure 2c. The isotropy of $|\psi(\vec{q})|^2$ is broken for our model due to the appearance of the preferable directions in the momentum space.

As one can see from Figure 2c, the existence of almost one-dimensional regions with sharp jump in $|\psi(\vec{q})|^2$ should probably lead to some anisotropy already in the configuration space for the carriers that we consider as manifestation of nematicity.

The approximation q^4 for the exchange operator expression presents a particular interest for systems with strong damping of quasi-particle excitations.

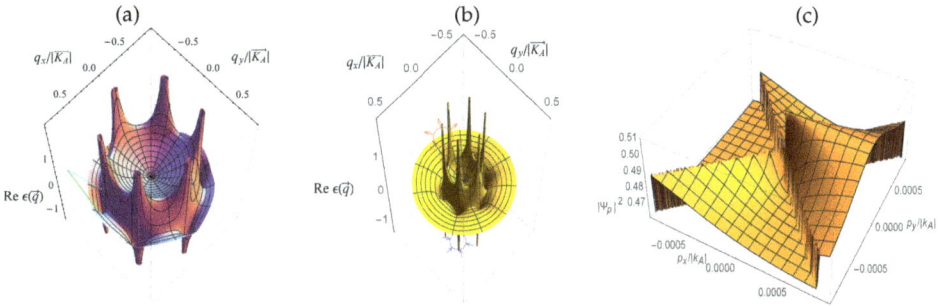

Figure 2. A splitting of Dirac cone replicas: for graphene (**a**) and Pb monolayer (**b**). One of the six pairs of Weyl-like nodes: source and sink are indicated; (**c**) the square of the absolute value of the upper spinor component $|\psi|^2$ of $\vec{q}$−eigenstate in the 2D semimetal model. $\vec{q} = \vec{p} - \vec{K}_A$. (in color)

The second approximation of the exchange, for which we can prove the harmonic origin of the proposed Majorana-like field, is the model exchange with full exponential factors taken into account, but with the phase-difference between $\pi(p)_z$-electrons wavefunction chosen to be identically zero (see Ref. [4] for detail). Numeric simulation of $\hat{\omega}^2(\vec{p})$ with this model exchange has been performed on a discrete lattice in the Brillouin zone. It has been demonstrated that the operator $\hat{\omega}^2(\vec{p})$ is always diagonal in this case.

Now, we perform the simulations with the exact expression for the exchange term.

In this general case, the exchange between $\pi(p_z)$-electron and its three nearest $\pi(p_z)$-electrons has been calculated based on the method proposed in [4]. Band structure of the 2D semimetal has the form of a degenerated Dirac cone in the neighborhood of the Dirac point. Then, the emergence of unfolding leads to replica appearance, and further splitting of these replicas gives the octagonal symmetry of the problem, as one can see in Figure 3. Hyperbolic points (saddle points) are located between nodes and at the apex of the Dirac cone (Van-Hove singularities) as one can see in Figure 2a,b [3,48–50]. Therefore, a fractal-like set of Fermi arcs which are shown in Figure 4, is formed in the absence of damping in the system. Contrary to the graphene case, the splitting of the Dirac bands for the Pb-monolayer occurs at sufficiently small q, and therefore, can be observed experimentally. In addition, for the Pb-monolayer, there exist regions with huge numbers of Fermi arcs, and, respectively, regions with strong fluctuations of antiferromagnetic ordering.

Thus, the secondary quantized field described by Equation (9) represents a field in which quanta manifest themselves as Dirac pseudo-fermions in the apex of the Dirac cone and as Weyl-like particles for sufficiently large q at the presence of the dumping in the system. For an ideal system ($\Im m\,\epsilon(\vec{q}) = 0$), such a behavior is similar to that of the mathematical pendulum in the vicinity of the separatrix [51,52].

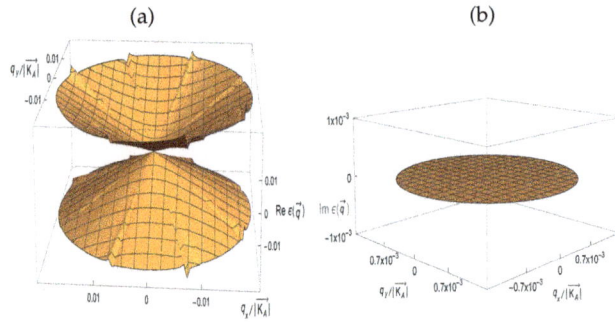

Figure 3. A band structure in the graphene model with partial unfolding of Dirac cone: real (a) and imaginary (b) parts of $\epsilon(\vec{q})$; range of high momenta. $\vec{q} = \vec{p} - \vec{K}_A$ (in color).

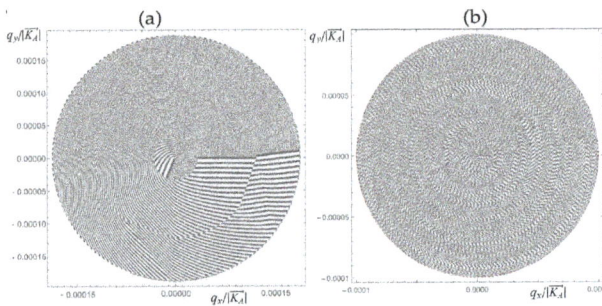

Figure 4. Density of Fermi arcs sets in graphene (a) and Pb-monolayer bands for values of momentum q in the range $0 \geq q/\left|\vec{K}_A\right| \leq 10^{-4}, \vec{q} = \vec{p} - \vec{K}_A$.

5. Discussion

Discussing the obtained results, we have to point out, firstly, that the excitations of the constructed secondary-quantized pseudo-fermionic field are Majorana-like massless quasiparticles.

The set of Fermi arcs in our model shows that the splitting of Dirac replicas on a huge number of Weyl-like states occurs in the momentum space except for the Dirac cone apex.

In contrast to known massless Dirac and Weyl models, in the proposed model, there is a partial removing of the degeneracy of the Dirac cone, and the octagonal symmetry of the bands emerges for sufficiently large q. Thus, Majorana particles in our model can be represented as a wave package of infinitely large number of Weyl-like states.

Secondly, the Dirac cone for the proposed 2D-semimetal model is degenerated in a very small neighborhood of the Dirac point $K_A(K_B)$ at $q \to 0$.

Thirdly, the first-approximation with damping demonstrates that sufficiently strong decay leads to diminishing the number of the Weyl states and formation of bands having hexagonal symmetry. In accordance with the obtained results, in the system with strong damping, only six pairs of Weyl nodes survive. In this case, each Dirac hole (electron) cone is surrounded by three electron (hole) bands relating to three Weyl pairs. Provided the lifetime of the Weyl-like states is sufficiently large (small but finite damping) to preserve the octagonal symmetry of the bands, each Dirac hole (electron) cone will be surrounded by four electron (hole) bands relating to four Weyl pairs.

Important features of the proposed model are that the fractal set of Fermi arches manifests pseudospin fluctuations and the phenomenon of nematicity is possible.

6. Conclusions

In conclusion, contrary to known Dirac and Weyl models, the constructed 2D-semimetal model allows for description, in a general formalism, the band structure of a wide class of existing strongly.

Acknowledgments: This work has been supported in part by Research grant No. 2.1.01.1 within the Basic Research Program "Microcosm and Universe" of the Republic of Belarus.

Author Contributions: Both authors equally contributed to this work.

Conflicts of Interest: The authors declare no conflict of interest.

References

1. Grushevskaya, H.V.; Hurski, L.I. Coherent charge transport in strongly correlated electron systems: Negatively charged exciton. *Quantum Matter* **2015**, *4*, 384–386.
2. Fefferman, C.L.; Weinstein, M.I. Honeycomb lattice potentials and Dirac points. *J. Am. Math. Soc.* **2012**, *25*, 1169–1220.
3. Grushevskaya, H.V.; Krylov, G. Quantum field theory of graphene with dynamical partial symmetry breaking. *J. Mod. Phys.* **2014**, *5*, 984–994.
4. Grushevskaya, H.V.; Krylov, G. Semimetals with Fermi Velocity Affected by Exchange Interactions: Two Dimensional Majorana Charge Carriers. *J. Nonlinear Phenom. Complex Syst.* **2015**, *18*, 266–283.
5. Semenoff, G.W.; Sodano, P. Stretched quantum states emerging from a Majorana medium. *J. Phys. B: At. Mol. Opt. Phys.* **2007**, *40*, 1479–1488.
6. Nadj-Perge, S.; Drozdov, I.K.; Li, J.; Chen, H.; Jeon, S.; Seo, J.; MacDonald, A.H.; Bernevig, A.; Yazdani, A. Observation of Majorana fermions in ferromagnetic atomic chains on a superconductor. *Science* **2014**, *346*, 602–607.
7. Gerber, S.; Bartkowiak, M.; Gavilano, J.L.; Ressouche, E.; Egetenmeyer, N.; Niedermayer, C.; Bianchi, A.D.; Movshovich, R.; Bauer, E.D.; Thompson, J.D.; *et al.* Switching of magnetic domains reveals spatially inhomogeneous superconductivity. *Nat. Phys.* **2014**, *10*, 126–129.
8. Shimojima, T.; Sakaguchi, F.; Ishizaka, K.; Ishida, Y.; Kiss, T.; Okawa, M.; Togashi, T.; Chen, C.-T.; Watanabe, S.; Arita, M.; et al. Orbital-independent superconducting gaps in iron-pnictides. *Science* **2011**, *332*, 564–567.
9. Davis, J.C.S.; Lee, D.-H. Concepts relating magnetic interactions, intertwined electronic orders, and strongly correlated superconductivity. *Proc. Natl. Acad. Sci. USA* **2013**, *110*, 17623–17630.
10. Borisenko, S.V.; Evtushinsky, D.V.; Liu, Z.-H.; Morozov, I.; Kappenberger, R.; Wurmehl, S.; Büchner, B.; Yaresko, A.N.; Kim, T.K.; Hoesch, M.; *et al.* Direct observation of spin-orbit coupling in iron-based superconductors. *Nat. Phys.* **2015**, doi:10.1038/nphys3594.
11. Hurski, L.I.; Grushevskaya, H.V.; Kalanda, N.A. Non-adiabatic paramagnetic model of pseudo-gap state in high-temperature cuprate superconductors. *Dokl. Nat. Acad. Sci. Belarus* **2010**, *54*, 55–62. (In Russian)
12. Diop, L.V.B.; Isnard, O.; Rodriguez-Carvajal, J. Ultrasharp magnetization steps in the antiferromagnetic itinerant-electron system $LaFe_{12}B_6$. *Phys. Rev.* **2016**, *B93*, 014440.
13. Kasahara, S.; Shi, H.J.; Hashimoto, K.; Tonegawa, S.; Mizukami, Y.; Shibauchi, T.; Sugimoto, K.; Fukuda, T.; Terashima, T.; Nevidomskyy, A.H.; *et al.* Electronic nematicity above the structural and superconducting transition in $BaFe_2(As_{1-x}P_x)_2$. *Nature* **2012**, *486*, 382–385.
14. Wang, Q.; Shen, Y.; Pan, B.; Hao, Y.; Ma, M.; Zhou, F.; Steffens, P.; Schmalzl, K.; Forrest, T.R.; Abdel-Hafiez, M.; *et al.* Strong interplay between stripe spin fluctuations, nematicity and superconductivity in FeSe. *Nat. Mater.* **2016**, *15*, 159–163.
15. Kushwaha, P.; Sunko, V.; Moll, Ph.J.W.; Bawden, L.; Riley, J.M.; Nandi, N.; Rosner, H.; Schmidt, M.P.; Arnold, F.; Hassinger, E.; et al. Nearly free electrons in a 5d delafossite oxide metal. *Sci. Adv.* **2015**, *1*, e1500692.
16. Lv, M.; Zhang, S.-C. Dielectric function, Friedel oscillation and plasmons in Weyl semimetals. *Int. J. Mod. Phys. B* **2013**, *27*, 1350177.
17. Xu, S.-Y.; Belopolski, I.; Alidoust, N.; Neupane, M.; Bian, G.; Zhang, C.; Sankar, R.; Chang, G.; Yuan, Z.; Lee, C.-C.; *et al.* Discovery of a Weyl Fermion semimetal and topological Fermi arcs. *Science* **2015**, *349*, 613–617.

18. Lv, B.Q.; Xu, N.; Weng, H.M.; Ma, J.Z.; Richard, P.; Huang, X.C.; Zhao, L.X.; Chen, G.F.; Matt, C.E.; Bisti, F.; *et al.* Observation of Weyl nodes in TaAs. *Nat. Phys.* **2015**, *11*, 724–727.

19. Huang, S.-M.; Xu, S.-Y.; Belopolski, I.; Lee, C.-C.; Chang, G.; Wang, B.K.; Alidoust, N.; Bian, G.; Neupane, M.; Zhang, C.; *et al.* A Weyl Fermion semimetal with surface Fermi arcs in the transition metal monopnictide TaAs class. *Nat. Commun.* **2015**, *6*, 7373.

20. Tan, B.S.; Hsu, Y.-T.; Zeng, B.; Ciomaga Hatnean, M.; Harrison, N.; Zhu, Z.; Hartstein, M.; Kiourlappou, M.; Srivastava, A.; Johannes, M.D.; *et al.* Unconventional Fermi surface in an insulating state. *Science* **2015**, *349*, 287–290.

21. Falkovsky, L.A. Optical properties of graphene and IV–VI semiconductors. *Phys.-Uspekhi* **2008**, *51*, 887–897.

22. Novoselov, K.S.; Jiang, D.; Schedin, F.; Booth, T.J.; Khotkevich, V.V.; Morozov, S.V.; Geim, A.K. Two-dimensional atomic crystals. *Proc. Natl. Acad. Sci. USA* **2005**, *102*, 10451–10453.

23. Keldysh, L.V. Coulomb interaction in thin semiconductor and semimetal films. *Lett. J. Exper. Theor. Phys.* **1979**, *29*, 716–719.

24. Dora, B.; Gulacsi, M.; Sodano, P. Majorana zero modes in graphene with trigonal warping. *Phys. Status Solidi RRL* **2009**, *3*, 169–171.

25. Elias, D.C.; Gorbachev, R.V.; Mayorov, A.S.; Morozov, S.V.; Zhukov, A.A.; Blake, P.; Ponomarenko, L.A.; Grigorieva, I.V.; Novoselov, K.S.; Guinea, F.; *et al.* Dirac cones reshaped by interaction effects in suspended graphene. *Nat. Phys.* **2012**, *8*, 172.

26. Du, X.; Skachko, I.; Barker, A.; Andrei, E.Y. Approaching ballistic transport in suspended graphene. *Nat. Nanotechnol.* **2008**, *3*, 491–495.

27. Cooper, D.R.; D'Anjou, B.; Ghattamaneni, N.A.; Harack, B.; Hilke, M.; Horth, A.; Majlis, N.; Massicotte, M.; Vandsburger, L.; Whiteway, E.; *et al.* Experimental Review of Graphene. *ISRN Condensed Matter Phys.* **2012**, *2012*, Article ID 501686.

28. San-Jose, P.; Lado, J. L.; Aguado, R.; Guinea, F.; Fernandez-Rossier, J. Majorana Zero Modes in Graphene. *Phys. Rev. X* **2015**, *5*, 041042.

29. Wang, J.R.; Liu, G.Z. Eliashberg theory of excitonic insulating transition in graphene. *J. Phys. Condensed Matter* **2011**, *23*, 155602.

30. Pessa, E. The Majorana Oscillator. *Electr. J. Theor. Phys.* **2006**, *3*, 285–292.

31. Majorana, E. Theory of Relativistic Particles with Arbitrary Intrinsic Moment. *Nuovo Cimento* **1932**, *9*, 335.

32. Peskin, M.E.; Schroeder, D.V. *An Introduction to Quantum Field Theory*; Addison-Wesley Publishing Company: Oxford, UK, 1995.

33. Simpao, V.A. Exact Solution of Majorana Equation via Heaviside Operational Ansatz. *Electr. J. Theor. Phys.* **2006**, *3*, 239–247.

34. Hainzl, C.; Lewin, M.; Sparber, C. Ground state properties of graphene in Hartree-Fock theory. *J. Math. Phys.* **2012**, *53*, 095220.

35. Grushevskaya, H.V.; Krylov, G.G. Charge Carriers Asymmetry and Energy Minigaps in Monolayer Graphene: Dirac–Hartree–Fock approach. *Int. J. Nonliner Phenom. Complex Syst.* **2013**, *16*, 189–208.

36. Grushevskaya, H.V.; Krylov, G.G. *Nanotechnology in the Security Systems, NATO Science for Peace and Security Series C: Environmental Security*; Bonča, J., Kruchinin, S., Eds.; Springer: Dordrecht, The Netherlands, 2015; Chapter 3.

37. Grushevskaya, H.V.; Krylov, G.G. Electronic Structure and Transport in Graphene: QuasiRelativistic Dirac–Hartree–Fock Self-Consistent Field Approximation. In *Graphene Science Handbook. Vol. 3: Electrical and Optical Properties*; Aliofkhazraei, M., Ali, N., Milne, W.I., Ozkan, C.S., Mitura, S., Gervasoni, J.L., Eds.; CRC Press—Taylor&Francis Group: Boca Raton, FL, USA, 2016.

38. Gribov, V.N. *Quantum Electrodynamics*; R & C Dynamics: Izhevsk, Russia, 2001. (In Russian)

39. Fock, V.A. *Principles of Quantum Mhechanics*; Science: Moscow, Russia, 1976. (In Russian)

40. Krylova, H.; Hursky, L. *Spin Polarization in Strong-Correlated Nanosystems*; LAP LAMBERT Academic Publishing, AV Akademikerverlag GmbH & Co.: Saarbrüken, Germany, 2013.

41. Semenoff, G.W. Condensed-matter simulation of a three-dimensional anomaly. *Phys. Rev. Lett.* **1984**, *53*, 2449.

42. Abergel, D.S.L.; Apalkov, V.; Berashevich, J.; Ziegler, K.; Chakraborty, T. Properties of graphene: A theoretical perspective. *Adv. Phys.* **2010**, *59*, 261.

43. Gusynin, V.P.; Sharapov, S.G.; Carbotte, J.P. AC Conductivity of Graphene: From Tight-binding model to 2 + 1-dimensional quantum electrodynamics. *Int. J. Mod. Phys. B* **2007**, *21*, 4611.

44. Wallace, P.R. The band theory of graphite. *Phys. Rev.* **1971**, *71*, 622–634.

45. Saito, R.; Dresselhaus, G.; Dresselhaus, M.S. *Physical Properties of Carbon Nanotubes*; Imperial: London, UK, 1998.

46. Reich, S.; Maultzsch, J.; Thomsen, C.; Ordejón, P. Tight-binding description of graphene. *Phys. Rev. B* **2002**, *66*, 035412.

47. Castro Neto, A.H.; Guinea, F.; Peres, N.M.; Novoselov, K.S.; Geim, A.K. The electronic properties of graphene. *Rev. Mod. Phys.* **2009**, *81*, 109.

48. Brihuega, I.; Mallet, P.; González-Herrero, H.; Trambly de Laissardière, G.; Ugeda, M.M.; Magaud, L.; Gomez-Rodríguez, J.M.; Ynduráin, F.; Veuillen, J.-Y. Unraveling the Intrinsic and Robust Nature of van Hove Singularities in Twisted Bilayer Graphene by Scanning Tunneling Microscopy and Theoretical Analysis. *Phys. Rev. Lett.* **2012**, *109*, 196802; Erratum in **2012**, *109*, 209905.

49. Andrei, E.Y.; Li, G.; Du, X. Electronic properties of graphene: A perspective from scanning tunneling microscopy and magnetotransport. *Rep. Prog. Phys.* **2012**, *75*, 056501.

50. Grushevskaya, H.V.; Krylov, G.; Gaisyonok, V.A.; Serow, D.V. Symmetry of Model N = 3 for Graphene with Charged Pseudo-Excitons. *J. Nonliner Phenom. Complex Sys.* **2015**, *18*, 81–98.

51. Zaslavsky, G. M.; Sagdeev, R.Z.; Usikov, D.A.; Chernikov, A.A. *Weak Chaos and Quasi-Regular Patterns*; Cambridge University Press: New York, NY, USA, 1991.

52. Guckenheimer, J.; Holmes, P. *Nonlinear Oscillations, Dynamical Systems, and Bifurcations of Vector Fields*; Springer-Verlag: New York, NY, USA, 1990; Volume 42.

© 2016 by the authors. Licensee MDPI, Basel, Switzerland. This article is an open access article distributed under the terms and conditions of the Creative Commons Attribution (CC BY) license (http://creativecommons.org/licenses/by/4.0/).

Chapter 3:
Papers Published by This Issue Editior in *Symmetry*

symmetry

[MDPI]

Article

Lorentz Harmonics, Squeeze Harmonics and Their Physical Applications

Young S. Kim [1,*] **and Marilyn E. Noz** [2]

[1] Center for Fundamental Physics, University of Maryland, College Park, MD 20742, USA
[2] Department of Radiology, New York University, New York, NY 10016, USA
* E-Mail: yskim@umd.edu; Tel.: 301-405-6024.

Received: 6 January 2011; in revised form: 7 February 2011 / Accepted: 11 February 2011 /
Published: 14 February 2011

Abstract: Among the symmetries in physics, the rotation symmetry is most familiar to us. It is known that the spherical harmonics serve useful purposes when the world is rotated. Squeeze transformations are also becoming more prominent in physics, particularly in optical sciences and in high-energy physics. As can be seen from Dirac's light-cone coordinate system, Lorentz boosts are squeeze transformations. Thus the squeeze transformation is one of the fundamental transformations in Einstein's Lorentz-covariant world. It is possible to define a complete set of orthonormal functions defined for one Lorentz frame. It is shown that the same set can be used for other Lorentz frames. Transformation properties are discussed. Physical applications are discussed in both optics and high-energy physics. It is shown that the Lorentz harmonics provide the mathematical basis for squeezed states of light. It is shown also that the same set of harmonics can be used for understanding Lorentz-boosted hadrons in high-energy physics. It is thus possible to transmit physics from one branch of physics to the other branch using the mathematical basis common to them.

Keywords: Lorentz harmonics; relativistic quantum mechanics; squeeze transformation; Dirac's efforts; hidden variables; Lorentz-covariant bound states; squeezed states of light

Classification: PACS 03.65.Ge, 03.65.Pm

1. Introduction

In this paper, we are concerned with symmetry transformations in two dimensions, and we are accustomed to the coordinate system specified by x and y variables. On the xy plane, we know how to make rotations and translations. The rotation in the xy plane is performed by the matrix algebra

$$\begin{pmatrix} x' \\ y' \end{pmatrix} = \begin{pmatrix} \cos\theta & -\sin\theta \\ \sin\theta & \cos\theta \end{pmatrix} \begin{pmatrix} x \\ y \end{pmatrix} \tag{1}$$

but we are not yet familiar with

$$\begin{pmatrix} z' \\ t' \end{pmatrix} = \begin{pmatrix} \cosh\eta & \sinh\eta \\ \sinh\eta & \cosh\eta \end{pmatrix} \begin{pmatrix} z \\ t \end{pmatrix} \tag{2}$$

Symmetry **2011**, 3, 16–36

We see this form when we learn Lorentz transformations, but there is a tendency in the literature to avoid this form, especially in high-energy physics. Since this transformation can also be written as

$$\begin{pmatrix} u' \\ v' \end{pmatrix} = \begin{pmatrix} \exp\left(\eta\right) & 0 \\ 0 & \exp\left(-\eta\right) \end{pmatrix} \begin{pmatrix} u \\ v \end{pmatrix}$$

(3)

with

$$u = \frac{z+t}{\sqrt{2}}, \qquad v = \frac{z-t}{\sqrt{2}}$$

(4)

where the variables u and v are expanded and contracted respectively, we call Equation (2) or Equation (3) **squeeze** transformations [1].

From the mathematical point of view, the symplectic group $Sp(2)$ contains both the rotation and squeeze transformations of Equations (1) and (2), and its mathematical properties have been extensively discussed in the literature [1,2]. This group has been shown to be one of the essential tools in quantum optics. From the mathematical point of view, the squeezed state in quantum optics is a harmonic oscillator representation of this $Sp(2)$ group [1].

We are interested in this paper in "squeeze transformations" of localized functions. We are quite familiar with the role of spherical harmonics in three dimensional rotations. We use there the same set of harmonics, but the rotated function has different linear combinations of those harmonics. Likewise, we are interested in a complete set of functions which will serve the same purpose for squeeze transformations. It will be shown that harmonic oscillator wave functions can serve the desired purpose. From the physical point of view, squeezed states define the squeeze or Lorentz harmonics.

In 2003, Giedke *et al.* used the Gaussian function to discuss the entanglement problems in information theory [3]. This paper allows us to use the oscillator wave functions to address many interesting current issues in quantum optics and information theory. In 2005, the present authors noted that the formalism of Lorentz-covariant harmonic oscillators leads to a space-time entanglement [4]. We developed the oscillator formalism to deal with hadronic phenomena observed in high-energy laboratories [5]. It is remarkable that the mathematical formalism of Giedke *et al.* is identical with that of our oscillator formalism.

While quantum optics or information theory is a relatively new branch of physics, the squeeze transformation has been the backbone of Einstein's special relativity. While Lorentz, Poincaré, and Einstein used the transformation of Equation (2) for Lorentz boosts, Dirac observed that the same equation can be written in the form of Equation (3) [6]. Unfortunately, this squeeze aspect of Lorentz boosts has not been fully addressed in high-energy physics dealing with particles moving with relativistic speeds.

Thus, we can call the same set of functions "squeeze harmonics" and "Lorentz harmonics" in quantum optics and high-energy physics respectively. This allows us to translate the physics of quantum optics or information theory into that of high-energy physics.

The physics of high-energy hadrons requires a Lorentz-covariant localized quantum system. This description requires one variable which is hidden in the present form of quantum mechanics. It is the time-separation variable between two constituent particles in a quantum bound system like the hydrogen atom, where the Bohr radius measures the separation between the proton and the electron. What happens to this quantity when the hydrogen atom is boosted and the time-separation variable starts playing its role? The Lorentz harmonics will allow us to address this question.

In Section 2, it is noted that the Lorentz boost of localized wave functions can be described in terms of one-dimensional harmonic oscillators. Thus, those wave functions constitute the Lorentz harmonics. It is also noted that the Lorentz boost is a squeeze transformation.

In Section 3, we examine Dirac's life-long efforts to make quantum mechanics consistent with special relativity, and present a Lorentz-covariant form of bound-state quantum mechanics. In Section 4, we construct a set of Lorentz-covariant harmonic oscillator wave functions, and show that they can be given a Lorentz-covariant probability interpretation.

In Section 5, the formalism is shown to constitute a mathematical basis for squeezed states of light, and for quantum entangled states. In Section 6, this formalism can serve as the language for Feynman's rest of the universe [7]. Finally, in Section 7, we show that the harmonic oscillator formalism can be applied to high-energy hadronic physics, and what we observe there can be interpreted in terms of what we learn from quantum optics.

2. Lorentz or Squeeze Harmonics

Let us start with the two-dimensional plane. We are quite familiar with rigid transformations such as rotations and translations in two-dimensional space. Things are different for non-rigid transformations such as a circle becoming an ellipse.

We start with the well-known one-dimensional harmonic oscillator eigenvalue equation

$$\frac{1}{2}\left[-\left(\frac{\partial}{\partial x}\right)^2 + x^2\right]\chi_n(x) = \left(n + \frac{1}{2}\right)\chi_n(x) \tag{5}$$

For a given value of integer n, the solution takes the form

$$\chi_n(x) = \left[\frac{1}{\sqrt{\pi}2^n n!}\right]^{1/2} H_n(x)\exp\left(\frac{-x^2}{2}\right) \tag{6}$$

where $H_n(x)$ is the Hermite polynomial of the n-th degree. We can then consider a set of functions with all integer values of n. They satisfy the orthogonality relation

$$\int \chi_n(x)\chi_{n'}(x) = \delta_{nn'} \tag{7}$$

This relation allows us to define $f(x)$ as

$$f(x) = \sum_n A_n \chi_n(x) \tag{8}$$

with

$$A_n = \int f(x)\chi_n(x)dx \tag{9}$$

Let us next consider another variable added to Equation (5), and the differential equation

$$\frac{1}{2}\left\{\left[-\left(\frac{\partial}{\partial x}\right)^2 + x^2\right] + \left[-\left(\frac{\partial}{\partial y}\right)^2 + y^2\right]\right\}\phi(x,y) = \lambda\phi(x,y) \tag{10}$$

This equation can be re-arranged to

$$\frac{1}{2}\left\{-\left(\frac{\partial}{\partial x}\right)^2 - \left(\frac{\partial}{\partial y}\right)^2 + x^2 + y^2\right\}\phi(x,y) = \lambda\phi(x,y) \tag{11}$$

This differential equation is invariant under the rotation defined in Equation (1). In terms of the polar coordinate system with

$$r = \sqrt{x^2 + y^2}, \qquad \tan\theta = \left(\frac{y}{x}\right) \tag{12}$$

this equation can be written:

$$\frac{1}{2}\left\{-\frac{\partial^2}{\partial r^2} - \frac{1}{r}\frac{\partial}{\partial r} - \frac{1}{r^2}\frac{\partial^2}{\partial\theta^2} + r^2\right\}\phi(r,\theta) = \lambda\phi(r,\theta) \tag{13}$$

and the solution takes the form

$$\phi(r,\theta) = e^{-r^2/2} R_{n,m}(r) \{ A_m \cos(m\theta) + B_n \sin(m\theta) \} \tag{14}$$

The radial equation should satisfy

$$\frac{1}{2} \left\{ -\frac{\partial^2}{\partial r^2} - \frac{1}{r} \frac{\partial}{\partial r} + \frac{m^2}{r^2} + r^2 \right\} R_{n,m}(r) = (n+m+1) R_{n,m}(r) \tag{15}$$

In the polar form of Equation (14), we can achieve the rotation of this function by changing the angle variable θ.

On the other hand, the differential equation of Equation (10) is separable in the x and y variables. The eigen solution takes the form

$$\phi_{n_x,n_y}(x,y) = \chi_{n_x}(x)\chi_{n_y}(y) \tag{16}$$

with

$$\lambda = n_x + n_y + 1 \tag{17}$$

If a function $f(x,y)$ is sufficiently localized around the origin, it can be expanded as

$$f(x,y) = \sum_{n_x,n_y} A_{n_x,n_y} \chi_{n_x}(x)\chi_{n_y}(y) \tag{18}$$

with

$$A_{n_x,n_y} = \int f(x,y)\chi_{n_x}(x)\chi_{n_y}(y)\,dx\,dy \tag{19}$$

If we rotate $f(x,y)$ according to Equation (1), it becomes $f(x^*,y^*)$, with

$$x^* = (\cos\theta)x - (\sin\theta)y, \qquad y^* = (\sin\theta)x + (\cos\theta)y \tag{20}$$

This rotated function can also be expanded in terms of $\chi_{n_x}(x)$ and $\chi_{n_y}(y)$:

$$f(x^*,y^*) = \sum_{n_x,n_y} A^*_{n_x,n_y} \chi_{n_x}(x)\chi_{n_y}(y) \tag{21}$$

with

$$A^*_{n_x,n_y} = \int f(x^*,y^*)\chi_{n_x}(x)\chi_{n_y}(y)\,dx\,dy \tag{22}$$

Next, let us consider the differential equation

$$\frac{1}{2} \left\{ -\left(\frac{\partial}{\partial z}\right)^2 + \left(\frac{\partial}{\partial t}\right)^2 + z^2 - t^2 \right\} \psi(z,t) = \lambda\psi(z,t) \tag{23}$$

Here we use the variables z and t, instead of x and y. Clearly, this equation can be also separated in the z and t coordinates, and the eigen solution can be written as

$$\psi_{n_z,n_t}(z,t) = \chi_{n_z}(z)\chi_{n_t}(z,t) \tag{24}$$

with

$$\lambda = n_z - n_t. \tag{25}$$

The oscillator equation is not invariant under coordinate rotations of the type given in Equation (1). It is however invariant under the squeeze transformation given in Equation (2).

The differential equation of Equation (23) becomes

$$\frac{1}{4}\left\{-\frac{\partial}{\partial u}\frac{\partial}{\partial v}+uv\right\}\psi(u,v)=\lambda\psi(u,v)\tag{26}$$

Both Equation (11) and Equation (23) are two-dimensional differential equations. They are invariant under rotations and squeeze transformations respectively. They take convenient forms in the polar and squeeze coordinate systems respectively as shown in Equation (13) and Equation (26).

The solutions of the rotation-invariant equation are well known, but the solutions of the squeeze-invariant equation are still strange to the physics community. Fortunately, both equations are separable in the Cartesian coordinate system. This allows us to study the latter in terms of the familiar rotation-invariant equation. This means that if the solution is sufficiently localized in the z and t plane, it can be written as

$$\psi(z,t)=\sum_{n_z,n_t}A_{n_z,n_t}\chi_{n_z}(z)\chi_{n_t}(t)\tag{27}$$

with

$$A_{n_z,n_t}=\int\psi(z,t)\chi_{n_z}(z)\chi_{n_t}(t)\,dz\,dt\tag{28}$$

If we squeeze the coordinate according to Equation (2),

$$\psi(z^*,t^*)=\sum_{n_z,n_t}A^*_{n_z,n_t}\chi_{n_z}(z)\chi_{n_t}(t)\tag{29}$$

with

$$A^*_{n_z,n_t}=\int\psi(z^*,t^*)\chi_{n_z}(z)\chi_{n_t}(t)\,dz\,dt\tag{30}$$

Here again both the original and transformed wave functions are linear combinations of the wave functions for the one-dimensional harmonic oscillator given in Equation (6).

The wave functions for the one-dimensional oscillator are well known, and they play important roles in many branches of physics. It is gratifying to note that they could play an essential role in squeeze transformations and Lorentz boosts, see Table (1). We choose to call them Lorentz harmonics or squeeze harmonics.

Table 1. Cylindrical and hyperbolic equations. The cylindrical equation is invariant under rotation while the hyperbolic equation is invariant under squeeze transformation

Equation	Invariant under	Eigenvalue
Cylindrical	Rotation	$\lambda=n_x+n_y+1$
Hyperbolic	Squeeze	$\lambda=n_x-n_y$

3. The Physical Origin of Squeeze Transformations

Paul A. M. Dirac made it his life-long effort to combine quantum mechanics with special relativity. We examine the following four of his papers.

- In 1927 [8], Dirac pointed out the time-energy uncertainty should be taken into consideration for efforts to combine quantum mechanics and special relativity.
- In 1945 [9], Dirac considered four-dimensional harmonic oscillator wave functions with

$$\exp\left\{-\frac{1}{2}\left(x^2+y^2+z^2+t^2\right)\right\}\tag{31}$$

and noted that this form is not Lorentz-covariant.

- In 1949 [6], Dirac introduced the light-cone variables of Equation (4). He also noted that the construction of a Lorentz-covariant quantum mechanics is equivalent to the construction of a representation of the Poncaré group.
- In 1963 [10], Dirac constructed a representation of the (3 + 2) deSitter group using two harmonic oscillators. This deSitter group contains three (3 + 1) Lorentz groups as its subgroups.

In each of these papers, Dirac presented the original ingredients which can serve as building blocks for making quantum mechanics relativistic. We combine those elements using Wigner's little groups [11] and and Feynman's observation of high-energy physics [12–14].

First of all, let us combine Dirac's 1945 paper and his light-cone coordinate system given in his 1949 paper. Since x and y variables are not affected by Lorentz boosts along the z direction in Equation (31), it is sufficient to study the Gaussian form

$$\exp\left\{-\frac{1}{2}\left(z^2 + t^2\right)\right\} \tag{32}$$

This form is certainly not invariant under Lorentz boost as Dirac noted. On the other hand, it can be written as

$$\exp\left\{-\frac{1}{2}\left(u^2 + v^2\right)\right\} \tag{33}$$

where u and v are the light-cone variables defined in Equation (4). If we make the Lorentz-boost or Lorentz squeeze according to Equation (3), this Gaussian form becomes

$$\exp\left\{-\frac{1}{2}\left(e^{-2\eta}u^2 + e^{2\eta}v^2\right)\right\} \tag{34}$$

If we write the Lorentz boost as

$$z' = \frac{z + \beta t}{\sqrt{1 - \beta^2}} \qquad t' = \frac{t + \beta z}{\sqrt{1 - \beta^2}} \tag{35}$$

where β is the the velocity parameter v/c, then β is related to η by

$$\beta = \tanh(\eta) \tag{36}$$

Let us go back to the Gaussian form of Equation (32), this expression is consistent with Dirac's earlier paper on the time-energy uncertainty relation [8]. According to Dirac, this is a c-number uncertainty relation without excitations. The existence of the time-energy uncertainty is illustrated in the first part of Figure 1.

In his 1927 paper, Dirac noted the space-time asymmetry in uncertainty relations. While there are no time-like excitations, quantum mechanics allows excitations along the z direction. How can we take care of problem?

If we suppress the excitations along the t coordinate, the normalized solution of this differential equation, Equation (24), is

$$\psi(z,t) = \left(\frac{1}{\pi 2^n n!}\right)^{1/2} H_n(z) \exp\left\{-\left(\frac{z^2 + t^2}{2}\right)\right\} \tag{37}$$

252

Symmetry **2011**, 3, 16–36

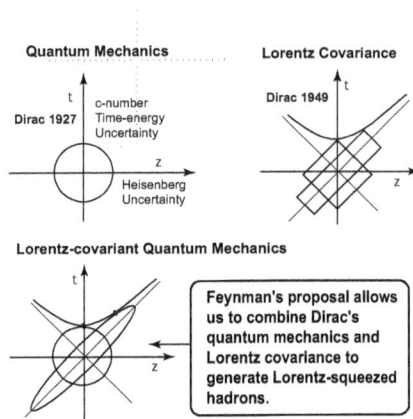

Figure 1. Space-time picture of quantum mechanics. In his 1927 paper, Dirac noted that there is a c-number time-energy uncertainty relation, in addition to Heisenberg's position-momentum uncertainty relations, with quantum excitations. This idea is illustrated in the first figure (upper left). In his 1949 paper, Dirac produced his light-cone coordinate system as illustrated in the second figure (upper right). It is then not difficult to produce the third figure, for a Lorentz-covariant picture of quantum mechanics. This Lorentz-squeeze property is observed in high-energy laboratories through Feynman's parton picture discussed in Section 7.

If we boost the coordinate system, the Lorentz-boosted wave functions should take the form

$$\psi_\eta^n(z,t) = \left(\frac{1}{\pi 2^n n!}\right)^{1/2} H_n\left(z\cosh\eta - t\sinh\eta\right)$$

$$\times \exp\left\{-\left[\frac{(\cosh 2\eta)(z^2+t^2) - 4(\sinh 2\eta)zt}{2}\right]\right\} \tag{38}$$

These are the solutions of the phenomenological equation of Feynman *et al.* [12] for internal motion of the quarks inside a hadron. In 1971, Feynman *et al.* wrote down a Lorentz-invariant differential equation of the form

$$\frac{1}{2}\left\{-\left(\frac{\partial}{\partial x_\mu}\right)^2 + x_\mu^2\right\}\psi\left(x_\mu\right) = (\lambda+1)\psi\left(x_\mu\right) \tag{39}$$

where x_μ is for the Lorentz-covariant space-time four vector. This oscillator equation is separable in the Cartesian coordinate system, and the transverse components can be seprated out. Thus, the differential of Equation (23) contains the essential element of the Lorentz-invariant Equation (39).

However, the solutions contained in Reference [12] are not normalizable and therefore cannot carry physical interpretations. It was shown later that there are normalizable solutions which constitute a representation of Wigner's $O(3)$-like little group [5,11,15]. The $O(3)$ group is the three-dimensional rotation group without a time-like direction or time-like excitations. This addresses Dirac's concern about the space-time asymmetry in uncertainty relations [8]. Indeed, the expression of Equation (37) is considered to be the representation of Wigner's little group for quantum bound states [11,15]. We shall return to more physical questions in Section 7.

4. Further Properties of the Lorentz Harmonics

Let us continue our discussion of quantum bound states using harmonic oscillators. We are interested in this section to see how the oscillator solution of Equation (37) would appear to a moving observer.

The variable z and t are the longitudinal and time-like separations between the two constituent particles. In terms of the light-cone variables defined in Equation (4), the solution of Equation (37) takes the form

$$\psi_0^n(z,t) = \left[\frac{1}{\pi n! 2^n}\right]^{1/2} H_n\left(\frac{u+v}{\sqrt{2}}\right) \exp\left\{-\left(\frac{u^2+v^2}{2}\right)\right\} \tag{40}$$

and

$$\psi_\eta^n(z,t) = \left[\frac{1}{\pi n! 2^n}\right]^{1/2} H_n\left(\frac{e^{-\eta}u + e^\eta v}{\sqrt{2}}\right) \exp\left\{-\left(\frac{e^{-2\eta}u^2 + e^{2\eta}v^2}{2}\right)\right\} \tag{41}$$

for the rest and moving hadrons respectively.

It is mathematically possible to expand this as [5,16]

$$\psi_\eta^n(z,t) = \left(\frac{1}{\cosh\eta}\right)^{(n+1)} \sum_k \left[\frac{(n+k)!}{n!k!}\right]^{1/2} (\tanh\eta)^k \chi_{n+k}(z)\chi_n(t) \tag{42}$$

where $\chi_n(z)$ is the n-th excited state oscillator wave function which takes the familiar form

$$\chi_n(z) = \left[\frac{1}{\sqrt{\pi} 2^n n!}\right]^{1/2} H_n(z) \exp\left(\frac{-z^2}{2}\right) \tag{43}$$

as given in Equation (6). This is an expansion of the Lorentz-boosted wave function in terms of the Lorentz harmonics.

If the hadron is at rest, there are no time-like oscillations. There are time-like oscillations for a moving hadron. This is the way in which the space and time variable mix covariantly. This also provides a resolution of the space-time asymmetry pointed out by Dirac in his 1927 paper [8]. We shall return to this question in Section 6. Our next question is whether those oscillator equations can be given a probability interpretation.

Even though we suppressed the excitations along the t direction in the hadronic rest frame, it is an interesting mathematical problem to start with the oscillator wave function with an excited state in the time variable. This problem was adressed by Rotbart in 1981 [17].

4.1. Lorentz-Invariant Orthogonality Relations

Let us consider two wave functions $\psi_\eta^n(z,t)$. If two covariant wave functions are in the same Lorentz frame and have thus the same value of η, the orthogonality relation

$$\left(\psi_\eta^{n'}, \psi_\eta^n\right) = \delta_{nn'} \tag{44}$$

is satisfied.

If those two wave functions have different values of η, we have to start with

$$\left(\psi_{\eta'}^{n'}, \psi_\eta^n\right) = \int \left(\psi_{\eta'}^{n'}(z,t)\right)^* \psi_\eta^n(z,t)\,dz\,dt \tag{45}$$

Without loss of generality, we can assume $\eta' = 0$ in the system where $\eta = 0$, and evaluate the integration. The result is [18]

$$\left(\psi_0^{n'}, \psi_\eta^n\right) = \int \left(\psi_0^{n'}(z,t)\right)^2 \psi_\eta^n(z,t)\,dx\,dt = \left(\sqrt{1-\beta^2}\right)^{(n+1)} \delta_{n,n'} \tag{46}$$

where $\beta = \tanh(\eta)$, as given in Equation (36). This is like the Lorentz-contraction property of a rigid rod. The ground state is like a single rod. Since we obtain the first excited state by applying a step-up operator, this state should behave like a multiplication of two rods, and a similar argument can be give to n rigid rods. This is illustrated in Figure 2.

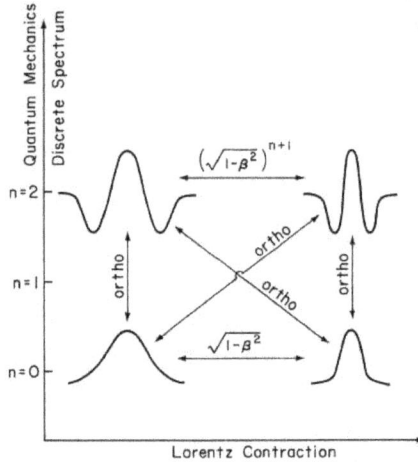

Figure 2. Orthogonality relations for the covariant harmonic oscillators. The orthogonality remains invariant. For the two wave functions in the orthogonality integral, the result is zero if they have different values of n. If both wave functions have the same value of n, the integral shows the Lorentz contraction property.

With these orthogonality properties, it is possible to give quantum probability interpretation in the Lorentz-covariant world, and it was so stated in our 1977 paper [19].

4.2. Probability Interpretations

Let us study the probability issue in terms of the one-dimensional oscillator solution of Equation (6) whose probability interpretation is indisputable. Let us also go back to the rotationally invariant differential equation of Equation (11). Then the product

$$\chi_{n_x}(x)\chi_{n_y}(y) \tag{47}$$

also has a probability interpretation with the eigen value $(n_x + n_y + 1)$. Thus the series of the form [1,5]

$$\phi_\eta^n(x,y) = \left(\frac{1}{\cosh\eta}\right)^{(n+1)} \sum_k \left[\frac{(n+k)!}{n!k!}\right]^{1/2} (\tanh\eta)^k \chi_{n+k}(x)\chi_n(y) \tag{48}$$

also has its probability interpretation, but it is not in an eigen state. Each term in this series has an eigenvalue $(2n + k + 1)$. The expectation value of Equation (11) is

$$\left(\frac{1}{\cosh\eta}\right)^{2(n+1)} \sum_k \frac{(2n+k+1)(n+k)!}{n!k!} (\tanh\eta)^{2k} \tag{49}$$

If we replace the variables x and y by z and t respectively in the above expression of Equation (48), it becomes the Lorentz-covariant wave function of Equation (42). Each term $\chi_{n+k}(z)\chi_k(t)$ in the series has the eigenvalue n. Thus the series is in the eigen state with the eigenvalue n.

This difference does not prevent us from importing the probability interpretation from that of Equation (48).

In the present covariant oscillator formalism, the time-separation variable can be separated from the rest of the wave function, and does not requite further interpretation. For a moving hadron, time-like excitations are mixed with longitudinal excitations. Is it possible to give a physical interpretation to those time-like excitations? To address this issue, we shall study in Section 5 two-mode squeezed states also based on the mathematics of Equation (48). There, both variables have their physical interpretations.

5. Two-Mode Squeezed States

Harmonic oscillators play the central role also in quantum optics. There the nth excited oscillator state corresponds to the n-photon state $|n>$. The ground state means the zero-photon or vacuum state $|0>$. The single-photon coherent state can be written as

$$|\alpha >= e^{-\alpha\alpha^*/2} \sum_n \frac{\alpha^n}{\sqrt{n!}}|n > \tag{50}$$

which can be written as [1]

$$|\alpha >= e^{-\alpha\alpha^*/2} \sum_n \frac{\alpha^n}{n!}\left(\hat{a}^\dagger\right)^n |0 >= \left\{e^{-\alpha\alpha^*/2}\right\} \exp\left\{\alpha\hat{a}^\dagger\right\}|0 > \tag{51}$$

This aspect of the single-photon coherent state is well known. Here we are dealing with one kind of photon, namely with a given momentum and polarization. The state $|n>$ means there are n photons of this kind.

Let us next consider a state of two kinds of photons, and write $|n_1, n_2 >$ as the state of n_1 photons of the first kind, and n_2 photons of the second kind [20]. We can then consider the form

$$\frac{1}{\cosh\eta} \exp\left\{(\tanh\eta)\hat{a}_1^\dagger\hat{a}_2^\dagger\right\}|0,0 > \tag{52}$$

The operator $\hat{a}_1^\dagger\hat{a}_2^\dagger$ was studied by Dirac in connection with his representation of the deSitter group, as we mentioned in Section 3. After making a Taylor expansion of Equation (52), we arrive at

$$\frac{1}{\cosh\eta} \sum_k (\tanh\eta)^k |k, k > \tag{53}$$

which is the squeezed vacuum state or two-photon coherent state [1,20]. This expression is the wave function of Equation (48) in a different notation. This form is also called the entangled Gaussian state of two photons [3] or the entangled oscillator state of space and time [4].

If we start with the n-particle state of the first photon, we obtain

$$\left[\frac{1}{\cosh\eta}\right]^{(n+1)} \exp\left\{(\tanh\eta)\hat{a}_1^\dagger\hat{a}_2^\dagger\right\}|n,0 >$$

$$= \left[\frac{1}{\cosh\eta}\right]^{(n+1)} \sum_k \left[\frac{(n+k)!}{n!k!}\right]^{1/2} (\tanh\eta)^k |k+n, k > \tag{54}$$

which is the wave function of Equation (42) in a different notation. This is the n-photon squeezed state [1].

Since the two-mode squeezed state and the covariant harmonic oscillators share the same set of mathematical formulas, it is possible to transmit physical interpretations from one to the other. For two-mode squeezed state, both photons carry physical interpretations, while the interpretation is yet to be given to the time-separation variable in the covariant oscillator formalism. It is clear from Equation (42) and Equation (54) that the time-like excitations are like the second-photon states.

What would happen if the second photon is not observed? This interesting problem was addressed by Yurke and Potasek [21] and by Ekert and Knight [22]. They used the density matrix formalism and integrated out the second-photon states. This increases the entropy and temperature of the system. We choose not to reproduce their mathematics, because we will be presenting the same mathematics in Section 6.

6. Time-Separation Variable in Feynman's Rest of the Universe

As was noted in the previous section, the time-separation variable has an important role in the covariant formulation of the harmonic oscillator wave functions. It should exist wherever the space separation exists. The Bohr radius is the measure of the separation between the proton and electron in the hydrogen atom. If this atom moves, the radius picks up the time separation, according to Einstein [23].

On the other hand, the present form of quantum mechanics does not include this time-separation variable. The best way we can interpret it at the present time is to treat this time-separation as a variable in Feynman's rest of the universe [24]. In his book on statistical mechanics [7], Feynman states

> When we solve a quantum-mechanical problem, what we really do is divide the universe into two
> parts - the system in which we are interested and the rest of the universe. We then usually act as if
> the system in which we are interested comprised the entire universe. To motivate the use of density
> matrices, let us see what happens when we include the part of the universe outside the system.

The failure to include what happens outside the system results in an increase of entropy. The entropy is a measure of our ignorance and is computed from the density matrix [25]. The density matrix is needed when the experimental procedure does not analyze all relevant variables to the maximum extent consistent with quantum mechanics [26]. If we do not take into account the time-separation variable, the result is an increase in entropy [27,28].

For the covariant oscillator wave functions defined in Equation (42), the pure-state density matrix is

$$\rho_\eta^n(z, t; z', t') = \psi_\eta^n(z, t)\psi_\eta^n(z', t') \tag{55}$$

which satisfies the condition $\rho^2 = \rho$:

$$\rho_\eta^n(z, t; x', t') = \int \rho_\eta^n(z, t; x'', t'')\rho_\eta^n(z'', t''; z', t')dz''dt'' \tag{56}$$

However, in the present form of quantum mechanics, it is not possible to take into account the time separation variables. Thus, we have to take the trace of the matrix with respect to the t variable. Then the resulting density matrix is

$$\rho_\eta^n(z, z') = \int \psi_\eta^n(z, t)\psi_\eta^n(z', t)dt$$

$$= \left(\frac{1}{\cosh \eta}\right)^{2(n+1)} \sum_k \frac{(n+k)!}{n!k!}(\tanh \eta)^{2k}\psi_{n+k}(z)\psi_{n+k}^*(z') \tag{57}$$

The trace of this density matrix is one, but the trace of ρ^2 is less than one, as

$$Tr\left(\rho^2\right) = \int \rho_\eta^n(z, z')\rho_\eta^n(z', z)dzdz'$$

$$= \left(\frac{1}{\cosh \eta}\right)^{4(n+1)} \sum_k \left[\frac{(n+k)!}{n!k!}\right]^2 (\tanh \eta)^{4k} \tag{58}$$

which is less than one. This is due to the fact that we do not know how to deal with the time-like separation in the present formulation of quantum mechanics. Our knowledge is less than complete.

The standard way to measure this ignorance is to calculate the entropy defined as

$$S = -Tr\left(\rho \ln(\rho)\right) \tag{59}$$

If we pretend to know the distribution along the time-like direction and use the pure-state density matrix given in Equation (55), then the entropy is zero. However, if we do not know how to deal with the distribution along t, then we should use the density matrix of Equation (57) to calculate the entropy, and the result is

$$S = 2(n+1)\left\{(\cosh\eta)^2\ln(\cosh\eta) - (\sinh\eta)\ln(\sinh\eta)\right\}$$

$$-\left(\frac{1}{\cosh\eta}\right)^{2(n+1)}\sum_k \frac{(n+k)!}{n!k!}\ln\left[\frac{(n+k)!}{n!k!}\right](\tanh\eta)^{2k} \tag{60}$$

In terms of the velocity v of the hadron,

$$S = -(n+1)\left\{\ln\left[1-\left(\frac{v}{c}\right)^2\right] + \frac{(v/c)^2\ln(v/c)^2}{1-(v/c)^2}\right\}$$

$$-\left[1-\left(\frac{1}{v}\right)^2\right]\sum_k \frac{(n+k)!}{n!k!}\ln\left[\frac{(n+k)!}{n!k!}\right]\left(\frac{v}{c}\right)^{2k} \tag{61}$$

Let us go back to the wave function given in Equation (41). As is illustrated in Figure 3, its localization property is dictated by the Gaussian factor which corresponds to the ground-state wave function. For this reason, we expect that much of the behavior of the density matrix or the entropy for the nth excited state will be the same as that for the ground state with $n = 0$. For this state, the density matrix and the entropy are

$$\rho(z,z') = \left(\frac{1}{\pi\cosh(2\eta)}\right)^{1/2}\exp\left\{-\frac{1}{4}\left[\frac{(z+z')^2}{\cosh(2\eta)} + (z-z')^2\cosh(2\eta)\right]\right\} \tag{62}$$

and

$$S = 2\left\{(\cosh\eta)^2\ln(\cosh\eta) - (\sinh\eta)^2\ln(\sinh\eta)\right\} \tag{63}$$

respectively. The quark distribution $\rho(z,z)$ becomes

$$\rho(z,z) = \left(\frac{1}{\pi\cosh(2\eta)}\right)^{1/2}\exp\left(\frac{-z^2}{\cosh(2\eta)}\right) \tag{64}$$

The width of the distribution becomes $\sqrt{\cosh\eta}$, and becomes wide-spread as the hadronic speed increases. Likewise, the momentum distribution becomes wide-spread [5,29]. This simultaneous increase in the momentum and position distribution widths is called the parton phenomenon in high-energy physics [13,14]. The position-momentum uncertainty becomes $\cosh\eta$. This increase in uncertainty is due to our ignorance about the physical but unmeasurable time-separation variable.

Let us next examine how this ignorance will lead to the concept of temperature. For the Lorentz-boosted ground state with $n = 0$, the density matrix of Equation (62) becomes that of the harmonic oscillator in a thermal equilibrium state if $(\tanh\eta)^2$ is identified as the Boltzmann factor [29]. For other states, it is very difficult, if not impossible, to describe them as thermal equilibrium states. Unlike the case of temperature, the entropy is clearly defined for all values of n. Indeed, the entropy in this case is derivable directly from the hadronic speed.

The time-separation variable exists in the Lorentz-covariant world, but we pretend not to know about it. It thus is in Feynman's rest of the universe. If we do not measure this time-separation, it becomes translated into the entropy.

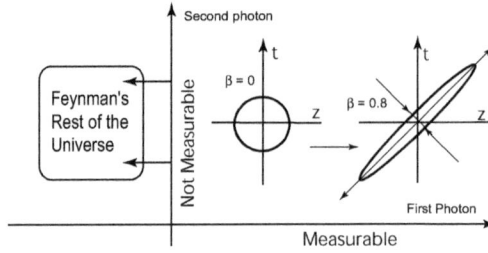

Figure 3. Localization property in the zt plane. When the hadron is at rest, the Gaussian form is concentrated within a circular region specified by $(z+t)^2 + (z-t)^2 = 1$. As the hadron gains speed, the region becomes deformed to $e^{-2\eta}(z+t)^2 + e^{2\eta}(z-t)^2 = 1$. Since it is not possible to make measurements along the t direction, we have to deal with information that is less than complete.

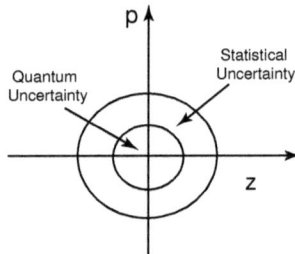

Figure 4. The uncertainty from the hidden time-separation coordinate. The small circle indicates the minimal uncertainty when the hadron is at rest. More uncertainty is added when the hadron moves. This is illustrated by a larger circle. The radius of this circle increases by $\sqrt{\cosh(2\eta)}$.

We can see the uncertainty in our measurement process from the Wigner function defined as

$$W(z,p) = \frac{1}{\pi} \int \rho(z+y, z-y)e^{2ipy} dy \tag{65}$$

After integration, this Wigner function becomes

$$W(z,p) = \frac{1}{\pi \cosh(2\eta)} \exp\left\{-\left(\frac{z^2 + p^2}{\cosh(2\eta)}\right)\right\} \tag{66}$$

This Wigner phase distribution is illustrated in Figure 4. The smaller inner circle corresponds to the minimal uncertainty of the single oscillator. The larger circle is for the total uncertainty including the statistical uncertainty from our failure to observe the time-separation variable. The two-mode squeezed state tells us how this happens. In the two-mode case, both the first and second photons are observable, but we can choose not to observe the second photon.

7. Lorentz-Covariant Quark Model

The hydrogen atom played the pivotal role while the present form of quantum mechanics was developed. At that time, the proton was in the absolute Galilean frame of reference, and it was thinkable that the proton could move with a speed close to that of light.

Also, at that time, both the proton and electron were point particles. However, the discovery of Hofstadter *et al.* changed the picture of the proton in 1955 [30]. The proton charge has its internal distribution. Within the framework of quantum electrodynamics, it is possible to calculate the Rutherford formula for the electron-proton scattering when both electron and proton are point particles. Because the proton is not a point particle, there is a deviation from the Rutherford formula. We describe this deviation using the formula called the "proton form factor" which depends on the momentum transfer during the electron-proton scattering.

Indeed, the study of the proton form factor has been and still is one of the central issues in high-energy physics. The form factor decreases as the momentum transfer increases. Its behavior is called the "dipole cut-off" meaning an inverse-square decrease, and it has been a challenging problem in quantum field theory and other theoretical models [31]. Since the emergence of the quark model in 1964 [32], the hadrons are regarded as quantum bound states of quarks with space-time wave functions. Thus, the quark model is responsible for explaining this form factor. There are indeed many papers written on this subject. We shall return to this problem in Subsection 7.2.

Another problem in high-energy physics is Feynman's parton picture [13,14]. If the hadron is at rest, we can approach this problem within the framework of bound-state quantum mechanics. If it moves with a speed close to that of light, it appears as a collection of an infinite number of partons, which interact with external signals incoherently. This phenomenon raises the question of whether the Lorentz boost destroys quantum coherence [33]. This leads to the concept of Feynman's decoherence [34]. We shall discuss this problem first.

7.1. Feynman's Parton Picture and Feynman's Decoherence

In 1969, Feynman observed that a fast-moving hadron can be regarded as a collection of many "partons" whose properties appear to be quite different from those of the quarks [5,14]. For example, the number of quarks inside a static proton is three, while the number of partons in a rapidly moving proton appears to be infinite. The question then is how the proton looking like a bound state of quarks to one observer can appear different to an observer in a different Lorentz frame? Feynman made the following systematic observations.

a. The picture is valid only for hadrons moving with velocity close to that of light.
b. The interaction time between the quarks becomes dilated, and partons behave as free independent particles.
c. The momentum distribution of partons becomes widespread as the hadron moves fast.
d. The number of partons seems to be infinite or much larger than that of quarks.

Because the hadron is believed to be a bound state of two or three quarks, each of the above phenomena appears as a paradox, particularly (b) and (c) together. How can a free particle have a wide-spread momentum distribution?

In order to address this question, let us go to Figure 5, which illustrates the Lorentz-squeeze property of the hadron as the hadron gains its speed. If we use the harmonic oscillator wave function, its momentum-energy wave function takes the same form as the space-time wave function. As the hadron gains its speed, both wave functions become squeezed.

As the wave function becomes squeezed, the distribution becomes wide-spread, the spring constant appear to become weaker. Consequently, the constituent quarks appear to become free particles.

If the constituent particles are confined in the narrow elliptic region, they become like massless particles. If those massless particles have a wide-spread momentum distribution, it is like a black-body radiation with infinite number of photon distributions.

We have addressed this question extensively in the literature, and concluded Gell-Mann's quark model and Feynman's parton model are two different manifestations of the same Lorentz-covariant quantity [19,35,36]. Thus coherent quarks and incoherent partons are perfectly consistent within the framework of quantum mechanics and special relativity [33]. Indeed, this defines Feynman's decoherence [34].

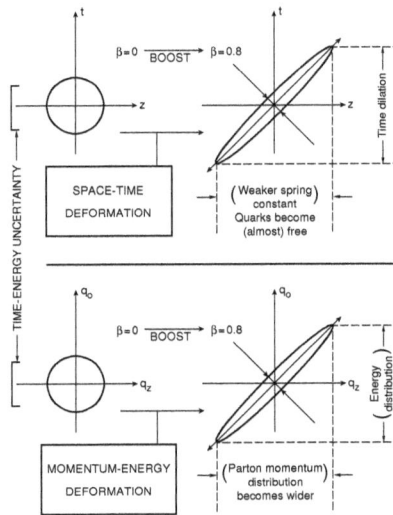

Figure 5. Lorentz-squeezed space-time and momentum-energy wave functions. As the hadron's speed approaches that of light, both wave functions become concentrated along their respective positive light-cone axes. These light-cone concentrations lead to Feynman's parton picture.

More recently, we were able to explain this decoherence problem in terms of the interaction time among the constituent quarks and the time required for each quark to interact with external signals [4].

7.2. Proton Form Factors and Lorentz Coherence

As early as in 1970, Fujimura *et al.* calculated the electromagnetic form factor of the proton using the wave functions given in this paper and obtained the so-called "dipole" cut-off of the form factor [37]. At that time, these authors did not have a benefit of the differential equation of Feynman and his co-authors [12]. Since their wave functions can now be given a bona-fide covariant probability interpretation, their calculation could be placed between the two limiting cases of quarks and partons.

Even before the calculation of Fujimura *et al.* in 1965, the covariant wave functions were discussed by various authors [38–40]. In 1970, Licht and Pagnamenta also discussed this problem with Lorentz-contracted wave functions [41].

In our 1973 paper [42], we attempted to explain the covariant oscillator wave function in terms of the coherence between the incoming signal and the width of the contracted wave function. This aspect was explained in terms of the overlap of the energy-momentum wave function in our book [5].

In this paper, we would like to go back to the coherence problem we raised in 1973, and follow-up on it. In the Lorentz frame where the momentum of the proton has the opposite signs before and after the collision, the four-momentum transfer is

$$(p, E) - (-p, E) = (2p, 0) \tag{67}$$

where the proton comes along the z direction with its momentum p, and its energy $\sqrt{p^2 + m^2}$.

Then the form factor becomes

$$F(p) = \int e^{2ipz} \left(\psi_\eta(z,t)\right)^* \psi_{-\eta}(z,t) \, dz \, dt \tag{68}$$

If we use the ground-state oscillator wave function, this integral becomes

$$\frac{1}{\pi} \int e^{2ipz} \exp\left\{-\cosh(2\eta)\left(z^2 + t^2\right)\right\} dz \, dt \tag{69}$$

After the t integration, this integral becomes

$$\frac{1}{\sqrt{\pi \cosh(2\eta)}} \int e^{2ipz} \exp\left\{-z^2 \cosh(2\eta)\right\} dz \tag{70}$$

The integrand is a product of a Gaussian factor and a sinusoidal oscillation. The width of the Gaussian factor shrinks by $1/\sqrt{\cosh(2\eta)}$, which becomes $\exp(-\eta)$ as η becomes large. The wave length of the sinusoidal factor is inversely proportional to the momentum p. The wave length decreases also at the rate of $\exp(-\eta)$. Thus, the rate of the shrinkage is the same for both the Gaussian and sinusoidal factors. For this reason, the cutoff rate of the form factor of Equation (68) should be less than that for

$$\int e^{2ipz} \left(\psi_0(z,t)\right)^* \psi_0(z,t) \, dz \, dt = \frac{1}{\sqrt{\pi}} \int e^{2ipz} \exp\left(-z^2\right) dz \tag{71}$$

which corresponds to the form factor without the squeeze effect on the wave function. The integration of this expression lead to $\exp(-p^2)$, which corresponds to an exponential cut-off as p^2 becomes large.

Let us go back to the form factor of Equation (68). If we complete the integral, it becomes

$$F(p) = \frac{1}{\cosh(2\eta)} \exp\left\{\frac{-p^2}{\cosh(2\eta)}\right\} \tag{72}$$

As p^2 becomes large, the Gaussian factor becomes a constant. However, the factor $1/\cosh(2\eta)$ leads the form factor decrease of $1/p^2$, which is a much slower decrease than the exponential cut-off without squeeze effect.

There still is a gap between this mathematical formula and the observed experimental data. Before looking at the experimental curve, we have to realize that there are three quarks inside the hadron with two oscillator mode. This will lead to a $\left(1/p^2\right)^2$ cut-off, which is commonly called the dipole cut-off in the literature.

There is still more work to be done. For instance, the effect of the quark spin should be addressed [43,44]. Also there are reports of deviations from the exact dipole cut-off [45]. There have been attempts to study the form factors based on the four-dimensional rotation group [46], and also on the lattice QCD [47].

Yet, it is gratifying to note that the effect of Lorentz squeeze lead to the polynomial decrease in the momentum transfer, thanks to the Lorentz coherence illustrated in Figure 6. We started our logic from the fundamental principles of quantum mechanics and relativity.

8. Conclusions

In this paper, we presented one mathematical formalism applicable both to the entanglement problems in quantum optics [3] and to high-energy hadronic physics [4]. The formalism is based on harmonic oscillators familiar to us. We have presented a complete orthonormal set with a Lorentz-covariant probability interpretation.

Since both branches of physics share the same mathematical base, it is possible to translate physics from one branch to the other. In this paper, we have given a physical interpretation to the

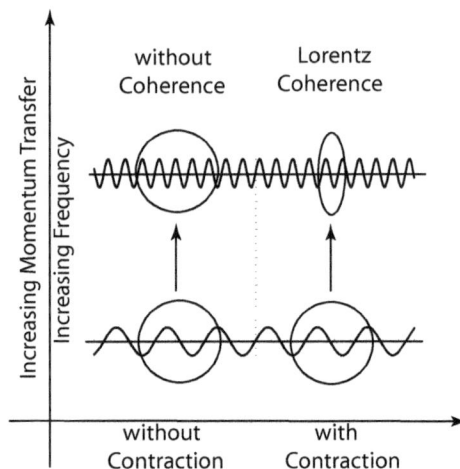

Figure 6. Coherence between the wavelength and the proton size. As the momentum transfer increases, the external signal sees Lorentz-contracting proton distribution. On the other hand, the wavelength of the signal also decreases. Thus, the cutoff is not as severe as the case where the proton distribution is not contracted.

time-separation variable as a hidden variable in Feynman's rest of the universe, in terms of the two-mode squeezed state where both photons are observable.

This paper is largely a review paper with an organization to suit the current interest in physics. For instance, the concepts of entanglement and decoherecne did not exist when those original papers were written. Furthermore, the probability interpretation given in Subsection 4.2 has not been published before.

The rotation symmetry plays its role in all branches of physics. We noted that the squeeze symmetry plays active roles in two different subjects of physics. It is possible that the squeeze transformation can serve useful purposes in many other fields, although we are not able to specify them at this time.

References

1. Kim, Y.S.; Noz, M.E. *Phase Space Picture of Quantum Mechanics*; World Scientific Publishing Company: Singapore, 1991.
2. Guillemin, V.; Sternberg, S. *Symplectic Techniques in Physics*; Cambridge University: Cambridge, UK, 1984.
3. Giedke, G.; Wolf, M.M.; Krger, O; Werner, R.F.; Cirac, J.J. Entanglement of formation for symmetric Gaussian states. *Phys. Rev. Lett.* **2003**, *91*, 107901-107904.
4. Kim, Y.S.; Noz, M.E. Coupled oscillators, entangled oscillators, and Lorentz-covariant harmonic oscillators. *J. Opt. B: Quantum Semiclass. Opt.* **2005**, *7*, S458-S467.
5. Kim, Y.S.; Noz, M.E. *Theory and Applications of the Poincaré Group* D; Reidel Publishing Company: Dordrecht, The Netherlands, 1986.
6. Dirac, P.A.M. Forms of Relativistic Dynamics. *Rev. Mod. Phys.* **1949**, *21*, 392-399.
7. Feynman, R.P. *Statistical Mechanics*; Benjamin/Cummings: Reading, MA, USA, 1972.
8. Dirac, P.A.M. The Quantum Theory of the Emission and Absorption of Radiation. *Proc. Roy. Soc. (London)* **1927**, *A114*, 243-265.
9. Dirac, P.A.M. Unitary Representations of the Lorentz Group. *Proc. Roy. Soc. (London)* **1945**, *A183*, 284-295.
10. Dirac, P.A.M. A Remarkable Representation of the 3 + 2 de Sitter Group. *J. Math. Phys.* **1963**, *4*, 901-909.
11. Wigner, E. On Unitary Representations of the Inhomogeneous Lorentz Group. *Ann. Math.* **1939**, *40*, 149-204.

12. Feynman, R.P.; Kislinger, M.; Ravndal F. Current Matrix Elements from a Relativistic Quark Model. *Phys. Rev. D* **1971**, *3*, 2706-2732.

13. Feynman, R.P. Very High-Energy Collisions of Hadrons. *Phys. Rev. Lett.* **1969**, *23*, 1415-1417.

14. Feynman, R.P. The Behavior of Hadron Collisions at Extreme Energies in High-Energy Collisions. In *Proceedings of the Third International Conference*; Gordon and Breach: New York, NY, USA, 1969; pp. 237-249.

15. Kim, Y.S.; Noz, M.E.; Oh, S.H. Representations of the Poincaré group for relativistic extended hadrons. *J. Math. Phys.* **1979**, *20*, 1341-1344.

16. Kim, Y.S.; Noz, M.E.; Oh, S.H.; A Simple Method for Illustrating the Difference between the Homogeneous and Inhomogeneous Lorentz Groups. *Am. J. Phys.* **1979**, *47*, 892-897.

17. Rotbart, F.C. Complete orthogonality relations for the covariant harmonic oscillator. *Phys. Rev. D* **1981**, *12*, 3078-3090.

18. Ruiz, M.J. Orthogonality relations for covariant harmonic oscillator wave functions. *Phys. Rev. D* **1974**, *10*, 4306-4307.

19. Kim, Y.S.; Noz, M.E. Covariant Harmonic Oscillators and the Parton Picture. *Phys. Rev. D* **1977**, *15*, 335-338.

20. Yuen, H.P. Two-photon coherent states of the radiation field. *Phys. Rev. A* **1976**, *13*, 2226-2243.

21. Yurke, B.; Potasek, M. Obtainment of Thermal Noise from a Pure State. *Phys. Rev. A* **1987**, *36*, 3464-3466.

22. Ekert, A.K.; Knight, P.L. Correlations and squeezing of two-mode oscillations. *Am. J. Phys.* **1989**, *57*, 692-697.

23. Kim, Y.S.; Noz, M.E. The Question of Simultaneity in Relativity and Quantum Mechanics. In *Quantum Theory: Reconsideration of Foundations–3*; Adenier, G., Khrennikov, A., Nieuwenhuizen, T.M., Eds.; AIP Conference Proceedings 180, American Institute of Physics, College Park, MD, USA, 2006; pp. 168-178.

24. Han, D.; Kim, Y.S.; Noz, M.E. Illustrative Example of Feynman's Rest of the Universe. *Am. J. Phys.* **1999**, *67*, 61-66.

25. von Neumann, J. *Die Mathematische Grundlagen der Quanten-mechanik*; Springer: Berlin, Germany, 1932.

26. Fano, U. Description of States in Quantum Mechanics by Density Matrix and Operator Techniques. *Rev. Mod. Phys.* **1967**, *29*, 74-93.

27. Kim, Y.S.; Wigner, E.P. Entropy and Lorentz Transformations. *Phys. Lett. A* **1990**, *147*, 343-347.

28. Kim, Y.S. Coupled oscillators and Feynman's three papers. *J. Phys. Conf. Ser.* **2007**, *70*, 012010: 1-19.

29. Han, D.; Kim, Y.S.; Noz, M.E. Lorentz-Squeezed Hadrons and Hadronic Temperature. *Phys. Lett. A* **1990**, *144*, 111-115.

30. Hofstadter, R.; McAllister, R.W. Electron Scattering from the Proton. *Phys. Rev.* **1955**, *98*, 217-218.

31. Frazer, W.; Fulco, J. Effect of a Pion-Pion Scattering Resonance on Nucleon Structure. *Phys. Rev. Lett.* **1960**, *2*, 365-368.

32. Gell-Mann, M. Nonleptonic Weak Decays and the Eightfold Way. *Phys. Lett.* **1964**, *12*, 155-156.

33. Kim, Y.S. Does Lorentz Boost Destroy Coherence? *Fortschr. der Physik* **1998**, *46*, 713-724.

34. Kim, Y.S.; Noz, M.E. Feynman's Decoherence. *Optics Spectro.* **2003**, *47*, 733-740.

35. Hussar, P.E. Valons and harmonic oscillators. *Phys. Rev. D* **1981**, *23*, 2781-2783.

36. Kim, Y.S. Observable gauge transformations in the parton picture. *Phys. Rev. Lett.* **1989**, *63*, 348-351.

37. Fujimura, K.; Kobayashi, T.; Namiki, M. Nucleon Electromagnetic Form Factors at High Momentum Transfers in an Extended Particle Model Based on the Quark Model. *Prog. Theor. Phys.* **1970**, *43*, 73-79.

38. Yukawa, H. Structure and Mass Spectrum of Elementary Particles. I. General Considerations. *Phys. Rev.* **1953**, *91*, 415-416.

39. Markov, M. On Dynamically Deformable Form Factors in the Theory Of Particles. *Suppl. Nuovo Cimento* **1956**, *3*, 760-772.

40. Ginzburg, V.L.; Man'ko, V.I. Relativistic oscillator models of elementary particles. *Nucl. Phys.* **1965**, *74*, 577-588.

41. Licht, A.L.; Pagnamenta, A. Wave Functions and Form Factors for Relativistic Composite Particles I. *Phys. Rev. D* **1970**, *2*, 1150-1156.

42. Kim, Y.S.; Noz, M.E. Covariant harmonic oscillators and the quark model. *Phys. Rev. D* **1973**, *8*, 3521-3627.

43. Lipes, R. Electromagnetic Excitations of the Nucleon in a Relativistic Quark Model. *Phys. Rev. D* **1972**, *5*, 2849-2863.

44. Henriques, A.B.; Keller, B.H.; Moorhouse, R.G. General three-spinor wave functions and the relativistic quark model. *Ann. Phys. (NY)* **1975**, *93*, 125-151.

45. Punjabi, V.; Perdrisat, C.F.; Aniol, K.A.; Baker, F.T.; Berthot, J.; Bertin, P.Y.; Bertozzi, W.; Besson, A.; Bimbot, L.; Boeglin, W.U.; *et al.* Proton elastic form factor ratios to Q2 = 3.5 GeV2 by polarization transfer. *Phys. Rev. C* **2005**, *71*, 055202-27.

46. Alkofer, R.; Holl, A.; Kloker, M.; Karssnigg A.; Roberts, C.D. On Nucleon Electromagnetic Form Factors. *Few-Body Sys.* **2005**, *37*, 1-31.

47. Matevosyan, H.H.; Thomas, A.W.; Miller, G.A. Study of lattice QCD form factors using the extended Gari-Krumpelmann model. *Phys. Rev. C* **2005**, *72*, 065204-5.

© 2011 by the authors. Licensee MDPI, Basel, Switzerland. This article is an open access article distributed under the terms and conditions of the Creative Commons Attribution (CC BY) license (http://creativecommons.org/licenses/by/4.0/).

symmetry

MDPI

Article

Dirac Matrices and Feynman's Rest of the Universe

Young S. Kim [1,*] **and Marilyn E. Noz** [2]

[1] Center for Fundamental Physics, University of Maryland, College Park, MD 20742, USA
[2] Department of Radiology, New York University, New York, NY 10016, USA; marilyne.noz@gmail.com
[*] Author to whom correspondence should be addressed; yskim@umd.edu; Tel.: +1-301-937-6306.

Received: 25 June 2012; in revised form: 6 October 2012; Accepted: 23 October 2012; Published: 30 October 2012

Abstract: There are two sets of four-by-four matrices introduced by Dirac. The first set consists of fifteen Majorana matrices derivable from his four γ matrices. These fifteen matrices can also serve as the generators of the group $SL(4,r)$. The second set consists of ten generators of the $Sp(4)$ group which Dirac derived from two coupled harmonic oscillators. It is shown possible to extend the symmetry of $Sp(4)$ to that of $SL(4,r)$ if the area of the phase space of one of the oscillators is allowed to become smaller without a lower limit. While there are no restrictions on the size of phase space in classical mechanics, Feynman's rest of the universe makes this $Sp(4)$-to-$SL(4,r)$ transition possible. The ten generators are for the world where quantum mechanics is valid. The remaining five generators belong to the rest of the universe. It is noted that the groups $SL(4,r)$ and $Sp(4)$ are locally isomorphic to the Lorentz groups $O(3,3)$ and $O(3,2)$ respectively. This allows us to interpret Feynman's rest of the universe in terms of space-time symmetry.

Keywords: Dirac gamma matrices; Feynman's rest of the universe; two coupled oscilators; Wigner's phase space; non-canonical transformations; group generators; $SL(4, r)$ isomorphic $O(3, 3)$; quantum mechanics interpretation

1. Introduction

In 1963, Paul A. M. Dirac published an interesting paper on the coupled harmonic oscillators [1]. Using step-up and step-down operators, Dirac was able to construct ten operators satisfying a closed set of commutation relations. He then noted that this set of commutation relations can also be used as the Lie algebra for the $O(3,2)$ de Sitter group applicable to three space and two time dimensions. He noted further that this is the same as the Lie algebra for the four-dimensional symplectic group $Sp(4)$.

His algebra later became the fundamental mathematical language for two-mode squeezed states in quantum optics [2–5]. Thus, Dirac's ten oscillator matrices play a fundamental role in modern physics.

In the Wigner phase-space representation, it is possible to write the Wigner function in terms of two position and two momentum variables. It was noted that those ten operators of Dirac can be translated into the operators with these four variables [4,6], which then can be written as four-by-four matrices. There are thus ten four-by-four matrices. We shall call them Dirac's oscillator matrices. They are indeed the generators of the symplectic group $Sp(4)$.

We are quite familiar with four Dirac matrices for the Dirac equation, namely $\gamma_1, \gamma_2, \gamma_3$, and γ_0. They all become imaginary in the Majorana representation. From them we can construct fifteen linearly independent four-by-four matrices. It is known that these four-by-four matrices can serve as the generators of the $SL(4,r)$ group [6,7]. It is also known that this $SL(4,r)$ group is locally isomorphic to the Lorentz group $O(3,3)$ applicable to the three space and three time dimensions [6,7].

There are now two sets of the four-by-four matrices constructed by Dirac. The first set consists of his ten oscillator matrices, and there are fifteen γ matrices coming from his Dirac equation. There is

Symmetry **2012**, *4*, 626–643

thus a difference of five matrices. The question is then whether this difference can be explained within the framework of the oscillator formalism with tangible physics.

It was noted that his original $O(3,2)$ symmetry can be extended to that of $O(3,3)$ Lorentz group applicable to the six dimensional space consisting of three space and three time dimensions. This requires the inclusion of non-canonical transformations in classical mechanics [6]. These non-canonical transformations cannot be interpreted in terms of the present form of quantum mechanics.

On the other hand, we can use this non-canonical effect to illustrate the concept of Feynman's rest of the universe. This oscillator system can serve as two different worlds. The first oscillator is the world in which we do quantum mechanics, and the second is for the rest of the universe. Our failure to observe the second oscillator results in the increase in the size of the Wigner phase space, thus increasing the entropy [8].

Instead of ignoring the second oscillator, it is of interest to see what happens to it. In this paper, it is shown that Planck's constant does not have a lower limit. This is allowed in classical mechanics, but not in quantum mechanics.

Indeed, Dirac's ten oscillator matrices explain the quantum world for both oscillators. The set of Dirac's fifteen γ matrices contains his ten oscillator matrices as a subset. We discuss in this paper the physics of this difference.

In Section 2, we start with Dirac's four γ matrices in the Majorana representation and construct all fifteen four-by-four matrices applicable to the Majorana form of the Dirac spinors. Section 3 reproduces Dirac's derivation of the $O(3,2)$ symmetry with ten generators from two coupled oscillators. This group is locally isomorphic to $Sp(4)$, which allows canonical transformations in classical mechanics.

In Section 4, we translate Dirac's formalism into the language of the Wigner phase space. This allows us to extend the $Sp(4)$ symmetry into the non-canonical region in classical mechanics. The resulting symmetry is that of $SL(4, r)$, isomorphic to that of the Lorentz group $O(3,3)$ with fifteen generators. This allows us to establish the correspondence between Dirac's Majorana matrices with those $SL(4, r)$ four-by-four matrices applicable to the two oscillator system, as well as the fifteen six-by-six matrices that serve as the generators of the $O(3,3)$ group.

Finally, in Section 5, it is shown that the difference between the ten oscillator matrices and the fifteen Majorana matrix can serve as an illustrative example of Feynman's rest of the universe [8,9].

2. Dirac Matrices in the Majorana Representation

Since all the generators for the two coupled oscillator system can be written as four-by-four matrices with imaginary elements, it is convenient to work with Dirac matrices in the Majorana representation, where the all the elements are imaginary [7,10,11]. In the Majorana representation, the four Dirac γ matrices are

$$\gamma_1 = i \begin{pmatrix} \sigma_3 & 0 \\ 0 & \sigma_3 \end{pmatrix}, \gamma_2 = \begin{pmatrix} 0 & -\sigma_2 \\ \sigma_2 & 0 \end{pmatrix}$$

$$\gamma_3 = -i \begin{pmatrix} \sigma_1 & 0 \\ 0 & \sigma_1 \end{pmatrix}, \gamma_0 = \begin{pmatrix} 0 & \sigma_2 \\ \sigma_2 & 0 \end{pmatrix} \tag{1}$$

where

$$\sigma_1 = \begin{pmatrix} 0 & 1 \\ 1 & 0 \end{pmatrix}, \sigma_2 = \begin{pmatrix} 0 & -i \\ i & 0 \end{pmatrix}, \sigma_1 = \begin{pmatrix} 1 & 0 \\ 0 & -1 \end{pmatrix}$$

These γ matrices are transformed like four-vectors under Lorentz transformations. From these four matrices, we can construct one pseudo-scalar matrix

$$\gamma_5 = i\gamma_0\gamma_1\gamma_2\gamma_3 = \begin{pmatrix} \sigma_2 & 0 \\ 0 & -\sigma_2 \end{pmatrix} \tag{2}$$

and a pseudo vector $i\gamma_5\gamma_\mu$ consisting of

$$i\gamma_5\gamma_1 = i\begin{pmatrix} -\sigma_1 & 0 \\ 0 & \sigma_1 \end{pmatrix}, i\gamma_5\gamma_2 = -i\begin{pmatrix} 0 & I \\ I & 0 \end{pmatrix}$$

$$i\gamma_5\gamma_0 = i\begin{pmatrix} 0 & I \\ -I & 0 \end{pmatrix}, i\gamma_5\gamma_3 = i\begin{pmatrix} -\sigma_3 & 0 \\ 0 & +\sigma_3 \end{pmatrix} \tag{3}$$

In addition, we can construct the tensor of the γ as

$$T_{\mu\nu} = \frac{i}{2}\left(\gamma_\mu\gamma_\nu - \gamma_\nu\gamma_\mu\right) \tag{4}$$

This antisymmetric tensor has six components. They are

$$i\gamma_0\gamma_1 = -i\begin{pmatrix} 0 & \sigma_1 \\ \sigma_1 & 0 \end{pmatrix}, i\gamma_0\gamma_2 = -i\begin{pmatrix} -I & 0 \\ 0 & I \end{pmatrix}, i\gamma_0\gamma_3 = -i\begin{pmatrix} 0 & \sigma_3 \\ \sigma_3 & 0 \end{pmatrix} \tag{5}$$

and

$$i\gamma_1\gamma_2 = i\begin{pmatrix} 0 & -\sigma_1 \\ \sigma_1 & 0 \end{pmatrix}, i\gamma_2\gamma_3 = -i\begin{pmatrix} 0 & -\sigma_3 \\ \sigma_3 & 0 \end{pmatrix}, i\gamma_3\gamma_1 = \begin{pmatrix} \sigma_2 & 0 \\ 0 & \sigma_2 \end{pmatrix} \tag{6}$$

There are now fifteen linearly independent four-by-four matrices. They are all traceless and their components are imaginary [7]. We shall call these Dirac's Majorana matrices.

In 1963 [1], Dirac constructed another set of four-by-four matrices from two coupled harmonic oscillators, within the framework of quantum mechanics. He ended up with ten four-by-four matrices. It is of interest to compare his oscillator matrices and his fifteen Majorana matrices.

3. Dirac's Coupled Oscillators

In his 1963 paper [1], Dirac started with the Hamiltonian for two harmonic oscillators. It can be written as

$$H = \frac{1}{2}\left(p_1^2 + x_1^2\right) + \frac{1}{2}\left(p_2^2 + x_2^2\right) \tag{7}$$

The ground-state wave function for this Hamiltonian is

$$\psi_0(x_1, x_2) = \frac{1}{\sqrt{\pi}} \exp\left\{-\frac{1}{2}\left(x_1^2 + x_2^2\right)\right\} \tag{8}$$

We can now consider unitary transformations applicable to the ground-state wave function of Equation (8), and Dirac noted that those unitary transformations are generated by [1]

$$
\begin{aligned}
L_1 &= \frac{1}{2}\left(a_1^\dagger a_2 + a_2^\dagger a_1\right), L_2 = \frac{1}{2i}\left(a_1^\dagger a_2 - a_2^\dagger a_1\right) \\
L_3 &= \frac{1}{2}\left(a_1^\dagger a_1 - a_2^\dagger a_2\right), S_3 = \frac{1}{2}\left(a_1^\dagger a_1 + a_2 a_2^\dagger\right) \\
K_1 &= -\frac{1}{4}\left(a_1^\dagger a_1^\dagger + a_1 a_1 - a_2^\dagger a_2^\dagger - a_2 a_2\right) \\
K_2 &= \frac{i}{4}\left(a_1^\dagger a_1^\dagger - a_1 a_1 + a_2^\dagger a_2^\dagger - a_2 a_2\right) \\
K_3 &= \frac{1}{2}\left(a_1^\dagger a_2^\dagger + a_1 a_2\right) \\
Q_1 &= -\frac{i}{4}\left(a_1^\dagger a_1^\dagger - a_1 a_1 - a_2^\dagger a_2^\dagger + a_2 a_2\right) \\
Q_2 &= -\frac{1}{4}\left(a_1^\dagger a_1^\dagger + a_1 a_1 + a_2^\dagger a_2^\dagger + a_2 a_2\right) \\
Q_3 &= \frac{i}{2}\left(a_1^\dagger a_2^\dagger - a_1 a_2\right)
\end{aligned} \tag{9}
$$

where $a^\dagger$ and a are the step-up and step-down operators applicable to harmonic oscillator wave functions. These operators satisfy the following set of commutation relations.

$$[L_i, L_j] = i\epsilon_{ijk}L_k, \ [L_i, K_j] = i\epsilon_{ijk}K_k, \ [L_i, Q_j] = i\epsilon_{ijk}Q_k$$

$$[K_i, K_j] = [Q_i, Q_j] = -i\epsilon_{ijk}L_k, \ [L_i, S_3] = 0$$

$$[K_i, Q_j] = -i\delta_{ij}S_3, \ [K_i, S_3] = -iQ_i, \ [Q_i, S_3] = iK_i \tag{10}$$

Dirac then determined that these commutation relations constitute the Lie algebra for the $O(3, 2)$ de Sitter group with ten generators. This de Sitter group is the Lorentz group applicable to three space coordinates and two time coordinates. Let us use the notation (x, y, z, t, s), with (x, y, z) as space coordinates and (t, s) as two time coordinates. Then the rotation around the z axis is generated by

$$L_3 = \begin{pmatrix} 0 & -i & 0 & 0 & 0 \\ i & 0 & 0 & 0 & 0 \\ 0 & 0 & 0 & 0 & 0 \\ 0 & 0 & 0 & 0 & 0 \\ 0 & 0 & 0 & 0 & 0 \end{pmatrix} \tag{11}$$

The generators L_1 and L_2 can be also be constructed. The K_3 and Q_3 will take the form

$$K_3 = \begin{pmatrix} 0 & 0 & 0 & 0 & 0 \\ 0 & 0 & 0 & 0 & 0 \\ 0 & 0 & 0 & i & 0 \\ 0 & 0 & i & 0 & 0 \\ 0 & 0 & 0 & 0 & 0 \end{pmatrix}, Q_3 = \begin{pmatrix} 0 & 0 & 0 & 0 & 0 \\ 0 & 0 & 0 & 0 & 0 \\ 0 & 0 & 0 & 0 & i \\ 0 & 0 & 0 & 0 & 0 \\ 0 & 0 & i & 0 & 0 \end{pmatrix} \tag{12}$$

From these two matrices, the generators K_1, K_2, Q_1, Q_2 can be constructed. The generator S_3 can be written as

$$S_3 = \begin{pmatrix} 0 & 0 & 0 & 0 & 0 \\ 0 & 0 & 0 & 0 & 0 \\ 0 & 0 & 0 & 0 & 0 \\ 0 & 0 & 0 & 0 & -i \\ 0 & 0 & 0 & i & 0 \end{pmatrix} \tag{13}$$

The last five-by-five matrix generates rotations in the two-dimensional space of (t, s).

In his 1963 paper [1], Dirac states that the Lie algebra of Equation (10) can serve as the four-dimensional symplectic group $Sp(4)$. In order to see this point, let us go to the Wigner phase-space picture of the coupled oscillators.

3.1. Wigner Phase-Space Representation

For this two-oscillator system, the Wigner function is defined as [4,6]

$$W(x_1, x_2; p_1, p_2) = \left(\tfrac{1}{\pi}\right)^2 \int \exp\{-2i(p_1 y_1 + p_2 y_2)\}$$

$$\times \psi^*(x_1 + y_1, x_2 + y_2)\psi(x_1 - y_1, x_2 - y_2)dy_1 dy_2 \tag{14}$$

Indeed, the Wigner function is defined over the four-dimensional phase space of (x_1, p_1, x_2, p_2) just as in the case of classical mechanics. The unitary transformations generated by the operators of Equation (9) are translated into linear canonical transformations of the Wigner function [4]. The canonical transformations are generated by the differential operators [4]:

$$L_1 = +\tfrac{i}{2}\left\{\left(x_1\tfrac{\partial}{\partial p_2} - p_2\tfrac{\partial}{\partial x_1}\right) + \left(x_2\tfrac{\partial}{\partial p_1} - p_1\tfrac{\partial}{\partial x_2}\right)\right\}$$

$$L_2 = -\frac{i}{2}\left\{\left(x_1\frac{\partial}{\partial x_2} - x_2\frac{\partial}{\partial x_1}\right) + \left(p_1\frac{\partial}{\partial p_2} - p_2\frac{\partial}{\partial p_1}\right)\right\}$$

$$L_3 = +\frac{i}{2}\left\{\left(x_1\frac{\partial}{\partial p_1} - p_1\frac{\partial}{\partial x_1}\right) - \left(x_2\frac{\partial}{\partial p_2} - p_2\frac{\partial}{\partial x_2}\right)\right\}$$

$$S_3 = -\frac{i}{2}\left\{\left(x_1\frac{\partial}{\partial p_1} - p_1\frac{\partial}{\partial x_1}\right) + \left(x_2\frac{\partial}{\partial p_2} - p_2\frac{\partial}{\partial x_2}\right)\right\} \tag{15}$$

and

$$K_1 = -\frac{i}{2}\left\{\left(x_1\frac{\partial}{\partial p_1} + p_1\frac{\partial}{\partial x_1}\right) - \left(x_2\frac{\partial}{\partial p_2} + p_2\frac{\partial}{\partial x_2}\right)\right\}$$

$$K_2 = -\frac{i}{2}\left\{\left(x_1\frac{\partial}{\partial x_1} - p_1\frac{\partial}{\partial p_1}\right) + \left(x_2\frac{\partial}{\partial x_2} - p_2\frac{\partial}{\partial p_2}\right)\right\}$$

$$K_3 = +\frac{i}{2}\left\{\left(x_1\frac{\partial}{\partial p_2} + p_2\frac{\partial}{\partial x_1}\right) + \left(x_2\frac{\partial}{\partial p_1} + p_1\frac{\partial}{\partial x_2}\right)\right\}$$

$$Q_1 = +\frac{i}{2}\left\{\left(x_1\frac{\partial}{\partial x_1} - p_1\frac{\partial}{\partial p_1}\right) - \left(x_2\frac{\partial}{\partial x_2} - p_2\frac{\partial}{\partial p_2}\right)\right\}$$

$$Q_2 = -\frac{i}{2}\left\{\left(x_1\frac{\partial}{\partial p_1} + p_1\frac{\partial}{\partial x_1}\right) + \left(x_2\frac{\partial}{\partial p_2} + p_2\frac{\partial}{\partial x_2}\right)\right\}$$

$$Q_3 = -\frac{i}{2}\left\{\left(x_2\frac{\partial}{\partial x_1} + x_1\frac{\partial}{\partial x_2}\right) - \left(p_2\frac{\partial}{\partial p_1} + p_1\frac{\partial}{\partial p_2}\right)\right\} \tag{16}$$

3.2. Translation into Four-by-Four Matrices

For a dynamical system consisting of two pairs of canonical variables x_1, p_1 and x_2, p_2, we can use the coordinate variables defined as [6]

$$(\eta_1, \eta_2, \eta_3, \eta_4) = (x_1, p_1, x_2, p_2) \tag{17}$$

Then the transformation of the variables from η_i to ξ_i is canonical if [12,13]

$$MJ\widetilde{M} = J \tag{18}$$

where M is a four-by-four matrix defined by

$$M_{ij} = \frac{\partial}{\partial \eta_j}\xi_i$$

and

$$J = \begin{pmatrix} 0 & 1 & 0 & 0 \\ -1 & 0 & 0 & 0 \\ 0 & 0 & 0 & 1 \\ 0 & 0 & -1 & 0 \end{pmatrix} \tag{19}$$

According to this form of the J matrix, the area of the phase space for x_1 and p_1 variables remains invariant, and the story is the same for the phase space of x_2 and p_2.

We can then write the generators of the $Sp(4)$ group as

$$L_1 = \frac{-1}{2}\begin{pmatrix} 0 & \sigma_2 \\ \sigma_2 & 0 \end{pmatrix}, L_2 = \frac{i}{2}\begin{pmatrix} 0 & -I \\ I & 0 \end{pmatrix}$$

$$L_3 = \frac{1}{2}\begin{pmatrix} -\sigma_2 & 0 \\ 0 & \sigma_2 \end{pmatrix}, S_3 = \frac{1}{2}\begin{pmatrix} \sigma_2 & 0 \\ 0 & \sigma_2 \end{pmatrix} \tag{20}$$

and

$$K_1 = \frac{i}{2}\begin{pmatrix} \sigma_1 & 0 \\ 0 & -\sigma_1 \end{pmatrix}, K_2 = \frac{i}{2}\begin{pmatrix} \sigma_3 & 0 \\ 0 & \sigma_3 \end{pmatrix}, K_3 = -\frac{i}{2}\begin{pmatrix} 0 & \sigma_1 \\ \sigma_1 & 0 \end{pmatrix}$$

and

$$Q_1 = \frac{i}{2}\begin{pmatrix} -\sigma_3 & 0 \\ 0 & \sigma_3 \end{pmatrix}, Q_2 = \frac{i}{2}\begin{pmatrix} \sigma_1 & 0 \\ 0 & \sigma_1 \end{pmatrix}, Q_3 = \frac{i}{2}\begin{pmatrix} 0 & \sigma_3 \\ \sigma_3 & 0 \end{pmatrix} \tag{21}$$

These four-by-four matrices satisfy the commutation relations given in Equation (10). Indeed, the de Sitter group $O(3,2)$ is locally isomorphic to the $Sp(4)$ group. The remaining question is whether these ten matrices can serve as the fifteen Dirac matrices given in Section 2. The answer is clearly no. How can ten matrices describe fifteen matrices? We should therefore add five more matrices.

4. Extension to $O(3,3)$ Symmetry

Unlike the case of the Schrödinger picture, it is possible to add five non-canonical generators to the above list. They are

$$S_1 = +\frac{i}{2}\left\{ \left(x_1\frac{\partial}{\partial x_2} - x_2\frac{\partial}{\partial x_1} \right) - \left(p_1\frac{\partial}{\partial p_2} - p_2\frac{\partial}{\partial p_1} \right) \right\}$$

$$S_2 = -\frac{i}{2}\left\{ \left(x_1\frac{\partial}{\partial p_2} - p_2\frac{\partial}{\partial x_1} \right) + \left(x_2\frac{\partial}{\partial p_1} - p_1\frac{\partial}{\partial x_2} \right) \right\} \tag{22}$$

as well as three additional squeeze operators:

$$G_1 = -\frac{i}{2}\left\{ \left(x_1\frac{\partial}{\partial x_2} + x_2\frac{\partial}{\partial x_1} \right) + \left(p_1\frac{\partial}{\partial p_2} + p_2\frac{\partial}{\partial p_1} \right) \right\}$$

$$G_2 = \frac{i}{2}\left\{ \left(x_1\frac{\partial}{\partial p_2} + p_2\frac{\partial}{\partial x_1} \right) - \left(x_2\frac{\partial}{\partial p_1} + p_1\frac{\partial}{\partial x_2} \right) \right\}$$

$$G_3 = -\frac{i}{2}\left\{ \left(x_1\frac{\partial}{\partial x_1} + p_1\frac{\partial}{\partial p_1} \right) + \left(x_2\frac{\partial}{\partial p_1} + p_1\frac{\partial}{\partial x_2} \right) \right\} \tag{23}$$

These five generators perform well-defined operations on the Wigner function. However, the question is whether these additional generators are acceptable in the present form of quantum mechanics.

In order to answer this question, let us note that the uncertainty principle in the phase-space picture of quantum mechanics is stated in terms of the minimum area in phase space for a given pair of conjugate variables. The minimum area is determined by Planck's constant. Thus we are allowed to expand the phase space, but are not allowed to contract it. With this point in mind, let us go back to G_3 of Equation (23), which generates transformations that simultaneously expand one phase space and contract the other. Thus, the G_3 generator is not acceptable in quantum mechanics even though it generates well-defined mathematical transformations of the Wigner function.

If the five generators of Equations (22) and (23) are added to the ten generators given in Equations (15) and (16), there are fifteen generators. They satisfy the following set of commutation relations.

$$[L_i, L_j] = i\epsilon_{ijk}L_k, \; [S_i, S_j] = i\epsilon_{ijk}S_k, \; [L_i, S_j] = 0$$

$$[L_i, K_j] = i\epsilon_{ijk}K_k, \; [L_i, Q_j] = i\epsilon_{ijk}Q_k, \; [L_i, G_j] = i\epsilon_{ijk}G_k$$

$$[K_i, K_j] = [Q_i, Q_j] = [Q_i, Q_j] = -i\epsilon_{ijk}L_k$$

$$[K_i, Q_j] = -i\delta_{ij}S_3, \; [Q_i, G_j] = -i\delta_{ij}S_1, \; [G_i, K_j] = -i\delta_{ij}S_2$$

$$[K_i, S_3] = -iQ_i, \; [Q_i, S_3] = iK_i, \; [G_i, S_3] = 0$$

$$[K_i, S_1] = 0, \; [Q_i, S_1] = -iG_i, \; [G_i, S_1] = iQ_i$$

$$[K_i, S_2] = iG_i, \; [Q_i, S_2] = 0, \; [G_i, S_2] = -iK_i \tag{24}$$

As we shall see in Section 4.2, this set of commutation relations serves as the Lie algebra for the group $SL(4, r)$ and also for the $O(3,3)$ Lorentz group.

These fifteen four-by-four matrices are written in terms of Dirac's fifteen Majorana matrices, and are tabulated in Table 1. There are six anti-symmetric and nine symmetric matrices. These anti-symmetric matrices were divided into two sets of three rotation generators in the four-dimensional phase space. The nine symmetric matrices can be divided into three sets of three squeeze generators. However, this classification scheme is easier to understand in terms the group $O(3,3)$, discussed in Section 4.2.

Table 1. $SL(4, r)$ and Dirac matrices. Two sets of rotation generators and three sets of boost generators. There are 15 generators.

	First component	Second component	Third component
Rotation	$L_1 = \frac{-i}{2}\gamma_0$	$L_2 = \frac{-i}{2}\gamma_5\gamma_0$	$L_3 = \frac{-1}{2}\gamma_5$
Rotation	$S_1 = \frac{i}{2}\gamma_2\gamma_3$	$S_2 = \frac{i}{2}\gamma_1\gamma_2$	$S_3 = \frac{i}{2}\gamma_3\gamma_1$
Boost	$K_1 = \frac{-i}{2}\gamma_5\gamma_1$	$K_2 = \frac{1}{2}\gamma_1$	$K_3 = \frac{i}{2}\gamma_0\gamma_1$
Boost	$Q_1 = \frac{i}{2}\gamma_5\gamma_3$	$Q_2 = \frac{-1}{2}\gamma_3$	$Q_3 = -\frac{i}{2}\gamma_0\gamma_3$
Boost	$G_1 = \frac{-i}{2}\gamma_5\gamma_2$	$G_2 = \frac{1}{2}\gamma_2$	$G_3 = \frac{i}{2}\gamma_0\gamma_2$

4.1. Non-Canonical Transformations in Classical Mechanics

In addition to Dirac's ten oscillator matrices, we can consider the matrix

$$G_3 = \frac{i}{2}\begin{pmatrix} I & 0 \\ 0 & -I \end{pmatrix} \tag{25}$$

which will generate a radial expansion of the phase space of the first oscillator, while contracting that of the second phase space [14], as illustrated in Figure 1. What is the physical significance of this operation? The expansion of phase space leads to an increase in uncertainty and entropy [8,14].

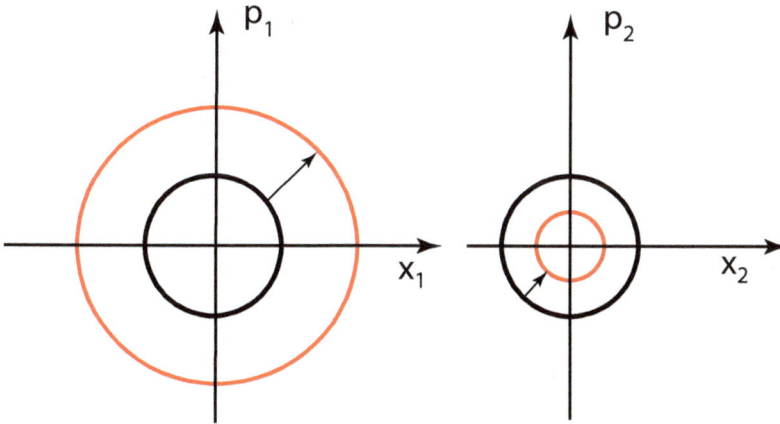

Figure 1. Expanding and contracting phase spaces. Canonical transformations leave the area of each phase space invariant. Non-canonical transformations can change them, yet the product of these two areas remains invariant.

The contraction of the second phase space has a lower limit in quantum mechanics, namely it cannot become smaller than Planck's constant. However, there is no such lower limit in classical mechanics. We shall go back to this question in Section 5.

In the meantime, let us study what happens when the matrix G_3 is introduced into the set of matrices given in Equations (20) and (21). It commutes with S_3, L_3, K_1, K_2, Q_1, and Q_2. However, its commutators with the rest of the matrices produce four more generators:

$$[G_3, L_1] = iG_2, [G_3, L_2] = -iG_1, [G_3, K_3] = iS_2, [G_3, Q_3] = -iS_1 \tag{26}$$

where

$$G_1 = \frac{i}{2}\begin{pmatrix} 0 & I \\ I & 0 \end{pmatrix}, G_2 = \frac{1}{2}\begin{pmatrix} 0 & -\sigma_2 \\ \sigma_2 & 0 \end{pmatrix}$$

$$S_1 = \frac{i}{2}\begin{pmatrix} 0 & \sigma_3 \\ -\sigma_3 & 0 \end{pmatrix}, S_2 = \frac{i}{2}\begin{pmatrix} 0 & -\sigma_1 \\ \sigma_1 & 0 \end{pmatrix} \tag{27}$$

If we take into account the above five generators in addition to the ten generators of $Sp(4)$, there are fifteen generators. These generators satisfy the set of commutation relations given in Equation (24).

Indeed, the ten $Sp(4)$ generators together with the five new generators form the Lie algebra for the group $SL(4, r)$. There are thus fifteen four-by-four matrices. They can be written in terms of the fifteen Majorana matrices, as given in Table 1.

4.2. Local Isomorphism between O(3,3) and SL(4,r)

It is now possible to write fifteen six-by-six matrices that generate Lorentz transformations on the three space coordinates and three time coordinates [6]. However, those matrices are difficult to handle and do not show existing regularities. In this section, we write those matrices as two-by-two matrices of three-by-three matrices.

For this purpose, we construct four sets of three-by-three matrices given in Table 2. There are two sets of rotation generators

$$L_i = \begin{pmatrix} A_i & 0 \\ 0 & 0 \end{pmatrix}, S_i = \begin{pmatrix} 0 & 0 \\ 0 & A_i \end{pmatrix} \tag{28}$$

applicable to the space and time coordinates respectively.

There are also three sets of boost generators. In the two-by-two representation of the matrices given in Table 2, they are

$$K_i = \begin{pmatrix} 0 & B_i \\ \widetilde{B_i} & 0 \end{pmatrix}, Q_i = \begin{pmatrix} 0 & C_i \\ \widetilde{C_i} & 0 \end{pmatrix}, G_i = \begin{pmatrix} 0 & D_i \\ \widetilde{D_i} & 0 \end{pmatrix} \tag{29}$$

where the three-by-three matrices A_i, B_i, C_i, and D_i are given in Table 2, and $\widetilde{A_i}, \widetilde{B_i}, \widetilde{C_i}, \widetilde{D_i}$ are their transposes respectively.

273

Table 2. Three-by-three matrices constituting the two-by-two representation of generators of the $O(3,3)$ group.

	i = 1	i = 2	i = 3
A_i	$\begin{pmatrix} 0 & 0 & 0 \\ 0 & 0 & -i \\ 0 & i & 0 \end{pmatrix}$	$\begin{pmatrix} 0 & 0 & i \\ 0 & 0 & 0 \\ -i & 0 & 0 \end{pmatrix}$	$\begin{pmatrix} 0 & -i & 0 \\ i & 0 & 0 \\ 0 & 0 & 0 \end{pmatrix}$
B_i	$\begin{pmatrix} i & 0 & 0 \\ 0 & 0 & 0 \\ 0 & 0 & 0 \end{pmatrix}$	$\begin{pmatrix} 0 & 0 & 0 \\ i & 0 & 0 \\ 0 & 0 & 0 \end{pmatrix}$	$\begin{pmatrix} 0 & 0 & 0 \\ 0 & 0 & 0 \\ i & 0 & 0 \end{pmatrix}$
C_i	$\begin{pmatrix} 0 & i & 0 \\ 0 & 0 & 0 \\ 0 & 0 & 0 \end{pmatrix}$	$\begin{pmatrix} 0 & 0 & 0 \\ 0 & i & 0 \\ 0 & 0 & 0 \end{pmatrix}$	$\begin{pmatrix} 0 & 0 & 0 \\ 0 & 0 & 0 \\ 0 & i & 0 \end{pmatrix}$
D_i	$\begin{pmatrix} 0 & 0 & i \\ 0 & 0 & 0 \\ 0 & 0 & 0 \end{pmatrix}$	$\begin{pmatrix} 0 & 0 & 0 \\ 0 & 0 & i \\ 0 & 0 & 0 \end{pmatrix}$	$\begin{pmatrix} 0 & 0 & 0 \\ 0 & 0 & 0 \\ 0 & 0 & i \end{pmatrix}$

There is a four-by-four Majorana matrix corresponding to each of these fifteen six-by-six matrices, as given in Table 1.

There are of course many interesting subgroups. The most interesting case is the $O(3,2)$ subgroup, and there are three of them. Another interesting feature is that there are three time dimensions. Thus, there are also $O(2,3)$ subgroups applicable to two space and three time coordinates. This symmetry between space and time coordinates could be an interesting future investigation.

5. Feynman's Rest of the Universe

In his book on statistical mechanics [9], Feynman makes the following statement. *When we solve a quantum-mechanical problem, what we really do is divide the universe into two parts - the system in which we are interested and the rest of the universe. We then usually act as if the system in which we are interested comprised the entire universe. To motivate the use of density matrices, let us see what happens when we include the part of the universe outside the system.*

We can use two coupled harmonic oscillators to illustrate what Feynman says about his rest of the universe. One of the oscillators can be used for the world in which we make physical measurements, while the other belongs to the rest of the universe [8].

Let us start with a single oscillator in its ground state. In quantum mechanics, there are many kinds of excitations of the oscillator, and three of them are familiar to us. First, it can be excited to a state with a definite energy eigenvalue. We obtain the excited-state wave functions by solving the eigenvalue problem for the Schrödinger equation, and this procedure is well known.

Second, the oscillator can go through coherent excitations. The ground-state oscillator can be excited to a coherent or squeezed state. During this process, the minimum uncertainty of the ground state is preserved. The coherent or squeezed state is not in an energy eigenstate. This kind of excited state plays a central role in coherent and squeezed states of light, which have recently become a standard item in quantum mechanics.

Third, the oscillator can go through thermal excitations. This is not a quantum excitation but a statistical ensemble. We cannot express a thermally excited state by making linear combinations of wave functions. We should treat this as a canonical ensemble. In order to deal with this thermal state, we need a density matrix.

For the thermally excited single-oscillator state, the density matrix takes the form [9,15,16].

$$\rho(x,y) = \left(1 - e^{-1/T}\right) \sum_{k} e^{-k/T} \phi_k(x) \phi_k^*(x) \tag{30}$$

where the absolute temperature T is measured in the scale of Boltzmann's constant, and $\phi_k(x)$ is the k-th excited state wave oscillator wave function. The index ranges from 0 to ∞.

We also use Wigner functions to deal with statistical problems in quantum mechanics. The Wigner function for this thermally excited state is [4,9,15]

$$W_T(x, p) = \frac{1}{\pi} \int e^{-2ipz} \rho(x - z, x + z) dz \tag{31}$$

which becomes

$$W_T = \left[\frac{\tanh(1/2T)}{\pi} \right] \exp\left[-\left(x^2 + p^2 \right) \tanh(1/2T) \right] \tag{32}$$

This Wigner function becomes

$$W_0 = \frac{1}{\pi} \exp\left[-\left(x^2 + p^2 \right) \right] \tag{33}$$

when $T = 0$. As the temperature increases, the radius of this Gaussian form increases from one to [14].

$$\frac{1}{\sqrt{\tanh(1/2T)}} \tag{34}$$

The question is whether we can derive this expanding Wigner function from the concept of Feynman's rest of the universe. In their 1999 paper [8], Han *et al.* used two coupled harmonic oscillators to illustrate what Feynman said about his rest of the universe. One of their two oscillators is for the world in which we do quantum mechanics and the other is for the rest of the universe. However, these authors did not use canonical transformations. In Section 5.1, we summarize the main point of their paper using the language of canonical transformations developed in the present paper.

Their work was motivated by the papers by Yurke *et al.* [17] and by Ekert *et al.* [18], and the Barnett–Phoenix version of information theory [19]. These authors asked the question of what happens when one of the photons is not observed in the two-mode squeezed state.

In Section 5.2, we introduce another form of Feynman's rest of the universe, based on non-canonical transformations discussed in the present paper. For a two-oscillator system, we can define a single-oscillator Wigner function for each oscillator. Then non-canonical transformations allow one Wigner function to expand while forcing the other to shrink. The shrinking Wigner function has a lower limit in quantum mechanics, while there is none in classical mechanics. Thus, Feynman's rest of the universe consists of classical mechanics where Planck's constant has no lower limit.

In Section 5.3, we translate the mathematics of the expanding Wigner function into the physical language of entropy.

5.1. Canonical Approach

Let us start with the ground-state wave function for the uncoupled system. Its Hamiltonian is given in Equation (7), and its wave function is

$$\psi_0(x_1, x_2) = \frac{1}{\sqrt{\pi}} \exp\left[-\frac{1}{2} \left(x_1^2 + x_2^2 \right) \right] \tag{35}$$

We can couple these two oscillators by making the following canonical transformations. First, let us rotate the coordinate system by $45°$ to get

$$\frac{1}{\sqrt{2}}(x_1 + x_2), \frac{1}{\sqrt{2}}(x_1 - x_2) \tag{36}$$

Let us then squeeze the coordinate system:

$$\frac{e^{\eta}}{\sqrt{2}}(x_1 + x_2), \frac{e^{-\eta}}{\sqrt{2}}(x_1 - x_2) \tag{37}$$

Likewise, we can transform the momentum coordinates to

$$\frac{e^{-\eta}}{\sqrt{2}}(p_1 + p_2), \frac{e^{\eta}}{\sqrt{2}}(p_1 - p_2) \tag{38}$$

Equations (37) and (38) constitute a very familiar canonical transformation. The resulting wave function for this coupled system becomes

$$\psi_\eta(x_1, x_2) = \frac{1}{\sqrt{\pi}} \exp\left\{-\frac{1}{4}\left[e^{\eta}(x_1 - x_2)^2 + e^{-\eta}(x_1 + x_2)^2\right]\right\} \tag{39}$$

This transformed wave function is illustrated in Figure 2.

As was discussed in the literature for several different purposes [4,20–22], this wave function can be expanded as

$$\psi_\eta(x_1, x_2) = \frac{1}{\cosh\eta} \sum_k \left(\tanh\frac{\eta}{2}\right)^k \phi_k(x_1)\phi_k(x_2) \tag{40}$$

where the wave function $\phi_k\phi(x)$ and the range of summation are defined in Equation (30). From this wave function, we can construct the pure-state density matrix

$$\rho(x_1, x_2; x_1', x_2') = \psi_\eta(x_1, x_2)\psi_\eta(x_1', x_2') \tag{41}$$

which satisfies the condition $\rho^2 = \rho$:

$$\rho(x_1, x_2; x_1', x_2') = \int \rho(x_1, x_2; x_1'', x_2'')\rho(x_1'', x_2''; x_1', x_2')dx_1''dx_2'' \tag{42}$$

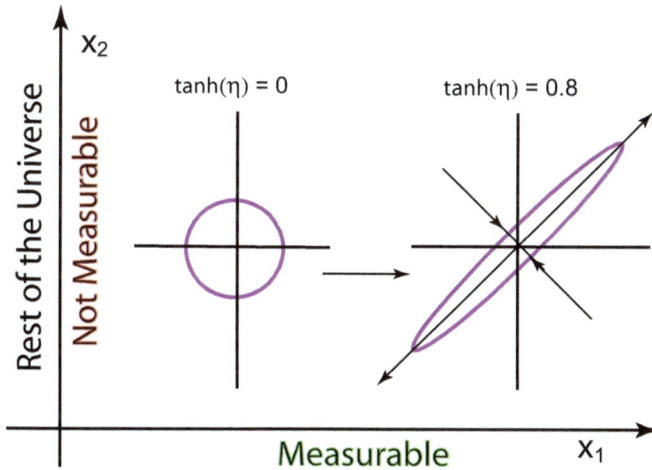

Figure 2. Two-dimensional Gaussian form for two-coupled oscillators. One of the variables is observable while the second variable is not observed. It belongs to Feynman's rest of the universe.

If we are not able to make observations on the x_2, we should take the trace of the ρ matrix with respect to the x_2 variable. Then the resulting density matrix is

$$\rho(x, x') = \int \psi_\eta(x, x_2)\{\psi_\eta(x', x_2)\}^* dx_2 \tag{43}$$

Here, we have replaced x_1 and x_1' by x and x' respectively. If we complete the integration over the x_2 variable,

$$\rho(x, x') = \left(\frac{1}{\pi \cosh \eta}\right)^{1/2} \exp\left\{-\left[\frac{(x + x')^2 + (x - x')^2 \cosh^2 \eta}{4 \cosh \eta}\right]\right\} \tag{44}$$

The diagonal elements of the above density matrix are

$$\rho(x, x) = \left(\frac{1}{\pi \cosh \eta}\right)^{1/2} \exp\left(-x^2 / \cosh \eta\right) \tag{45}$$

With this expression, we can confirm the property of the density matrix: $Tr(\rho) = 1$.

As for the trace of ρ^2, we can perform the integration

$$Tr\left(\rho^2\right) = \int \rho(x, x')\rho(x', x)dx'dx = \frac{1}{\cosh \eta} \tag{46}$$

which is less than one for nonzero values of η.

The density matrix can also be calculated from the expansion of the wave function given in Equation (40). If we perform the integral of Equation (43), the result is

$$\rho(x, x') = \left(\frac{1}{\cosh(\eta/2)}\right)^2 \sum_k \left(\tanh \frac{\eta}{2}\right)^{2k} \phi_k(x)\phi_k^*(x') \tag{47}$$

which leads to $Tr(\rho) = 1$. It is also straightforward to compute the integral for $Tr(\rho^2)$. The calculation leads to

$$Tr\left(\rho^2\right) = \left(\frac{1}{\cosh(\eta/2)}\right)^4 \sum_k \left(\tanh \frac{\eta}{2}\right)^{4k} \tag{48}$$

The sum of this series becomes to $(1/\cosh \eta)$, as given in Equation (46).

We can approach this problem using the Wigner function. The Wigner function for the two oscillator system is [4]

$$W_0(x_1, p_1; x_2, p_2) = \left(\frac{1}{\pi}\right)^2 \exp\left[-\left(x_1^2 + p_1^2 + x_2^2 + p_2^2\right)\right] \tag{49}$$

If we pretend not to make measurement on the second oscillator coordinate, the x_2 and p_2 variables have to be integrated out [8]. The net result becomes the Wigner function for the first oscillator.

The canonical transformation of Equations (37) and (38) changes this Wigner function to

$$W(x_1, x_2; p_1, p_2) = \left(\frac{1}{\pi}\right)^2 \exp\left\{-\frac{1}{2}\left[e^{\eta}(x_1 - x_2)^2 + e^{-\eta}(x_1 + x_2)^2\right.\right.$$
$$\left.\left. + e^{-\eta}(p_1 - p_2)^2 + e^{\eta}(p_1 + p_2)^2\right]\right\} \tag{50}$$

If we do not observe the second pair of variables, we have to integrate this function over x_2 and p_2:

$$W_\eta(x_1, p_1) = \int W(x_1, x_2; p_1, p_2)dx_2dp_2 \tag{51}$$

and the evaluation of this integration leads to [8]

$$W_\eta(x, p) = \frac{1}{\pi \cosh \eta} \exp\left[-\left(\frac{x^2 + p^2}{\cosh \eta}\right)\right] \tag{52}$$

where we use x and p for x_1 and p_1 respectively.

This Wigner function is of the form given in Equation (32) for the thermal excitation, if we identify the squeeze parameter η as [23]

$$\cosh \eta = \frac{1}{\tanh(1/2T)} \tag{53}$$

The failure to make measurement on the second oscillator leads to the radial expansion of the Wigner phase space as in the case of the thermal excitation.

5.2. Non-Canonical Approach

As we noted before, among the fifteen Dirac matrices, ten of them can be used for canonical transformations in classical mechanics, and thus in quantum mechanics. They play a special role in quantum optics [2–5].

The remaining five of them can have their roles if the change in the phase space area is allowed. In quantum mechanics, the area can be increased, but it has a lower limit called Plank's constant. In classical mechanics, this constraint does not exist. The mathematical formalism given in this paper allows us to study this aspect of the system of coupled oscillators.

Let us choose the following three matrices from those in Equations (20) and (21).

$$S_3 = \frac{1}{2} \begin{pmatrix} \sigma_2 & 0 \\ 0 & \sigma_2 \end{pmatrix}, K_2 = \frac{i}{2} \begin{pmatrix} \sigma_3 & 0 \\ 0 & \sigma_3 \end{pmatrix}, Q_2 = \frac{i}{2} \begin{pmatrix} \sigma_1 & 0 \\ 0 & \sigma_1 \end{pmatrix} \tag{54}$$

They satisfy the closed set of commutation relations:

$$[S_3, K_2] = iQ_2, [S_3, Q_2] = -iQ_3, [K_2, Q_2] = -iS_3 \tag{55}$$

This is the Lie algebra for the $Sp(2)$ group, This is the symmetry group applicable to the single-oscillator phase space [4], with one rotation and two squeezes. These matrices generate the same transformation for the first and second oscillators.

We can choose three other sets with similar properties. They are

$$S_3 = \frac{1}{2} \begin{pmatrix} \sigma_2 & 0 \\ 0 & \sigma_2 \end{pmatrix}, Q_1 = \frac{i}{2} \begin{pmatrix} \sigma_3 & 0 \\ 0 & -\sigma_3 \end{pmatrix}, K_1 = \frac{i}{2} \begin{pmatrix} \sigma_1 & 0 \\ 0 & -\sigma_1 \end{pmatrix} \tag{56}$$

$$L_3 = \frac{1}{2} \begin{pmatrix} -\sigma_2 & 0 \\ 0 & \sigma_2 \end{pmatrix}, K_2 = \frac{i}{2} \begin{pmatrix} \sigma_3 & 0 \\ 0 & \sigma_3 \end{pmatrix}, K_1 = \frac{i}{2} \begin{pmatrix} -\sigma_1 & 0 \\ 0 & \sigma_1 \end{pmatrix} \tag{57}$$

and

$$L_3 = \frac{1}{2} \begin{pmatrix} -\sigma_2 & 0 \\ 0 & \sigma_2 \end{pmatrix}, -Q_2 = \frac{i}{2} \begin{pmatrix} -\sigma_3 & 0 \\ 0 & \sigma_3 \end{pmatrix}, Q_2 = \frac{i}{2} \begin{pmatrix} \sigma_1 & 0 \\ 0 & \sigma_1 \end{pmatrix} \tag{58}$$

These matrices also satisfy the commutation relations given in Equation (55). In this case, the squeeze transformations take opposite directions in the second phase space.

Since all these transformations are canonical, they leave the area of each phase space invariant. However, let us look at the non-canonical generator G_3 of Equation (25). It generates the transformation matrix of the form

$$\begin{pmatrix} e^\eta & 0 \\ 0 & e^{-\eta} \end{pmatrix} \tag{59}$$

If η is positive, this matrix expands the first phase space while contracting the second. This contraction of the second phase space is allowed in classical mechanics, but it has a lower limit in quantum mechanics.

The expansion of the first phase space is exactly like the thermal expansion resulting from our failure to observe the second oscillator that belongs to the rest of the universe. If we expand the system of Dirac's ten oscillator matrices to the world of his fifteen Majorana matrices, we can expand and contract the first and second phase spaces without mixing them up. We can thus construct a model where the observed world and the rest of the universe remain separated. In the observable world, quantum mechanics remains valid with thermal excitations. In the rest of the universe, since the area of the phase space can decrease without lower limit, only classical mechanics is valid.

During the expansion/contraction process, the product of the areas of the two phase spaces remains constant. This may or may not be an extended interpretation of the uncertainty principle, but we choose not to speculate further on this issue.

Let us turn our attention to the fact that the groups $SL(4, r)$ and $Sp(4)$ are locally isomorphic to $O(3, 3)$ and $O(3, 2)$ respectively. This means that we can do quantum mechanics in one of the $O(3, 2)$ subgroups of $O(3, 3)$, as Dirac noted in his 1963 paper [1]. The remaining generators belong to Feynman's rest of the universe.

5.3. Entropy and the Expanding Wigner Phase Space

We have seen how Feynman's rest of the universe increases the radius of the Wigner function. It is important to note that the entropy of the system also increases.

Let us go back to the density matrix. The standard way to measure this ignorance is to calculate the entropy defined as [16,24–27].

$$S = -Tr(\rho \ln(\rho)) \tag{60}$$

where S is measured in units of Boltzmann's constant. If we use the density matrix given in Equation (44), the entropy becomes

$$S = 2\left\{ \cosh^2\left(\frac{\eta}{2}\right) \ln\left(\cosh\frac{\eta}{2}\right) - \sinh^2\left(\frac{\eta}{2}\right) \ln\left(\sinh\frac{\eta}{2}\right) \right\} \tag{61}$$

In order to express this equation in terms of the temperature variable T, we write Equation (53) as

$$\cosh \eta = \frac{1 + e^{-1/T}}{1 - e^{-1/T}} \tag{62}$$

which leads to

$$\cosh^2\left(\frac{\eta}{2}\right) = \frac{1}{1 + e^{-1/T}}, \sinh^2\left(\frac{\eta}{2}\right) = \frac{e^{-1/T}}{1 + e^{-1/T}} \tag{63}$$

Then the entropy of Equation (61) takes the form [8]

$$S = \left(\frac{1}{T}\right) \left\{ \frac{1}{\exp(1/T) - 1} \right\} - \ln\left(1 - e^{-1/T}\right) \tag{64}$$

This familiar expression is for the entropy of an oscillator state in thermal equilibrium. Thus, for this oscillator system, we can relate our ignorance of the Feynman's rest of the universe, measured by of the coupling parameter η, to the temperature.

6. Concluding Remarks

In this paper, we started with the fifteen four-by-four matrices for the Majorana representation of the Dirac matrices, and the ten generators of the $Sp(4)$ group corresponding to Dirac's oscillator matrices. Their explicit forms are given in the literature [6,7], and their roles in modern physics are well-known [3,4,11]. We re-organized them into tables.

The difference between these two representations consists of five matrices. The physics of this difference is discussed in terms of Feynman's rest of the universe [9]. According to Feynman, this universe consists of the world in which we do quantum mechanics, and the rest of the universe. In the rest of the universe, our physical laws may or may not be respected. In the case of coupled oscillators, without the lower limit on Planck's constant, we can do classical mechanics but not quantum mechanics in the rest of the universe.

Symmetry **2012**, *4*, 626–643

In 1971, Feynman *et al.* [28] published a paper on the oscillator model of hadrons, where the proton consists of three quarks linked up by oscillator springs. In order to treat this problem, they use a three-particle symmetry group formulated by Dirac in his book on quantum mechanics [29,30]. An interesting problem could be to see what happens to the two quarks when one of them is not observed. Another interesting question could be to see what happens to one of the quarks when two of them are not observed.

Finally, we note here that group theory is a very powerful tool in approaching problems in modern physics. Different groups can share the same set of commutation relations for their generators. Recently, the group $SL(2, c)$ through its correspondence with the $SO(3, 1)$ has been shown to be the underlying language for classical and modern optics [4,31]. In this paper, we exploited the correspondence between $SL(4, r)$ and $O(3, 3)$, as well as the correspondence between $Sp(4)$ and $O(3, 2)$, which was first noted by Paul A. M. Dirac [1].

There could be more applications of group isomorphisms in the future. A comprehensive list of those correspondences is given in Gilmore's book on Lie groups [32].

Acknowledgments: We would like to thank Christian Baumgarten for telling us about the $Sp(2)$ symmetry in classical mechanics.

References

1. Dirac, P.A.M. A remarkable representation of the 3 + 2 de Sitter Group. *J. Math. Phys.* **1963**, *4*, 901–909. [CrossRef]
2. Yuen, H.P. Two-photon coherent states of the radiation field. *Phys. Rev. A* **1976**, *13*, 2226–2243. [CrossRef]
3. Yurke, B.S.; McCall, S.L.; Klauder, J.R. SU(2) and SU(1,1) interferometers. *Phys. Rev. A* **1986**, *33*, 4033–4054. [CrossRef] [PubMed]
4. Kim, Y.S.; Noz, M.E. *Phase Space Picture of Quantum Mechanics*; World Scientific Publishing Company: Singapore, 1991.
5. Han, D.; Kim, Y.S.; Noz, M.E.; Yeh, L. Symmetries of two-mode squeezed states. *J. Math. Phys.* **1993**, *34*, 5493–5508. [CrossRef]
6. Han, D.; Kim, Y.S.; Noz, M.E. O(3,3)-like symmetries of coupled harmonic oscillators. *J. Math. Phys.* **1995**, *36*, 3940–3954. [CrossRef]
7. Lee, D.-G. The Dirac gamma matrices as "relics" of a hidden symmetry?: As fundamental representation of the algebra Sp(4,r). *J. Math. Phys.* **1995**, *36*, 524–530. [CrossRef]
8. Han, D.; Kim, Y.S.; Noz, M.E. Illustrative example of Feynman's rest of the universe. *Am. J. Phys.* **1999**, *67*, 61–66. [CrossRef]
9. Feynman, R.P. *Statistical Mechanics*; Benjamin/Cummings: Reading, MA, USA, 1972.
10. Majorana, E. Relativistic theory of particles with arbitrary intrinsic angular momentum. *Nuovo Cimento* **1932**, *9*, 335–341. [CrossRef]
11. Itzykson, C.; Zuber, J.B. *Quantum Field Theory*; MaGraw-Hill: New York, NY, USA, 1980.
12. Goldstein, H. *Classical Mechanics*, 2nd ed.; Addison-Wesley: Reading, MA, USA, 1980.
13. Abraham, R.; Marsden, J.E. *Foundations of Mechanics*, 2nd ed.; Benjamin/Cummings: Reading, MA, USA, 1978.
14. Kim, Y.S.; Li, M. Squeezed states and thermally excited states in the Wigner phase-space picture of quantum mechanics. *Phys. Lett. A* **1989**, *139*, 445–448. [CrossRef]
15. Davies, R.W.; Davies, K.T.R. On the Wigner distribution function for an oscillator. *Ann. Phys.* **1975**, *89*, 261–273. [CrossRef]
16. Landau, L.D.; Lifshitz, E.M. *Statistical Physics*; Pergamon Press: London, UK, 1958.
17. Yurke, B.; Potasek, M. Obtainment of thermal noise from a pure state. *Phys. Rev. A* **1987**, *36*, 3464–3466. [CrossRef] [PubMed]
18. Ekert, A.K.; Knight, P.L. Correlations and squeezing of two-mode oscillations. *Am. J. Phys.* **1989**, *57*, 692–697. [CrossRef]
19. Barnett, S.M.; Phoenix, S.J.D. Information theory, squeezing and quantum correlations. *Phys. Rev. A* **1991**, *44*, 535–545. [CrossRef] [PubMed]

20. Kim, Y.S.; Noz, M.E.; Oh, S.H. A simple method for illustrating the difference between the homogeneous and inhomogeneous Lorentz Groups. *Am. J. Phys.* **1979**, *47*, 892–897. [CrossRef]
21. Kim, Y.S.; Noz, M.E. *Theory and Applications of the Poincaré Group*; Reidel: Dordrecht, the Netherlands, 1986.
22. Giedke, G.; Wolf, M.M.; Krueger, O.; Werner, R.F.; Cirac, J.J. Entanglement of formation for symmetric Gaussian states. *Phys. Rev. Lett.* **2003**, *91*, 107901–107904. [CrossRef] [PubMed]
23. Han, D.; Kim, Y.S.; Noz, M.E. Lorentz-squeezed hadrons and hadronic temperature. *Phys. Lett. A* **1990**, *144*, 111–115. [CrossRef]
24. von Neumann, J. *Mathematical Foundation of Quantum Mechanics*; Princeton University: Princeton, NJ, USA, 1955.
25. Fano, U. Description of states in quantum mechanics by density matrix and operator techniques. *Rev. Mod. Phys.* **1957**, *29*, 74–93. [CrossRef]
26. Blum, K. *Density Matrix Theory and Applications*; Plenum: New York, NY, USA, 1981.
27. Kim, Y.S.; Wigner, E.P. Entropy and Lorentz transformations. *Phys. Lett. A* **1990**, *147*, 343–347. [CrossRef]
28. Feynman, R.P.; Kislinger, M.; Ravndal, F. Current matrix elements from a relativistic Quark Model. *Phys. Rev. D* **1971**, *3*, 2706–2732. [CrossRef]
29. Dirac, P.A.M. *Principles of Quantum Mechanics*, 4th ed.; Oxford University: London, UK, 1958.
30. Hussar, P.E.; Kim, Y.S.; Noz, M.E. Three-particle symmetry classifications according to the method of Dirac. *Am. J. Phys.* **1980**, *48*, 1038–1042. [CrossRef]
31. Başkal, S.; Kim, Y.S. Lorentz Group in ray and polarization optics. In *Mathematical Optics: Classical, Quantum and Imaging Methods*; Lakshminarayanan, V., Calvo, M.L., Alieva, T., Eds.; CRC Press: New York, NY, USA, 2012.
32. Gilmore, R. *Lie Groups, Lie Algebras, and Some of Their Applications*; Wiley: New York, NY, USA, 1974.

© 2012 by the authors. Licensee MDPI, Basel, Switzerland. This article is an open access article distributed under the terms and conditions of the Creative Commons Attribution (CC BY) license (http://creativecommons.org/licenses/by/4.0/).

symmetry

Article

Symmetries Shared by the Poincaré Group and the Poincaré Sphere

Young S. Kim [1],* and Marilyn E. Noz [2]

[1] Center for Fundamental Physics, University of Maryland, College Park, MD 20742, USA
[2] Department of Radiology, New York University, New York, NY 10016, USA; marilyne.noz@gmail.com
* Author to whom correspondence should be addressed; yskim@umd.edu; Tel.: +1-301-937-1306.

Received: 29 May 2013; in revised form: 9 June 2013; Accepted: 9 June 2013; Published: 27 June 2013

Abstract: Henri Poincaré formulated the mathematics of Lorentz transformations, known as the Poincaré group. He also formulated the Poincaré sphere for polarization optics. It is shown that these two mathematical instruments can be derived from the two-by-two representations of the Lorentz group. Wigner's little groups for internal space-time symmetries are studied in detail. While the particle mass is a Lorentz-invariant quantity, it is shown to be possible to address its variations in terms of the decoherence mechanism in polarization optics.

Keywords: Poincaré group; Poincaré sphere; Wigner's little groups; particle mass; decoherence mechanism; two-by-two representations; Lorentz group

1. Introduction

It was Henri Poincaré who worked out the mathematics of Lorentz transformations before Einstein and Minkowski, and the Poincaré group is the underlying language for special relativity. In order to analyze the polarization of light, Poincaré also constructed a graphic illustration known as the Poincaré sphere [1–3].

It is of interest to see whether the Poincaré sphere can also speak the language of special relativity. In that case, we can study the physics of relativity in terms of what we observe in optical laboratories. For that purpose, we note first that the Lorentz group starts as a group of four-by-four matrices, while the Poincaré sphere is based on the two-by-two matrix consisting of four Stokes parameters. Thus, it is essential to find a two-by-two representation of the Lorentz group. Fortunately, this representation exists in the literature [4,5], and we shall use it in this paper.

As for the problems in relativity, we shall discuss here Wigner's little groups dictating the internal space-time symmetries of relativistic particles [6]. In his original paper of 1939 [7], Wigner considered the subgroups of the Lorentz group, whose transformations leave the four-momentum of a given particle invariant. While this problem has been extensively discussed in the literature, we propose here to study it using Naimark's two-by-two representation of the Lorentz group [4,5].

This two-by-two representation is useful for communicating with the symmetries of the Poincaré sphere based on the four Stokes parameters, which can take the form of two-by-two matrices. We shall prove here that the Poincaré sphere shares the same symmetry property as that of the Lorentz group, particularly in approaching Wigner's little groups. By doing this, we can study the Lorentz symmetries of elementary particles from what we observe in optical laboratories.

The present paper starts from an unpublished note based on an invited paper presented by one of the authors (YSK) at the Fedorov Memorial Symposium: Spins and Photonic Beams at Interface held in Minsk (2011) [8]. To this, we have added a detailed discussion of how the decoherence mechanism in polarization optics is mathematically equivalent to a massless particle gaining mass to become a massive particle. We are particularly interested in how the variation of mass can be accommodated in the study of internal space-time symmetries.

Symmetry **2013**, *5*, 233–252

In Section 2, we define the symmetry problem we propose to study in this paper. We are interested in the subgroups of the Lorentz group, whose transformations leave the four-momentum of a given particle invariant. This is an old problem and has been repeatedly discussed in the literature [6,7,9]. In this paper, we discuss this problem using the two-by-two formulation of the Lorentz group. This two-by-two language is directly applicable to polarization optics and the Poincaré sphere.

While Wigner formulated his little groups for particles in their given Lorentz frames, we give a formalism applicable to all Lorentz frames. In his 1939 paper, Wigner pointed out that his little groups are different for massive, massless and imaginary-particles. In Section 3, we discuss the possibility of deriving the symmetry properties for massive and imaginary-mass particles from that of the massless particle.

In Section 4, we assemble the variables in polarization optics, and define the matrix operators corresponding to transformations applicable to those variables. We write the Stokes parameters in the form of a two-by-two matrix. The Poincaré sphere can be constructed from this two-by-two Stokes matrix. In Section 5, we note that there can be two radii for the Poincaré sphere. Poincaré's original sphere has one fixed radius, but this radius can change, depending on the degree of coherence. Based on what we studied in Section 3, we can associate this change of the radius to the change in mass of the particle.

2. Poincaré Group and Wigner's Little Groups

Poincaré formulated the group theory of Lorentz transformations applicable to four-dimensional space consisting of three space coordinates and one time variable. There are six generators for this group consisting of three rotation and three boost generators.

In addition, Poincaré considered translations applicable to those four space-time variables, with four generators. If we add these four generators to the six generators for the homogenous Lorentz group, the result is the inhomogeneous Lorentz group [7] with ten generators. This larger group is called the Poincaré group in the literature.

The four translation generators produce space-time four-vectors consisting of the energy and momentum. Thus, within the framework of the Poincaré group, we can consider the subgroup of the Lorentz group for a fixed value of momentum [7]. This subgroup defines the internal space-time symmetry of the particle. Let us consider a particle at rest. Its momentum consists of its mass as its time-like variable and zero for the three momentum components.

$$(m, 0, 0, 0) \tag{1}$$

For convenience, we use the four-vector convention, (t, z, x, y) and (E, p_z, p_x, p_y).

This four-momentum of Equation (1) is invariant under three-dimensional rotations applicable only to the z, x, y coordinates. The dynamical variable associated with this rotational degree of freedom is called the spin of the particle.

We are then interested in what happens when the particle moves with a non-zero momentum. If it moves along the z direction, the four-momentum takes the value:

$$m(\cosh \eta, \sinh \eta, 0, 0) \tag{2}$$

which means:

$$p_0 = m(\cosh \eta) \quad p_z = m(\sinh \eta) e^\eta = \sqrt{\frac{p_0 + p_z}{p_0 - p_z}} \tag{3}$$

Accordingly, the little group consists of Lorentz-boosted rotation matrices. This aspect of the little group has been discussed in the literature [6,9]. The question then is whether we could carry out the same logic using two-by-two matrices

Of particular interest is what happens when the transformation parameter, η, becomes very large and the four-momentum becomes that of a massless particle. This problem has also been discussed in the literature within the framework of four-dimensional Minkowski space. The η parameter becomes large when the momentum becomes large, but it can also become large when the mass becomes very small. The two-by-two formulation allows us to study these two cases separately, as we will do in Section 3.

If the particle has an imaginary mass, it moves faster than light and is not observable. Yet, particles of this kind play important roles in Feynman diagrams, and their space-time symmetry should also be studied. In his original paper [7], Wigner studied the little group as the subgroup of the Lorentz group whose transformations leave the four-momentum invariant of the form:

$$(0, k, 0, 0) \tag{4}$$

Wigner observed that this four-momentum remains invariant under the Lorentz boost along the x or y direction.

If we boost this four-momentum along the z direction, the four-momentum becomes:

$$k(\sinh \eta, \cosh \eta, 0, 0) \tag{5}$$

with:

$$e^{\eta} = \sqrt{\frac{p_0 + p_z}{p_z - p_0}} \tag{6}$$

The two-by-two formalism also allows us to study this problem.

In Section 2.1, we shall present the two-by-two representation of the Lorentz group. In Section 2.2, we shall present Wigner's little groups in this two-by-two representation. While Wigner's analysis was based on particles in their fixed Lorentz frames, we are interested in what happens when they start moving. We shall deal with this problem in Section 3.

2.1. Two-by-Two Representation of the Lorentz Groups

The Lorentz group starts with a group of four-by-four matrices performing Lorentz transformations on the Minkowskian vector space of (t, z, x, y), leaving the quantity:

$$t^2 - z^2 - x^2 - y^2 \tag{7}$$

invariant. It is possible to perform this transformation using two-by-two representations [4,5]. This mathematical aspect is known as $SL(2, c)$, the universal covering group for the Lorentz group.

In this two-by-two representation, we write the four-vector as a matrix:

$$X = \begin{pmatrix} t + z & x - iy \\ x + iy & t - z \end{pmatrix} \tag{8}$$

Then, its determinant is precisely the quantity given in Equation (7). Thus, the Lorentz transformation on this matrix is a determinant-preserving transformation. Let us consider the transformation matrix as:

$$G = \begin{pmatrix} \alpha & \beta \\ \gamma & \delta \end{pmatrix} \quad G^{\dagger} = \begin{pmatrix} \alpha^* & \gamma^* \\ \beta^* & \delta^* \end{pmatrix} \tag{9}$$

with:

$$\det(G) = 1 \tag{10}$$

The G matrix starts with four complex numbers. Due to the above condition on its determinant, it has six independent parameters. The group of these G matrices is known to be locally isomorphic to the group of four-by-four matrices performing Lorentz transformations on the four-vector (t, z, x, y). In other words, for each G matrix, there is a corresponding four-by-four Lorentz-transform matrix, as is illustrated in the Appendix A.

The matrix, G, is not a unitary matrix, because its Hermitian conjugate is not always its inverse. The group can have a unitary subgroup, called $SU(2)$, performing rotations on electron spins. As far as we can see, this G-matrix formalism was first presented by Naimark in 1954 [4]. Thus, we call this formalism the Naimark representation of the Lorentz group. We shall see first that this representation is convenient for studying space-time symmetries of particles. We shall then note that this Naimark representation is the natural language for the Stokes parameters in polarization optics.

With this point in mind, we can now consider the transformation:

$$X' = GXG^\dagger \tag{11}$$

Since G is not a unitary matrix, it is not a unitary transformation. In order to tell this difference, we call this the "Naimark transformation". This expression can be written explicitly as:

$$\begin{pmatrix} t' + z' & x' - iy' \\ x + iy & t' - z' \end{pmatrix} = \begin{pmatrix} \alpha & \beta \\ \gamma & \delta \end{pmatrix} \begin{pmatrix} t + z & x - iy \\ x + iy & t - z \end{pmatrix} \begin{pmatrix} \alpha^* & \gamma^* \\ \beta^* & \delta^* \end{pmatrix} \tag{12}$$

For this transformation, we have to deal with four complex numbers. However, for all practical purposes, we may work with two Hermitian matrices:

$$Z(\delta) = \begin{pmatrix} e^{i\delta/2} & 0 \\ 0 & e^{-i\delta/2} \end{pmatrix} R(\delta) = \begin{pmatrix} \cos(\theta/2) & -\sin(\theta/2) \\ \sin(\theta/2) & \cos(\theta/2) \end{pmatrix} \tag{13}$$

and two symmetric matrices:

$$B(\eta) = \begin{pmatrix} e^{\eta/2} & 0 \\ 0 & e^{-\eta/2} \end{pmatrix} S(\lambda) = \begin{pmatrix} \cosh(\lambda/2) & \sinh(\lambda/2) \\ \sinh(\lambda/2) & \cosh(\lambda/2) \end{pmatrix} \tag{14}$$

whose Hermitian conjugates are not their inverses. The two Hermitian matrices in Equation (13) lead to rotations around the z and y axes, respectively. The symmetric matrices in Equation (14) perform Lorentz boosts along the z and x directions, respectively.

Repeated applications of these four matrices will lead to the most general form of the G matrix of Equation (9) with six independent parameters. For each two-by-two Naimark transformation, there is a four-by-four matrix performing the corresponding Lorentz transformation on the four-component four-vector. In the Appendix A, the four-by-four equivalents are given for the matrices of Equations (13) and (14).

It was Einstein who defined the energy-momentum four-vector and showed that it also has the same Lorentz-transformation law as the space-time four-vector. We write the energy-momentum four-vector as:

$$P = \begin{pmatrix} E + p_z & p_x - ip_y \\ p_x + ip_y & E - p_z \end{pmatrix} \tag{15}$$

with:

$$\det(P) = E^2 - p_x^2 - p_y^2 - p_z^2 \tag{16}$$

which means:

$$\det(P) = m^2 \tag{17}$$

where m is the particle mass.

Now, Einstein's transformation law can be written as:

$$P' = GPG^\dagger \tag{18}$$

or explicitly:

$$\begin{pmatrix} E' + p'_z & p'_x - ip'_y \\ p'_x + ip'_y & E' - p'_z \end{pmatrix} = \begin{pmatrix} \alpha & \beta \\ \gamma & \delta \end{pmatrix} \begin{pmatrix} E + p_z & p_x - ip_y \\ p_x + ip_y & E - p_z \end{pmatrix} \begin{pmatrix} \alpha^* & \gamma^*\beta^* \\ \delta^* & \end{pmatrix} \tag{19}$$

2.2. Wigner's Little Groups

Later in 1939 [7], Wigner was interested in constructing subgroups of the Lorentz group whose transformations leave a given four-momentum invariant. He called these subsets "little groups". Thus, Wigner's little group consists of two-by-two matrices satisfying:

$$P = WPW^\dagger \tag{20}$$

This two-by-two W matrix is not an identity matrix, but tells about the internal space-time symmetry of a particle with a given energy-momentum four-vector. This aspect was not known when Einstein formulated his special relativity in 1905. The internal space-time symmetry was not an issue at that time.

If its determinant is a positive number, the P matrix can be brought to a form proportional to:

$$P = \begin{pmatrix} 1 & 0 \\ 0 & 1 \end{pmatrix} \tag{21}$$

corresponding to a massive particle at rest.

If the determinant is negative, it can be brought to a form proportional to:

$$P = \begin{pmatrix} 1 & 0 \\ 0 & -1 \end{pmatrix} \tag{22}$$

corresponding to an imaginary-mass particle moving faster than light along the z direction, with its vanishing energy component.

If the determinant is zero, we may write P as:

$$P = \begin{pmatrix} 1 & 0 \\ 0 & 0 \end{pmatrix} \tag{23}$$

which is proportional to the four-momentum matrix for a massless particle moving along the z direction.

For all three of the above cases, the matrix of the form:

$$Z(\delta) = \begin{pmatrix} e^{i\delta/2} & 0 \\ 0 & e^{-i\delta/2} \end{pmatrix} \tag{24}$$

will satisfy the Wigner condition of Equation (20). This matrix corresponds to rotations around the z axis, as is shown in the Appendix A.

For the massive particle with the four-momentum of Equation (21), the Naimark transformations with the rotation matrix of the form:

$$R(\theta) = \begin{pmatrix} \cos(\theta/2) & -\sin(\theta/2) \\ \sin(\theta/2) & \cos(\theta/2) \end{pmatrix} \tag{25}$$

also leave the P matrix of Equation (21) invariant. Together with the $Z(\delta)$ matrix, this rotation matrix leads to the subgroup consisting of the unitary subset of the G matrices. The unitary subset of G is $SU(2)$, corresponding to the three-dimensional rotation group dictating the spin of the particle [9].

For the massless case, the transformations with the triangular matrix of the form:

$$\begin{pmatrix} 1 & \gamma \\ 0 & 1 \end{pmatrix} \tag{26}$$

leave the momentum matrix of Equation (23) invariant. The physics of this matrix has a stormy history, and the variable, γ, leads to gauge transformation applicable to massless particles [6,10].

For a particle with its imaginary mass, the W matrix of the form:

$$S(\lambda) = \begin{pmatrix} \cosh(\lambda/2) & \sinh(\lambda/2) \\ \sinh(\lambda/2) & \cosh(\lambda/2) \end{pmatrix} \tag{27}$$

will leave the four-momentum of Equation (22) invariant. This unobservable particle does not appear to have observable internal space-time degrees of freedom.

Table 1 summarizes the transformation matrices for Wigner's subgroups for massive, massless and imaginary-mass particles. Of course, it is a challenging problem to have one expression for all those three cases, and this problem has been addressed in the literature [11].

Table 1. Wigner's Little Groups. The little groups are the subgroups of the Lorentz group, whose transformations leave the four-momentum of a given particle invariant. Thus, the little groups define the internal space-time symmetries of particles. The four-momentum remains invariant under the rotation around it. In addition, the four-momentum remains invariant under the following transformations. These transformations are different for massive, massless and imaginary-mass particles.

Particle mass	Four-momentum	Transform matrices
Massive	$\begin{pmatrix} 1 & 0 \\ 0 & 1 \end{pmatrix}$	$\begin{pmatrix} \cos(\theta/2) & -\sin(\theta/2) \\ \sin(\theta/2) & \cos(\theta/2) \end{pmatrix}$
Massless	$\begin{pmatrix} 1 & 0 \\ 0 & 0 \end{pmatrix}$	$\begin{pmatrix} 1 & \gamma \\ 0 & 1 \end{pmatrix}$
Imaginary mass	$\begin{pmatrix} 1 & 0 \\ 0 & -1 \end{pmatrix}$	$\begin{pmatrix} \cosh(\lambda/2) & \sinh(\lambda/2) \\ \sinh(\lambda/2) & \cosh(\lambda/2) \end{pmatrix}$

3. Lorentz Completion of Wigner's Little Groups

In his original paper [7], Wigner worked out his little groups for specific Lorentz frames. For the massive particle, he constructed his little group in the frame where the particle is at rest. For the imaginary-mass particle, the energy-component of his frame is zero.

For the massless particle, it moves along the z direction with a nonzero momentum. There are no specific frames particularly convenient for us. Thus, the specific frame can be chosen for an arbitrary value of the momentum, and the triangular matrix of Equation (26) should remain invariant under Lorentz boosts along the z direction.

For the massive particle, let us Lorentz-boost the four-momentum matrix of Equation (21) by performing a Naimark transformation:

$$\begin{pmatrix} e^{\eta/2} & 0 \\ 0 & e^{-\eta/2} \end{pmatrix} \begin{pmatrix} 1 & 0 \\ 0 & 1 \end{pmatrix} \begin{pmatrix} e^{\eta/2} & 0 \\ 0 & e^{-\eta/2} \end{pmatrix} \tag{28}$$

which leads to:

$$\begin{pmatrix} e^{\eta} & 0 \\ 0 & e^{-\eta} \end{pmatrix} \tag{29}$$

This resulting matrix corresponds to the Lorentz-boosted four-momentum given in Equation (2). For simplicity, we let $m = 1$ hereafter in this paper. The Lorentz transformation applicable to the four-momentum matrix is not a similarity transformation, but it is a Naimark transformation, as defined in Equation (11).

On the other hand, the rotation matrix of Equation (25) is Lorentz-boosted as a similarity transformation:

$$\begin{pmatrix} e^{\eta/2} & 0 \\ 0 & e^{-\eta/2} \end{pmatrix} \begin{pmatrix} \cos(\theta/2) & -\sin(\theta/2) \\ \sin(\theta/2) & \cos(\theta/2) \end{pmatrix} \begin{pmatrix} e^{-\eta/2} & 0 \\ 0 & e^{\eta/2} \end{pmatrix} \qquad (30)$$

and it becomes:

$$\begin{pmatrix} \cos(\theta/2) & -e^{\eta}\sin(\theta/2) \\ e^{-\eta}\sin(\theta/2) & \cos(\theta/2) \end{pmatrix} \qquad (31)$$

If we perform the Naimark transformation of the four-momentum matrix of Equation (29) with this Lorentz-boosted rotation matrix:

$$\begin{pmatrix} \cos(\theta/2) & -e^{\eta}\sin(\theta/2) \\ e^{-\eta/2}\sin(\theta/2) & \cos(\theta/2) \end{pmatrix} \begin{pmatrix} e^{\eta} & 0 \\ 0 & e^{-\eta} \end{pmatrix} \begin{pmatrix} \cos(\theta/2) & e^{\eta}\sin(\theta/2) \\ -e^{-\eta}\sin(\theta/2) & \cos(\theta/2) \end{pmatrix} \qquad (32)$$

the result is the four-momentum matrix of Equation (29). This means that the Lorentz-boosted rotation matrix of Equation (31) represents the little group, whose transformations leave the four-momentum matrix of Equation (29) invariant.

For the imaginary-mass case, the Lorentz boosted four-momentum matrix becomes:

$$\begin{pmatrix} e^{\eta} & 0 \\ 0 & -e^{-\eta} \end{pmatrix} \qquad (33)$$

The little group matrix is:

$$\begin{pmatrix} \cosh(\lambda/2) & e^{\eta}\sinh(\lambda/2) \\ e^{-\eta}\sinh(\lambda/2) & \cosh(\lambda/2) \end{pmatrix} \qquad (34)$$

where η is given in Equation (6).

For the massless case, if we boost the four-momentum matrix of Equation (23), the result is:

$$e^{\eta}\begin{pmatrix} 1 & 0 \\ 0 & 0 \end{pmatrix} \qquad (35)$$

Here, the η parameter is an independent variable and cannot be defined in terms of the momentum or energy.

The remaining problem is to see whether the massive and imaginary-mass cases collapse to the massless case in the large η limit. This variable becomes large when the momentum becomes large or the mass becomes small. We shall discuss these two cases separately.

3.1. Large-Momentum Limit

While Wigner defined his little group for the massive particle in its rest frame in his original paper [7], the little group represented by Equation (31) is applicable to the moving particle, whose four-momentum is given in Equation (29). This matrix can also be written as:

$$e^{\eta}\begin{pmatrix} 1 & 0 \\ 0 & e^{-2\eta} \end{pmatrix} \qquad (36)$$

In the limit of large η, we can change the above expression into:

$$e^{\eta}\begin{pmatrix} 1 & 0 \\ 0 & 0 \end{pmatrix} \tag{37}$$

This process is continuous, but not necessarily analytic [11]. After making this transition, we can come back to the original frame to obtain the four momentum matrix of Equation (23).

The remaining problem is the Lorentz-boosted rotation matrix of Equation (31). If this matrix is going to remain finite as η approaches infinity, the upper-right element should be finite for large values of η. Let it be γ. Then:

$$-e^{\eta}\sin(\theta/2) = \gamma \tag{38}$$

This means that angle θ has to become zero. As a consequence, the little group matrix of Equation (31) becomes the triangular matrix given in Equation (26) for massless particles.

Imaginary-mass particles move faster than light, and they are not observable. On the other hand, the mathematics applicable to Wigner's little group for this particle has been useful in the two-by-two beam transfer matrix in ray and polarization optics [12].

Let us go back to the four-momentum matrix of Equation (22). If we boost this matrix, it becomes:

$$\begin{pmatrix} e^{\eta} & 0 \\ 0 & -e^{-\eta} \end{pmatrix} \tag{39}$$

which can be written as:

$$e^{\eta}\begin{pmatrix} 1 & 0 \\ 0 & -e^{-2\eta} \end{pmatrix} \tag{40}$$

This matrix can be changed to form Equation (37) in the limit of large η.

Indeed, the little groups for massive, massless and imaginary cases coincide in the large-η limit. Thus, it is possible to jump from one little group to another, and it is a continuous process, but not necessarily analytic [12].

The η parameter can become large as the momentum becomes large or the mass becomes small. In this subsection, we considered the case for large momentum. However, it is of interest to see the limiting process when the mass becomes small, especially in view of the fact that neutrinos have small masses.

3.2. Small-Mass Limit

Let us start with a massive particle with fixed energy, E. Then, $p_0 = E$, and $p_z = E\cos\chi$. The four-momentum matrix is

$$E\begin{pmatrix} 1+\cos\chi & 0 \\ 0 & 1-\cos\chi \end{pmatrix} \tag{41}$$

The determinant of this matrix is $E^2(\sin\chi)^2$. In the regime of the Lorentz group, this is the $(mass)^2$ and is a Lorentz-invariant quantity. There are no Lorentz transformations that change the angle, χ. Thus, with this extra variable, it is possible to study the little groups for variable masses, including the small-mass limit and the zero-mass case.

If $\chi = 0$, the matrix of Equation (41) becomes that of the four-momentum matrix for a massless particle. As it becomes a positive small number, the matrix of Equation (41) can be written as:

$$E(\sin\chi)\begin{pmatrix} e^{\eta} & 0 \\ 0 & e^{-\eta} \end{pmatrix} \tag{42}$$

with

$$e^{\eta} = \sqrt{\frac{1 + \cos\chi}{1 - \cos\chi}} \tag{43}$$

Here, again, the determinant of Equation (42) is $E^2(\sin\chi)^2$. With this matrix, we can construct Wigner's little group for each value of the angle, χ. If χ is not zero, even if it is very small, the little group is $O(3)$-like, as in the case of all massive particles. As the angle, χ, varies continuously from zero to $90°$, the mass increases from zero to its maximum value.

It is important to note that the little groups are different for the small-mass limit and for the zero-mass case. In this section, we studied the internal space-time symmetries dictated by Wigner's little groups, and we are able to present their Lorentz-covariant picture in Table 2.

Table 2. Covariance of the energy-momentum relation and covariance of the internal space-time symmetry groups. The γ parameter for the massless case has been studied in earlier papers in the four-by-four matrix formulation [6]. It corresponds to a gauge transformation. Among the three spin components, S_3 is along the direction of the momentum and remains invariant. It is called the "helicity".

Massive, Slow	Covariance	Massless, Fast
$E = p^2/2m$	Einstein's $E = mc^2$	$E = cp$
S_3		Helicity
	Wigner's Little Group	
S_1, S_2		Gauge Transformation

4. Jones Vectors and Stokes Parameters

In studying polarized light propagating along the z direction, the traditional approach is to consider the x and y components of the electric fields. Their amplitude ratio and the phase difference determine the state of polarization. Thus, we can change the polarization either by adjusting the amplitudes, by changing the relative phase or both. For convenience, we call the optical device that changes amplitudes an "attenuator" and the device that changes the relative phase a "phase shifter".

The traditional language for this two-component light is the Jones-vector formalism, which is discussed in standard optics textbooks [13]. In this formalism, the above two components are combined into one column matrix, with the exponential form for the sinusoidal function:

$$\begin{pmatrix} \psi_1(z,t) \\ \psi_2(z,t) \end{pmatrix} = \begin{pmatrix} a\exp\{i(kz - \omega t + \phi_1)\} \\ b\exp\{i(kz - \omega t + \phi_2)\} \end{pmatrix} \tag{44}$$

This column matrix is called the Jones vector.

When the beam goes through a medium with different values of indexes of refraction for the x and y directions, we have to apply the matrix:

$$\begin{pmatrix} e^{i\delta_1} & 0 \\ 0 & e^{i\delta_2} \end{pmatrix} = e^{i(\delta_1 + \delta_2)/2}\begin{pmatrix} e^{-i\delta/2} & 0 \\ 0 & e^{i\delta/2} \end{pmatrix} \tag{45}$$

with $\delta = \delta_1 - \delta_2$. In measurement processes, the overall phase factor, $e^{i(\delta_1 + \delta_2)/2}$, cannot be detected and can therefore be deleted. The polarization effect of the filter is solely determined by the matrix:

$$Z(\delta) = \begin{pmatrix} e^{i\delta/2} & 0 \\ 0 & e^{-i\delta/2} \end{pmatrix} \tag{46}$$

which leads to a phase difference of δ between the x and y components. The form of this matrix is given in Equation (13), which serves as the rotation around the z axis in the Minkowski space and time.

Also along the x and y directions, the attenuation coefficients could be different. This will lead to the matrix [14]:

$$\begin{pmatrix} e^{-\eta_1} & 0 \\ 0 & e^{-\eta_2} \end{pmatrix} = e^{-(\eta_1+\eta_2)/2} \begin{pmatrix} e^{\eta/2} & 0 \\ 0 & e^{-\eta/2} \end{pmatrix} \tag{47}$$

with $\eta = \eta_2 - \eta_1$. If $\eta_1 = 0$ and $\eta_2 = \infty$, the above matrix becomes:

$$\begin{pmatrix} 1 & 0 \\ 0 & 0 \end{pmatrix} \tag{48}$$

which eliminates the y component. This matrix is known as a polarizer in the textbooks [13] and is a special case of the attenuation matrix of Equation (47).

This attenuation matrix tells us that the electric fields are attenuated at two different rates. The exponential factor, $e^{-(\eta_1+\eta_2)/2}$, reduces both components at the same rate and does not affect the state of polarization. The effect of polarization is solely determined by the squeeze matrix [14]:

$$B(\eta) = \begin{pmatrix} e^{\eta/2} & 0 \\ 0 & e^{-\eta/2} \end{pmatrix} \tag{49}$$

This diagonal matrix is given in Equation (14). In the language of space-time symmetries, this matrix performs a Lorentz boost along the z direction.

The polarization axes are not always the x and y axes. For this reason, we need the rotation matrix:

$$R(\theta) = \begin{pmatrix} \cos(\theta/2) & -\sin(\theta/2) \\ \sin(\theta/2) & \cos(\theta/2) \end{pmatrix} \tag{50}$$

which, according to Equation (13), corresponds to the rotation around the y axis in the space-time symmetry.

Among the rotation angles, the angle of $45°$ plays an important role in polarization optics. Indeed, if we rotate the squeeze matrix of Equation (49) by $45°$, we end up with the squeeze matrix:

$$R(\theta) = \begin{pmatrix} \cosh(\lambda/2) & \sinh(\lambda/2) \\ \sinh(\lambda/2) & \cosh(\lambda/2) \end{pmatrix} \tag{51}$$

which is also given in Equation (14). In the language of space-time physics, this matrix leads to a Lorentz boost along the x axis.

Indeed, the G matrix of Equation (9) is the most general form of the transformation matrix applicable to the Jones vector. Each of the above four matrices plays its important role in special relativity, as we discussed in Section 2. Their respective roles in optics and particle physics are given in Table 3.

However, the Jones vector alone cannot tell us whether the two components are coherent with each other. In order to address this important degree of freedom, we use the coherency matrix [1,2]:

$$C = \begin{pmatrix} S_{11} & S_{12} \\ S_{21} & S_{22} \end{pmatrix} \tag{52}$$

with:

$$< \psi_i^* \psi_j > = \frac{1}{T} \int_0^T \psi_i^*(t+\tau)\psi_j(t)dt \tag{53}$$

where T, for a sufficiently long time interval, is much larger than τ. Then, those four elements become [15]:

$$S_{11} =< \psi_1^*\psi_1 >= a^2 \qquad S_{12} =< \psi_1^*\psi_2 >= abe^{-(\sigma+i\delta)}$$
$$S_{21} =< \psi_2^*\psi_1 >= abe^{-(\sigma-i\delta)} \quad S_{22} =< \psi_2^*\psi_2 >= b^2 \tag{54}$$

The diagonal elements are the absolute values of ψ_1 and ψ_2, respectively. The off-diagonal elements could be smaller than the product of ψ_1 and ψ_2, if the two beams are not completely coherent. The σ parameter specifies the degree of coherency.

This coherency matrix is not always real, but it is Hermitian. Thus, it can be diagonalized by a unitary transformation. If this matrix is normalized so that its trace is one, it becomes a density matrix [16,17].

Table 3. Polarization optics and special relativity sharing the same mathematics. Each matrix has its clear role in both optics and relativity. The determinant of the Stokes or the four-momentum matrix remains invariant under Lorentz transformations. It is interesting to note that the decoherence parameter (least fundamental) in optics corresponds to the mass (most fundamental) in particle physics.

Polarization Optics	Transformation Matrix	Particle Symmetry
Phase shift δ	$\begin{pmatrix} e^{\delta/2} & 0 \\ 0 & e^{-i\delta/2} \end{pmatrix}$	Rotation around z
Rotation around z	$\begin{pmatrix} \cos(\theta/2) & -\sin(\theta/2) \\ \sin(\theta/2) & \cos(\theta/2) \end{pmatrix}$	Rotation around y
Squeeze along x and y	$\begin{pmatrix} e^{\eta/2} & 0 \\ 0 & e^{-\eta/2} \end{pmatrix}$	Boost along z
Squeeze along $45°$	$\begin{pmatrix} \cosh(\lambda/2) & \sinh(\lambda/2) \\ \sinh(\lambda/2) & \cosh(\lambda/2) \end{pmatrix}$	Boost along x
$(ab)^2 \sin^2\chi$	Determinant	$(mass)^2$

If we start with the Jones vector of the form of Equation (44), the coherency matrix becomes:

$$C = \begin{pmatrix} a^2 & ab\, e^{-(\sigma+i\delta)} \\ ab\, e^{-(\sigma-i\delta)} & b^2 \end{pmatrix} \tag{55}$$

We are interested in the symmetry properties of this matrix. Since the transformation matrix applicable to the Jones vector is the two-by-two representation of the Lorentz group, we are particularly interested in the transformation matrices applicable to this coherency matrix.

The trace and the determinant of the above coherency matrix are:

$$\det(C) = (ab)^2(1 - e^{-2\sigma})$$
$$\operatorname{tr}(C) = a^2 + b^2 \tag{56}$$

Since $e^{-\sigma}$ is always smaller than one, we can introduce an angle, χ, defined as:

$$\cos\chi = e^{-\sigma} \tag{57}$$

and call it the "decoherence angle". If $\chi = 0$, the decoherence is minimum, and it becomes maximum when $\chi = 90°$. We can then write the coherency matrix of Equation (55) as:

$$C = \begin{pmatrix} a^2 & ab(\cos\chi)e^{-i\delta} \\ ab(\cos\chi)e^{i\delta} & b^2 \end{pmatrix} \tag{58}$$

The degree of polarization is defined as [13]:

$$f = \sqrt{1 - \frac{4\det(C)}{(\mathrm{tr}(C))^2}} = \sqrt{1 - \frac{4(ab)^2 \sin^2 \chi}{(a^2 + b^2)^2}} \tag{59}$$

This degree is one if $\chi = 0$. When $\chi = 90°$, it becomes:

$$\frac{a^2 - b^2}{a^2 + b^2} \tag{60}$$

Without loss of generality, we can assume that a is greater than b. If they are equal, this minimum degree of polarization is zero.

Under the influence of the Naimark transformation given in Equation (11), this coherency matrix is transformed as:

$$\tag{61}$$

It is more convenient to make the following linear combinations:

$$S_0 = \frac{S_{11} + S_{22}}{2} \quad S_3 = \frac{S_{11} - S_{22}}{2}$$
$$S_1 = \frac{S_{12} + S_{21}}{2} \quad S_2 = \frac{S_{12} - S_{21}}{2i} \tag{62}$$

These four parameters are called Stokes parameters, and four-by-four transformations applicable to these parameters are widely known as Mueller matrices [1,3]. However, if the Naimark transformation given in Equation (61) is translated into the four-by-four Lorentz transformations according to the correspondence given in the Appendix A, the Mueller matrices constitute a representation of the Lorentz group.

Another interesting aspect of the two-by-two matrix formalism is that the coherency matrix can be formulated in terms of quarternions [18–20]. The quarternion representation can be translated into rotations in four-dimensional space. There is a long history between the Lorentz group and the four-dimensional rotation group. It would be interesting to see what the quarternion representation of polarization optics will add to this history between those two similar, but different, groups.

As for earlier applications of the two-by-two representation of the Lorentz group, we note the vector representation by Fedorov [21,22]. Fedorov showed that it is easier to carry out kinematical calculations using his two-by-two representation. For instance, the computation of the Wigner rotation angle is possible in the two-by-two representation [23]. Earlier papers on group theoretical approaches to polarization optics include also those on Mueller matrices [24] and on relativistic kinematics and polarization optics [25].

5. Geometry of the Poincaré Sphere

We now have the four-vector, (S_0, S_3, S_1, S_2), which is Lorentz-transformed like the space-time four-vector, (t, z, x, y), or the energy-momentum four-vector of Equation (15). This Stokes four-vector has a three-component subspace, (S_3, S_1, S_2), which is like the three-dimensional Euclidean subspace in the four-dimensional Minkowski space. In this three-dimensional subspace, we can introduce the spherical coordinate system with:

$$R = \sqrt{S_3^2 + S_1^2 + S_2^2}$$
$$S_3 = R \cos \zeta$$
$$S_1 = R(\sin \zeta) \cos \delta \quad S_2 = R(\sin \zeta) \sin \delta$$

(63)

The radius, R, is the radius of this sphere, and is:

$$R = \frac{1}{2}\sqrt{(a^2 - b^2)^2 + 4(ab)^2 \cos^2 \chi}$$

(64)

with:

$$S_3 = \frac{a^2 - b^2}{2}$$

(65)

This spherical picture is traditionally known as the Poincaré sphere [1–3]. Without loss of generality, we assume a is greater than b, and S_3 is non-negative. In addition, we can consider another sphere with its radius:

$$S_0 = \frac{a^2 + b^2}{2}$$

(66)

according to Equation (62).

The radius, R, takes its maximum value, S_0, when $\chi = 0°$. It decreases and reaches its minimum value, S_3, when $\chi = 90°$. In terms of R, the degree of polarization given in Equation (59) is:

$$f = \frac{R}{S_0}$$

(67)

This aspect of the radius R is illustrated in Figure 1a. The minimum value of R is S_3 of Equation (64).

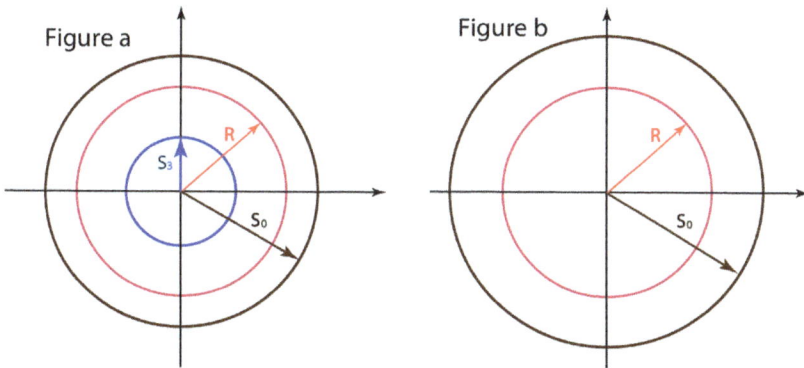

Figure 1. Radius of the Poincaré sphere. The radius, R, takes its maximum value, S_0, when the decoherence angle, χ, is zero. It becomes smaller as χ increases. It becomes minimum when the angle reaches 90°. Its minimum value is S_3, as is illustrated in Figure 1a. The degree of polarization is maximum when $R = S_0$ and is minimum when $R = S_3$. According to Equation (65), S_3 becomes zero when $a = b$, and the minimum value of R becomes zero, as is indicated in Figure 1b. Its maximum value is still S_0. This maximum radius can become larger because b becomes larger to make $a = b$.

Let us go back to the four-momentum matrix of Equation (15). Its determinant is m^2 and remains invariant. Likewise, the determinant of the coherency matrix of Equation (58) should also remain invariant. The determinant in this case is:

$$S_0^2 - R^2 = (ab)^2 \sin^2 \chi \tag{68}$$

This quantity remains invariant. This aspect is shown on the last row of Table 3.

Let us go back to Equation (49). This matrix changes the relative magnitude of the amplitudes, a and b. Thus, without loss of generality, we can study the Stokes parameters with $a = b$. The coherency matrix then becomes:

$$C = a^2 \begin{pmatrix} 1 & (\cos \chi)e^{-i\delta} \\ (\cos \chi)e^{i\delta} & 1 \end{pmatrix} \tag{69}$$

Since the angle, δ, does not play any essential roles, we can let $\delta = 0$ and write the coherency matrix as:

$$C = a^2 \begin{pmatrix} 1 & \cos \chi \\ \cos \chi & 1 \end{pmatrix} \tag{70}$$

Then, the minimum radius, $S_3 = 0$, and S_0 of Equation (62) and R of Equation (64) become:

$$S_0 = a^2 R = a^2(\cos \chi) \tag{71}$$

respectively. The Poincaré sphere becomes simplified to that of Figure 1b. This Poincaré sphere allows R to decrease to zero.

The determinant of the above two-by-two matrix is:

$$a^4 \left(1 - \cos^2 \chi\right) = a^4 \sin^2 \chi \tag{72}$$

Since the Lorentz transformation leaves the determinant invariant, the change in this χ variable is not a Lorentz transformation. It is of course possible to construct a larger group in which this variable plays a role in a group transformation [23], but in this paper, we are more interested in its role in a particle gaining a mass. With this point in mind, let us diagonalize the coherency matrix of Equation (69). Then it takes the form:

$$a^2 \begin{pmatrix} 1 + \cos \chi & 0 \\ 0 & 1 - \cos \chi \end{pmatrix} \tag{73}$$

This form is the same as the four-momentum matrix given in Equation (41). There, we were not able to associate the variable, χ, with any known physical process or symmetry operations of the Lorentz group. Fortunately, in this section, we noted that this variable comes from the degree of decoherence in polarization optics.

6. Concluding Remarks

In this paper, we noted first that the group of Lorentz transformations can be formulated in terms of two-by-two matrices. This two-by-two formalism can also be used for transformations of the coherency matrix in polarization optics consisting of four Stokes parameters.

Thus, this set of the four parameters is like a Minkowskian four-vector under four-by-four Lorentz transformations. In order to accommodate all four Stokes parameters, we noted that the radius of the Poincaré sphere should be allowed to vary from its maximum value to its minimum, corresponding to the fully and minimal coherent cases.

As in the case of the particle mass, the decoherence parameter in the Stokes formalism is invariant under Lorentz transformations. However, the Poincaré sphere, with a variable radius, provides the

Symmetry **2013**, 5, 233–252

mechanism for the variations of the decoherence parameter. It was noted that this variation gives a physical process whose mathematics correspond to that of the mass variable in particle physics.

As for polarization optics, the traditional approach has been to work with two polarizer matrices, like:

$$\begin{pmatrix} 1 & 0 \\ 0 & 0 \end{pmatrix} \begin{pmatrix} 0 & 0 \\ 0 & 1 \end{pmatrix} \tag{74}$$

We have replaced these two matrices by one attenuation matrix of Equation (47). This replacement enables us to formulate the Lorentz group for the Stokes parameters [15]. Furthermore, this attenuation matrix makes it possible to make a continuous transformation from one matrix to another by adjusting the attenuation parameters in optical media. It could be interesting to design optical experiments along this direction.

Acknowledgments: This paper is in part based on an invited paper presented by one of the authors (YSK) at the Fedorov Memorial Symposium: International Conference "Spins and Photonic Beams at Interface", dedicated to the 100th anniversary of F.I. Fedorov (1911–1994) (Minsk, Belarus, 2011). He would like to thank Sergei Kilin for inviting him to the conference.

In addition to numerous original contributions in optics, Fedorov wrote a book on two-by-two representations of the Lorentz group based on his own research on this subject. It was, therefore, quite appropriate for him (YSK) to present a paper on applications of the Lorentz group to optical science. He would like to thank V. A. Dluganovich and M. Glaynskii for bringing the papers and the book written by Academician Fedorov, as well as their own papers to his attention.

Conflicts of Interest: The authors declare no conflict of interest.

Appendix Appendix

In Section 2, we listed four two-by-two matrices whose repeated applications lead to the most general form of the two-by-two matrix, G. It is known that every G matrix can be translated into a four-by-four Lorentz transformation matrix through [4,9,15]:

$$\begin{pmatrix} t'+z' \\ x'-iy' \\ x'+iy' \\ t'-z' \end{pmatrix} = \begin{pmatrix} \alpha\alpha^* & \alpha\beta^* & \beta\alpha^* & \beta\beta^* \\ \alpha\gamma^* & \alpha\delta^* & \beta\gamma^* & \beta\delta^* \\ \gamma\alpha^* & \gamma\beta^* & \delta\alpha^* & \delta\beta^* \\ \gamma\gamma^* & \gamma\delta^* & \delta\gamma^* & \delta\delta^* \end{pmatrix} \begin{pmatrix} t+z \\ x-iy \\ x+iy \\ t-z \end{pmatrix} \tag{75}$$

and:

$$\begin{pmatrix} t \\ z \\ x \\ y \end{pmatrix} = \frac{1}{2} \begin{pmatrix} 1 & 0 & 0 & 1 \\ 1 & 0 & 0 & -1 \\ 0 & 1 & 1 & 0 \\ 0 & i & -i & 0 \end{pmatrix} \begin{pmatrix} t+z \\ x-iy \\ x+iy \\ t-z \end{pmatrix} \tag{76}$$

These matrices appear to be complicated, but it is enough to study the matrices of Equation (13) and Equation (14) to cover all the matrices in this group. Thus, we give their four-by-four equivalents in this Appendix A:

$$Z(\delta) = \begin{pmatrix} e^{i\delta/2} & 0 \\ 0 & e^{-i\delta/2} \end{pmatrix} \tag{77}$$

leads to the four-by-four matrix:

$$\begin{pmatrix} 1 & 0 & 0 & 0 \\ 1 & 0 & 0 & 0 \\ 0 & 1 & \cos\delta & -\sin\delta \\ 0 & 0 & \sin\delta & \cos\delta \end{pmatrix} \tag{78}$$

Symmetry **2013**, *5*, 233–252

Likewise:

$$B(\eta) = \begin{pmatrix} e^{\eta/2} & 0 \\ 0 & e^{-\eta/2} \end{pmatrix} \rightarrow \begin{pmatrix} \cosh\eta & \sinh\eta & 0 & 0 \\ \sinh\eta & \cosh\eta & 0 & 0 \\ 0 & 0 & 1 & 0 \\ 0 & 0 & 0 & 1 \end{pmatrix} \tag{79}$$

$$R(\theta) = \begin{pmatrix} \cos(\theta/2) & -\sin(\theta/2) \\ \sin(\theta/2) & \sin(\theta/2) \end{pmatrix} \rightarrow \begin{pmatrix} 1 & 0 & 0 & 0 \\ 0 & \cos\theta & -\sin\theta & 0 \\ 0 & \sin\theta & \cos\theta & 0 \\ 0 & 0 & 0 & 1 \end{pmatrix} \tag{80}$$

and:

$$S(\lambda) = \begin{pmatrix} \cosh(\lambda/2) & \sinh(\lambda/2) \\ \sinh(\lambda/2) & \sinh(\lambda/2) \end{pmatrix} \rightarrow \begin{pmatrix} \cosh\lambda & 0 & \sinh\lambda & 0 \\ 0 & 1 & 0 & 0 \\ \sinh\lambda & 0 & \cosh\lambda & 0 \\ 0 & 0 & 0 & 1 \end{pmatrix} \tag{81}$$

References

1. Azzam, R.A.M.; Bashara, I. *Ellipsometry and Polarized Light*; North-Holland: Amsterdam, The Netherlands, 1977.
2. Born, M.; Wolf, E. *Principles of Optics*, 6th ed.; Pergamon: Oxford, NY, USA, 1980.
3. Brosseau, C. *Fundamentals of Polarized Light: A Statistical Optics Approach*; John Wiley: New York, NY, USA, 1998.
4. Naimark, M.A. Linear representation of the Lorentz group. *Uspekhi Mater. Nauk* **1954**, *9*, 19–93, Translated by Atkinson, F.V., American Mathematical Society Translations, Series 2, **1957**, 6, 379–458.
5. Naimark, M.A. *Linear Representations of the Lorentz Group*; Pergamon Press: Oxford, NY, USA, 1958; Translated by Swinfen, A.; Marstrand, O.J., 1964.
6. Kim, Y.S.; Wigner, E.P. Space-time geometry of relativistic particles. *J. Math. Phys.* **1990**, *31*, 55–60. [CrossRef]
7. Wigner, E. On unitary representations of the inhomogeneous Lorentz group. *Ann. Math.* **1939**, *40*, 149–204. [CrossRef]
8. Kim, Y.S. Poincaré Sphere and Decoherence Problems. Available online: http://arxiv.org/abs/1203.4539 (accessed on 17 June 2013).
9. Kim, Y.S.; Noz, M.E. *Theory and Applications of the Poincaré Group*; Reidel: Dordrecht, The Netherlands, 1986.
10. Han, D.; Kim, Y.S.; Son, D. E(2)-like little group for massless particles and polarization of neutrinos. *Phys. Rev. D* **1982**, *26*, 3717–3725.
11. Başkal, S.; Kim, Y.S. One analytic form for four branches of the ABCD matrix. *J. Mod. Opt.* **2010**, *57*, 1251–1259. [CrossRef]
12. Başkal, S.; Kim, Y.S. Lorentz Group in Ray and Polarization Optics. In *Mathematical Optics: Classical, Quantum and Computational Methods*; Lakshminarayanan, V., Calvo, M.L., Alieva, T., Eds.; CRC Taylor and Francis: New York, NY, USA, 2013; Chapter 9; pp. 303–349.
13. Saleh, B.E.A.; Teich, M.C. *Fundamentals of Photonics*, 2nd ed.; John Wiley: Hoboken, NJ, USA, 2007.
14. Han, D.; Kim, Y.S.; Noz, M.E. Jones-vector formalism as a representation of the Lorentz group. *J. Opt. Soc. Am. A* **1997**, *14*, 2290–2298.
15. Han, D.; Kim, Y.S.; Noz, M.E. Stokes parameters as a Minkowskian four-vector. *Phys. Rev. E* **1997**, *56*, 6065–6076.
16. Feynman, R.P. *Statistical Mechanics*; Benjamin/Cummings: Reading, MA, USA, 1972.
17. Han, D.; Kim, Y.S.; Noz, M.E. Illustrative example of Feynman's rest of the universe. *Am. J. Phys.* **1999**, *67*, 61–66. [CrossRef]
18. Pellat-Finet, P. Geometric approach to polarization optics. II. Quarternionic representation of polarized light. *Optik* **1991**, *87*, 68–76.
19. Dlugunovich, V.A.; Kurochkin, Y.A. Vector parameterization of the Lorentz group transformations and polar decomposition of Mueller matrices. *Opt. Spectrosc.* **2009**, *107*, 312–317. [CrossRef]

Symmetry **2013**, *5*, 233–252

20. Tudor, T. Vectorial Pauli algebraic approach in polarization optics. I. Device and state operators. *Optik* **2010**, *121*, 1226–1235. [CrossRef]
21. Fedorov, F.I. Vector parametrization of the Lorentz group and relativistic kinematics. *Theor. Math. Phys.* **1970**, *2*, 248–252. [CrossRef]
22. Fedorov, F.I. *Lorentz Group*; [in Russian]; Global Science, Physical-Mathematical Literature: Moscow, Russia, 1979.
23. Başkal, S.; Kim, Y.S. De Sitter group as a symmetry for optical decoherence. *J. Phys. A* **2006**, *39*, 7775–7788.
24. Dargys, A. Optical Mueller matrices in terms of geometric algebra. *Opt. Commun.* **2012**, *285*, 4785–4792. [CrossRef]
25. Pellat-Finet, P.; Basset, M. What is common to both polarization optics and relativistic kinematics? *Optik* **1992**, *90*, 101–106.

© 2013 by the authors. Licensee MDPI, Basel, Switzerland. This article is an open access article distributed under the terms and conditions of the Creative Commons Attribution (CC BY) license (http://creativecommons.org/licenses/by/4.0/).

symmetry

MDPI

Article

Wigner's Space-Time Symmetries Based on the Two-by-Two Matrices of the Damped Harmonic Oscillators and the Poincaré Sphere

Sibel Başkal [1], Young S. Kim [2],* and Marilyn E. Noz [3]

[1] Department of Physics, Middle East Technical University, Ankara 06800, Turkey; E-Mail: baskal@newton.physics.metu.edu.tr

[2] Center for Fundamental Physics, University of Maryland, College Park, MD 20742, USA

[3] Department of Radiology, New York University, New York, NY 10016, USA; E-Mail: marilyne.noz@gmail.com

* E-Mail: yskim@umd.edu; Tel.: +1-301-937-1306.

Received: 28 February 2014; in revised form: 28 May 2014 / Accepted: 9 June 2014 / Published: 25 June 2014

Abstract: The second-order differential equation for a damped harmonic oscillator can be converted to two coupled first-order equations, with two two-by-two matrices leading to the group $Sp(2)$. It is shown that this oscillator system contains the essential features of Wigner's little groups dictating the internal space-time symmetries of particles in the Lorentz-covariant world. The little groups are the subgroups of the Lorentz group whose transformations leave the four-momentum of a given particle invariant. It is shown that the damping modes of the oscillator correspond to the little groups for massive and imaginary-mass particles respectively. When the system makes the transition from the oscillation to damping mode, it corresponds to the little group for massless particles. Rotations around the momentum leave the four-momentum invariant. This degree of freedom extends the $Sp(2)$ symmetry to that of $SL(2,c)$ corresponding to the Lorentz group applicable to the four-dimensional Minkowski space. The Poincaré sphere contains the $SL(2,c)$ symmetry. In addition, it has a non-Lorentzian parameter allowing us to reduce the mass continuously to zero. It is thus possible to construct the little group for massless particles from that of the massive particle by reducing its mass to zero. Spin-1/2 particles and spin-1 particles are discussed in detail.

Keywords: damped harmonic oscillators; coupled first-order equations; unimodular matrices; Wigner's little groups; Poincaré sphere; $Sp(2)$ group; $SL(2,c)$ group; gauge invariance; neutrinos; photons

PACS: 03.65.Fd, 03.67.-a, 05.30.-d

1. Introduction

We are quite familiar with the second-order differential equation

$$m\frac{d^2y}{dt^2} + b\frac{dy}{dt} + Ky = 0 \tag{1}$$

for a damped harmonic oscillator. This equation has the same mathematical form as

$$L\frac{d^2Q}{dt^2} + R\frac{dQ}{dt} + \frac{1}{C}Q = 0 \tag{2}$$

for electrical circuits, where $L, R,$ and C are the inductance, resistance, and capacitance respectively. These two equations play fundamental roles in physical and engineering sciences. Since they start from the same set of mathematical equations, one set of problems can be studied in terms of the other. For instance, many mechanical phenomena can be studied in terms of electrical circuits.

In Equation (1), when $b = 0$, the equation is that of a simple harmonic oscillator with the frequency $\omega = \sqrt{K/m}$. As b increases, the oscillation becomes damped. When b is larger than $2\sqrt{Km}$, the oscillation disappears, as the solution is a damping mode.

Consider that increasing b continuously, while difficult mechanically, can be done electrically using Equation (2) by adjusting the resistance R. The transition from the oscillation mode to the damping mode is a continuous physical process.

This b term leads to energy dissipation, but is not regarded as a fundamental force. It is inconvenient in the Hamiltonian formulation of mechanics and troublesome in transition to quantum mechanics, yet, plays an important role in classical mechanics. In this paper this term will help us understand the fundamental space-time symmetries of elementary particles.

We are interested in constructing the fundamental symmetry group for particles in the Lorentz-covariant world. For this purpose, we transform the second-order differential equation of Equation (1) to two coupled first-order equations using two-by-two matrices. Only two linearly independent matrices are needed. They are the anti-symmetric and symmetric matrices

$$A = \begin{pmatrix} 0 & -i \\ i & 0 \end{pmatrix}, \quad \text{and} \quad S = \begin{pmatrix} 0 & i \\ i & 0 \end{pmatrix} \tag{3}$$

respectively. The anti-symmetric matrix A is Hermitian and corresponds to the oscillation part, while the symmetric S matrix corresponds to the damping.

These two matrices lead to the $Sp(2)$ group consisting of two-by-two unimodular matrices with real elements. This group is isomorphic to the three-dimensional Lorentz group applicable to two space-like and one time-like coordinates. This group is commonly called the $O(2,1)$ group.

This $O(2,1)$ group can explain all the essential features of Wigner's little groups dictating internal space-time symmetries of particles [1]. Wigner defined his little groups as the subgroups of the Lorentz group whose transformations leave the four-momentum of a given particle invariant. He observed that the little groups are different for massive, massless, and imaginary-mass particles. It has been a challenge to design a mathematical model which will combine those three into one formalism, but we show that the damped harmonic oscillator provides the desired mathematical framework.

For the two space-like coordinates, we can assign one of them to the direction of the momentum, and the other to the direction perpendicular to the momentum. Let the direction of the momentum be along the z axis, and let the perpendicular direction be along the x axis. We therefore study the kinematics of the group within the zx plane, then see what happens when we rotate the system around the z axis without changing the momentum [2].

The Poincaré sphere for polarization optics contains the $SL(2,c)$ symmetry isomorphic to the four-dimensional Lorentz group applicable to the Minkowski space [3–7]. Thus, the Poincaré sphere extends Wigner's picture into the three space-like and one time-like coordinates. Specifically, this extension adds rotations around the given momentum which leaves the four-momentum invariant [2].

While the particle mass is a Lorentz-invariant variable, the Poincaré sphere contains an extra variable which allows the mass to change. This variable allows us to take the mass-limit of the symmetry operations. The transverse rotational degrees of freedom collapse into one gauge degree of freedom and polarization of neutrinos is a consequence of the requirement of gauge invariance [8,9].

The $SL(2,c)$ group contains symmetries not seen in the three-dimensional rotation group. While we are familiar with two spinors for a spin-1/2 particle in nonrelativistic quantum mechanics, there are two additional spinors due to the reflection properties of the Lorentz group. There are thus 16 bilinear combinations of those four spinors. This leads to two scalars, two four-vectors, and one antisymmetric four-by-four tensor. The Maxwell-type electromagnetic field tensor can be obtained as a massless limit of this tensor [10].

In Section 2, we review the damped harmonic oscillator in classical mechanics, and note that the solution can be either in the oscillation mode or damping mode depending on the magnitude of the damping parameter. The translation of the second order equation into a first order differential equation with two-by-two matrices is possible. This first-order equation is similar to the Schrödinger equation for a spin-1/2 particle in a magnetic field.

Section 3 shows that the two-by-two matrices of Section 2 can be formulated in terms of the $Sp(2)$ group. These matrices can be decomposed into the Bargmann and Wigner decompositions. Furthermore, this group is isomorphic to the three-dimensional Lorentz group with two space and one time-like coordinates.

In Section 4, it is noted that this three-dimensional Lorentz group has all the essential features of Wigner's little groups which dictate the internal space-time symmetries of the particles in the Lorentz-covariant world. Wigner's little groups are the subgroups of the Lorentz group whose transformations leave the four-momentum of a given particle invariant. The Bargmann Wigner decompositions are shown to be useful tools for studying the little groups.

In Section 5, we note that the given momentum is invariant under rotations around it. The addition of this rotational degree of freedom extends the $Sp(2)$ symmetry to the six-parameter $SL(2, c)$ symmetry. In the space-time language, this extends the three dimensional group to the Lorentz group applicable to three space and one time dimensions.

Section 6 shows that the Poincaré sphere contains the symmetries of $SL(2, c)$ group. In addition, it contains an extra variable which allows us to change the mass of the particle, which is not allowed in the Lorentz group.

In Section 7, the symmetries of massless particles are studied in detail. In addition to rotation around the momentum, Wigner's little group generates gauge transformations. While gauge transformations on spin-1 photons are well known, the gauge invariance leads to the polarization of massless spin-1/2 particles, as observed in neutrino polarizations.

In Section 8, it is noted that there are four spinors for spin-1/2 particles in the Lorentz-covariant world. It is thus possible to construct 16 bilinear forms, applicable to two scalars, and two vectors, and one antisymmetric second-rank tensor. The electromagnetic field tensor is derived as the massless limit. This tensor is shown to be gauge-invariant.

2. Classical Damped Oscillators

For convenience, we write Equation (1) as

$$\frac{d^2y}{dt^2} + 2\mu \frac{dy}{dt} + \omega^2 y = 0 \tag{4}$$

with

$$\omega = \sqrt{\frac{K}{m}}, \quad \text{and} \quad \mu = \frac{b}{2m} \tag{5}$$

The damping parameter μ is positive when there are no external forces. When ω is greater than μ, the solution takes the form

$$y = e^{-\mu t} \left[C_1 \cos(\omega' t) + C_2 \sin(\omega' t) \right] \tag{6}$$

where

$$\omega' = \sqrt{\omega^2 - \mu^2} \tag{7}$$

and C_1 and C_2 are the constants to be determined by the initial conditions. This expression is for a damped harmonic oscillator. Conversely, when μ is greater than ω, the quantity inside the square-root sign is negative, then the solution becomes

$$y = e^{-\mu t} \left[C_3 \cosh(\mu' t) + C_4 \sinh(\mu' t) \right] \tag{8}$$

with

$$\mu' = \sqrt{\mu^2 - \omega^2} \tag{9}$$

If $\omega = \mu$, both Equations (6) and (8) collapse into one solution

$$y(t) = e^{-\mu t}[C_5 + C_6\, t] \tag{10}$$

These three different cases are treated separately in textbooks. Here we are interested in the transition from Equation (6) to Equation (8), via Equation (10). For convenience, we start from μ greater than ω with μ' given by Equation (9).

For a given value of μ, the square root becomes zero when ω equals μ. If ω becomes larger, the square root becomes imaginary and divides into two branches.

$$\pm i\sqrt{\omega^2 - \mu^2} \tag{11}$$

This is a continuous transition, but not an analytic continuation. To study this in detail, we translate the second order differential equation of Equation (4) into the first-order equation with two-by-two matrices.

Given the solutions of Equations (6) and (10), it is convenient to use $\psi(t)$ defined as

$$\psi(t) = e^{\mu t}y(t), \quad \text{and} \quad y = e^{-\mu t}\psi(t) \tag{12}$$

Then $\psi(t)$ satisfies the differential equation

$$\frac{d^2\psi(t)}{dt^2} + (\omega^2 - \mu^2)\psi(t) = 0 \tag{13}$$

2.1. Two-by-Two Matrix Formulation

In order to convert this second-order equation to a first-order system, we introduce $\psi_1(t)$ and $\psi_2(t)$ satisfying two coupled differential equations

$$\frac{d\psi_1(t)}{dt} = (\mu - \omega)\psi_2(t) \tag{14}$$

$$\frac{d\psi_2(t)}{dt} = (\mu + \omega)\psi_1(t) \tag{15}$$

which can be written in matrix form as

$$\frac{d}{dt}\begin{pmatrix} \psi_1 \\ \psi_2 \end{pmatrix} = \begin{pmatrix} 0 & \mu - \omega \\ \mu + \omega & 0 \end{pmatrix}\begin{pmatrix} \psi_1 \\ \psi_2 \end{pmatrix} \tag{16}$$

Using the Hermitian and anti-Hermitian matrices of Equation (3) in Section 1, we construct the linear combination

$$H = \omega\begin{pmatrix} 0 & -i \\ i & 0 \end{pmatrix} + \mu\begin{pmatrix} 0 & i \\ i & 0 \end{pmatrix} \tag{17}$$

We can then consider the first-order differential equation

$$i\frac{\partial}{\partial t}\psi(t) = H\psi(t) \tag{18}$$

While this equation is like the Schrödinger equation for an electron in a magnetic field, the two-by-two matrix is not Hermitian. Its first matrix is Hermitian, but the second matrix is anti-Hermitian. It is of course an interesting problem to give a physical interpretation to this non-Hermitian matrix in connection with quantum dissipation [11], but this is beyond the scope of the present paper. The solution of Equation (18) is

$$\psi(t) = \exp\left\{ \begin{pmatrix} 0 & -\omega + \mu \\ \omega + \mu & 0 \end{pmatrix} t \right\} \begin{pmatrix} C_7 \\ C_8 \end{pmatrix} \tag{19}$$

where $C_7 = \psi_1(0)$ and $C_8 = \psi_2(0)$ respectively.

2.2. Transition from the Oscillation Mode to Damping Mode

It appears straight-forward to compute this expression by a Taylor expansion, but it is not. This issue was extensively discussed in the earlier papers by two of us [12,13]. The key idea is to write the matrix

$$\begin{pmatrix} 0 & -\omega + \mu \\ \omega + \mu & 0 \end{pmatrix} \tag{20}$$

as a similarity transformation of

$$\omega' \begin{pmatrix} 0 & -1 \\ 1 & 0 \end{pmatrix} \qquad (\omega > \mu) \tag{21}$$

and as that of

$$\mu' \begin{pmatrix} 0 & 1 \\ 1 & 0 \end{pmatrix} \qquad (\mu > \omega) \tag{22}$$

with ω' and μ' defined in Equations (7) and (9), respectively.
Then the Taylor expansion leads to

$$\begin{pmatrix} \cos(\omega't) & -\sqrt{(\omega - \mu)/(\omega + \mu)} \, \sin(\omega't) \\ \sqrt{(\omega + \mu)/(\omega - \mu)} \, \sin(\omega't) & \cos(\omega't) \end{pmatrix} \tag{23}$$

when ω is greater than μ. The solution $\psi(t)$ takes the form

$$\begin{pmatrix} C_7 \cos(\omega't) - C_8\sqrt{(\omega - \mu)/(\omega + \mu)} \, \sin(\omega't) \\ C_7\sqrt{(\omega + \mu)/(\omega - \mu)} \, \sin(\omega't) + C_8 \cos(\omega't) \end{pmatrix} \tag{24}$$

If μ is greater than ω, the Taylor expansion becomes

$$\begin{pmatrix} \cosh(\mu't) & \sqrt{(\mu - \omega)/(\mu + \omega)} \, \sinh(\mu't) \\ \sqrt{(\mu + \omega)/(\mu - \omega)} \, \sinh(\mu't) & \cosh(\mu't) \end{pmatrix} \tag{25}$$

When ω is equal to μ, both Equations (23) and (25) become

$$\begin{pmatrix} 1 & 0 \\ 2\omega t & 1 \end{pmatrix} \tag{26}$$

If ω is sufficiently close to but smaller than μ, the matrix of Equation (25) becomes

$$\begin{pmatrix} 1 + (\epsilon/2)(2\omega t)^2 & +\epsilon(2\omega t) \\ (2\omega t) & 1 + (\epsilon/2)(2\omega t)^2 \end{pmatrix} \tag{27}$$

with

$$\epsilon = \frac{\mu - \omega}{\mu + \omega} \tag{28}$$

If ω is sufficiently close to μ, we can let

$$\mu + \omega = 2\omega, \quad \text{and} \quad \mu - \omega = 2\mu\epsilon \tag{29}$$

If ω is greater than μ, ϵ defined in Equation (28) becomes negative, the matrix of Equation (23) becomes

$$\begin{pmatrix} 1 - (-\epsilon/2)(2\omega t)^2 & -(-\epsilon)(2\omega t) \\ 2\omega t & 1 - (-\epsilon/2)(2\omega t)^2 \end{pmatrix} \tag{30}$$

We can rewrite this matrix as

$$\begin{pmatrix} 1 - (1/2)\left[\left(2\omega\sqrt{-\epsilon}\right)t\right]^2 & -\sqrt{-\epsilon}\left[\left(2\omega\sqrt{-\epsilon}\right)t\right] \\ 2\omega t & 1 - (1/2)\left[\left(2\omega\sqrt{-\epsilon}\right)t\right]^2 \end{pmatrix} \tag{31}$$

If ϵ becomes positive, Equation (27) can be written as

$$\begin{pmatrix} 1 + (1/2)\left[\left(2\omega\sqrt{\epsilon}\right)t\right]^2 & \sqrt{\epsilon}\left[\left(2\omega\sqrt{\epsilon}\right)t\right] \\ 2\omega t & 1 + (1/2)\left[\left(2\omega\sqrt{\epsilon}\right)t\right]^2 \end{pmatrix} \tag{32}$$

The transition from Equation (31) to Equation (32) is continuous as they become identical when $\epsilon = 0$. As ϵ changes its sign, the diagonal elements of above matrices tell us how $\cos(\omega' t)$ becomes $\cosh(\mu' t)$. As for the upper-right element element, $-\sin(\omega' t)$ becomes $\sinh(\mu' t)$. This non-analytic continuity is discussed in detail in one of the earlier papers by two of us on lens optics [13]. This type of continuity was called there "tangential continuity." There, the function and its first derivative are continuous while the second derivative is not.

2.3. Mathematical Forms of the Solutions

In this section, we use the Heisenberg approach to the problem, and obtain the solutions in the form of two-by-two matrices. We note that

1. For the oscillation mode, the trace of the matrix is smaller than 2. The solution takes the form of

$$\begin{pmatrix} \cos(x) & -e^{-\eta}\sin(x) \\ e^{\eta}\sin(x) & \cos(x) \end{pmatrix} \tag{33}$$

   with trace $2\cos(x)$. The trace is independent of η.
2. For the damping mode, the trace of the matrix is greater than 2.

$$\begin{pmatrix} \cosh(x) & e^{-\eta}\sinh(x) \\ e^{\eta}\sinh(x) & \cosh(x) \end{pmatrix} \tag{34}$$

   with trace $2\cosh(x)$. Again, the trace is independent of η.
3. For the transition mode, the trace is equal to 2, and the matrix is triangular and takes the form of

$$\begin{pmatrix} 1 & 0 \\ \gamma & 1 \end{pmatrix} \tag{35}$$

When x approaches zero, the Equations (33) and (34) take the form

$$\begin{pmatrix} 1 - x^2/2 & -xe^{-\eta} \\ xe^{\eta} & 1 - x^2/2 \end{pmatrix}, \quad \text{and} \quad \begin{pmatrix} 1 + x^2/2 & xe^{-\eta} \\ xe^{\eta} & 1 + x^2/2 \end{pmatrix} \tag{36}$$

respectively. These two matrices have the same lower-left element. Let us fix this element to be a positive number γ. Then

$$x = \gamma e^{-\eta} \tag{37}$$

Then the matrices of Equation (36) become

$$\begin{pmatrix} 1 - \gamma^2 e^{-2\eta}/2 & -\gamma e^{-2\eta} \\ \gamma & 1 - \gamma^2 e^{-2\eta}/2 \end{pmatrix}, \quad \text{and} \quad \begin{pmatrix} 1 + \gamma^2 e^{-2\eta}/2 & \gamma e^{-2\eta} \\ \gamma & 1 + \gamma^2 e^{-2\eta}/2 \end{pmatrix} \tag{38}$$

If we introduce a small number ϵ defined as

$$\epsilon = \sqrt{\gamma} e^{-\eta} \tag{39}$$

the matrices of Equation (38) become

$$\begin{pmatrix} e^{-\eta/2} & 0 \\ 0 & e^{\eta/2} \end{pmatrix} \begin{pmatrix} 1 - \gamma\epsilon^2/2 & \sqrt{\gamma}\epsilon \\ \sqrt{\gamma}\epsilon & 1 - \gamma\epsilon^2/2 \end{pmatrix} \begin{pmatrix} e^{\eta/2} & 0 \\ 0 & e^{-\eta/2} \end{pmatrix}$$

$$\begin{pmatrix} e^{-\eta/2} & 0 \\ 0 & e^{\eta/2} \end{pmatrix} \begin{pmatrix} 1 + \gamma\epsilon^2/2 & \sqrt{\gamma}\epsilon \\ \sqrt{\gamma}\epsilon & 1 + \gamma\epsilon^2/2 \end{pmatrix} \begin{pmatrix} e^{\eta/2} & 0 \\ 0 & e^{-\eta/2} \end{pmatrix} \tag{40}$$

respectively, with $e^{-\eta} = \epsilon/\sqrt{\gamma}$.

3. Groups of Two-by-Two Matrices

If a two-by-two matrix has four complex elements, it has eight independent parameters. If the determinant of this matrix is one, it is known as an unimodular matrix and the number of independent parameters is reduced to six. The group of two-by-two unimodular matrices is called $SL(2,c)$. This six-parameter group is isomorphic to the Lorentz group applicable to the Minkowski space of three space-like and one time-like dimensions [14].

We can start with two subgroups of $SL(2,c)$.

1. While the matrices of $SL(2,c)$ are not unitary, we can consider the subset consisting of unitary matrices. This subgroup is called $SU(2)$, and is isomorphic to the three-dimensional rotation group. This three-parameter group is the basic scientific language for spin-$1/2$ particles.

2. We can also consider the subset of matrices with real elements. This three-parameter group is called $Sp(2)$ and is isomorphic to the three-dimensional Lorentz group applicable to two space-like and one time-like coordinates.

In the Lorentz group, there are three space-like dimensions with $x, y,$ and z coordinates. However, for many physical problems, it is more convenient to study the problem in the two-dimensional (x, z) plane first and generalize it to three-dimensional space by rotating the system around the z axis. This process can be called Euler decomposition and Euler generalization [2].

First, we study $Sp(2)$ symmetry in detail, and achieve the generalization by augmenting the two-by-two matrix corresponding to the rotation around the z axis. In this section, we study in detail properties of $Sp(2)$ matrices, then generalize them to $SL(2,c)$ in Section 5.

There are three classes of $Sp(2)$ matrices. Their traces can be smaller or greater than two, or equal to two. While these subjects are already discussed in the literature [15–17] our main interest is what happens as the trace goes from less than two to greater than two. Here we are guided by the model we have discussed in Section 2, which accounts for the transition from the oscillation mode to the damping mode.

3.1. Lie Algebra of Sp(2)

The two linearly independent matrices of Equation (3) can be written as

$$K_1 = \frac{1}{2} \begin{pmatrix} 0 & i \\ i & 0 \end{pmatrix}, \quad \text{and} \quad J_2 = \frac{1}{2} \begin{pmatrix} 0 & -i \\ i & 0 \end{pmatrix} \tag{41}$$

However, the Taylor series expansion of the exponential form of Equation (23) or Equation (25) requires an additional matrix

$$K_3 = \frac{1}{2} \begin{pmatrix} i & 0 \\ 0 & -i \end{pmatrix} \tag{42}$$

These matrices satisfy the following closed set of commutation relations.

$$[K_1, J_2] = iK_3, \qquad [J_2, K_3] = iK_1, \qquad [K_3, K_1] = -iJ_2 \tag{43}$$

These commutation relations remain invariant under Hermitian conjugation, even though K_1 and K_3 are anti-Hermitian. The algebra generated by these three matrices is known in the literature as the group $Sp(2)$ [17]. Furthermore, the closed set of commutation relations is commonly called the Lie algebra. Indeed, Equation (43) is the Lie algebra of the $Sp(2)$ group.

The Hermitian matrix J_2 generates the rotation matrix

$$R(\theta) = \exp\left(-i\theta J_2\right) = \begin{pmatrix} \cos(\theta/2) & -\sin(\theta/2) \\ \sin(\theta/2) & \cos(\theta/2) \end{pmatrix} \tag{44}$$

and the anti-Hermitian matrices K_1 and K_2, generate the following squeeze matrices.

$$S(\lambda) = \exp\left(-i\lambda K_1\right) = \begin{pmatrix} \cosh(\lambda/2) & \sinh(\lambda/2) \\ \sinh(\lambda/2) & \cosh(\lambda/2) \end{pmatrix} \tag{45}$$

and

$$B(\eta) = \exp\left(-i\eta K_3\right) = \begin{pmatrix} \exp\left(\eta/2\right) & 0 \\ 0 & \exp\left(-\eta/2\right) \end{pmatrix} \tag{46}$$

respectively.

Returning to the Lie algebra of Equation (43), since K_1 and K_3 are anti-Hermitian, and J_2 is Hermitian, the set of commutation relation is invariant under the Hermitian conjugation. In other words, the commutation relations remain invariant, even if we change the sign of K_1 and K_3, while keeping that of J_2 invariant. Next, let us take the complex conjugate of the entire system. Then both the J and K matrices change their signs.

3.2. Bargmann and Wigner Decompositions

Since the $Sp(2)$ matrix has three independent parameters, it can be written as [15]

$$\begin{pmatrix} \cos\left(\alpha_1/2\right) & -\sin\left(\alpha_1/2\right) \\ \sin\left(\alpha_1/2\right) & \cos\left(\alpha_1/2\right) \end{pmatrix} \begin{pmatrix} \cosh\chi & \sinh\chi \\ \sinh\chi & \cosh\chi \end{pmatrix} \begin{pmatrix} \cos\left(\alpha_2/2\right) & -\sin\left(\alpha_2/2\right) \\ \sin\left(\alpha_2/2\right) & \cos\left(\alpha_2/2\right) \end{pmatrix} \tag{47}$$

This matrix can be written as

$$\begin{pmatrix} \cos(\delta/2) & -\sin(\delta/2) \\ \sin(\delta/2) & \cos(\delta/2) \end{pmatrix} \begin{pmatrix} a & b \\ c & d \end{pmatrix} \begin{pmatrix} \cos(\delta/2) & \sin(\delta/2) \\ -\sin(\delta/2) & \cos(\delta/2) \end{pmatrix} \tag{48}$$

where

$$\begin{pmatrix} a & b \\ c & d \end{pmatrix} = \begin{pmatrix} \cos(\alpha/2) & -\sin(\alpha/2) \\ \sin(\alpha/2) & \cos(\alpha/2) \end{pmatrix} \begin{pmatrix} \cosh\chi & \sinh\chi \\ \sinh\chi & \cosh\chi \end{pmatrix} \begin{pmatrix} \cos(\alpha/2) & -\sin(\alpha/2) \\ \sin(\alpha/2) & \cos(\alpha/2) \end{pmatrix} \tag{49}$$

with

$$\delta = \frac{1}{2}(\alpha_1 - \alpha_2), \quad \text{and} \quad \alpha = \frac{1}{2}(\alpha_1 + \alpha_2) \tag{50}$$

If we complete the matrix multiplication of Equation (49), the result is

$$\begin{pmatrix} (\cosh\chi)\cos\alpha & \sinh\chi - (\cosh\chi)\sin\alpha \\ \sinh\chi + (\cosh\chi)\sin\alpha & (\cosh\chi)\cos\alpha \end{pmatrix} \tag{51}$$

We shall call hereafter the decomposition of Equation (49) the Bargmann decomposition. This means that every matrix in the $Sp(2)$ group can be brought to the Bargmann decomposition by a similarity transformation of rotation, as given in Equation (48). This decomposition leads to an equidiagonal matrix with two independent parameters.

For the matrix of Equation (49), we can now consider the following three cases. Let us assume that χ is positive, and the angle θ is less than 90°. Let us look at the upper-right element.

1. If it is negative with $[\sinh\chi < (\cosh\chi)\sin\alpha]$, then the trace of the matrix is smaller than 2, and the matrix can be written as

$$\begin{pmatrix} \cos(\theta/2) & -e^{-\eta}\sin(\theta/2) \\ e^{\eta}\sin(\theta/2) & \cos(\theta/2) \end{pmatrix} \tag{52}$$

with

$$\cos(\theta/2) = (\cosh\chi)\cos\alpha, \quad \text{and} \quad e^{-2\eta} = \frac{(\cosh\chi)\sin\alpha - \sinh\chi}{(\cosh\chi)\sin\alpha + \sinh\chi} \tag{53}$$

2. If it is positive with $[\sinh\chi > (\cosh\chi)\sin\alpha)]$, then the trace is greater than 2, and the matrix can be written as

$$\begin{pmatrix} \cosh(\lambda/2) & e^{-\eta}\sinh(\lambda/2) \\ e^{\eta}\sinh(\lambda/2) & \cosh(\lambda/2) \end{pmatrix} \tag{54}$$

with

$$\cosh(\lambda/2) = (\cosh\chi)\cos\alpha, \quad \text{and} \quad e^{-2\eta} = \frac{\sinh\chi - (\cosh\chi)\sin\alpha}{(\cosh\chi)\sin\alpha + \sinh\chi} \tag{55}$$

3. If it is zero with $[(\sinh\chi = (\cosh\chi)\sin\alpha)]$, then the trace is equal to 2, and the matrix takes the form

$$\begin{pmatrix} 1 & 0 \\ 2\sinh\chi & 1 \end{pmatrix} \tag{56}$$

The above repeats the mathematics given in Section 2.3.

Returning to Equations (52) and (53), they can be decomposed into

$$M(\theta, \eta) = \begin{pmatrix} e^{\eta/2} & 0 \\ 0 & e^{-\eta/2} \end{pmatrix} \begin{pmatrix} \cos(\theta/2) & -\sin(\theta/2) \\ \sin(\theta/2) & \cos(\theta/2) \end{pmatrix} \begin{pmatrix} e^{-\eta/2} & 0 \\ 0 & e^{\eta/2} \end{pmatrix} \tag{57}$$

and

$$M(\lambda, \eta) = \begin{pmatrix} e^{\eta/2} & 0 \\ 0 & e^{-\eta/2} \end{pmatrix} \begin{pmatrix} \cosh(\lambda/2) & \sinh(\lambda/2) \\ \sinh(\lambda/2) & \cos(\lambda/2) \end{pmatrix} \begin{pmatrix} e^{-\eta/2} & 0 \\ 0 & e^{\eta/2} \end{pmatrix} \tag{58}$$

respectively. In view of the physical examples given in Section 6, we shall call this the "Wigner decomposition." Unlike the Bargmann decomposition, the Wigner decomposition is in the form of a similarity transformation.

We note that both Equations (57) and (58) are written as similarity transformations. Thus

$$[M(\theta, \eta)]^n = \begin{pmatrix} \cos(n\theta/2) & -e^{-\eta}\sin(n\theta/2) \\ e^{\eta}\sin(n\theta/2) & \cos(n\theta/2) \end{pmatrix} \tag{59}$$

$$[M(\lambda, \eta)]^n = \begin{pmatrix} \cosh(n\lambda/2) & e^{\eta}\sinh(n\lambda/2) \\ e^{-\eta}\sinh(n\lambda/2) & \cosh(n\lambda/2) \end{pmatrix} \tag{60}$$

$$[M(\gamma)]^n = \begin{pmatrix} 1 & 0 \\ n\gamma & 1 \end{pmatrix} \tag{61}$$

These expressions are useful for studying periodic systems [18].

The question is what physics these decompositions describe in the real world. To address this, we study what the Lorentz group does in the real world, and study isomorphism between the $Sp(2)$ group and the Lorentz group applicable to the three-dimensional space consisting of one time and two space coordinates.

3.3. Isomorphism with the Lorentz Group

The purpose of this section is to give physical interpretations of the mathematical formulas given in Section 3.2. We will interpret these formulae in terms of the Lorentz transformations which are normally described by four-by-four matrices. For this purpose, it is necessary to establish a correspondence between the two-by-two representation of Section 3.2 and the four-by-four representations of the Lorentz group.

Let us consider the Minkowskian space-time four-vector

$$(t, z, x, y) \tag{62}$$

where $(t^2 - z^2 - x^2 - y^2)$ remains invariant under Lorentz transformations. The Lorentz group consists of four-by-four matrices performing Lorentz transformations in the Minkowski space.

In order to give physical interpretations to the three two-by-two matrices given in Equations (44)–(46), we consider rotations around the y axis, boosts along the x axis, and boosts along the z axis. The transformation is restricted in the three-dimensional subspace of (t, z, x). It is then straight-forward to construct those four-by-four transformation matrices where the y coordinate remains invariant. They are given in Table 1. Their generators also given. Those four-by-four generators satisfy the Lie algebra given in Equation (43).

Table 1. Matrices in the two-by-two representation, and their corresponding four-by-four generators and transformation matrices.

Matrices	Generators	Four-by-Four	Transform matrices
$R(\theta)$	$J_2 = \frac{1}{2}\begin{pmatrix} 0 & -i \\ i & 0 \end{pmatrix}$	$\begin{pmatrix} 0 & 0 & 0 & 0 \\ 0 & 0 & -i & 0 \\ 0 & i & 0 & 0 \\ 0 & 0 & 0 & 0 \end{pmatrix}$	$\begin{pmatrix} 1 & 0 & 0 & 0 \\ 0 & \cos\theta & -\sin\theta & 0 \\ 0 & \sin\theta & \cos\theta & 0 \\ 0 & 0 & 0 & 1 \end{pmatrix}$
$B(\eta)$	$K_3 = \frac{1}{2}\begin{pmatrix} i & 0 \\ 0 & -i \end{pmatrix}$	$\begin{pmatrix} 0 & i & 0 & 0 \\ i & 0 & 0 & 0 \\ 0 & 0 & 0 & 0 \\ 0 & 0 & 0 & 0 \end{pmatrix}$	$\begin{pmatrix} \cosh\eta & \sinh\eta & 0 & 0 \\ \sinh\eta & \cosh\eta & 0 & 0 \\ 0 & 0 & 1 & 0 \\ 0 & 0 & 0 & 1 \end{pmatrix}$
$S(\lambda)$	$K_1 = \frac{1}{2}\begin{pmatrix} 0 & i \\ i & 0 \end{pmatrix}$	$\begin{pmatrix} 0 & 0 & i & 0 \\ 0 & 0 & 0 & 0 \\ i & 0 & 0 & 0 \\ 0 & 0 & 0 & 0 \end{pmatrix}$	$\begin{pmatrix} \cosh\lambda & 0 & \sinh\lambda & 0 \\ 0 & 1 & 0 & 0 \\ \sinh\lambda & 0 & \cosh\lambda & 0 \\ 0 & 0 & 0 & 1 \end{pmatrix}$

4. Internal Space-Time Symmetries

We have seen that there corresponds a two-by-two matrix for each four-by-four Lorentz transformation matrix. It is possible to give physical interpretations to those four-by-four matrices. It must thus be possible to attach a physical interpretation to each two-by-two matrix.

Since 1939 [1] when Wigner introduced the concept of the little groups many papers have been published on this subject, but most of them were based on the four-by-four representation. In this section, we shall give the formalism of little groups in the language of two-by-two matrices. In so doing, we provide physical interpretations to the Bargmann and Wigner decompositions introduced in Section 3.2.

4.1. Wigner's Little Groups

In [1], Wigner started with a free relativistic particle with momentum, then constructed subgroups of the Lorentz group whose transformations leave the four-momentum invariant. These subgroups thus define the internal space-time symmetry of the given particle. Without loss of generality, we assume that the particle momentum is along the z direction. Thus rotations around the momentum leave the momentum invariant, and this degree of freedom defines the helicity, or the spin parallel to the momentum.

We shall use the word "Wigner transformation" for the transformation which leaves the four-momentum invariant:

1. For a massive particle, it is possible to find a Lorentz frame where it is at rest with zero momentum. The four-momentum can be written as $m(1,0,0,0)$, where m is the mass. This four-momentum is invariant under rotations in the three-dimensional (z, x, y) space.
2. For an imaginary-mass particle, there is the Lorentz frame where the energy component vanishes. The momentum four-vector can be written as $p(0,1,0,0)$, where p is the magnitude of the momentum.
3. If the particle is massless, its four-momentum becomes $p(1,1,0,0)$. Here the first and second components are equal in magnitude.

The constant factors in these four-momenta do not play any significant roles. Thus we write them as $(1,0,0,0), (0,1,0,0)$, and $(1,1,0,0)$ respectively. Since Wigner worked with these three specific four-momenta [1], we call them Wigner four-vectors.

All of these four-vectors are invariant under rotations around the z axis. The rotation matrix is

$$Z(\phi) = \begin{pmatrix} 1 & 0 & 0 & 0 \\ 0 & 1 & 0 & 0 \\ 0 & 0 & \cos\phi & -\sin\phi \\ 0 & 0 & \sin\phi & \cos\phi \end{pmatrix} \tag{63}$$

In addition, the four-momentum of a massive particle is invariant under the rotation around the y axis, whose four-by-four matrix was given in Table 1. The four-momentum of an imaginary particle is invariant under the boost matrix $S(\lambda)$ given in Table 1. The problem for the massless particle is more complicated, but will be discussed in detail in Section 7. See Table 2.

Table 2. Wigner four-vectors and Wigner transformation matrices applicable to two space-like and one time-like dimensions. Each Wigner four-vector remains invariant under the application of its Wigner matrix.

Mass	Wigner Four-Vector	Wigner Transformation
Massive	$(1,0,0,0)$	$\begin{pmatrix} 1 & 0 & 0 & 0 \\ 0 & \cos\theta & -\sin\theta & 0 \\ 0 & \sin\theta & \cos\theta & 0 \\ 0 & 0 & 0 & 1 \end{pmatrix}$
Massless	$(1,1,0,0)$	$\begin{pmatrix} 1+\gamma^2/2 & -\gamma^2/2 & \gamma & 0 \\ \gamma^2/2 & 1-\gamma^2/2 & \gamma & 0 \\ -\gamma & \gamma & 1 & 0 \\ 0 & 0 & 0 & 1 \end{pmatrix}$
Imaginary mass	$(0,1,0,0)$	$\begin{pmatrix} \cosh\lambda & 0 & \sinh\lambda & 0 \\ 0 & 1 & 0 & 0 \\ \sinh\lambda & 0 & \cosh\lambda & 0 \\ 0 & 0 & 0 & 1 \end{pmatrix}$

4.2. Two-by-Two Formulation of Lorentz Transformations

The Lorentz group is a group of four-by-four matrices performing Lorentz transformations on the Minkowskian vector space of (t, z, x, y), leaving the quantity

$$t^2 - z^2 - x^2 - y^2 \tag{64}$$

invariant. It is possible to perform the same transformation using two-by-two matrices [7,14,19].

In this two-by-two representation, the four-vector is written as

$$X = \begin{pmatrix} t+z & x-iy \\ x+iy & t-z \end{pmatrix} \tag{65}$$

where its determinant is precisely the quantity given in Equation (64) and the Lorentz transformation on this matrix is a determinant-preserving, or unimodular transformation. Let us consider the transformation matrix as [7,19]

$$G = \begin{pmatrix} \alpha & \beta \\ \gamma & \delta \end{pmatrix}, \quad \text{and} \quad G^\dagger = \begin{pmatrix} \alpha^* & \gamma^* \\ \beta^* & \delta^* \end{pmatrix} \tag{66}$$

with

$$\det(G) = 1 \tag{67}$$

and the transformation

$$X' = GXG^\dagger \tag{68}$$

Since G is not a unitary matrix, Equation (68) not a unitary transformation, but rather we call this the "Hermitian transformation". Equation (68) can be written as

$$\begin{pmatrix} t'+z' & x'-iy' \\ x+iy & t'-z' \end{pmatrix} = \begin{pmatrix} \alpha & \beta \\ \gamma & \delta \end{pmatrix} \begin{pmatrix} t+z & x-iy \\ x+iy & t-z \end{pmatrix} \begin{pmatrix} \alpha^* & \gamma^* \\ \beta^* & \delta^* \end{pmatrix} \tag{69}$$

It is still a determinant-preserving unimodular transformation, thus it is possible to write this as a four-by-four transformation matrix applicable to the four-vector (t, z, x, y) [7,14].

Since the G matrix starts with four complex numbers and its determinant is one by Equation (67), it has six independent parameters. The group of these G matrices is known to be locally isomorphic to the group of four-by-four matrices performing Lorentz transformations on the four-vector (t, z, x, y). In other words, for each G matrix there is a corresponding four-by-four Lorentz-transform matrix [7].

The matrix G is not a unitary matrix, because its Hermitian conjugate is not always its inverse. This group has a unitary subgroup called $SU(2)$ and another consisting only of real matrices called $Sp(2)$. For this later subgroup, it is sufficient to work with the three matrices $R(\theta), S(\lambda)$, and $B(\eta)$ given in Equations (44)–(46) respectively. Each of these matrices has its corresponding four-by-four matrix applicable to the (t, z, x, y). These matrices with their four-by-four counterparts are tabulated in Table 1.

The energy-momentum four vector can also be written as a two-by-two matrix. It can be written as

$$P = \begin{pmatrix} p_0 + p_z & p_x - ip_y \\ p_x + ip_y & p_0 - p_z \end{pmatrix} \tag{70}$$

with

$$\det(P) = p_0^2 - p_x^2 - p_y^2 - p_z^2 \tag{71}$$

which means

$$\det(P) = m^2 \tag{72}$$

where m is the particle mass.

The Lorentz transformation can be written explicitly as

$$P' = GPG^\dagger \tag{73}$$

or

$$\begin{pmatrix} p_0' + p_z' & p_x' - ip_y' \\ p_x' + ip_y' & E' - p_z' \end{pmatrix} = \begin{pmatrix} \alpha & \beta \\ \gamma & \delta \end{pmatrix} \begin{pmatrix} p_0 + p_z & p_x - ip_y \\ p_x + ip_y & p_0 - p_z \end{pmatrix} \begin{pmatrix} \alpha^* & \gamma^* \\ \beta^* & \delta^* \end{pmatrix} \tag{74}$$

This is an unimodular transformation, and the mass is a Lorentz-invariant variable. Furthermore, it was shown in [7] that Wigner's little groups for massive, massless, and imaginary-mass particles can be explicitly defined in terms of two-by-two matrices.

Wigner's little group consists of two-by-two matrices satisfying

$$P = WPW^\dagger \tag{75}$$

The two-by-two W matrix is not an identity matrix, but tells about the internal space-time symmetry of a particle with a given energy-momentum four-vector. This aspect was not known when Einstein formulated his special relativity in 1905, hence the internal space-time symmetry was not an issue at that time. We call the two-by-two matrix W the Wigner matrix, and call the condition of Equation (75) the Wigner condition.

If determinant of W is a positive number, then P is proportional to

$$P = \begin{pmatrix} 1 & 0 \\ 0 & 1 \end{pmatrix} \tag{76}$$

corresponding to a massive particle at rest, while if the determinant is negative, it is proportional to

$$P = \begin{pmatrix} 1 & 0 \\ 0 & -1 \end{pmatrix} \tag{77}$$

corresponding to an imaginary-mass particle moving faster than light along the z direction, with a vanishing energy component. If the determinant is zero, P is

$$P = \begin{pmatrix} 1 & 0 \\ 0 & 0 \end{pmatrix} \tag{78}$$

which is proportional to the four-momentum matrix for a massless particle moving along the z direction.

For all three cases, the matrix of the form

$$Z(\phi) = \begin{pmatrix} e^{-i\phi/2} & 0 \\ 0 & e^{i\phi/2} \end{pmatrix} \tag{79}$$

will satisfy the Wigner condition of Equation (75). This matrix corresponds to rotations around the z axis.

For the massive particle with the four-momentum of Equation (76), the transformations with the rotation matrix of Equation (44) leave the P matrix of Equation (76) invariant. Together with the $Z(\phi)$ matrix, this rotation matrix leads to the subgroup consisting of the unitary subset of the G matrices. The unitary subset of G is $SU(2)$ corresponding to the three-dimensional rotation group dictating the spin of the particle [14].

For the massless case, the transformations with the triangular matrix of the form

$$\begin{pmatrix} 1 & \gamma \\ 0 & 1 \end{pmatrix} \tag{80}$$

leave the momentum matrix of Equation (78) invariant. The physics of this matrix has a stormy history, and the variable γ leads to a gauge transformation applicable to massless particles [8,9,20,21].

For a particle with an imaginary mass, a W matrix of the form of Equation (45) leaves the four-momentum of Equation (77) invariant.

Table 3 summarizes the transformation matrices for Wigner's little groups for massive, massless, and imaginary-mass particles. Furthermore, in terms of their traces, the matrices given in this subsection can be compared with those given in Section 2.3 for the damped oscillator. The comparisons are given in Table 4.

Of course, it is a challenging problem to have one expression for all three classes. This problem has been discussed in the literature [12], and the damped oscillator case of Section 2 addresses the continuity problem.

Table 3. Wigner vectors and Wigner matrices in the two-by-two representation. The trace of the matrix tells whether the particle m^2 is positive, zero, or negative.

Particle Mass	Four-Momentum	Transform Matrix	Trace
Massive	$\begin{pmatrix} 1 & 0 \\ 0 & 1 \end{pmatrix}$	$\begin{pmatrix} \cos(\theta/2) & -\sin(\theta/2) \\ \sin(\theta/2) & \cos(\theta/2) \end{pmatrix}$	less than 2
Massless	$\begin{pmatrix} 1 & 0 \\ 0 & 0 \end{pmatrix}$	$\begin{pmatrix} 1 & \gamma \\ 0 & 1 \end{pmatrix}$	equal to 2
Imaginary mass	$\begin{pmatrix} 1 & 0 \\ 0 & -1 \end{pmatrix}$	$\begin{pmatrix} \cosh(\lambda/2) & \sinh(\lambda/2) \\ \sinh(\lambda/2) & \cosh(\lambda/2) \end{pmatrix}$	greater than 2

Table 4. Damped Oscillators and Space-time Symmetries. Both share $Sp(2)$ as their symmetry group.

Trace	Damped Oscillator	Particle Symmetry
Smaller than 2	Oscillation Mode	Massive Particles
Equal to 2	Transition Mode	Massless Particles
Larger than 2	Damping Mode	Imaginary-mass Particles

5. Lorentz Completion of Wigner's Little Groups

So far we have considered transformations applicable only to (t, z, x) space. In order to study the full symmetry, we have to consider rotations around the z axis. As previously stated, when a particle moves along this axis, this rotation defines the helicity of the particle.

In [1], Wigner worked out the little group of a massive particle at rest. When the particle gains a momentum along the z direction, the single particle can reverse the direction of momentum, the spin, or both. What happens to the internal space-time symmetries is discussed in this section.

5.1. Rotation around the z Axis

In Section 3, our kinematics was restricted to the two-dimensional space of z and x, and thus includes rotations around the y axis. We now introduce the four-by-four matrix of Equation (63) performing rotations around the z axis. Its corresponding two-by-two matrix was given in Equation (79). Its generator is

$$J_3 = \frac{1}{2} \begin{pmatrix} 1 & 0 \\ 0 & -1 \end{pmatrix} \tag{81}$$

If we introduce this additional matrix for the three generators we used in Sections 3 and 3.2, we end up the closed set of commutation relations

$$[J_i, J_j] = i\epsilon_{ijk}J_k, \qquad [J_i, K_j] = i\epsilon_{ijk}K_k, \qquad [K_i, K_j] = -i\epsilon_{ijk}J_k \tag{82}$$

with

$$J_i = \frac{1}{2}\sigma_i, \quad \text{and} \quad K_i = \frac{i}{2}\sigma_i \tag{83}$$

where σ_i are the two-by-two Pauli spin matrices.

For each of these two-by-two matrices there is a corresponding four-by-four matrix generating Lorentz transformations on the four-dimensional Lorentz group. When these two-by-two matrices are imaginary, the corresponding four-by-four matrices were given in Table 1. If they are real, the corresponding four-by-four matrices were given in Table 5.

Table 5. Two-by-two and four-by-four generators not included in Table 1. The generators given there and given here constitute the set of six generators for $SL(2,c)$ or of the Lorentz group given in Equation (82).

Generator	Two-by-Two	Four-by-Four
J_3	$\frac{1}{2}\begin{pmatrix} 1 & 0 \\ 0 & -1 \end{pmatrix}$	$\begin{pmatrix} 0 & 0 & 0 & 0 \\ 0 & 0 & 0 & 0 \\ 0 & 0 & 0 & -i \\ 0 & 0 & i & 0 \end{pmatrix}$
J_1	$\frac{1}{2}\begin{pmatrix} 0 & 1 \\ 1 & 0 \end{pmatrix}$	$\begin{pmatrix} 0 & 0 & 0 & 0 \\ 0 & 0 & 0 & i \\ 0 & 0 & 0 & 0 \\ 0 & -i & 0 & 0 \end{pmatrix}$
K_2	$\frac{1}{2}\begin{pmatrix} 0 & 1 \\ -1 & 0 \end{pmatrix}$	$\begin{pmatrix} 0 & 0 & 0 & i \\ 0 & 0 & 0 & 0 \\ 0 & 0 & 0 & 0 \\ i & 0 & 0 & 0 \end{pmatrix}$

This set of commutation relations is known as the Lie algebra for the $SL(2,c)$, namely the group of two-by-two elements with unit determinants. Their elements are complex. This set is also the Lorentz group performing Lorentz transformations on the four-dimensional Minkowski space.

This set has many useful subgroups. For the group $SL(2,c)$, there is a subgroup consisting only of real matrices, generated by the two-by-two matrices given in Table 1. This three-parameter subgroup is precisely the $Sp(2)$ group we used in Sections 3 and 3.2. Their generators satisfy the Lie algebra given in Equation (43).

In addition, this group has the following Wigner subgroups governing the internal space-time symmetries of particles in the Lorentz-covariant world [1]:

1. The J_i matrices form a closed set of commutation relations. The subgroup generated by these Hermitian matrices is $SU(2)$ for electron spins. The corresponding rotation group does not change the four-momentum of the particle at rest. This is Wigner's little group for massive particles.

   If the particle is at rest, the two-by-two form of the four-vector is given by Equation (76). The Lorentz transformation generated by J_3 takes the form

   $$\begin{pmatrix} e^{i\phi/2} & 0 \\ 0 & e^{-i\phi/2} \end{pmatrix} \begin{pmatrix} 1 & 0 \\ 0 & 1 \end{pmatrix} \begin{pmatrix} e^{-i\phi/2} & 0 \\ 0 & e^{i\phi/2} \end{pmatrix} = \begin{pmatrix} 1 & 0 \\ 0 & 1 \end{pmatrix} \tag{84}$$

   Similar computations can be carried out for J_1 and J_2.
2. There is another $Sp(2)$ subgroup, generated by $K_1, K_2,$ and J_3. They satisfy the commutation relations

   $$[K_1, K_2] = -iJ_3, \qquad [J_3, K_1] = iK_2, \qquad [K_2, J_3] = iK_1. \tag{85}$$

The Wigner transformation generated by these two-by-two matrices leave the momentum four-vector of Equation (77) invariant. For instance, the transformation matrix generated by K_2 takes the form

$$\exp\left(-i\xi K_2\right) = \begin{pmatrix} \cosh(\xi/2) & i\sinh(\xi/2) \\ i\sinh(\xi/2) & \cosh(\xi/2) \end{pmatrix} \tag{86}$$

and the Wigner transformation takes the form

$$\begin{pmatrix} \cosh(\xi/2) & i\sinh(\xi/2) \\ -i\sinh(\xi/2) & \cosh(\xi/2) \end{pmatrix} \begin{pmatrix} 1 & 0 \\ 0 & -1 \end{pmatrix} \begin{pmatrix} \cosh(\xi/2) & i\sinh(\xi/2) \\ -i\sinh(\xi/2) & \cosh(\xi/2) \end{pmatrix} = \begin{pmatrix} 1 & 0 \\ 0 & -1 \end{pmatrix} \tag{87}$$

Computations with K_2 and J_3 lead to the same result.

Since the determinant of the four-momentum matrix is negative, the particle has an imaginary mass. In the language of the four-by-four matrix, the transformation matrices leave the four-momentum of the form $(0, 1, 0, 0)$ invariant.

3. Furthermore, we can consider the following combinations of the generators:

$$N_1 = K_1 - J_2 = \begin{pmatrix} 0 & i \\ 0 & 0 \end{pmatrix}, \quad \text{and} \quad N_2 = K_2 + J_1 = \begin{pmatrix} 0 & 1 \\ 0 & 0 \end{pmatrix} \tag{88}$$

Together with J_3, they satisfy the the following commutation relations.

$$[N_1, N_2] = 0, \qquad [N_1, J_3] = -iN_2, \qquad [N_2, J_3] = iN_1 \tag{89}$$

In order to understand this set of commutation relations, we can consider an $x\,y$ coordinate system in a two-dimensional space. Then rotation around the origin is generated by

$$J_3 = -i \left(x \frac{\partial}{\partial y} - y \frac{\partial}{\partial x} \right) \tag{90}$$

and the two translations are generated by

$$N_1 = -i \frac{\partial}{\partial x}, \quad \text{and} \quad N_2 = -i \frac{\partial}{\partial y} \tag{91}$$

for the x and y directions respectively. These operators satisfy the commutations relations given in Equation (89).

The two-by-two matrices of Equation (88) generate the following transformation matrix.

$$G(\gamma, \phi) = \exp\left[-i\gamma \left(N_1 \cos\phi + N_2 \sin\phi\right)\right] = \begin{pmatrix} 1 & \gamma e^{-i\phi} \\ 0 & 1 \end{pmatrix} \tag{92}$$

The two-by-two form for the four-momentum for the massless particle is given by Equation (78). The computation of the Hermitian transformation using this matrix is

$$\begin{pmatrix} 1 & \gamma e^{-i\phi} \\ 0 & 1 \end{pmatrix} \begin{pmatrix} 1 & 0 \\ 0 & 0 \end{pmatrix} \begin{pmatrix} 1 & 0 \\ \gamma e^{i\phi} & 1 \end{pmatrix} = \begin{pmatrix} 1 & 0 \\ 0 & 0 \end{pmatrix} \tag{93}$$

confirming that N_1 and N_2, together with J_3, are the generators of the $E(2)$-like little group for massless particles in the two-by-two representation. The transformation that does this in the physical world is described in the following section.

5.2. E(2)-Like Symmetry of Massless Particles

From the four-by-four generators of $K_{1,2}$ and $J_{1,2}$, we can write

$$N_1 = \begin{pmatrix} 0 & 0 & i & 0 \\ 0 & 0 & i & 0 \\ i & -i & 0 & 0 \\ 0 & 0 & 0 & 0 \end{pmatrix}, \quad \text{and} \quad N_2 = \begin{pmatrix} 0 & 0 & 0 & i \\ 0 & 0 & 0 & i \\ 0 & 0 & 0 & 0 \\ i & -i & 0 & 0 \end{pmatrix} \tag{94}$$

These matrices lead to the transformation matrix of the form

$$G(\gamma, \phi) = \begin{pmatrix} 1 + \gamma^2/2 & -\gamma^2/2 & \gamma \cos \phi & \gamma \sin \phi \\ \gamma^2/2 & 1 - \gamma^2/2 & \gamma \cos \phi & \gamma \sin \phi \\ -\gamma \cos \phi & \gamma \cos \phi & 1 & 0 \\ -\gamma \sin \phi & \gamma \sin \phi & 0 & 1 \end{pmatrix} \tag{95}$$

This matrix leaves the four-momentum invariant, as we can see from

$$G(\gamma, \phi) \begin{pmatrix} 1 \\ 1 \\ 0 \\ 0 \end{pmatrix} = \begin{pmatrix} 1 \\ 1 \\ 0 \\ 0 \end{pmatrix} \tag{96}$$

When it is applied to the photon four-potential

$$G(\gamma, \phi) \begin{pmatrix} A_0 \\ A_3 \\ A_1 \\ A_2 \end{pmatrix} = \begin{pmatrix} A_0 \\ A_3 \\ A_1 \\ A_2 \end{pmatrix} + \gamma \left(A_1 \cos \phi + A_2 \sin \phi \right) \begin{pmatrix} 1 \\ 1 \\ 0 \\ 0 \end{pmatrix} \tag{97}$$

with the Lorentz condition which leads to $A_3 = A_0$ in the zero mass case. Gauge transformations are well known for electromagnetic fields and photons. Thus Wigner's little group leads to gauge transformations.

In the two-by-two representation, the electromagnetic four-potential takes the form

$$\begin{pmatrix} 2A_0 & A_1 - iA_2 \\ A_1 + iA_2 & 0 \end{pmatrix} \tag{98}$$

with the Lorentz condition $A_3 = A_0$. Then the two-by-two form of Equation (97) is

$$\begin{pmatrix} 1 & \gamma e^{-i\phi} \\ 0 & 1 \end{pmatrix} \begin{pmatrix} 2A_0 & A_1 - iA_2 \\ A_1 + iA_2 & 0 \end{pmatrix} \begin{pmatrix} 1 & 0 \\ \gamma e^{i\phi} & 1 \end{pmatrix} \tag{99}$$

which becomes

$$\begin{pmatrix} A_0 & A_1 - iA_2 \\ A_1 + iA_2 & 0 \end{pmatrix} + \begin{pmatrix} 2\gamma \left(A_1 \cos \phi - A_2 \sin \phi \right) & 0 \\ 0 & 0 \end{pmatrix} \tag{100}$$

This is the two-by-two equivalent of the gauge transformation given in Equation (97).

For massless spin-1/2 particles starting with the two-by-two expression of $G(\gamma, \phi)$ given in Equation (92), and considering the spinors

$$u = \begin{pmatrix} 1 \\ 0 \end{pmatrix}, \quad \text{and} \quad v = \begin{pmatrix} 0 \\ 1 \end{pmatrix} \tag{101}$$

for spin-up and spin-down states respectively,

$$Gu = u, \quad \text{and} \quad Gv = v + \gamma e^{-i\phi} u \tag{102}$$

This means that the spinor u for spin up is invariant under the gauge transformation while v is not. Thus, the polarization of massless spin-1/2 particle, such as neutrinos, is a consequence of the gauge invariance. We shall continue this discussion in Section 7.

5.3. Boosts along the z Axis

In Sections 4.1 and 5.1, we studied Wigner transformations for fixed values of the four-momenta. The next question is what happens when the system is boosted along the z direction, with the transformation

$$\begin{pmatrix} t' \\ z' \end{pmatrix} = \begin{pmatrix} \cosh \eta & \sinh \eta \\ \sinh \eta & \cosh \eta \end{pmatrix} \begin{pmatrix} t \\ z \end{pmatrix} \tag{103}$$

Then the four-momenta become

$$(\cosh \eta, \sinh \eta, 0, 0), \quad (\sinh \eta, \cosh \eta, 0, 0), \quad e^\eta (1, 1, 0, 0) \tag{104}$$

respectively for massive, imaginary, and massless particles cases. In the two-by-two representation, the boost matrix is

$$\begin{pmatrix} e^{\eta/2} & 0 \\ 0 & e^{-\eta/2} \end{pmatrix} \tag{105}$$

and the four-momenta of Equation (104) become

$$\begin{pmatrix} e^\eta & 0 \\ 0 & e^{-\eta} \end{pmatrix}, \quad \begin{pmatrix} e^\eta & 0 \\ 0 & -e^{-\eta} \end{pmatrix}, \quad \begin{pmatrix} e^\eta & 0 \\ 0 & 0 \end{pmatrix} \tag{106}$$

respectively. These matrices become Equations (76)–(78) respectively when $\eta = 0$.

We are interested in Lorentz transformations which leave a given non-zero momentum invariant. We can consider a Lorentz boost along the direction preceded and followed by identical rotation matrices, as described in Figure 1 and the transformation matrix as

$$\begin{pmatrix} \cos(\alpha/2) & -\sin(\alpha/2) \\ \sin(\alpha/2) & \cos(\alpha/2) \end{pmatrix} \begin{pmatrix} \cosh \chi & -\sinh \chi \\ -\sinh \chi & \cosh \chi \end{pmatrix} \begin{pmatrix} \cos(\alpha/2) & -\sin(\alpha/2) \\ \sin(\alpha/2) & \cos(\alpha/2) \end{pmatrix} \tag{107}$$

which becomes

$$\begin{pmatrix} (\cos \alpha) \cosh \chi & -\sinh \chi - (\sin \alpha) \cosh \chi \\ -\sinh \chi + (\sin \alpha) \cosh \chi & (\cos \alpha) \cosh \chi \end{pmatrix} \tag{108}$$

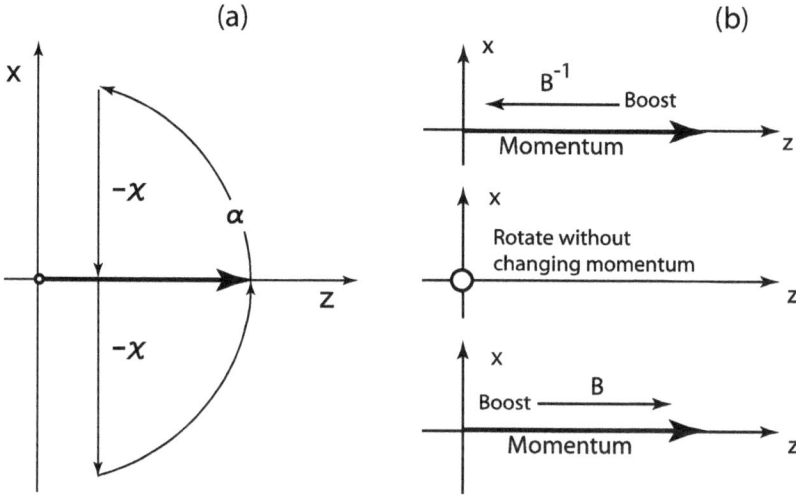

Figure 1. Bargmann and Wigner decompositions. (a) Bargmann decomposition; (b) Wigner decomposition. In the Bargmann decomposition, we start from a momentum along the *z* direction. We can rotate, boost, and rotate to bring the momentum to the original position. The resulting matrix is the product of one boost and two rotation matrices. In the Wigner decomposition, the particle is boosted back to the frame where the Wigner transformation can be applied. Make a Wigner transformation there and come back to the original state of the momentum. This process also can also be written as the product of three simple matrices.

Except the sign of χ, the two-by-two matrices of Equations (107) and (108) are identical with those given in Section 3.2. The only difference is the sign of the parameter χ. We are thus ready to interpret this expression in terms of physics.

1. If the particle is massive, the off-diagonal elements of Equation (108) have opposite signs, and this matrix can be decomposed into

$$\begin{pmatrix} e^{\eta/2} & 0 \\ 0 & e^{-\eta/2} \end{pmatrix} \begin{pmatrix} \cos(\theta/2) & -\sin(\theta/2) \\ \sin(\theta/2) & \cos(\theta/2) \end{pmatrix} \begin{pmatrix} e^{\eta/2} & 0 \\ 0 & e^{-\eta/2} \end{pmatrix} \tag{109}$$

with

$$\cos(\theta/2) = (\cosh \chi) \cos \alpha, \quad \text{and} \quad e^{2\eta} = \frac{\cosh(\chi) \sin \alpha + \sinh \chi}{\cosh(\chi) \sin \alpha - \sinh \chi} \tag{110}$$

and

$$e^{2\eta} = \frac{p_0 + p_z}{p_0 - p_z} \tag{111}$$

According to Equation (109) the first matrix (far right) reduces the particle momentum to zero. The second matrix rotates the particle without changing the momentum. The third matrix boosts the particle to restore its original momentum. This is the extension of Wigner's original idea to moving particles.

2. If the particle has an imaginary mass, the off-diagonal elements of Equation (108) have the same sign,

$$\begin{pmatrix} e^{\eta/2} & 0 \\ 0 & e^{-\eta/2} \end{pmatrix} \begin{pmatrix} \cosh(\lambda/2) & -\sinh(\lambda/2) \\ \sinh(\lambda/2) & \cosh(\lambda/2) \end{pmatrix} \begin{pmatrix} e^{\eta/2} & 0 \\ 0 & e^{-\eta/2} \end{pmatrix} \tag{112}$$

with

$$\cosh(\lambda/2) = (\cosh\chi)\cos\alpha, \quad \text{and} \quad e^{2\eta} = \frac{\sinh\chi + \cosh(\chi)\sin\alpha}{\cosh(\chi)\sin\alpha - \sinh\chi} \tag{113}$$

and

$$e^{2\eta} = \frac{p_0 + p_z}{p_z - p_0} \tag{114}$$

This is also a three-step operation. The first matrix brings the particle momentum to the zero-energy state with $p_0 = 0$. Boosts along the x or y direction do not change the four-momentum. We can then boost the particle back to restore its momentum. This operation is also an extension of the Wigner's original little group. Thus, it is quite appropriate to call the formulas of Equations (109) and (112) Wigner decompositions.

3. If the particle mass is zero with

$$\sinh\chi = (\cosh\chi)\sin\alpha \tag{115}$$

the η parameter becomes infinite, and the Wigner decomposition does not appear to be useful. We can then go back to the Bargmann decomposition of Equation (107). With the condition of Equations (115) and (108) becomes

$$\begin{pmatrix} 1 & -\gamma \\ 0 & 1 \end{pmatrix} \tag{116}$$

with

$$\gamma = 2\sinh\chi \tag{117}$$

The decomposition ending with a triangular matrix is called the Iwasawa decomposition [16,22] and its physical interpretation was given in Section 5.2. The γ parameter does not depend on η.

Thus, we have given physical interpretations to the Bargmann and Wigner decompositions given in Section (3.2). Consider what happens when the momentum becomes large. Then η becomes large for nonzero mass cases. All three four-momenta in Equation (106) become

$$e^{\eta}\begin{pmatrix} 1 & 0 \\ 0 & 0 \end{pmatrix} \tag{118}$$

As for the Bargmann-Wigner matrices, they become the triangular matrix of Equation (116), with $\gamma = \sin(\theta/2)e^{\eta}$ and $\gamma = \sinh(\lambda/2)e^{\eta}$, respectively for the massive and imaginary-mass cases.

In Section 5.2, we concluded that the triangular matrix corresponds to gauge transformations. However, particles with imaginary mass are not observed. For massive particles, we can start with the three-dimensional rotation group. The rotation around the z axis is called helicity, and remains invariant under the boost along the z direction. As for the transverse rotations, they become gauge transformation as illustrated in Table 6.

Table 6. Covariance of the energy-momentum relation, and covariance of the internal space-time symmetry. Under the Lorentz boost along the z direction, J_3 remains invariant, and this invariant component of the angular momentum is called the helicity. The transverse component J_1 and J_2 collapse into a gauge transformation. The γ parameter for the massless case has been studied in earlier papers in the four-by-four matrix formulation of Wigner's little groups [8,21].

Massive, Slow	Covariance	Massless, Fast
$E = p^2/2m$	Einstein's $E = mc^2$	$E = cp$
J_3		Helicity
	Wigner's Little Group	
J_1, J_2		Gauge Transformation

5.4. Conjugate Transformations

The most general form of the $SL(2,c)$ matrix is given in Equation (66). Transformation operators for the Lorentz group are given in exponential form as:

$$D = \exp\left\{-i\sum_{i=1}^{3}(\theta_i J_i + \eta_i K_i)\right\} \tag{119}$$

where the J_i are the generators of rotations and the K_i are the generators of proper Lorentz boosts. They satisfy the Lie algebra given in Equation (43). This set of commutation relations is invariant under the sign change of the boost generators K_i. Thus, we can consider "dot conjugation" defined as

$$\dot{D} = \exp\left\{-i\sum_{i=1}^{3}(\theta_i J_i - \eta_i K_i)\right\} \tag{120}$$

Since K_i are anti-Hermitian while J_i are Hermitian, the Hermitian conjugate of the above expression is

$$D^\dagger = \exp\left\{-i\sum_{i=1}^{3}(-\theta_i J_i + \eta_i K_i)\right\} \tag{121}$$

while the Hermitian conjugate of G is

$$\dot{D}^\dagger = \exp\left\{-i\sum_{i=1}^{3}(-\theta_i J_i - \eta_i K_i)\right\} \tag{122}$$

Since we understand the rotation around the z axis, we can now restrict the kinematics to the zt plane, and work with the $Sp(2)$ symmetry. Then the D matrices can be considered as Bargmann decompositions. First, D and $\dot{D}$, and their Hermitian conjugates are

$$D(\alpha, \chi) = \begin{pmatrix} (\cos\alpha)\cosh\chi & \sinh\chi - (\sin\alpha)\cosh\chi \\ \sinh\chi + (\sin\alpha)\cosh\chi & (\cos\alpha)\cosh\chi \end{pmatrix} \tag{123}$$

$$\dot{D}(\alpha, \chi) = \begin{pmatrix} (\cos\alpha)\cosh\chi & -\sinh\chi - (\sin\alpha)\cosh\chi \\ -\sinh\chi + (\sin\alpha)\cosh\chi & (\cos\alpha)\cosh\chi \end{pmatrix} \tag{124}$$

These matrices correspond to the "D loops" given in Figure 2a,b respectively. The "dot" conjugation changes the direction of boosts. The dot conjugation leads to the inversion of the space which is called the parity operation.

We can also consider changing the direction of rotations. Then they result in the Hermitian conjugates. We can write their matrices as

$$D^\dagger(\alpha, \chi) = \begin{pmatrix} (\cos\alpha)\cosh\chi & \sinh\chi + (\sin\alpha)\cosh\chi \\ \sinh\chi - (\sin\alpha)\cosh\chi & (\cos\alpha)\cosh\chi \end{pmatrix} \tag{125}$$

$$\dot{D}^\dagger(\alpha, \chi) = \begin{pmatrix} (\cos\alpha)\cosh\chi & -\sinh\chi + (\sin\alpha)\cosh\chi \\ -\sinh\chi - (\sin\alpha)\cosh\chi & (\cos\alpha)\cosh\chi \end{pmatrix} \tag{126}$$

From the exponential expressions from Equation (119) to Equation (122), it is clear that

$$D^\dagger = \dot{D}^{-1}, \quad \text{and} \quad \dot{D}^\dagger = D^{-1} \tag{127}$$

The D loop given in Figure 1 corresponds to $\dot{D}$. We shall return to these loops in Section 7.

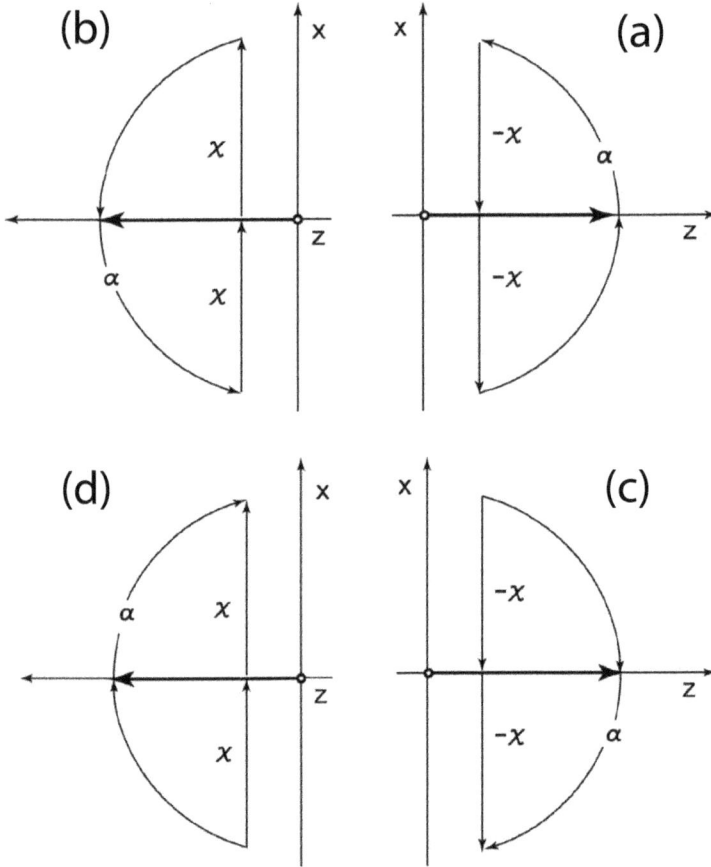

Figure 2. Four D-loops resulting from the Bargmann decomposition. (**a**) Bargmann decomposition from Figure 1; (**b**) Direction of the Lorentz boost is reversed; (**c**) Direction of rotation is reversed; (**d**) Both directions are reversed. These operations correspond to the space-inversion, charge conjugation, and the time reversal respectively.

6. Symmetries Derivable from the Poincaré Sphere

The Poincaré sphere serves as the basic language for polarization physics. Its underlying language is the two-by-two coherency matrix. This coherency matrix contains the symmetry of $SL(2,c)$ isomorphic to the the the Lorentz group applicable to three space-like and one time-like dimensions [4,6,7].

For polarized light propagating along the z direction, the amplitude ratio and phase difference of electric field x and y components traditionally determine the state of polarization. Hence, the polarization can be changed by adjusting the amplitude ratio or the phase difference or both. Usually, the optical device which changes amplitude is called an "attenuator" (or "amplifier") and the device which changes the relative phase a "phase shifter".

Let us start with the Jones vector:

$$\begin{pmatrix} \psi_1(z,t) \\ \psi_2(z,t) \end{pmatrix} = \begin{pmatrix} a \exp\left[i(kz - \omega t)\right] \\ a \exp\left[i(kz - \omega t)\right] \end{pmatrix} \tag{128}$$

To this matrix, we can apply the phase shift matrix of Equation (79) which brings the Jones vector to

$$\begin{pmatrix} \psi_1(z,t) \\ \psi_2(z,t) \end{pmatrix} = \begin{pmatrix} a\exp\left[i(kz-\omega t - i\phi/2)\right] \\ a\exp\left[i(kz-\omega t + i\phi/2)\right] \end{pmatrix} \tag{129}$$

The generator of this phase-shifter is J_3 given Table 5.

The optical beam can be attenuated differently in the two directions. The resulting matrix is

$$e^{-\mu}\begin{pmatrix} e^{\eta/2} & 0 \\ 0 & e^{-\eta/2} \end{pmatrix} \tag{130}$$

with the attenuation factor of $\exp\left(-\mu_0 + \eta/2\right)$ and $\exp\left(-\mu - \eta/2\right)$ for the x and y directions respectively. We are interested only the relative attenuation given in Equation (46) which leads to different amplitudes for the x and y component, and the Jones vector becomes

$$\begin{pmatrix} \psi_1(z,t) \\ \psi_2(z,t) \end{pmatrix} = \begin{pmatrix} ae^{\mu/2}\exp\left[i(kz-\omega t - i\phi/2)\right] \\ ae^{-\mu/2}\exp\left[i(kz-\omega t + i\phi/2)\right] \end{pmatrix} \tag{131}$$

The squeeze matrix of Equation (46) is generated by K_3 given in Table 1.

The polarization is not always along the x and y axes, but can be rotated around the z axis using Equation (79) generated by J_2 given in Table 1.

Among the rotation angles, the angle of $45°$ plays an important role in polarization optics. Indeed, if we rotate the squeeze matrix of Equation (46) by $45°$, we end up with the squeeze matrix of Equation (45) generated by K_1 given also in Table 1.

Each of these four matrices plays an important role in special relativity, as we discussed in Sections 3.2 and 6. Their respective roles in optics and particle physics are given in Table 7.

Table 7. Polarization optics and special relativity share the same mathematics. Each matrix has its clear role in both optics and relativity. The determinant of the Stokes or the four-momentum matrix remains invariant under Lorentz transformations. It is interesting to note that the decoherence parameter (least fundamental) in optics corresponds to the $(mass)^2$ (most fundamental) in particle physics.

Polarization Optics	Transformation Matrix	Particle Symmetry
Phase shift by ϕ	$\begin{pmatrix} e^{-i\phi/2} & 0 \\ 0 & e^{i\phi/2} \end{pmatrix}$	Rotation around z.
Rotation around z	$\begin{pmatrix} \cos(\theta/2) & -\sin(\theta/2) \\ \sin(\theta/2) & \cos(\theta/2) \end{pmatrix}$	Rotation around y.
Squeeze along x and y	$\begin{pmatrix} e^{\eta/2} & 0 \\ 0 & e^{-\eta/2} \end{pmatrix}$	Boost along z.
Squeeze along $45°$	$\begin{pmatrix} \cosh(\lambda/2) & \sinh(\lambda/2) \\ \sinh(\lambda/2) & \cosh(\lambda/2) \end{pmatrix}$	Boost along x.
$a^4\,(\sin\xi)^2$	Determinant	$(mass)^2$

The most general form for the two-by-two matrix applicable to the Jones vector is the G matrix of Equation (66). This matrix is of course a representation of the $SL(2,c)$ group. It brings the simplest Jones vector of Equation (128) to its most general form.

6.1. Coherency Matrix

However, the Jones vector alone cannot tell us whether the two components are coherent with each other. In order to address this important degree of freedom, we use the coherency matrix defined as [3,23]

$$C = \begin{pmatrix} S_{11} & S_{12} \\ S_{21} & S_{22} \end{pmatrix} \tag{132}$$

where

$$< \psi_i^* \psi_j >= \frac{1}{T} \int_0^T \psi_i^*(t + \tau)\psi_j(t)dt \tag{133}$$

where T is a sufficiently long time interval. Then, those four elements become [4]

$$S_{11} =< \psi_1^* \psi_1 >= a^2, \qquad S_{12} =< \psi_1^* \psi_2 >= a^2(\cos \xi)e^{-i\phi} \tag{134}$$

$$S_{21} =< \psi_2^* \psi_1 >= a^2(\cos \xi)e^{+i\phi}, \qquad S_{22} =< \psi_2^* \psi_2 >= a^2 \tag{135}$$

The diagonal elements are the absolute values of ψ_1 and ψ_2 respectively. The angle ϕ could be different from the value of the phase-shift angle given in Equation (79), but this difference does not play any role in the reasoning. The off-diagonal elements could be smaller than the product of ψ_1 and ψ_2, if the two polarizations are not completely coherent.

The angle ξ specifies the degree of coherency. If it is zero, the system is fully coherent, while the system is totally incoherent if ξ is $90°$. This can therefore be called the "decoherence angle."

While the most general form of the transformation applicable to the Jones vector is G of Equation (66), the transformation applicable to the coherency matrix is

$$C' = G C G^\dagger \tag{136}$$

The determinant of the coherency matrix is invariant under this transformation, and it is

$$\det(C) = a^4(\sin \xi)^2 \tag{137}$$

Thus, angle ξ remains invariant. In the language of the Lorentz transformation applicable to the four-vector, the determinant is equivalent to the $(mass)^2$ and is therefore a Lorentz-invariant quantity.

6.2. Two Radii of the Poincaré Sphere

Let us write explicitly the transformation of Equation (136) as

$$\begin{pmatrix} S'_{11} & S'_{12} \\ S'_{21} & S'_{22} \end{pmatrix} = \begin{pmatrix} \alpha & \beta \\ \gamma & \delta \end{pmatrix} \begin{pmatrix} S_{11} & S_{12} \\ S_{21} & S_{22} \end{pmatrix} \begin{pmatrix} \alpha^* & \gamma^* \\ \beta^* & \delta^* \end{pmatrix} \tag{138}$$

It is then possible to construct the following quantities,

$$S_0 = \frac{S_{11} + S_{22}}{2}, \qquad S_3 = \frac{S_{11} - S_{22}}{2} \tag{139}$$

$$S_1 = \frac{S_{12} + S_{21}}{2}, \qquad S_2 = \frac{S_{12} - S_{21}}{2i} \tag{140}$$

These are known as the Stokes parameters, and constitute a four-vector (S_0, S_3, S_1, S_2) under the Lorentz transformation.

In the Jones vector of Equation (128), the amplitudes of the two orthogonal components are equal. Thus, the two diagonal elements of the coherency matrix are equal. This leads to $S_3 = 0$, and the problem is reduced from the sphere to a circle. In the resulting two-dimensional subspace, we can introduce the polar coordinate system with

$$R = \sqrt{S_1^2 + S_2^2} \tag{141}$$

$$S_1 = R \cos \phi \tag{142}$$

$$S_2 = R \sin \phi \tag{143}$$

The radius R is the radius of this circle, and is

$$R = a^2 \cos \xi \tag{144}$$

The radius R takes its maximum value S_0 when $\xi = 0°$. It decreases as ξ increases and vanishes when $\xi = 90°$. This aspect of the radius R is illustrated in Figure 3.

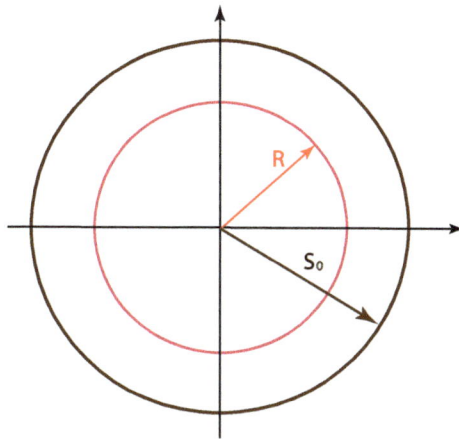

Figure 3. Radius of the Poincaré sphere. The radius R takes its maximum value S_0 when the decoherence angle ξ is zero. It becomes smaller as ξ increases. It becomes zero when the angle reaches 90°.

In order to see its implications in special relativity, let us go back to the four-momentum matrix of $m(1,0,0,0)$. Its determinant is m^2 and remains invariant. Likewise, the determinant of the coherency matrix of Equation (132) should also remain invariant. The determinant in this case is

$$S_0^2 - R^2 = a^4 \sin^2 \xi \tag{145}$$

This quantity remains invariant under the Hermitian transformation of Equation (138), which is a Lorentz transformation as discussed in Sections 3.2 and 6. This aspect is shown on the last row of Table 7.

The coherency matrix then becomes

$$C = a^2 \begin{pmatrix} 1 & (\cos \xi) e^{-i\phi} \\ (\cos \xi) e^{i\phi} & 1 \end{pmatrix} \tag{146}$$

Symmetry **2014**, *6*, 473–515

Since the angle ϕ does not play any essential role, we can let $\phi = 0$, and write the coherency matrix as

$$C = a^2 \begin{pmatrix} 1 & \cos \xi \\ \cos \xi & 1 \end{pmatrix} \tag{147}$$

The determinant of the above two-by-two matrix is

$$a^4 \left(1 - \cos^2 \xi \right) = a^4 \sin^2 \xi \tag{148}$$

Since the Lorentz transformation leaves the determinant invariant, the change in this ξ variable is not a Lorentz transformation. It is of course possible to construct a larger group in which this variable plays a role in a group transformation [6], but here we are more interested in its role in a particle gaining a mass from zero or the mass becoming zero.

6.3. Extra-Lorentzian Symmetry

The coherency matrix of Equation (146) can be diagonalized to

$$a^2 \begin{pmatrix} 1 + \cos \xi & 0 \\ 0 & 1 - \cos \xi \end{pmatrix} \tag{149}$$

by a rotation. Let us then go back to the four-momentum matrix of Equation (70). If $p_x = p_y = 0$, and $p_z = p_0 \cos \xi$, we can write this matrix as

$$p_0 \begin{pmatrix} 1 + \cos \xi & 0 \\ 0 & 1 - \cos \xi \end{pmatrix} \tag{150}$$

Thus, with this extra variable, it is possible to study the little groups for variable masses, including the small-mass limit and the zero-mass case.

For a fixed value of p_0, the $(mass)^2$ becomes

$$(mass)^2 = (p_0 \sin \xi)^2, \quad \text{and} \quad (momentum)^2 = (p_0 \cos \xi)^2 \tag{151}$$

resulting in

$$(energy)^2 = (mass)^2 + (momentum)^2 \tag{152}$$

This transition is illustrated in Figure 4. We are interested in reaching a point on the light cone from mass hyperbola while keeping the energy fixed. According to this figure, we do not have to make an excursion to infinite-momentum limit. If the energy is fixed during this process, Equation (152) tells the mass and momentum relation, and Figure 5 illustrates this relation.

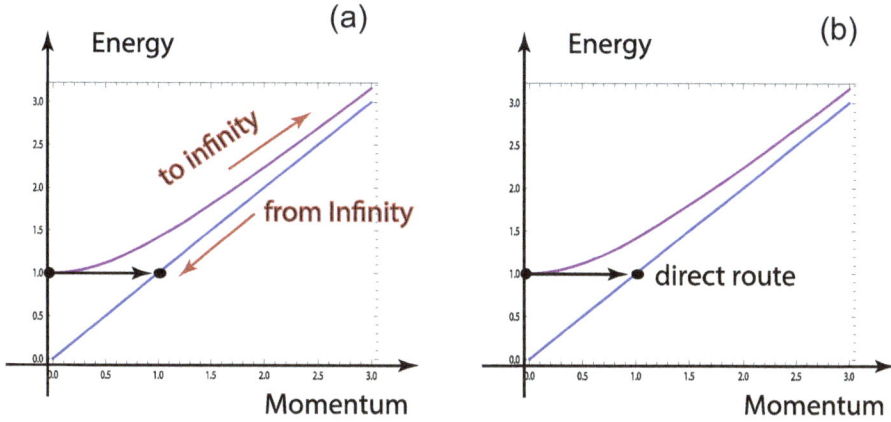

Figure 4. Transition from the massive to massless case. (**a**) Transition within the framework of the Lorentz group; (**b**) Trasnsition allowed in the symmetry of the Poincaré sphere. Within the framework of the Lorentz group, it is not possible to go from the massive to massless case directly, because it requires the change in the mass which is a Lorentz-invariant quantity. The only way is to move to infinite momentum and jump from the hyperbola to the light cone, and come back. The extra symmetry of the Poincaré sphere allows a direct transition

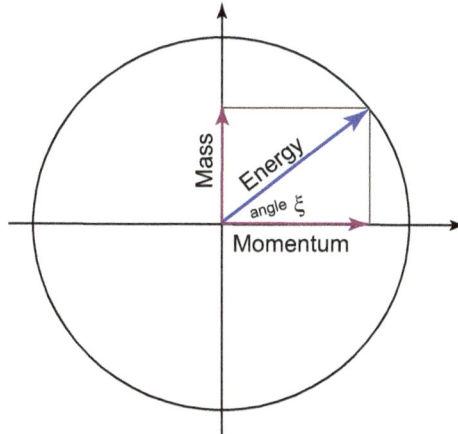

Figure 5. Energy-momentum-mass relation. This circle illustrates the case where the energy is fixed, while the mass and momentum are related according to the triangular rule. The value of the angle ξ changes from zero to 180°. The particle mass is negative for negative values of this angle. However, in the Lorentz group, only $(mass)^2$ is a relevant variable, and negative masses might play a role for theoretical purposes.

Within the framework of the Lorentz group, it is possible, by making an excursion to infinite momentum where the mass hyperbola coincides with the light cone, to then come back to the desired point. On the other hand, the mass formula of Equation (151) allows us to go there directly. The decoherence mechanism of the coherency matrix makes this possible.

7. Small-Mass and Massless Particles

We now have a mathematical tool to reduce the mass of a massive particle from its positive value to zero. During this process, the Lorentz-boosted rotation matrix becomes a gauge transformation for the spin-1 particle, as discussed Section 5.2. For spin-1/2 particles, there are two issues.

1. It was seen in Section 5.2 that the requirement of gauge invariance lead to a polarization of massless spin-1/2 particle, such as neutrinos. What happens to anti-neutrinos?
2. There are strong experimental indications that neutrinos have a small mass. What happens to the $E(2)$ symmetry?

7.1. Spin-1/2 Particles

Let us go back to the two-by-two matrices of Section 5.4, and the two-by-two D matrix. For a massive particle, its Wigner decomposition leads to

$$D = \begin{pmatrix} \cos(\theta/2) & -e^{-\eta}\sin(\theta/2) \\ e^{\eta}\sin(\theta/2) & \cos(\theta/2) \end{pmatrix} \tag{153}$$

This matrix is applicable to the spinors u and v defined in Equation (101) respectively for the spin-up and spin-down states along the z direction.

Since the Lie algebra of $SL(2,c)$ is invariant under the sign change of the K_i matrices, we can consider the "dotted" representation, where the system is boosted in the opposite direction, while the direction of rotations remain the same. Thus, the Wigner decomposition leads to

$$\dot{D} = \begin{pmatrix} \cos(\theta/2) & -e^{\eta}\sin(\theta/2) \\ e^{-\eta}\sin(\theta/2) & \cos(\theta/2) \end{pmatrix} \tag{154}$$

with its spinors

$$\dot{u} = \begin{pmatrix} 1 \\ 0 \end{pmatrix}, \quad \text{and} \quad \dot{v} = \begin{pmatrix} 0 \\ 1 \end{pmatrix} \tag{155}$$

For anti-neutrinos, the helicity is reversed but the momentum is unchanged. Thus, $D^{\dagger}$ is the appropriate matrix. However, $D^{\dagger} = \dot{D}^{-1}$ as was noted in Section 5.4. Thus, we shall use $\dot{D}$ for anti-neutrinos.

When the particle mass becomes very small,

$$e^{-\eta} = \frac{m}{2p} \tag{156}$$

becomes small. Thus, if we let

$$e^{\eta}\sin(\theta/2) = \gamma, \quad \text{and} \quad e^{-\eta}\sin(\theta/2) = \epsilon^2 \tag{157}$$

then the D matrix of Equation (153) and the $\dot{D}$ of Equation (154) become

$$\begin{pmatrix} 1 - \gamma\epsilon^2/2 & -\epsilon^2 \\ \gamma & 1 - \gamma\epsilon^2 \end{pmatrix}, \quad \text{and} \quad \begin{pmatrix} 1 - \gamma\epsilon^2/2 & -\gamma \\ \epsilon^2 & 1 - \gamma\epsilon^2 \end{pmatrix} \tag{158}$$

respectively where γ is an independent parameter and

$$\epsilon^2 = \gamma \left(\frac{m}{2p} \right)^2 \tag{159}$$

When the particle mass becomes zero, they become

$$\begin{pmatrix} 1 & 0 \\ \gamma & 1 \end{pmatrix}, \quad \text{and} \quad \begin{pmatrix} 1 & -\gamma \\ 0 & 1 \end{pmatrix} \tag{160}$$

respectively, applicable to the spinors (u, v) and $(\dot{u}, \dot{v})$ respectively.

For neutrinos,

$$\begin{pmatrix} 1 & 0 \\ \gamma & 1 \end{pmatrix}\begin{pmatrix} 1 \\ 0 \end{pmatrix} = \begin{pmatrix} 1 \\ \gamma \end{pmatrix}, \quad \text{and} \quad \begin{pmatrix} 1 & 0 \\ \gamma & 1 \end{pmatrix}\begin{pmatrix} 0 \\ 1 \end{pmatrix} = \begin{pmatrix} 0 \\ 1 \end{pmatrix} \tag{161}$$

For anti-neutrinos,

$$\begin{pmatrix} 1 & -\gamma \\ 0 & 1 \end{pmatrix}\begin{pmatrix} 1 \\ 0 \end{pmatrix} = \begin{pmatrix} 1 \\ 0 \end{pmatrix}, \quad \text{and} \quad \begin{pmatrix} 1 & -\gamma \\ 0 & 1 \end{pmatrix}\begin{pmatrix} 0 \\ 1 \end{pmatrix} = \begin{pmatrix} -\gamma \\ 1 \end{pmatrix} \tag{162}$$

It was noted in Section 5.2 that the triangular matrices of Equation (160) perform gauge transformations. Thus, for Equations (161) and (162) the requirement of gauge invariance leads to the polarization of neutrinos. The neutrinos are left-handed while the anti-neutrinos are right-handed. Since, however, nature cannot tell the difference between the dotted and undotted representations, the Lorentz group cannot tell which neutrino is right handed. It can say only that the neutrinos and anti-neutrinos are oppositely polarized.

If the neutrino has a small mass, the gauge invariance is modified to

$$\begin{pmatrix} 1 - \gamma\epsilon^2/2 & -\epsilon^2 \\ \gamma & 1 - \gamma\epsilon^2/2 \end{pmatrix}\begin{pmatrix} 0 \\ 1 \end{pmatrix} = \begin{pmatrix} 0 \\ 1 \end{pmatrix} - \epsilon^2\begin{pmatrix} 1 \\ \gamma/2 \end{pmatrix} \tag{163}$$

and

$$\begin{pmatrix} 1 - \gamma\epsilon^2/2 & -\gamma \\ \epsilon^2 & 1 - \gamma\epsilon^2 \end{pmatrix}\begin{pmatrix} 1 \\ 0 \end{pmatrix} = \begin{pmatrix} 1 \\ 0 \end{pmatrix} + \epsilon^2\begin{pmatrix} -\gamma/2 \\ 1 \end{pmatrix} \tag{164}$$

respectively for neutrinos and anti-neutrinos. Thus the violation of the gauge invariance in both cases is proportional to ϵ^2 which is $m^2/4p^2$.

7.2. Small-Mass Neutrinos in the Real World

Whether neutrinos have mass or not and the consequences of this relative to the Standard Model and lepton number is the subject of much theoretical speculation [24,25], and of cosmology [26], nuclear reactors [27], and high energy experimentations [28,29]. Neutrinos are fast becoming an important component of the search for dark matter and dark radiation [30]. Their importance within the Standard Model is reflected by the fact that they are the only particles which seem to exist with only one direction of chirality, *i.e.*, only left-handed neutrinos have been confirmed to exist so far.

It was speculated some time ago that neutrinos in constant electric and magnetic fields would acquire a small mass, and that right-handed neutrinos would be trapped within the interaction field [31]. Solving generalized electroweak models using left- and right-handed neutrinos has been discussed recently [32]. Today these right-handed neutrinos which do not participate in weak interactions are called "sterile" neutrinos [33]. A comprehensive discussion of the place of neutrinos in the scheme of physics has been given by Drewes [30]. We should note also that the three different neutrinos, namely ν_e, ν_μ, and ν_τ, may have different masses [34].

8. Scalars, Four-Vectors, and Four-Tensors

In Sections 5 and 7, our primary interest has been the two-by-two matrices applicable to spinors for spin-1/2 particles. Since we also used four-by-four matrices, we indirectly studied the four-component particle consisting of spin-1 and spin-zero components.

If there are two spin 1/2 states, we are accustomed to construct one spin-zero state, and one spin-one state with three degeneracies.

In this paper, we are confronted with two spinors, but each spinor can also be dotted. For this reason, there are 16 orthogonal states consisting of spin-one and spin-zero states. How many spin-zero states? How many spin-one states?

For particles at rest, it is known that the addition of two one-half spins result in spin-zero and spin-one states. In this paper, we have two different spinors behaving differently under the Lorentz boost. Around the z direction, both spinors are transformed by

$$Z(\phi) = \exp(-i\phi J_3) = \begin{pmatrix} e^{-i\phi/2} & 0 \\ 0 & e^{i\phi/2} \end{pmatrix} \tag{165}$$

However, they are boosted by

$$B(\eta) = \exp(-i\eta K_3) = \begin{pmatrix} e^{\eta/2} & 0 \\ 0 & e^{-\eta/2} \end{pmatrix} \tag{166}$$

$$\dot{B}(\eta) = \exp(i\eta K_3) = \begin{pmatrix} e^{-\eta/2} & 0 \\ 0 & e^{\eta/2} \end{pmatrix} \tag{167}$$

applicable to the undotted and dotted spinors respectively. These two matrices commute with each other, and also with the rotation matrix $Z(\phi)$ of Equation (165). Since K_3 and J_3 commute with each other, we can work with the matrix $Q(\eta, \phi)$ defined as

$$Q(\eta, \phi) = B(\eta)Z(\phi) = \begin{pmatrix} e^{(\eta-i\phi)/2} & 0 \\ 0 & e^{-(\eta-i\phi)/2} \end{pmatrix} \tag{168}$$

$$\dot{Q}(\eta, \phi) = \dot{B}(\eta)\dot{Z}(\phi) = \begin{pmatrix} e^{-(\eta+i\phi)/2} & 0 \\ 0 & e^{(\eta+i\phi)/2} \end{pmatrix} \tag{169}$$

When this combined matrix is applied to the spinors,

$$Q(\eta, \phi)u = e^{(\eta-i\phi)/2}u, \qquad Q(\eta, \phi)v = e^{-(\eta-i\phi)/2}v \tag{170}$$

$$\dot{Q}(\eta, \phi)\dot{u} = e^{-(\eta+i\phi)/2}\dot{u}, \qquad \dot{Q}(\eta, \phi)\dot{v} = e^{(\eta+i\phi)/2}\dot{v} \tag{171}$$

If the particle is at rest, we can construct the combinations

$$uu, \qquad \frac{1}{\sqrt{2}}(uv + vu), \qquad vv \tag{172}$$

to construct the spin-1 state, and

$$\frac{1}{\sqrt{2}}(uv - vu) \tag{173}$$

for the spin-zero state. There are four bilinear states. In the $SL(2, c)$ regime, there are two dotted spinors. If we include both dotted and undotted spinors, there are 16 independent bilinear combinations. They are given in Table 8. This table also gives the effect of the operation of $Q(\eta, \phi)$.

Table 8. Sixteen combinations of the $SL(2,c)$ spinors. In the $SU(2)$ regime, there are two spinors leading to four bilinear forms. In the $SL(2,c)$ world, there are two undotted and two dotted spinors. These four spinors lead to 16 independent bilinear combinations.

Spin 1			Spin 0
$uu,$	$\frac{1}{\sqrt{2}}(uv+vu),$	$vv,$	$\frac{1}{\sqrt{2}}(uv-vu)$
$\dot{u}\dot{u},$	$\frac{1}{\sqrt{2}}(\dot{u}\dot{v}+\dot{v}\dot{u}),$	$\dot{v}\dot{v},$	$\frac{1}{\sqrt{2}}(\dot{u}\dot{v}-\dot{v}\dot{u})$
$u\dot{u},$	$\frac{1}{\sqrt{2}}(u\dot{v}+v\dot{u}),$	$v\dot{v},$	$\frac{1}{\sqrt{2}}(u\dot{v}-v\dot{u})$
$\dot{u}u,$	$\frac{1}{\sqrt{2}}(\dot{u}v+\dot{v}u),$	$\dot{v}v,$	$\frac{1}{\sqrt{2}}(\dot{u}v-\dot{v}u)$
After the Operation of $Q(\eta,\phi)$ and $\dot{Q}(\eta,\phi)$			
$e^{-i\phi}e^{\eta}uu,$	$\frac{1}{\sqrt{2}}(uv+vu),$	$e^{i\phi}e^{-\eta}vv,$	$\frac{1}{\sqrt{2}}(uv-vu)$
$e^{-i\phi}e^{-\eta}\dot{u}\dot{u},$	$\frac{1}{\sqrt{2}}(\dot{u}\dot{v}+\dot{v}\dot{u}),$	$e^{i\phi}e^{\eta}\dot{v}\dot{v},$	$\frac{1}{\sqrt{2}}(\dot{u}\dot{v}-\dot{v}\dot{u})$
$e^{-i\phi}u\dot{u},$	$\frac{1}{\sqrt{2}}(e^{\eta}u\dot{v}+e^{-\eta}v\dot{u}),$	$e^{i\phi}v\dot{v},$	$\frac{1}{\sqrt{2}}(e^{\eta}u\dot{v}-e^{-\eta}v\dot{u})$
$e^{-i\phi}\dot{u}u,$	$\frac{1}{\sqrt{2}}(\dot{u}v+\dot{v}u),$	$e^{i\phi}\dot{v}v,$	$\frac{1}{\sqrt{2}}(e^{-\eta}\dot{u}v-e^{\eta}\dot{v}u)$

Among the bilinear combinations given in Table 8, the following two are invariant under rotations and also under boosts.

$$S = \frac{1}{\sqrt{2}}(uv-vu), \quad \text{and} \quad \dot{S} = -\frac{1}{\sqrt{2}}(\dot{u}\dot{v}-\dot{v}\dot{u}) \tag{174}$$

They are thus scalars in the Lorentz-covariant world. Are they the same or different? Let us consider the following combinations

$$S_+ = \frac{1}{\sqrt{2}}\left(S+\dot{S}\right), \quad \text{and} \quad S_- = \frac{1}{\sqrt{2}}\left(S-\dot{S}\right) \tag{175}$$

Under the dot conjugation, S_+ remains invariant, but S_- changes its sign.

Under the dot conjugation, the boost is performed in the opposite direction. Therefore it is the operation of space inversion, and S_+ is a scalar while S_- is called the pseudo-scalar.

8.1. Four-Vectors

Let us consider the bilinear products of one dotted and one undotted spinor as $u\dot{u}, u\dot{v}, \dot{u}v, v\dot{v}$, and construct the matrix

$$U = \begin{pmatrix} u\dot{v} & v\dot{v} \\ u\dot{u} & v\dot{u} \end{pmatrix} \tag{176}$$

Under the rotation $Z(\phi)$ and the boost $B(\eta)$ they become

$$\begin{pmatrix} e^{\eta}u\dot{v} & e^{-i\phi}v\dot{v} \\ e^{i\phi}u\dot{u} & e^{-\eta}v\dot{u} \end{pmatrix} \tag{177}$$

Indeed, this matrix is consistent with the transformation properties given in Table 8, and transforms like the four-vector

$$\begin{pmatrix} t+z & x-iy \\ x+iy & t-z \end{pmatrix} \tag{178}$$

This form was given in Equation (65), and played the central role throughout this paper. Under the space inversion, this matrix becomes

$$\begin{pmatrix} t-z & -(x-iy) \\ -(x+iy) & t+z \end{pmatrix} \tag{179}$$

This space inversion is known as the parity operation.

The form of Equation (176) for a particle or field with four-components, is given by (V_0, V_z, V_x, V_y). The two-by-two form of this four-vector is

$$U = \begin{pmatrix} V_0 + V_z & V_x - iV_y \\ V_x + iV_y & V_0 - V_z \end{pmatrix} \tag{180}$$

If boosted along the z direction, this matrix becomes

$$\begin{pmatrix} e^{\eta} (V_0 + V_z) & V_x - iV_y \\ V_x + iV_y & e^{-\eta} (V_0 - V_z) \end{pmatrix} \tag{181}$$

In the mass-zero limit, the four-vector matrix of Equation (181) becomes

$$\begin{pmatrix} 2A_0 & A_x - iA_y \\ A_x + iA_y & 0 \end{pmatrix} \tag{182}$$

with the Lorentz condition $A_0 = A_z$. The gauge transformation applicable to the photon four-vector was discussed in detail in Section 5.2.

Let us go back to the matrix of Equation (180), we can construct another matrix $\dot{U}$. Since the dot conjugation leads to the space inversion,

$$\dot{U} = \begin{pmatrix} \dot{u}v & \dot{v}v \\ \dot{u}u & \dot{v}u \end{pmatrix} \tag{183}$$

Then

$$\dot{u}v \simeq (t - z), \qquad \dot{v}u \simeq (t + z) \tag{184}$$

$$\dot{v}v \simeq -(x - iy), \qquad \dot{u}u \simeq -(x + iy) \tag{185}$$

where the symbol $\simeq$ means "transforms like".

Thus, U of Equation (176) and $\dot{U}$ of Equation (183) used up 8 of the 16 bilinear forms. Since there are two bilinear forms in the scalar and pseudo-scalar as given in Equation (175), we have to give interpretations to the six remaining bilinear forms.

8.2. Second-Rank Tensor

In this subsection, we are studying bilinear forms with both spinors dotted and undotted. In Section 8.1, each bilinear spinor consisted of one dotted and one undotted spinor. There are also bilinear spinors which are both dotted or both undotted. We are interested in two sets of three quantities satisfying the $O(3)$ symmetry. They should therefore transform like

$$(x + iy)/\sqrt{2}, \qquad (x - iy)/\sqrt{2}, \qquad z \tag{186}$$

which are like

$$uu, \qquad vv, \qquad (uv + vu)/\sqrt{2} \tag{187}$$

respectively in the $O(3)$ regime. Since the dot conjugation is the parity operation, they are like

$$-\dot{u}\dot{u}, \qquad -\dot{v}\dot{v}, \qquad -(\dot{u}\dot{v} + \dot{v}\dot{u})/\sqrt{2} \tag{188}$$

In other words,

$$(uu\dot{)} = -\dot{u}\dot{u}, \quad \text{and} \quad (vv\dot{)} = -\dot{v}\dot{v} \tag{189}$$

We noticed a similar sign change in Equation (184).

In order to construct the z component in this $O(3)$ space, let us first consider

$$f_z = \frac{1}{2}\left[(uv+vu)-(\dot{u}\dot{v}+\dot{v}\dot{u})\right], \qquad g_z = \frac{1}{2i}\left[(uv+vu)+(\dot{u}\dot{v}+\dot{v}\dot{u})\right] \tag{190}$$

where f_z and g_z are respectively symmetric and anti-symmetric under the dot conjugation or the parity operation. These quantities are invariant under the boost along the z direction. They are also invariant under rotations around this axis, but they are not invariant under boost along or rotations around the x or y axis. They are different from the scalars given in Equation (174).

Next, in order to construct the x and y components, we start with $g_\pm$ as

$$f_+ = \frac{1}{\sqrt{2}}\left(uu - \dot{u}\dot{u}\right) \qquad g_+ = \frac{1}{\sqrt{2i}}\left(uu + \dot{u}\dot{u}\right) \tag{191}$$

$$f_- = \frac{1}{\sqrt{2}}\left(vv - \dot{v}\dot{v}\right) \qquad g_- = \frac{1}{\sqrt{2i}}\left(vv + \dot{v}\dot{v}\right) \tag{192}$$

Then

$$f_x = \frac{1}{\sqrt{2}}\left(f_+ + f_-\right) = \frac{1}{2}\left[(uu - \dot{u}\dot{u}) + (vv - \dot{v}\dot{v})\right] \tag{193}$$

$$f_y = \frac{1}{\sqrt{2i}}\left(f_+ - f_-\right) = \frac{1}{2i}\left[(uu - \dot{u}\dot{u}) - (vv - \dot{v}\dot{v})\right] \tag{194}$$

and

$$g_x = \frac{1}{\sqrt{2}}\left(g_+ + g_-\right) = \frac{1}{2i}\left[(uu + \dot{u}\dot{u}) + (vv + \dot{v}\dot{v})\right] \tag{195}$$

$$g_y = \frac{1}{\sqrt{2i}}\left(g_+ - g_-\right) = -\frac{1}{2}\left[(uu + \dot{u}\dot{u}) - (vv + \dot{v}\dot{v})\right] \tag{196}$$

Here f_x and f_y are symmetric under dot conjugation, while g_x and g_y are anti-symmetric.

Furthermore, f_z, f_x, and f_y of Equations (190) and (193) transform like a three-dimensional vector. The same can be said for g_i of Equations (190) and (195). Thus, they can grouped into the second-rank tensor

$$T = \begin{pmatrix} 0 & -g_z & -g_x & -g_y \\ g_z & 0 & -f_y & f_x \\ g_x & f_y & 0 & -f_z \\ g_y & -f_x & f_z & 0 \end{pmatrix} \tag{197}$$

whose Lorentz-transformation properties are well known. The g_i components change their signs under space inversion, while the f_i components remain invariant. They are like the electric and magnetic fields respectively.

If the system is Lorentz-booted, f_i and g_i can be computed from Table 8. We are now interested in the symmetry of photons by taking the massless limit. According to the procedure developed in Section 6, we can keep only the terms which become larger for larger values of η. Thus,

$$f_x \to \frac{1}{2}\left(uu - \dot{v}\dot{v}\right), \qquad f_y \to \frac{1}{2i}\left(uu + \dot{v}\dot{v}\right) \tag{198}$$

$$g_x \to \frac{1}{2i}\left(uu + \dot{v}\dot{v}\right), \qquad g_y \to -\frac{1}{2}\left(uu - \dot{v}\dot{v}\right) \tag{199}$$

in the massless limit.

Then the tensor of Equation (197) becomes

$$
F = \begin{pmatrix} 0 & 0 & -E_x & -E_y \\ 0 & 0 & -B_y & B_x \\ E_x & B_y & 0 & 0 \\ E_y & -B_x & 0 & 0 \end{pmatrix}
\tag{200}
$$

with

$$
B_x \simeq \frac{1}{2}\left(uu - \dot{v}\dot{v}\right), \qquad B_y \simeq \frac{1}{2i}\left(uu + \dot{v}\dot{v}\right)
\tag{201}
$$

$$
E_x = \frac{1}{2i}\left(uu + \dot{v}\dot{v}\right), \qquad E_y = -\frac{1}{2}\left(uu - \dot{v}\dot{v}\right)
\tag{202}
$$

The electric and magnetic field components are perpendicular to each other. Furthermore,

$$
E_x = B_y, \qquad E_y = -B_x
\tag{203}
$$

In order to address this question, let us go back to Equation (191). In the massless limit,

$$
B_+ \simeq E_+ \simeq uu, \qquad B_- \simeq E_- \simeq \dot{v}\dot{v}
\tag{204}
$$

The gauge transformation applicable to u and $\dot{v}$ are the two-by-two matrices

$$
\begin{pmatrix} 1 & -\gamma \\ 0 & 1 \end{pmatrix}, \quad \text{and} \quad \begin{pmatrix} 1 & 0 \\ -\gamma & 1 \end{pmatrix}
\tag{205}
$$

respectively as noted in Sections 5.2 and 7.1. Both u and $\dot{v}$ are invariant under gauge transformations, while $\dot{u}$ and v do not.

The B_+ and E_+ are for the photon spin along the z direction, while B_- and E_- are for the opposite direction. In 1964 [35], Weinberg constructed gauge-invariant state vectors for massless particles starting from Wigner's 1939 paper [1]. The bilinear spinors uu and and $\dot{v}\dot{v}$ correspond to Weinberg's state vectors.

8.3. Possible Symmetry of the Higgs Mechanism

In this section, we discussed how the two-by-two formalism of the group $SL(2,c)$ leads the scalar, four-vector, and tensor representations of the Lorentz group. We discussed in detail how the four-vector for a massive particle can be decomposed into the symmetry of a two-component massless particle and one gauge degree of freedom. This aspect was studied in detail by Kim and Wigner [20,21], and their results are illustrated in Figure 6. This decomposition is known in the literature as the group contraction.

The four-dimensional Lorentz group can be contracted to the Euclidean and cylindrical groups. These contraction processes could transform a four-component massive vector meson into a massless spin-one particle with two spin components, and one gauge degree of freedom.

Since this contraction procedure is spelled out detail in [21], as well as in the present paper, its reverse process is also well understood. We start with one two-component massless particle with one gauge degree of freedom, and end up with a massive vector meson with its four components.

The mathematics of this process is not unlike the Higgs mechanism [36,37], where one massless field with two degrees of freedom absorbs one gauge degree freedom to become a quartet of bosons, namely that of $W, Z^{\pm}$ plus the Higgs boson. As is well known, this mechanism is the basis for the theory of electro-weak interaction formulated by Weinberg and Salam [38,39].

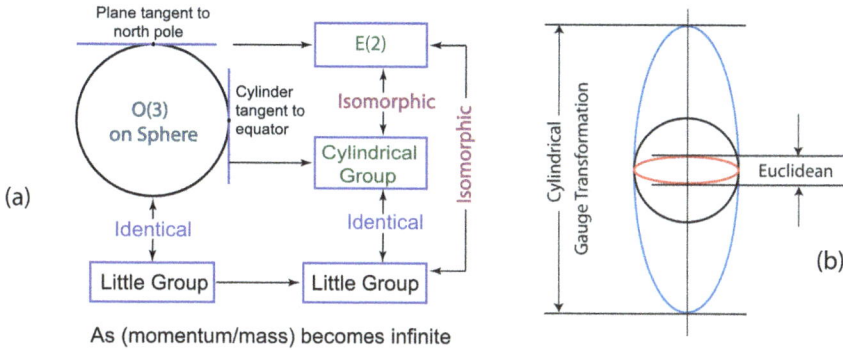

Figure 6. Contractions of the three-dimensional rotation group. (**a**) Contraction in terms of the tangential plane and the tangential cylinder [20]; (**b**) Contraction in terms of the expansion and contraction of the longitudinal axis [21]. In both cases, the symmetry ends up with one rotation around the longitudinal direction and one translational degree along the longitudinal axis. The rotation and translation corresponds to the helicity and gauge degrees of freedom.

The word "spontaneous symmetry breaking" is used for the Higgs mechanism. It could be an interesting problem to see that this symmetry breaking for the two Higgs doublet model can be formulated in terms of the Lorentz group and its contractions. In this connection, we note an interesting recent paper by Dée and Ivanov [40].

9. Conclusions

The damped harmonic oscillator, Wigner'e little groups, and the Poincaré sphere belong to the three different branches of physics. In this paper, it was noted that they are based on the same mathematical framework, namely the algebra of two-by-two matrices.

The second-order differential equation for damped harmonic oscillators can be formulated in terms of two-by-two matrices. These matrices produce the algebra of the group $Sp(2)$. While there are three trace classes of the two-by-two matrices of this group, the damped oscillator tells us how to make transitions from one class to another.

It is shown that Wigner's three little groups can be defined in terms of the trace classes of the $Sp(2)$ group. If the trace is smaller than two, the little group is for massive particles. If greater than two, the little group is for imaginary-mass particles. If the trace is equal to two, the little group is for massless particles. Thus, the damped harmonic oscillator provides a procedure for transition from one little group to another.

The Poincaré sphere contains the symmetry of the six-parameter $SL(2, c)$ group. Thus, the sphere provides the procedure for extending the symmetry of the little group defined within the Lorentz group of three-dimensional Minkowski space to its full Lorentz group in the four-dimensional space-time. In addition, the Poincaré sphere offers the variable which allows us to change the symmetry of a massive particle to that of a massless particle by continuously decreasing the mass.

In this paper, we extracted the mathematical properties of Wigner's little groups from the damped harmonic oscillator and the Poincaré sphere. In so doing, we have shown that the transition from one little group to another is tangentially continuous.

This subject was initiated by Inönü and Wigner in 1953 as the group contraction [41]. In their paper, they discussed the contraction of the three-dimensional rotation group becoming contracted to the two-dimensional Euclidean group with one rotational and two translational degrees of freedom. While the $O(3)$ rotation group can be illustrated by a three-dimensional sphere, the plane tangential at the north pole is for the $E(2)$ Euclidean group. However, we can also consider a cylinder tangential at the equatorial belt. The resulting cylindrical group is isomorphic to the Euclidean group [20]. While the rotational degree of freedom of this cylinder is for the photon spin, the up and down translations on the surface of the cylinder correspond to the gauge degree of freedom of the photon, as illustrated in Figure 6.

It was noted also that the Bargmann decomposition of two-by-two matrices, as illustrated in Figure 1 and Figure 2, allows us to study more detailed properties of the little groups, including space and time reflection reflection properties. Also in this paper, we have discussed how the scalars, four-vectors, and four-tensors can be constructed from the two-by-two representation in the Lorentz-covariant world.

In addition, it should be noted that the symmetry of the Lorentz group is also contained in the squeezed state of light [14] and the $ABCD$ matrix for optical beam transfers [18]. We also mentioned the possibility of understanding the mathematics of the Higgs mechanism in terms of the Lorentz group and its contractions.

Acknowledgements

In his 1939 paper [1], Wigner worked out the subgroups of the Lorentz group whose transformations leave the four momentum of a given particle invariant. In so doing, he worked out their internal space-time symmetries. In spite of its importance, this paper remains as one of the most difficult papers to understand. Wigner was eager to make his paper understandable to younger physicists.

While he was the pioneer in introducing the mathematics of group theory to physics, he was also quite fond of using two-by-two matrices to explain group theoretical ideas. He asked one of the present authors (Young S. Kim) to rewrite his 1939 paper [1] using the language of those matrices. This is precisely what we did in the present paper.

We are grateful to Eugene Paul Wigner for this valuable suggestion.

Author Contributions

This paper is largely based on the earlier papers by Young S. Kim and Marilyn E. Noz, and those by Sibel Başkal and Young S. Kim. The two-by-two formulation of the damped oscillator in Section 2 was jointly developed by Sibel Başkal and Yound S. Kim during the summer of 2012. Marilyn E. Noz developed the idea of the symmetry of small-mass neutrinos in Section 7. The limiting process in the symmetry of the Poincaré sphere was formulated by Young S. Kim. Sibel Başkal initially constructed the four-by-four tensor representation in Section 8.

The initial organization of this paper was conceived by Young S. Kim in his attempt to follow Wigner's suggestion to translate his 1939 paper into the language of two-by-two matrices. Sibel Başkal and Marilyn E. Noz tightened the organization and filled in the details.

Conflicts of Interest

The authors declare no conflicts of interest.

References

1. Wigner, E. On unitary representations of the inhomogeneous Lorentz Group. *Ann. Math.* **1939**, *40*, 149–204.
2. Han, D.; Kim, Y.S.; Son, D. Eulerian parametrization of Wigner little groups and gauge transformations in terms of rotations in 2-component spinors. *J. Math. Phys.* **1986**, *27*, 2228–2235.
3. Born, M.; Wolf, E. *Principles of Optics*, 6th ed.; Pergamon: Oxford, UK, 1980.

4. Han, D.; Kim, Y.S.; Noz, M.E. Stokes parameters as a Minkowskian four-vector. *Phys. Rev. E* **1997**, *56*, 6065–6076.

5. Brosseau, C. *Fundamentals of Polarized Light: A Statistical Optics Approach*; John Wiley: New York, NY, USA, 1998.

6. Başkal, S.; Kim, Y.S. De Sitter group as a symmetry for optical decoherence. *J. Phys. A* **2006**, *39*, 7775–7788.

7. Kim, Y.S.; Noz, M.E. Symmetries shared by the Poincaré Group and the Poincaré Sphere. *Symmetry* **2013**, *5*, 233–252.

8. Han, D.; Kim, Y.S.; Son, D. E(2)-like little group for massless particles and polarization of neutrinos. *Phys. Rev. D* **1982**, *26*, 3717–3725.

9. Han, D.; Kim, Y.S.; Son, D. Photons, neutrinos and gauge transformations. *Am. J. Phys.* **1986**, *54*, 818–821.

10. Başkal, S.; Kim, Y.S. Little groups and Maxwell-type tensors for massive and massless particles. *Europhys. Lett.* **1997**, *40*, 375–380.

11. Leggett, A.; Chakravarty, S.; Dorsey, A.; Fisher, M.; Garg, A.; Zwerger, W. Dynamics of the dissipative 2-state system. *Rev. Mod. Phys.* **1987**, *59*, 1–85.

12. Başkal, S.; Kim, Y.S. One analytic form for four branches of the ABCD matrix. *J. Mod. Opt.* **2010**, *57*, 1251–1259.

13. Başkal, S.; Kim, Y.S. Lens optics and the continuity problems of the ABCD matrix. *J. Mod. Opt.* **2014**, *61*, 161–166.

14. Kim, Y.S.; Noz, M.E. *Theory and Applications of the Poincaré Group*; Reidel: Dordrecht, The Netherlands, 1986.

15. Bargmann, V. Irreducible unitary representations of the Lorentz group. *Ann. Math.* **1947**, *48*, 568–640.

16. Iwasawa, K. On some types of topological groups. *Ann. Math.* **1949**, *50*, 507–558.

17. Guillemin, V.; Sternberg, S. *Symplectic Techniques in Physics*; Cambridge University Press: Cambridge, UK, 1984.

18. Başkal, S.; Kim, Y.S. Lorentz Group in Ray and Polarization Optics. In *Mathematical Optics: Classical, Quantum and Computational Methods*; Lakshminarayanan, V., Calvo, M.L., Alieva, T., Eds.; CRC Taylor and Francis: New York, NY, USA, 2013; Chapter 9, pp. 303–340.

19. Naimark, M.A. *Linear Representations of the Lorentz Group*; Pergamon: Oxford, UK, 1964.

20. Kim, Y.S.; Wigner, E.P. Cylindrical group and masless particles. *J. Math. Phys.* **1987**, *28*, 1175–1179.

21. Kim, Y.S.; Wigner, E.P. Space-time geometry of relativistic particles. *J. Math. Phys.* **1990**, *31*, 55–60.

22. Georgieva, E.; Kim, Y.S. Iwasawa effects in multilayer optics. *Phys. Rev. E* **2001**, *64*, doi:10.1103/PhysRevE.64.026602.

23. Saleh, B.E.A.; Teich, M.C. *Fundamentals of Photonics*, 2nd ed.; John Wiley: Hoboken, NJ, USA, 2007.

24. Papoulias, D.K.; Kosmas, T.S. Exotic Lepton Flavour Violating Processes in the Presence of Nuclei. *J. Phys.: Conf. Ser.* **2013**, *410*, 012123:1–012123:5.

25. Dinh, D.N.; Petcov, S.T.; Sasao, N.; Tanaka, M.; Yoshimura, M. Observables in neutrino mass spectroscopy using atoms. *Phys. Lett. B.* **2013**, *719*, 154–163.

26. Miramonti, L.; Antonelli, V. Advancements in Solar Neutrino physics. *Int. J. Mod. Phys. E.* **2013**, *22*, 1–16.

27. Li, Y.-F.; Cao, J.; Jun, Y.; Wang, Y.; Zhan, L. Unambiguous determination of the neutrino mass hierarchy using reactor neutrinos. *Phys. Rev. D.* **2013**, *88*, 013008:1–013008:9.

28. Bergstrom, J. Combining and comparing neutrinoless double beta decay experiments using different 584 nuclei. *J. High Energy Phys.* **2013**, *02*, 093:1–093:27.

29. Han, T.; Lewis, I.; Ruiz, R.; Si, Z.-G. Lepton number violation and W' chiral couplings at the LHC. *Phys. Rev. D* **2013**, *87*, 035011:1–035011:25.

30. Drewes, M. The phenomenology of right handed neutrinos. *Int. J. Mod. Phys. E* **2013**, *22*, 1330019:1–1330019:75.

31. Barut, A.O.; McEwan, J. The four states of the massless neutrino with pauli coupling by spin-gauge invariance. *Lett. Math. Phys.* **1986**, *11*, 67–72.

32. Palcu, A. Neutrino Mass as a consequence of the exact solution of 3-3-1 gauge models without exotic electric charges. *Mod. Phys. Lett. A* **2006**, *21*, 1203–1217.

33. Bilenky, S.M. Neutrino. *Phys. Part. Nucl.* **2013**, *44*, 1–46.

34. Alhendi, H. A.; Lashin, E. I.; Mudlej, A. A. Textures with two traceless submatrices of the neutrino mass matrix. *Phys.Rev. D* **2008**, *77*, 013009.1–013009.1-13.

35. Weinberg, S. Photons and gravitons in S-Matrix theory: Derivation of charge conservation and equality of gravitational and inertial mass. *Phys. Rev.* **1964**, *135*, B1049–B1056.

36. Higgs, P.W. Broken symmetries and the masses of gauge bosons. *Phys. Rev. Lett.* **1964**, *13*, 508–509.

37. Guralnik, G.S.; Hagen, C.R.; Kibble, T.W.B. Global conservation laws and massless particles. *Phys. Rev. Lett.* **1964**, *13*, 585–587.

38. Weinberg, S. A model of leptons. *Phys. Rev. Lett.* **1967**, *19*, 1265–1266.

39. Weinberg, S. *Quantum Theory of Fields, Volume II, Modern Applications*; Cambridge University Press: Cambridge, UK, 1996.

40. Dée, A.; Ivanov, I.P. Higgs boson masses of the general two-Higgs-doublet model in the Minkowski-space formalism. *Phys. Rev. D* **2010**, *81*, 015012:1–015012:8.

41. Inönü, E.; Wigner, E.P. On the contraction of groups and their representations. *Proc. Natl. Acad. Sci. USA* **1953**, *39*, 510–524.

© 2014 by the authors. Licensee MDPI, Basel, Switzerland. This article is an open access article distributed under the terms and conditions of the Creative Commons Attribution (CC BY) license (http://creativecommons.org/licenses/by/4.0/).

symmetry

MDPI

Article

Loop Representation of Wigner's Little Groups

Sibel Başkal [1], Young S. Kim [2,*] and Marilyn E. Noz [3]

[1] Department of Physics, Middle East Technical University, 06800 Ankara, Turkey;
baskal@newton.physics.metu.edu.tr

[2] Center for Fundamental Physics, University of Maryland College Park, Maryland, MD 20742, USA

[3] Department of Radiology, New York University, New York, NY 10016, USA; marilyn.noz@med.nyu.edu

* Correspondence: yskim@umd.edu; Tel.: +1-301-937-6306

Academic Editor: Sergei D. Odintsov

Received: 12 May 2017; Accepted: 15 June 2017; Published: 23 June 2017

Abstract: Wigner's little groups are the subgroups of the Lorentz group whose transformations leave the momentum of a given particle invariant. They thus define the internal space-time symmetries of relativistic particles. These symmetries take different mathematical forms for massive and for massless particles. However, it is shown possible to construct one unified representation using a graphical description. This graphical approach allows us to describe vividly parity, time reversal, and charge conjugation of the internal symmetry groups. As for the language of group theory, the two-by-two representation is used throughout the paper. While this two-by-two representation is for spin-1/2 particles, it is shown possible to construct the representations for spin-0 particles, spin-1 particles, as well as for higher-spin particles, for both massive and massless cases. It is shown also that the four-by-four Dirac matrices constitute a two-by-two representation of Wigner's little group.

Keywords: Wigner's little groups; Lorentz group; unified picture of massive and massless particles; two-by-two representations; graphical approach to internal space-time symmetries

PACS: 02.10.Yn; 02.20.Uw; 03.65.Fd

1. Introduction

In his 1939 paper [1], Wigner introduced subgroups of the Lorentz group whose transformations leave the momentum of a given particle invariant. These subgroups are called Wigner's little groups in the literature and are known as the symmetry groups for internal space-time structure.

For instance, a massive particle at rest can have spin that can be rotated in three-dimensional space. The little group in this case is the three-dimensional rotation group. For a massless particle moving along the z direction, Wigner noted that rotations around the z axis do not change the momentum. In addition, he found two more degrees of freedom, which together with the rotation, constitute a subgroup locally isomorphic to the two-dimensional Euclidean group.

However, Wigner's 1939 paper did not deal with the following critical issues.

1. As for the massive particle, Wigner worked out his little group in the Lorentz frame where the particle is at rest with zero momentum, resulting in the three-dimensional rotation group. He could have Lorentz-boosted the $O(3)$-like little group to make the little group for a moving particle.

2. While the little group for a massless particle is like $E(2)$, it is not difficult to associate the rotational degree of freedom to the helicity. However, Wigner did not give physical interpretations to the two translation-like degrees of freedom.

3. While the Lorentz group does not allow mass variations, particles with infinite momentum should behave like massless particles. The question is whether the Lorentz-boosted $O(3)$-like little group becomes the $E(2)$-like little group for particles with infinite momentum.

These issues have been properly addressed since then [2–5]. The translation-like degrees of freedom for massless particles collapse into one gauge degree of freedom, and the $E(2)$-like little group can be obtained as the infinite-momentum limit of the $O(3)$-like little group. This history is summarized in Figure 1.

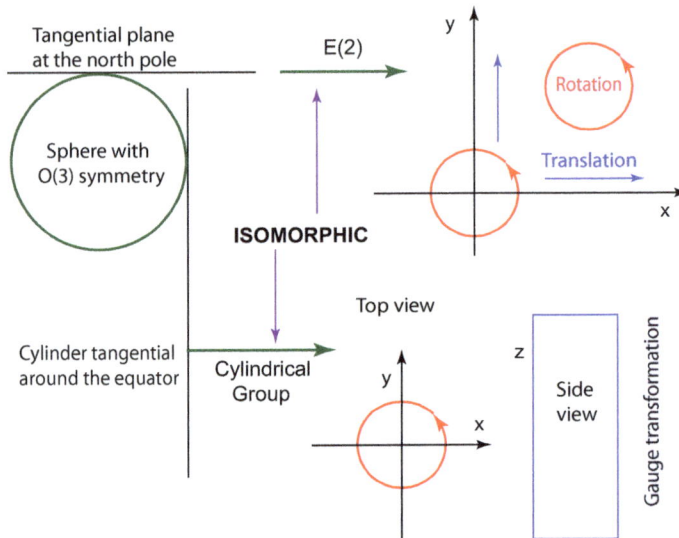

Figure 1. $O(3)$-like and $E(2)$-like internal space-time symmetries of massive and massless particles. The sphere corresponds to the $O(3)$-like little group for the massive particle. There is a plane tangential to the sphere at its north pole, which is $E(2)$. There is also a cylinder tangent to the sphere at its equatorial belt. This cylinder gives one helicity and one gauge degree of freedom. This figure thus gives a unified picture of the little groups for massive and massless particles [5].

In this paper, we shall present these developments using a mathematical language more transparent than those used in earlier papers.

1. In his original paper [1], Wigner worked out his little group for the massive particle when its momentum is zero. How about moving massive particles? In this paper, we start with a moving particle with non-zero momentum. We then perform rotations and boosts whose net effect does not change the momentum [6–8]. This procedure can be applied to the massive, massless, and imaginary-mass cases.

2. By now, we have a clear understanding of the group $SL(2, c)$ as the universal covering group of the Lorentz group. The logic with two-by-two matrices is far more transparent than the mathematics based on four-by-four matrices. We shall thus use the two-by-two representation of the Lorentz group throughout the paper [5,9–11].

The purpose of this paper is to make the physics contained in Wigner's original paper more transparent. In Section 2, we give the six generators of the Lorentz group. It is possible to write them in terms of coordinate transformations, four-by-four matrices, and two-by-two matrices. In Section 3, we introduce Wigner's little groups in terms of two-by-two matrices. In Section 4, it is shown possible to construct transformation matrices of the little group by performing rotations and a boost resulting in a non-trivial matrix, which leaves the given momentum invariant.

Since we are more familiar with Dirac matrices than the Lorentz group, it is shown in Section 5 that Dirac matrices are a representation of the Lorentz group, and his four-by-four matrices are two-by-two representations of the two-by-two representation of Wigner's little groups. In Section 6, we construct spin-0 and spin-1 particles for the SL(2,c) spinors. We also discuss massless higher spin particles.

2. Lorentz Group and Its Representations

The group of four-by-four matrices, which performs Lorentz transformations on the four-dimensional Minkowski space leaving invariant the quantity $(t^2 - z^2 - x^2 - y^2)$, forms the starting point for the Lorentz group. As there are three rotation and three boost generators, the Lorentz group is a six-parameter group.

Einstein, by observing that this Lorentz group also leaves invariant (E, p_z, p_x, p_y), was able to derive his Lorentz-covariant energy-momentum relation commonly known as $E = mc^2$. Thus, the particle mass is a Lorentz-invariant quantity.

The Lorentz group is generated by the three rotation operators:

$$J_i = -i \left(x_j \frac{\partial}{\partial x_k} - x_k \frac{\partial}{\partial x_j} \right),$$

(1)

where $i, j, k = 1, 2, 3$, and three boost operators:

$$K_i = -i \left(t \frac{\partial}{\partial x_i} + x_i \frac{\partial}{\partial t} \right).$$

(2)

These generators satisfy the closed set of commutation relations:

$$[J_i, J_j] = i\epsilon_{ijk} J_k, \qquad [J_i, K_j] = i\epsilon_{ijk} K_k, \qquad [K_i, K_j] = -i\epsilon_{ijk} J_k,$$

(3)

which are known as the Lie algebra for the Lorentz group.

Under the space inversion, $x_i \rightarrow -x_i$, or the time reflection, $t \rightarrow -t$, the boost generators K_i change sign. However, the Lie algebra remains invariant, which means that the commutation relations remain invariant under Hermitian conjugation.

In terms of four-by-four matrices applicable to the Minkowskian coordinate of (t, z, x, y), the generators can be written as:

$$J_3 = \begin{pmatrix} 0 & 0 & 0 & 0 \\ 0 & 0 & 0 & 0 \\ 0 & 0 & 0 & -i \\ 0 & 0 & i & 0 \end{pmatrix}, \qquad K_3 = \begin{pmatrix} 0 & i & 0 & 0 \\ i & 0 & 0 & 0 \\ 0 & 0 & 0 & 0 \\ 0 & 0 & 0 & 0 \end{pmatrix},$$

(4)

for rotations around and boosts along the z direction, respectively. Similar expressions can be written for the x and y directions. We see here that the rotation generators J_i are Hermitian, but the boost generators K_i are anti-Hermitian.

We can also consider the two-by-two matrices:

$$J_i = \frac{1}{2}\sigma_i, \quad \text{and} \quad K_i = \frac{i}{2}\sigma_i,$$

(5)

where σ_i are the Pauli spin matrices. These matrices also satisfy the commutation relations given in Equation (3).

There are interesting three-parameter subgroups of the Lorentz group. In 1939 [1], Wigner considered the subgroups whose transformations leave the four-momentum of a given particle invariant. First of all, consider a massive particle at rest. The momentum of this particle is invariant under rotations in three-dimensional space. What happens for the massless particle that cannot be brought to a rest frame? In this paper we shall consider this and other problems using the two-by-two representation of the Lorentz group.

3. Two-by-Two Representation of Wigner's Little Groups

The six generators of Equation (5) lead to the group of two-by-two unimodular matrices of the form:

$$G = \begin{pmatrix} \alpha & \beta \\ \gamma & \delta \end{pmatrix}, \tag{6}$$

with $\det(G) = 1$, where the matrix elements are complex numbers. There are thus six independent real numbers to accommodate the six generators given in Equation (5). The groups of matrices of this form are called $SL(2, c)$ in the literature. Since the generators K_i are not Hermitian, the matrix G is not always unitary. Its Hermitian conjugate is not necessarily the inverse.

The space-time four-vector can be written as [5,9,11]:

$$\begin{pmatrix} t + z & x - iy \\ x + iy & t - z \end{pmatrix}, \tag{7}$$

whose determinant is $t^2 - z^2 - x^2 - z^2$, and remains invariant under the Hermitian transformation:

$$X' = G \, X \, G^\dagger. \tag{8}$$

This is thus a Lorentz transformation. This transformation can be explicitly written as:

$$\begin{pmatrix} t' + z' & x' - iy' \\ x' + iy' & t' - z' \end{pmatrix} = \begin{pmatrix} \alpha & \beta \\ \gamma & \delta \end{pmatrix} \begin{pmatrix} t + z & x - iy \\ x + iy & t - z \end{pmatrix} \begin{pmatrix} \alpha^* & \gamma^* \\ \beta^* & \delta^* \end{pmatrix}. \tag{9}$$

With these six independent real parameters, it is possible to construct four-by-four matrices for Lorentz transformations applicable to the four-dimensional Minkowskian space [5,12]. For the purpose of the present paper, we need some special cases, and they are given in Table 1.

Table 1. Two-by-two and four-by-four representations of the Lorentz group.

Generators	Two-by-Two	Four-by-Four
$J_3 = \frac{1}{2}\begin{pmatrix} 1 & 0 \\ 0 & -1 \end{pmatrix}$	$\begin{pmatrix} \exp(i\phi/2) & 0 \\ 0 & \exp(-i\phi/2) \end{pmatrix}$	$\begin{pmatrix} 1 & 0 & 0 & 0 \\ 0 & 1 & 0 & 0 \\ 0 & 0 & \cos\phi & -\sin\phi \\ 0 & 0 & \sin\phi & \cos\phi \end{pmatrix}$
$K_3 = \frac{1}{2}\begin{pmatrix} i & 0 \\ 0 & -i \end{pmatrix}$	$\begin{pmatrix} \exp(\eta/2) & 0 \\ 0 & \exp(-\eta/2) \end{pmatrix}$	$\begin{pmatrix} \cosh\eta & \sinh\eta & 0 & 0 \\ \sinh\eta & \cosh\eta & 0 & 0 \\ 0 & 0 & 1 & 0 \\ 0 & 0 & 0 & 1 \end{pmatrix}$
$J_1 = \frac{1}{2}\begin{pmatrix} 0 & 1 \\ 1 & 0 \end{pmatrix}$	$\begin{pmatrix} \cos(\theta/2) & i\sin(\theta/2) \\ i\sin(\theta/2) & \cos(\theta/2) \end{pmatrix}$	$\begin{pmatrix} 1 & 0 & 0 & 0 \\ 0 & \cos\theta & 0 & \sin\theta \\ 0 & 0 & 1 & 0 \\ 0 & -\sin\theta & 0 & \cos\theta \end{pmatrix}$
$K_1 = \frac{1}{2}\begin{pmatrix} 0 & i \\ i & 0 \end{pmatrix}$	$\begin{pmatrix} \cosh(\lambda/2) & \sinh(\lambda/2) \\ \sinh(\lambda/2) & \cosh(\lambda/2) \end{pmatrix}$	$\begin{pmatrix} \cosh\lambda & 0 & \sinh\lambda & 0 \\ 0 & 1 & 0 & 0 \\ \sinh\lambda & 0 & \cosh\lambda & 0 \\ 0 & 0 & 0 & 1 \end{pmatrix}$
$J_2 = \frac{1}{2}\begin{pmatrix} 0 & -i \\ i & 0 \end{pmatrix}$	$\begin{pmatrix} \cos(\theta/2) & -\sin(\theta/2) \\ \sin(\theta/2) & \cos(\theta/2) \end{pmatrix}$	$\begin{pmatrix} 1 & 0 & 0 & 0 \\ 0 & \cos\theta & -\sin\theta & 0 \\ 0 & \sin\theta & \cos\theta & 0 \\ 0 & 0 & 0 & 1 \end{pmatrix}$
$K_2 = \frac{1}{2}\begin{pmatrix} 0 & 1 \\ -1 & 0 \end{pmatrix}$	$\begin{pmatrix} \cosh(\lambda/2) & -i\sinh(\lambda/2) \\ i\sinh(\lambda/2) & \cosh(\lambda/2) \end{pmatrix}$	$\begin{pmatrix} \cosh\lambda & 0 & 0 & \sinh\lambda \\ 0 & 1 & 0 & 0 \\ 0 & 0 & 1 & 0 \\ \sinh\lambda & 0 & 0 & \cosh\lambda \end{pmatrix}$

Likewise, the two-by-two matrix for the four-momentum takes the form:

$$P = \begin{pmatrix} p_0 + p_z & p_x - ip_y \\ p_x + ip_y & p_0 - p_z \end{pmatrix}, \tag{10}$$

with $p_0 = \sqrt{m^2 + p_z^2 + p_x^2 + p_y^2}$. The transformation property of Equation (9) is applicable also to this energy-momentum four-vector.

In 1939 [1], Wigner considered the following three four-vectors.

$$P_+ = \begin{pmatrix} 1 & 0 \\ 0 & 1 \end{pmatrix}, \qquad P_0 = \begin{pmatrix} 1 & 0 \\ 0 & 0 \end{pmatrix}, \qquad P_- = \begin{pmatrix} 1 & 0 \\ 0 & -1 \end{pmatrix}. \tag{11}$$

whose determinants are 1, 0, and -1, respectively, corresponding to the four-momenta of massive, massless, and imaginary-mass particles, as shown in Table 2.

Table 2. The Wigner momentum vectors in the two-by-two matrix representation together with the corresponding transformation matrix. These four-momentum matrices have determinants that are positive, zero, and negative for massive, massless, and imaginary-mass particles, respectively.

Particle Mass	Four-Momentum	Transform Matrix
Massive	$\begin{pmatrix} 1 & 0 \\ 0 & 1 \end{pmatrix}$	$\begin{pmatrix} \cos(\theta/2) & -\sin(\theta/2) \\ \sin(\theta/2) & \cos(\theta/2) \end{pmatrix}$
Massless	$\begin{pmatrix} 1 & 0 \\ 0 & 0 \end{pmatrix}$	$\begin{pmatrix} 1 & -\gamma \\ 0 & 1 \end{pmatrix}$
Imaginary mass	$\begin{pmatrix} 1 & 0 \\ 0 & -1 \end{pmatrix}$	$\begin{pmatrix} \cosh(\lambda/2) & \sinh(\lambda/2) \\ \sinh(\lambda/2) & \cosh(\lambda/2) \end{pmatrix}$

He then constructed the subgroups of the Lorentz group whose transformations leave these four-momenta invariant. These subgroups are called Wigner's little groups in the literature. Thus, the matrices of these little groups should satisfy:

$$W P_i W^\dagger = P_i, \tag{12}$$

where $i = +, 0, -$. Since the momentum of the particle is fixed, these little groups define the internal space-time symmetries of the particle. For all three cases, the momentum is invariant under rotations around the z axis, as can be seen from the expression given for the rotation matrix generated by J_3 given in Table 1.

For the first case corresponding to a massive particle at rest, the requirement of the subgroup is:

$$W P_+ W^\dagger = P_+. \tag{13}$$

This requirement tells that the subgroup is the rotation subgroup with the rotation matrix around the y direction:

$$R(\theta) = \begin{pmatrix} \cos(\theta/2) & -\sin(\theta/2) \\ \sin(\theta/2) & \cos(\theta/2) \end{pmatrix}. \tag{14}$$

For the second case of P_0, the triangular matrix of the form:

$$\Gamma(\xi) = \begin{pmatrix} 1 & -\xi \\ 0 & 1 \end{pmatrix}, \tag{15}$$

satisfies the Wigner condition of Equation (12). If we allow rotations around the z axis, the expression becomes:

$$\Gamma(\xi, \phi) = \begin{pmatrix} 1 & -\xi \exp(-i\phi) \\ 0 & 1 \end{pmatrix}. \tag{16}$$

This matrix is generated by:

$$N_1 = J_2 - K_1 = \begin{pmatrix} 0 & -i \\ 0 & 0 \end{pmatrix}, \quad \text{and} \quad N_2 = J_1 + K_2 = \begin{pmatrix} 0 & 1 \\ 0 & 0 \end{pmatrix}. \tag{17}$$

Thus, the little group is generated by J_3, N_1, and N_2. They satisfy the commutation relations:

$$[N_1, N_2] = 0, \qquad [J_3, N_1] = iN_2, \qquad [J_3, K_2] = -iN_1. \tag{18}$$

Wigner in 1939 [1] observed that this set is the same as that of the two-dimensional Euclidean group with one rotation and two translations. The physical interpretation of the rotation is easy to understand. It is the helicity of the massless particle. On the other hand, the physics of the N_1 and N_2 matrices has a stormy history, and the issue was not completely settled until 1990 [4]. They generate gauge transformations.

For the third case of P_-, the matrix of the form:

$$S(\lambda) = \begin{pmatrix} \cosh(\lambda/2) & \sinh(\lambda/2) \\ \sinh(\lambda/2) & \cosh(\lambda/2) \end{pmatrix}, \tag{19}$$

satisfies the Wigner condition of Equation (12). This corresponds to the Lorentz boost along the x direction generated by K_1 as shown in Table 1. Because of the rotation symmetry around the z axis, the Wigner condition is satisfied also by the boost along the y axis. The little group is thus generated by J_3, K_1, and K_2. These three generators:

$$[J_3, K_1] = iK_2, \qquad [J_3, K_2] = -iK_1, \qquad [K_1, K_2] = -iJ_3 \tag{20}$$

form the little group $O(2,1)$, which is the Lorentz group applicable to two space-like and one time-like dimensions.

Of course, we can add rotations around the z axis. Let us Lorentz-boost these matrices along the z direction with the diagonal matrix:

$$B(\eta) = \begin{pmatrix} \exp(\eta/2) & 0 \\ 0 & \exp(-\eta/2) \end{pmatrix}. \tag{21}$$

Then, the matrices of Equations (14), (15), and (19) become:

$$B(\eta)R(\theta)B(-\eta) = \begin{pmatrix} \cos(\theta/2) & -e^{\eta}\sin(\theta/2) \\ e^{-\eta}\sin(\theta/2) & \cos(\theta/2) \end{pmatrix}, \tag{22}$$

$$B(\eta)\Gamma(\xi)B(-\eta) = \begin{pmatrix} 1 & -e^{\eta}\xi \\ 0 & 1 \end{pmatrix}, \tag{23}$$

$$B(\eta)S(-\lambda)B(-\eta) = \begin{pmatrix} \cosh(\lambda/2) & -e^{\eta}\sinh(\lambda/2) \\ -e^{-\eta}\sinh(\lambda/2) & \cosh(\lambda/2) \end{pmatrix}, \tag{24}$$

respectively. We have changed the sign of λ for future convenience.

When η becomes large, θ, ξ, and λ should become small if the upper-right elements of the these three matrices are to remain finite. In that case, the diagonal elements become one, and all three matrices become like the triangular matrix:

$$\begin{pmatrix} 1 & -\gamma \\ 0 & 1 \end{pmatrix}. \tag{25}$$

Here comes the question of whether the matrix of Equation (24) can be continued from Equation (22), via Equation (23). For this purpose, let us write Equation (22) as:

$$\begin{pmatrix} 1 - (\gamma\epsilon)^2/2 & -\gamma \\ \gamma\epsilon^2 & 1 - (\gamma\epsilon)^2/2 \end{pmatrix}, \tag{26}$$

for small $\theta = 2\gamma\epsilon$, with $\epsilon = e^{-\eta}$. For Equation (24), we can write:

$$\begin{pmatrix} 1 + (\gamma\epsilon)^2/2 & -\gamma \\ -\gamma\epsilon^2 & 1 + (\gamma\epsilon)^2/2 \end{pmatrix}, \tag{27}$$

with $\lambda = -2\gamma\epsilon$. Both of these expressions become the triangular matrix of Equation (25) when $\epsilon = 0$.

For small values of ϵ, the diagonal elements change from $\cos(\theta/2)$ to $\cosh(\lambda/2)$ while $\sin(\theta/2)$ becomes $-\sinh(\lambda/2)$. Thus, it is possible to continue from Equation (22) to Equation (24). The mathematical details of this process have been discussed in our earlier paper on this subject [13].

We are then led to the question of whether there is one expression that will take care of all three cases. We shall discuss this issue in Section 4.

4. Loop Representation of Wigner's Little Groups

It was noted in Section 3 that matrices of Wigner's little group take different forms for massive, massless, and imaginary-mass particles. In this section, we construct one two-by-two matrix that works for all three different cases.

In his original paper [1], Wigner constructs those matrices in specific Lorentz frames. For instance, for a moving massive particle with a non-zero momentum, Wigner brings it to the rest frame and works out the $O(3)$ subgroup of the Lorentz group as the little group for this massive particle. In order to complete the little group, we should boost this $O(3)$ to the frame with the original non-zero momentum [4].

In this section, we construct transformation matrices without changing the momentum. Let us assume that the momentum is along the z direction; the rotation around the z axis leaves the momentum invariant. According to the Euler decomposition, the rotation around the y axis, in addition, will accommodate rotations along all three directions. For this reason, it is enough to study what happens in transformations within the xz plane [14].

It was Kupersztych [6] who showed in 1976 that it is possible to construct a momentum-preserving transformation by a rotation followed by a boost as shown in Figure 2. In 1981 [7], Han and Kim showed that the boost can be decomposed into two components as illustrated in Figure 2. In 1988 [8], Han and Kim showed that the same purpose can be achieved by one boost preceded and followed by the same rotation matrix, as shown also in Figure 2. We choose to call this loop the "D loop" and write the transformation matrix as:

$$D(\alpha, \chi) = R(\alpha)S(-2\chi)R(\alpha). \tag{28}$$

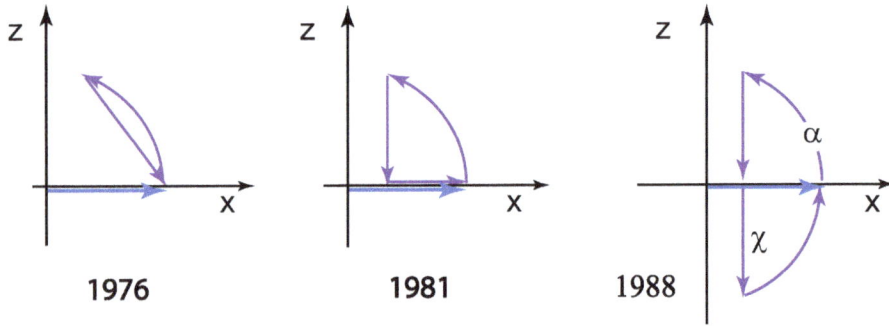

Figure 2. Evolution of the Wigner loop. In 1976 [6], Kupersztych considered a rotation followed by a boost whose net result will leave the momentum invariant. In 1981 [7], Han and Kim considered the same problem with simpler forms for boost matrices. In 1988, Han and Kim [8] constructed the Lorentz kinematics corresponding to the Bargmann decomposition [10] consisting of one boost matrix sandwiched by two rotation matrices. In the present case, the two rotation matrices are identical.

The D matrix can now be written as three matrices. This form is known in the literature as the Bargmann decomposition [10]. This form gives additional convenience. When we take the inverse or the Hermitian conjugate, we have to reverse the order of matrices. However, this particular form does not require re-ordering.

The D matrix of Equation (28) becomes:

$$D(\alpha, \chi) = \begin{pmatrix} (\cos \alpha) \cosh \chi & -\sinh \chi - (\sin \alpha) \cosh \chi \\ -\sinh \chi + (\sin \alpha) \cosh \chi & (\cos \alpha) \cosh \chi \end{pmatrix}. \tag{29}$$

If the diagonal element is smaller than one with $((\cos \alpha) \cosh \chi) < 1$, the off-diagonal elements have opposite signs. Thus, this D matrix can serve as the Wigner matrix of Equation (22) for massive particles. If the diagonal elements are one, one of the off-diagonal elements vanishes, and this matrix becomes triangular like Equation (23). If the diagonal elements are greater than one with $((\cos \alpha) \cosh \chi) > 1$, this matrix can become Equation (24). In this way, the matrix of Equation (28) can accommodate the three different expressions given in Equations (22)–(24).

4.1. Continuity Problems

Let us go back to the three separate formulas given in Equations (22)–(24). If η becomes infinity, all three of them become triangular. For the massive particle, $\tanh \eta$ is the particle speed, and:

$$\tanh \eta = \frac{p}{p_0}, \tag{30}$$

where p and p_0 are the momentum and energy of the particle, respectively.

When the particle is massive with $m^2 > 0$, the ratio:

$$\frac{\text{lower-left element}}{\text{upper-right element}}, \tag{31}$$

is negative and is:

$$-e^{-2\eta} = \frac{1 - \sqrt{1 + m^2/p^2}}{1 + \sqrt{1 + m^2/p^2}}. \tag{32}$$

If the mass is imaginary with $m^2 < 0$, the ratio is positive and:

$$e^{-2\eta} = \frac{1 - \sqrt{1 + m^2/p^2}}{1 + \sqrt{1 + m^2/p^2}}.$$

(33)

This ratio is zero for massless particles. This means that when m^2 changes from positive to negative, the ratio changes from $-e^{-2\eta}$ to $e^{-2\eta}$. This transition is continuous, but not analytic. This aspect of non-analytic continuity has been discussed in one of our earlier papers [13].

The D matrix of Equation (29) combines all three matrices given in Equations (22)–(24) into one matrix. For this matrix, the ratio of Equation (31) becomes:

$$\frac{\tanh \chi - \sin \alpha}{\tanh \chi + \sin \alpha} = \frac{1 - \sqrt{1 + (m/p)^2}}{1 + \sqrt{1 + (m/p^2)}}.$$

(34)

Thus,

$$\frac{m^2}{p^2} = \left(\frac{\sin \alpha}{\tanh \chi} \right)^2 - 1.$$

(35)

For the D loop of Figure 2, both $\tanh \chi$ and $\sin \alpha$ range from 0–1, as illustrated in Figure 3. For small values of the mass for a fixed value of the momentum, this expression becomes:

$$-\frac{m^2}{4p^2}.$$

(36)

Thus, the change from positive values of m^2 to negative values is continuous and analytic. For massless particles, m^2 is zero, while it is negative for imaginary-mass particles.

We realize that the mass cannot be changed within the frame of the Lorentz group and that both α and η are parameters of the Lorentz group. On the other hand, their combinations according to the D loop of Figure 2 can change the value of m^2 according to Equation (35) and Figure 3.

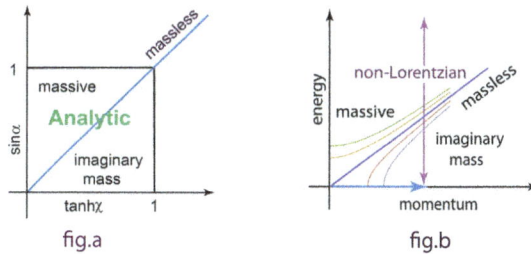

Figure 3. Non-Lorentzian transformations allowing mass variations. The D matrix of Equation (29) allows us to change the χ and α analytically within the square region in (**a**). These variations allow the mass variations illustrated in (**b**), not allowed in Lorentz transformations. The Lorentz transformations are possible along the hyperbolas given in this figure.

4.2. Parity, Time Reversal, and Charge Conjugation

Space inversion leads to the sign change in χ:

$$D(\alpha, -\chi) = \begin{pmatrix} (\cos \alpha) \cosh \chi & \sinh \chi - (\sin \alpha) \cosh \chi \\ \sinh \chi + (\sin \alpha) \cosh \chi & (\cos \alpha) \cosh \chi \end{pmatrix},$$

(37)

and time reversal leads to the sign change in both α and χ:

$$D(-\alpha, -\chi) = \begin{pmatrix} (\cos \alpha) \cosh \chi & \sinh \chi + (\sin \alpha) \cosh \chi \\ \sinh \chi - (\sin \alpha) \cosh \chi & (\cos \alpha) \cosh \chi \end{pmatrix}. \tag{38}$$

If we space-invert this expression, the result is a change only in the direction of rotation,

$$D(-\alpha, \chi) = \begin{pmatrix} (\cos \alpha) \cosh \chi & -\sinh \chi + (\sin \alpha) \cosh \chi \\ -\sinh \chi - (\sin \alpha) \cosh \chi & (\cos \alpha) \cosh \chi \end{pmatrix}. \tag{39}$$

The combined transformation of space inversion and time reversal is known as the "charge conjugation". All of these transformations are illustrated in Figure 4.

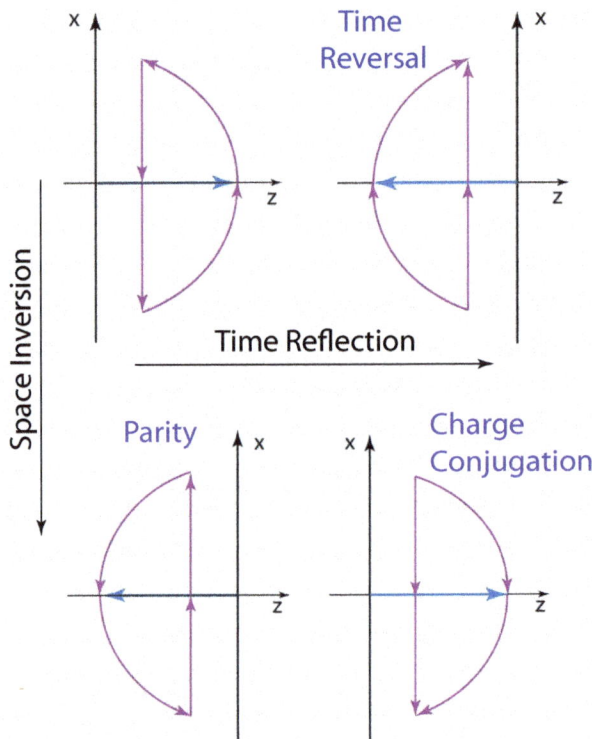

Figure 4. Parity, time reversal, and charge conjugation of Wigner's little groups in the loop representation.

Let us go back to the Lie algebra of Equation (3). This algebra is invariant under Hermitian conjugation. This means that there is another set of commutation relations,

$$[J_i, J_j] = i\epsilon_{ijk} J_k, \qquad [J_i, \dot{K}_j] = i\epsilon_{ijk} \dot{K}_k, \qquad [\dot{K}_i, \dot{K}_j] = -i\epsilon_{ijk} J_k, \tag{40}$$

where K_i is replaced with $\dot{K}_i = -K_i$. Let us go back to the expression of Equation (2). This transition to the dotted representation is achieved by the space inversion or by the parity operation.

On the other hand, the complex conjugation of the Lie algebra of Equation (3) leads to:

$$\left[J_i^*, J_j^*\right] = -i\epsilon_{ijk} J_k^*, \qquad \left[J_i^*, K_j^*\right] = -i\epsilon_{ijk} K_k^*, \qquad \left[K_i^*, K_j^*\right] = i\epsilon_{ijk} J_k^*. \tag{41}$$

It is possible to restore this algebra to that of the original form of Equation (3) if we replace J_i^* by $-J_i$ and K_i^* by $-K_i$. This corresponds to the time-reversal process. This operation is known as the anti-unitary transformation in the literature [15,16].

Since the algebras of Equations (3) and (41) are invariant under the sign change of K_i and K_i^*, respectively, there is another Lie algebra with J_i^* replaced by $-Ji$ and K_i^* by $-\dot{K}_i$. This is the parity operation followed by time reversal, resulting in charge conjugation. With the four-by-four matrices for spin-1 particles, this complex conjugation is trivial, and $J_i^* = -J_i$, as well as $K_i^* = -K_i$.

On the other hand, for spin 1/2 particles, we note that:

$$J_1^* = J_1, \qquad J_2^* = -J_2, \qquad J_3^* = J_3,$$

$$K_1^* = -K_1, \qquad K_2^* = K_2, \qquad K_3^* = -K_3. \tag{42}$$

Thus, J_i^* should be replaced by $\sigma_2 J_i \sigma_2$, and K_i^* by $-\sigma_2 K_i \sigma_2$.

5. Dirac Matrices as a Representation of the Little Group

The Dirac equation, Dirac matrices, and Dirac spinors constitute the basic language for spin-1/2 particles in physics. Yet, they are not widely recognized as the package for Wigner's little group. Yes, the little group is for spins, so are the Dirac matrices.

Let us write the Dirac equation as:

$$(p \cdot \gamma - m)\psi(\vec{x}, t) = \lambda \psi(\vec{x}, t). \tag{43}$$

This equation can be explicitly written as:

$$\left(-i\gamma_0 \frac{\partial}{\partial t} - i\gamma_1 \frac{\partial}{\partial x} - i\gamma_2 \frac{\partial}{\partial y} - i\gamma_3 \frac{\partial}{\partial z} - m \right) \psi(\vec{x}, t) = \lambda \psi(\vec{x}, t), \tag{44}$$

where:

$$\gamma_0 = \begin{pmatrix} 0 & I \\ I & 0 \end{pmatrix}, \quad \gamma_1 = \begin{pmatrix} 0 & \sigma_1 \\ -\sigma_1 & 0 \end{pmatrix}, \quad \gamma_2 = \begin{pmatrix} 0 & \sigma_2 \\ -\sigma_2 & 0 \end{pmatrix}, \quad \gamma_3 = \begin{pmatrix} 0 & \sigma_3 \\ -\sigma_3 & 0 \end{pmatrix}, \tag{45}$$

where I is the two-by-two unit matrix. We use here the Weyl representation of the Dirac matrices.

The Dirac spinor has four components. Thus, we write the wave function for a free particle as:

$$\psi(\vec{x}, t) = U_{\pm} \exp\left[i\left(\vec{p} \cdot \vec{x} - p_0 t \right) \right], \tag{46}$$

with the Dirac spinor:

$$U_+ = \begin{pmatrix} u \\ \dot{u} \end{pmatrix}, \qquad U_- = \begin{pmatrix} v \\ \dot{v} \end{pmatrix}, \tag{47}$$

where:

$$u = \dot{u} = \begin{pmatrix} 1 \\ 0 \end{pmatrix}, \quad \text{and} \quad v = \dot{v} = \begin{pmatrix} 0 \\ 1 \end{pmatrix}. \tag{48}$$

In Equation (46), the exponential form $\exp\left[i\left(\vec{p} \cdot \vec{x} - p_0 t \right) \right]$ defines the particle momentum, and the column vector $U_{\pm}$ is for the representation space for Wigner's little group dictating the internal space-time symmetries of spin-1/2 particles.

In this four-by-four representation, the generators for rotations and boosts take the form:

$$J_i = \frac{1}{2} \begin{pmatrix} \sigma_i & 0 \\ 0 & \sigma_i \end{pmatrix}, \quad \text{and} \quad K_i = \frac{i}{2} \begin{pmatrix} \sigma_i & 0 \\ 0 & -\sigma_i \end{pmatrix}. \tag{49}$$

This means that both dotted and undotted spinor are transformed in the same way under rotation, while they are boosted in the opposite directions.

When this γ_0 matrix is applied to $U_\pm$:

$$\gamma_0 U_+ = \begin{pmatrix} 0 & I \\ I & 0 \end{pmatrix} \begin{pmatrix} u \\ \dot{u} \end{pmatrix} = \begin{pmatrix} \dot{u} \\ u \end{pmatrix}, \quad \text{and} \quad \gamma_0 U_- = \begin{pmatrix} 0 & I \\ I & 0 \end{pmatrix} \begin{pmatrix} v \\ \dot{v} \end{pmatrix} = \begin{pmatrix} \dot{v} \\ v \end{pmatrix}. \tag{50}$$

Thus, the γ_0 matrix interchanges the dotted and undotted spinors.

The four-by-four matrix for the rotation around the y axis is:

$$R_{44}(\theta) = \begin{pmatrix} R(\theta) & 0 \\ 0 & R(\theta) \end{pmatrix}, \tag{51}$$

while the matrix for the boost along the z direction is:

$$B_{44}(\eta) = \begin{pmatrix} B(\eta) & 0 \\ 0 & B(-\eta) \end{pmatrix}, \tag{52}$$

with:

$$B(\pm\eta) = \begin{pmatrix} e^{\pm\eta/2} & 0 \\ 0 & e^{\mp\eta/2} \end{pmatrix}. \tag{53}$$

These γ matrices satisfy the anticommutation relations:

$$\{\gamma_\mu, \gamma_\nu\} = 2g_{\mu\nu}, \tag{54}$$

where:

$$g_{00} = 1, \quad g_{11} = g_{22} = g_{22} = -1,$$

$$g_{\mu\nu} = 0 \quad \text{if} \quad \mu \neq \nu. \tag{55}$$

Let us consider space inversion with the exponential form changing to $\exp\left[i\left(-\vec{p}\cdot\vec{x} - p_0 t\right)\right]$. For this purpose, we can change the sign of x in the Dirac equation of Equation (44). It then becomes:

$$\left(-i\gamma_0\frac{\partial}{\partial t} + i\gamma_1\frac{\partial}{\partial x} + i\gamma_2\frac{\partial}{\partial y} + i\gamma_3\frac{\partial}{\partial z} - m\right)\psi(-\vec{x}, t) = \lambda\psi(-\vec{x}, t). \tag{56}$$

Since $\gamma_0\gamma_i = -\gamma_i\gamma_0$ for $i = 1, 2, 3$,

$$\left(-i\gamma_0\frac{\partial}{\partial t} - i\gamma_1\frac{\partial}{\partial x} - i\gamma_2\frac{\partial}{\partial y} - i\gamma_3\frac{\partial}{\partial z} - m\right)[\gamma_0\psi(-\vec{x}\cdot\vec{p}, p_0 t)] = \lambda[\gamma_0\psi(-\vec{x}\cdot\vec{p}, p_0 t)]. \tag{57}$$

This is the Dirac equation for the wave function under the space inversion or the parity operation. The Dirac spinor $U_\pm$ becomes $\gamma_0 U_\pm$, according to Equation (50). This operation is illustrated in Table 3 and Figure 4.

Table 3. Parity, charge conjugation, and time reversal in the loop representation.

	Start	Time Reflection
Start	Start with $R(\alpha)S(-2\chi)R(\alpha)$	Time Reversal $R(-\alpha)S(2\chi)R(-\alpha)$
Space Inversion	Parity $R(\alpha)S(2\chi)R(\alpha)$	Charge Conjugation $R(-\alpha)S(-2\chi)R(-\alpha)$

We are interested in changing the sign of t. First, we can change both space and time variables, and then, we can change the space variable. We can take the complex conjugate of the equation first. Since γ_2 is imaginary, while all others are real, the Dirac equation becomes:

$$\left(i\gamma_0 \frac{\partial}{\partial t} + i\gamma_1 \frac{\partial}{\partial x} - i\gamma_2 \frac{\partial}{\partial y} + i\gamma_3 \frac{\partial}{\partial z} - m \right) \psi^*(\vec{x}, t) = \lambda \psi^*(\vec{x}, t). \tag{58}$$

We are now interested in restoring this equation to the original form of Equation (44). In order to achieve this goal, let us consider $(\gamma_1 \gamma_3)$. This form commutes with γ_0 and γ_2 and anti-commutes with γ_1 and γ_3. Thus,

$$\left(-i\gamma_0 \frac{\partial}{\partial t} - i\gamma_1 \frac{\partial}{\partial x} - i\gamma_2 \frac{\partial}{\partial y} - i\gamma_3 \frac{\partial}{\partial z} - m \right) (\gamma_1 \gamma_3) \psi^*(\vec{x}, -t) = \lambda (\gamma_1 \gamma_3) \psi^*(\vec{x}, -t). \tag{59}$$

Furthermore, since:

$$\gamma_1 \gamma_3 = \begin{pmatrix} i\sigma_2 & 0 \\ 0 & i\sigma_2 \end{pmatrix}, \tag{60}$$

this four-by-four matrix changes the direction of the spin. Indeed, this form of time reversal is consistent with Table 3 and Figure 4.

Finally, let us change the signs of both $\vec{x}$ and t. For this purpose, we go back to the complex-conjugated Dirac equation of Equation (43). Here, γ_2 anti-commutes with all others. Thus, the wave function:

$$\gamma_2 \psi(-\vec{x} \cdot \vec{p}, -p_0 t), \tag{61}$$

should satisfy the Dirac equation. This form is known as the charge-conjugated wave function, and it is also illustrated in Table 3 and Figure 4.

5.1. Polarization of Massless Neutrinos

For massless neutrinos, the little group consists of rotations around the z axis, in addition to N_i and $\dot{N}_i$ applicable to the upper and lower components of the Dirac spinors. Thus, the four-by-four matrix for these generators is:

$$N_{44(i)} = \begin{pmatrix} N_i & 0 \\ 0 & \dot{N}_i \end{pmatrix}. \tag{62}$$

The transformation matrix is thus:

$$D_{44}(\alpha, \beta) = \exp\left(-i\alpha N_{44(1)} - i\beta N_{44(2)} \right) = \begin{pmatrix} D(\alpha, \beta) & 0 \\ 0 & \dot{D}(\alpha, \beta) \end{pmatrix}, \tag{63}$$

with:

$$D(\alpha, \beta) = \begin{pmatrix} 1 & \alpha - i\beta \\ 0 & 1 \end{pmatrix}, \qquad \dot{D}(\alpha, \beta) \begin{pmatrix} 1 & 0 \\ -\alpha - i\beta & 1 \end{pmatrix}. \tag{64}$$

As is illustrated in Figure 1, the D transformation performs the gauge transformation on massless photons. Thus, this transformation allows us to extend the concept of gauge transformations to massless spin-1/2 particles. With this point in mind, let us see what happens when this D transformation is applied to the Dirac spinors.

$$D(\alpha, \beta)u = u, \qquad \dot{D}(\alpha, \beta)\dot{v} = \dot{v}. \tag{65}$$

Thus, u and $\dot{v}$ are invariant gauge transformations.

What happens to v and $\dot{u}$?

$$D(\alpha, \beta)v = v + (\alpha - i\beta)u, \qquad \dot{D}(\alpha, \beta)\dot{u} = \dot{u} - (\alpha + i\beta)\dot{v}. \tag{66}$$

These spinors are not invariant under gauge transformations [17,18].
Thus, the Dirac spinor:

$$U_{inv} = \begin{pmatrix} u \\ \dot{v} \end{pmatrix}, \tag{67}$$

is gauge-invariant while the spinor

$$U_{non} = \begin{pmatrix} v \\ \dot{u} \end{pmatrix}, \tag{68}$$

is not. Thus, gauge invariance leads to the polarization of massless spin-1/2 particles. Indeed, this is what we observe in the real world.

5.2. Small-Mass Neutrinos

Neutrino oscillation experiments presently suggest that neutrinos have a small, but finite mass [19]. If neutrinos have mass, there should be a Lorentz frame in which they can be brought to rest with an $O(3)$-like $SU(2)$ little group for their internal space-time symmetry. However, it is not likely that at-rest neutrinos will be found anytime soon. In the meantime, we have to work with the neutrino with a fixed momentum and a small mass [20]. Indeed, the present loop representation is suitable for this problem.

Since the mass is so small, it is appropriate to approach this small-mass problem as a departure from the massless case. In Section 5.1, it was noted that the polarization of massless neutrinos is a consequence of gauge invariance. Let us start with a left-handed massless neutrino with the spinor:

$$\dot{v} = \begin{pmatrix} 0 \\ 1 \end{pmatrix}, \tag{69}$$

and the gauge transformation applicable to this spinor:

$$\dot{\Gamma}(\gamma) = \begin{pmatrix} 1 & 0 \\ \gamma & 1 \end{pmatrix}. \tag{70}$$

Since:

$$\begin{pmatrix} 1 & 0 \\ \gamma & 1 \end{pmatrix} \begin{pmatrix} 0 \\ 1 \end{pmatrix} = \begin{pmatrix} 0 \\ 1 \end{pmatrix}, \tag{71}$$

the spinor of Equation (69) is invariant under the gauge transformation of Equation (70).

If the neutrino has a small mass, the transformation matrix is for a rotation. However, for a small non-zero mass, the deviation from the triangular form is small. The procedure for deriving the Wigner matrix for this case is given toward the end of Section 3. The matrix in this case is:

$$\dot{D}(\gamma) = \begin{pmatrix} 1 - (\gamma\epsilon)^2/2 & -\gamma\epsilon^2 \\ \gamma & 1 - (\gamma\epsilon)^2/2 \end{pmatrix}, \tag{72}$$

with $\epsilon^2 = m/p$, where m and p are the mass and momentum of the neutrino, respectively. This matrix becomes the gauge transformation of Equation (70) for $\epsilon = 0$. If this matrix is applied to the spinor of Equation (69), it becomes:

$$D(\gamma)\dot{v} = \begin{pmatrix} -\gamma\epsilon^2 \\ 1 \end{pmatrix}. \tag{73}$$

In this way, the left-handed neutrino gains a right-handed component. We took into account that $(\gamma\epsilon)^2$ is much smaller than one.

Since massless neutrinos are gauge independent, we cannot measure the value of γ. For the small-mass case, we can determine this value from the measured values of m/p and the density of right-handed neutrinos.

6. Scalars, Vectors, and Tensors

We are quite familiar with the process of constructing three spin-1 states and one spin-0 state from two spinors. Since each spinor has two states, there are four states if combined.

In the Lorentz-covariant world, for each spin-1/2 particle, there are two additional two-component spinors coming from the dotted representation [12,21–23]. There are thus four states. If two spinors are combined, there are 16 states. In this section, we show that they can be partitioned into

1. scalar with one state,
2. pseudo-scalar with one state,
3. four-vector with four states,
4. axial vector with four states,
5. second-rank tensor with six states.

These quantities contain sixteen states. We made an attempt to construct these quantities in our earlier publication [5], but this earlier version is not complete. There, we did not take into account the parity operation properly. We thus propose to complete the job in this section.

For particles at rest, it is known that the addition of two one-half spins result in spin-zero and spin-one states. Hence, we have two different spinors behaving differently under the Lorentz boost. Around the z direction, both spinors are transformed by:

$$Z(\phi) = \exp\left(-i\phi J_3\right) = \begin{pmatrix} e^{-i\phi/2} & 0 \\ 0 & e^{i\phi/2} \end{pmatrix}. \tag{74}$$

However, they are boosted by:

$$B(\eta) = \exp\left(-i\eta K_3\right) = \begin{pmatrix} e^{\eta/2} & 0 \\ 0 & e^{-\eta/2} \end{pmatrix},$$

$$\dot{B}(\eta) = \exp\left(i\eta K_3\right), = \begin{pmatrix} e^{-\eta/2} & 0 \\ 0 & e^{\eta/2} \end{pmatrix}, \tag{75}$$

which are applicable to the undotted and dotted spinors, respectively. These two matrices commute with each other and also with the rotation matrix $Z(\phi)$ of Equation (74). Since K_3 and J_3 commute with each other, we can work with the matrix $Q(\eta, \phi)$ defined as:

$$Q(\eta, \phi) = B(\eta)Z(\phi) = \begin{pmatrix} e^{(\eta-i\phi)/2} & 0 \\ 0 & e^{-(\eta-i\phi)/2} \end{pmatrix},$$

$$\dot{Q}(\eta, \phi) = \dot{B}(\eta)\dot{Z}(\phi) = \begin{pmatrix} e^{-(\eta+i\phi)/2} & 0 \\ 0 & e^{(\eta+i\phi)/2} \end{pmatrix}. \tag{76}$$

When this combined matrix is applied to the spinors,

$$Q(\eta, \phi)u = e^{(\eta-i\phi)/2}u, \qquad Q(\eta, \phi)v = e^{-(\eta-i\phi)/2}v,$$

$$\dot{Q}(\eta, \phi)\dot{u} = e^{-(\eta+i\phi)/2}\dot{u}, \qquad \dot{Q}(\eta, \phi)\dot{v} = e^{(\eta+i\phi)/2}\dot{v}. \tag{77}$$

If the particle is at rest, we can explicitly construct the combinations:

$$uu, \qquad \frac{1}{\sqrt{2}}(uv + vu), \qquad vv, \tag{78}$$

to obtain the spin-1 state and:

$$\frac{1}{\sqrt{2}}(uv - vu), \tag{79}$$

for the spin-zero state. This results in four bilinear states. In the $SL(2,c)$ regime, there are two dotted spinors, which result in four more bilinear states. If we include both dotted and undotted spinors, there are sixteen independent bilinear combinations. They are given in Table 4. This table also gives the effect of the operation of $Q(\eta, \phi)$.

Table 4. Sixteen combinations of the $SL(2,c)$ spinors. In the $SU(2)$ regime, there are two spinors leading to four bilinear forms. In the $SL(2,c)$ world, there are two undotted and two dotted spinors. These four-spinors lead to sixteen independent bilinear combinations.

Spin 1			Spin 0
$uu,$	$\frac{1}{\sqrt{2}}(uv + vu),$	$vv,$	$\frac{1}{\sqrt{2}}(uv - vu)$
$\dot{u}\dot{u},$	$\frac{1}{\sqrt{2}}(\dot{u}\dot{v} + \dot{v}\dot{u}),$	$\dot{v}\dot{v},$	$\frac{1}{\sqrt{2}}(\dot{u}\dot{v} - \dot{v}\dot{u})$
$u\dot{u},$	$\frac{1}{\sqrt{2}}(u\dot{v} + v\dot{u}),$	$v\dot{v},$	$\frac{1}{\sqrt{2}}(u\dot{v} - v\dot{u})$
$\dot{u}u,$	$\frac{1}{\sqrt{2}}(\dot{u}v + \dot{v}u),$	$\dot{v}v,$	$\frac{1}{\sqrt{2}}(\dot{u}v - \dot{v}u)$
After the operation of $Q(\eta, \phi)$ and $\dot{Q}(\eta, \phi)$			
$e^{-i\phi}e^{\eta}uu,$	$\frac{1}{\sqrt{2}}(uv + vu),$	$e^{i\phi}e^{-\eta}vv,$	$\frac{1}{\sqrt{2}}(uv - vu)$
$e^{-i\phi}e^{-\eta}\dot{u}\dot{u},$	$\frac{1}{\sqrt{2}}(\dot{u}\dot{v} + \dot{v}\dot{u}),$	$e^{i\phi}e^{\eta}\dot{v}\dot{v},$	$\frac{1}{\sqrt{2}}(\dot{u}\dot{v} - \dot{v}\dot{u})$
$e^{-i\phi}u\dot{u},$	$\frac{1}{\sqrt{2}}(e^{\eta}u\dot{v} + e^{-\eta}v\dot{u}),$	$e^{i\phi}v\dot{v},$	$\frac{1}{\sqrt{2}}(e^{\eta}u\dot{v} - e^{-\eta}v\dot{u})$
$e^{-i\phi}\dot{u}u,$	$\frac{1}{\sqrt{2}}(\dot{u}v + \dot{v}u),$	$e^{i\phi}\dot{v}v,$	$\frac{1}{\sqrt{2}}(e^{-\eta}\dot{u}v - e^{\eta}\dot{v}u)$

Among the bilinear combinations given in Table 4, the following two equations are invariant under rotations and also under boosts:

$$S = \frac{1}{\sqrt{2}}(uv - vu), \quad \text{and} \quad \dot{S} = -\frac{1}{\sqrt{2}}(\dot{u}\dot{v} - \dot{v}\dot{u}). \tag{80}$$

They are thus scalars in the Lorentz-covariant world. Are they the same or different? Let us consider the following combinations

$$S_+ = \frac{1}{\sqrt{2}}\left(S + \dot{S}\right), \quad \text{and} \quad S_- = \frac{1}{\sqrt{2}}\left(S - \dot{S}\right). \tag{81}$$

Under the dot conjugation, S_+ remains invariant, but S_- changes sign. The boost is performed in the opposite direction and therefore is the operation of space inversion. Thus, S_+ is a scalar, while S_- is called a pseudo-scalar.

6.1. Four-Vectors

Let us go back to Equation (78) and make a dot-conjugation on one of the spinors.

$$u\dot{u}, \qquad \frac{1}{\sqrt{2}}(u\dot{v} + v\dot{u}), \qquad v\dot{v}, \qquad \frac{1}{\sqrt{2}}(u\dot{v} - v\dot{u}),$$

$$\dot{u}u, \qquad \frac{1}{\sqrt{2}}(\dot{u}v + \dot{v}u), \qquad \dot{v}v, \qquad \frac{1}{\sqrt{2}}(\dot{u}v - \dot{v}u). \tag{82}$$

We can make symmetric combinations under dot conjugation, which lead to:

$$\frac{1}{\sqrt{2}}\left(u\dot{u}+\dot{u}u\right), \quad \frac{1}{2}[(u\dot{v}+v\dot{u})+(\dot{u}v+\dot{v}u)], \quad \frac{1}{\sqrt{2}}(v\dot{v}+\dot{v}v), \quad \text{for spin 1,}$$

$$\frac{1}{2}[(u\dot{v}-v\dot{u})+(\dot{u}v-\dot{v}u)], \quad \text{for spin 0,} \tag{83}$$

and anti-symmetric combinations, which lead to:

$$\frac{1}{\sqrt{2}}\left(u\dot{u}-\dot{u}u\right), \quad \frac{1}{2}[(u\dot{v}+v\dot{u})-(\dot{u}v+\dot{v}u)], \quad \frac{1}{\sqrt{2}}(v\dot{v}-\dot{v}v), \quad \text{for spin 1,}$$

$$\frac{1}{2}[(u\dot{v}-v\dot{u})-(\dot{u}v-\dot{v}u)], \quad \text{for spin 0.} \tag{84}$$

Let us rewrite the expression for the space-time four-vector given in Equation (7) as:

$$\begin{pmatrix} t+z & x-iy \\ x+iy & t-z \end{pmatrix}, \tag{85}$$

which, under the parity operation, becomes

$$\begin{pmatrix} t-z & -x+iy \\ -x-iy & t+z \end{pmatrix}. \tag{86}$$

If the expression of Equation (85) is for an axial vector, the parity operation leads to:

$$\begin{pmatrix} -t+z & x-iy \\ x+iy & -t-z \end{pmatrix}, \tag{87}$$

where only the sign of t is changed. The off-diagonal elements remain invariant, while the diagonal elements are interchanged with sign changes.

We note here that the parity operation corresponds to dot conjugation. Then, from the expressions given in Equations (83) and (84), it is possible to construct the four-vector as:

$$V = \begin{pmatrix} u\dot{v}-\dot{v}u & v\dot{v}-\dot{v}v \\ u\dot{u}-\dot{u}u & \dot{u}v-v\dot{u} \end{pmatrix}, \tag{88}$$

where the off-diagonal elements change their signs under the dot conjugation, while the diagonal elements are interchanged.

The axial vector can be written as:

$$A = \begin{pmatrix} u\dot{v}+\dot{v}u & v\dot{v}+\dot{v}v \\ u\dot{u}+\dot{u}u & -\dot{u}v-v\dot{u} \end{pmatrix}. \tag{89}$$

Here, the off-diagonal elements do not change their signs under dot conjugation, and the diagonal elements become interchanged with a sign change. This matrix thus represents an axial vector.

6.2. Second-Rank Tensor

There are also bilinear spinors, which are both dotted or both undotted. We are interested in two sets of three quantities satisfying the $O(3)$ symmetry. They should therefore transform like:

$$(x+iy)/\sqrt{2}, \quad (x-iy)/\sqrt{2}, \quad z, \tag{90}$$

which are like:

$$uu, \qquad vv, \qquad (uv + vu)/\sqrt{2}, \tag{91}$$

respectively, in the $O(3)$ regime. Since the dot conjugation is the parity operation, they are like:

$$-\dot{u}\dot{u}, \qquad -\dot{v}\dot{v}, \qquad -(\dot{u}\dot{v} + \dot{v}\dot{u})/\sqrt{2}. \tag{92}$$

In other words,

$$(uu\dot{)} = -\dot{u}\dot{u}, \quad \text{and} \quad (vv\dot{)} = -\dot{v}\dot{v}. \tag{93}$$

We noticed a similar sign change in Equation (86).

In order to construct the z component in this $O(3)$ space, let us first consider:

$$f_z = \frac{1}{2}\left[(uv + vu) - (\dot{u}\dot{v} + \dot{v}\dot{u})\right], \qquad g_z = \frac{1}{2i}\left[(uv + vu) + (\dot{u}\dot{v} + \dot{v}\dot{u})\right]. \tag{94}$$

Here, f_z and g_z are respectively symmetric and anti-symmetric under the dot conjugation or the parity operation. These quantities are invariant under the boost along the z direction. They are also invariant under rotations around this axis, but they are not invariant under boosts along or rotations around the x or y axis. They are different from the scalars given in Equation (80).

Next, in order to construct the x and y components, we start with $f_\pm$ and $g_\pm$ as:

$$f_+ = \frac{1}{\sqrt{2}}\left(uu - \dot{u}\dot{u}\right), \qquad f_- = \frac{1}{\sqrt{2}}\left(vv - \dot{v}\dot{v}\right),$$

$$g_+ = \frac{1}{\sqrt{2}i}\left(uu + \dot{u}\dot{u}\right), \qquad g_- = \frac{1}{\sqrt{2}i}\left(vv + \dot{v}\dot{v}\right). \tag{95}$$

Then:

$$f_x = \frac{1}{\sqrt{2}}\left(f_+ + f_-\right) = \frac{1}{2}\left[(uu + vv) - (\dot{u}\dot{u} + \dot{v}\dot{v})\right],$$

$$f_y = \frac{1}{\sqrt{2}i}\left(f_+ - f_-\right) = \frac{1}{2i}\left[(uu - vv) - (\dot{u}\dot{u} - \dot{v}\dot{v})\right], \tag{96}$$

and:

$$g_x = \frac{1}{\sqrt{2}}\left(g_+ + g_-\right) = \frac{1}{2}\left[(uu + vv) + (\dot{u}\dot{u} + \dot{v}\dot{v})\right],$$

$$g_y = \frac{1}{\sqrt{2}i}\left(g_+ - g_-\right) = \frac{1}{2i}\left[(uu - vv) + (\dot{u}\dot{u} - \dot{v}\dot{v})\right]. \tag{97}$$

Here, f_x and f_y are symmetric under dot conjugation, while g_x and g_y are anti-symmetric.

Furthermore, f_z, f_x and f_y of Equations (94) and (96) transform like a three-dimensional vector. The same can be said for g_i of Equations (94) and (97). Thus, they can be grouped into the second-rank tensor:

$$\begin{pmatrix} 0 & -f_z & -f_x & -f_y \\ f_z & 0 & -g_y & g_x \\ f_x & g_y & 0 & -g_z \\ f_y & -g_x & g_z & 0 \end{pmatrix}, \tag{98}$$

whose Lorentz-transformation properties are well known. The g_i components change their signs under space inversion, while the f_i components remain invariant. They are like the electric and magnetic fields, respectively.

If the system is Lorentz-boosted, f_i and g_i can be computed from Table 4. We are now interested in the symmetry of photons by taking the massless limit. Thus, we keep only the terms that become larger for larger values of η. Thus,

$$f_x \to \frac{1}{2}(uu - \dot{v}\dot{v}), \qquad f_y \to \frac{1}{2i}(uu + \dot{v}\dot{v}),$$

$$g_x \to \frac{1}{2i}(uu + \dot{v}\dot{v}), \qquad g_y \to -\frac{1}{2}(uu - \dot{v}\dot{v}), \tag{99}$$

in the massless limit.

Then, the tensor of Equation (98) becomes:

$$\begin{pmatrix} 0 & 0 & -E_x & -E_y \\ 0 & 0 & -B_y & B_x \\ E_x & B_y & 0 & 0 \\ E_y & -B_x & 0 & 0 \end{pmatrix}, \tag{100}$$

with:

$$E_x \simeq \frac{1}{2}(uu - \dot{v}\dot{v}), \qquad E_y \simeq \frac{1}{2i}(uu + \dot{v}\dot{v}),$$

$$B_x = \frac{1}{2i}(uu + \dot{v}\dot{v}), \qquad B_y = -\frac{1}{2}(uu - \dot{v}\dot{v}). \tag{101}$$

The electric and magnetic field components are perpendicular to each other. Furthermore,

$$B_x = E_y, \qquad B_y = -E_x. \tag{102}$$

In order to address symmetry of photons, let us go back to Equation (95). In the massless limit,

$$B_+ \simeq E_+ \simeq uu, \qquad B_- \simeq E_- \simeq \dot{v}\dot{v}. \tag{103}$$

The gauge transformations applicable to u and $\dot{v}$ are the two-by-two matrices:

$$\begin{pmatrix} 1 & -\gamma \\ 0 & 1 \end{pmatrix}, \quad \text{and} \quad \begin{pmatrix} 1 & 0 \\ \gamma & 1 \end{pmatrix}, \tag{104}$$

respectively. Both u and $\dot{v}$ are invariant under gauge transformations, while $\dot{u}$ and v are not.

The B_+ and E_+ are for the photon spin along the z direction, while B_- and E_- are for the opposite direction.

6.3. Higher Spins

Since Wigner's original book of 1931 [24,25], the rotation group, without Lorentz transformations, has been extensively discussed in the literature [22,26,27]. One of the main issues was how to construct the most general spin state from the two-component spinors for the spin-1/2 particle.

Since there are two states for the spin-1/2 particle, four states can be constructed from two spinors, leading to one state for the spin-0 state and three spin-1 states. With three spinors, it is possible to construct four spin-3/2 states and two spin-1/2 states, resulting in six states. This partition process is much more complicated [28,29] for the case of three spinors. Yet, this partition process is possible for all higher spin states.

In the Lorentz-covariant world, there are four states for each spin-1/2 particle. With two spinors, we end up with sixteen (4×4) states, and they are tabulated in Table 4. There should be 64 states for three spinors and 256 states for four spinors. We now know how to Lorentz-boost those spinors. We also know that the transverse rotations become gauge transformations in the limit of zero-mass or infinite-η. It is thus possible to bundle all of them into the table given in Figure 5.

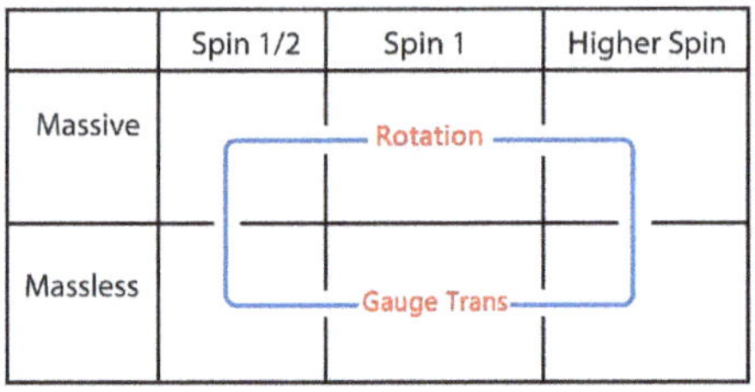

	Spin 1/2	Spin 1	Higher Spin
Massive		Rotation	
Massless		Gauge Trans	

Figure 5. Unified picture of massive and massless particles. The gauge transformation is a Lorentz-boosted rotation matrix and is applicable to all massless particles. It is possible to construct higher-spin states starting from the four states of the spin-1/2 particle in the Lorentz-covariant world.

In the relativistic regime, we are interested in photons and gravitons. As was noted in Sections 6.1 and 6.2, the observable components are invariant under gauge transformations. They are also the terms that become largest for large values of η.

We have seen in Section 6.2 that the photon state consists of uu and $\dot{v}\dot{v}$ for those whose spins are parallel and anti-parallel to the momentum, respectively. Thus, for spin-2 gravitons, the states must be $uuuu$ and $\dot{v}\dot{v}\dot{v}\dot{v}$, respectively.

In his effort to understand photons and gravitons, Weinberg constructed his states for massless particles [30], especially photons and gravitons [31]. He started with the conditions:

$$N_1|\text{state} >= 0, \quad \text{and} \quad N_2|\text{state} >= 0, \tag{105}$$

where N_1 and N_2 are defined in Equation (17). Since they are now known as the generators of gauge transformations, Weinberg's states are gauge-invariant states. Thus, uu and $\dot{v}\dot{v}$ are Weinberg's states for photons, and $uuuu$ are $\dot{v}\dot{v}\dot{v}\dot{v}$ are Weinberg's states for gravitons.

7. Concluding Remarks

Since the publication of Wigner's original paper [1], there have been many papers written on the subject. The issue is how to construct subgroups of the Lorentz group whose transformations do not change the momentum of a given particle. The traditional approach to this problem has been to work with a fixed mass, which remains invariant under Lorentz transformation.

In this paper, we have presented a different approach. Since, we are interested in transformations that leave the momentum invariant, we do not change the momentum throughout mathematical processes. Figure 3 tells the difference. In our approach, we fix the momentum, and we allow transitions from one hyperbola to another analytically with one transformation matrix. It is an interesting future problem to see what larger group can accommodate this process.

Since the purpose of this paper is to provide a simpler mathematics for understanding the physics of Wigner's little groups, we used the two-by-two $SL(2,c)$ representation, instead of four-by-four matrices, for the Lorentz group throughout the paper. During this process, it was noted in Section 5 that the Dirac equation is a representation of Wigner's little group.

We also discussed how to construct higher-spin states starting from four-component spinors for the spin-1/2 particle. We studied how the spins can be added in the Lorentz-covariant world, as illustrated in Figure 5.

Author Contributions: Each of the authors participated in developing the material presented in this paper and in writing the manuscript.

Conflicts of Interest: The authors declare no conflict of interest.

References

1. Wigner, E. On unitary representations of the inhomogeneous Lorentz group. *Ann. Math.* **1939**, *40*, 149–204.
2. Han, D.; Kim, Y.S.; Son, D. Gauge transformations as Lorentz-boosted rotations. *Phys. Lett. B* **1983**, *131*, 327–329.
3. Kim, Y.S.; Wigner, E.P. Cylindrical group and massless particles. *J. Math. Phys.* **1987**, *28*, 1175–1179.
4. Kim, Y.S.; Wigner, E.P. Space-time geometry of relativistic-particles. *J. Math. Phys.* **1990**, *31*, 55–60.
5. Başkal, S.; Kim, Y.S.; Noz, M.E. *Physics of the Lorentz Group, IOP Concise Physics*; Morgan & Claypool Publishers: San Rafael, CA, USA, 2015.
6. Kupersztych, J. Is there a link between gauge invariance, relativistic invariance and Electron Spin? *Nuovo Cimento* **1976**, *31B*, 1–11.
7. Han, D.; Kim, Y.S. Little group for photons and gauge transformations. *Am. J. Phys.* **1981**, *49*, 348–351.
8. Han, D.; Kim, Y.S. Special relativity and interferometers. *Phys. Rev. A* **1988**, *37*, 4494–4496.
9. Dirac, P.A.M. Applications of quaternions to Lorentz transformations. *Proc. R. Irish Acad.* **1945**, *A50*, 261–270.
10. Bargmann, V. Irreducible unitary representations of the Lorentz group. *Ann. Math.* **1947**, *48*, 568–640.
11. Naimark, M.A. *Linear Representations of the Lorentz Group*; Pergamon Press: Oxford, UK, 1954.
12. Kim, Y.S.; Noz, M.E. *Theory and Applications of the Poincaré Group*; Reidel: Dordrecht, The Netherlands, 1986.
13. Başkal, S.; Kim, Y.S.; Noz, M.E. Wigner's space-time symmetries based on the two-by-two matrices of the damped harmonic oscillators and the poincaré sphere. *Symmetry* **2014**, *6*, 473–515.
14. Han, D.; Kim, Y.S.; Son, D. Eulerian parametrization of Wigner little groups and gauge transformations in terms of rotations in 2-component spinors. *J. Math. Phys.* **1986**, *27*, 2228–2235.
15. Wigner, E.P. Normal form of antiunitary operators. *J. Math. Phys.* **1960**, *1*, 409–413.
16. Wigner, E.P. Phenomenological distinction between unitary and antiunitary symmetry operators. *J. Math. Phys.* **1960**, *1*, 413–416.
17. Han, D.; Kim, Y.S.; Son, D. E(2)-like little group for massless particles and polarization of neutrinos. *Phys. Rev. D* **1982**, *26*, 3717–3725.
18. Han, D.; Kim, Y.S.; Son, D. Photons, neutrinos, and gauge transformations. *Am. J. Phys.* **1986**, *54*, 818–821.
19. Mohapatra, R.N.; Smirnov, A.Y. Neutrino mass and new physics. *Ann. Rev. Nucl. Part. Sci.* **2006**, *56*, 569–628.
20. Kim, Y.S.; Maguire, G.Q., Jr.; Noz, M.E. Do small-mass neutrinos participate in gauge transformations? *Adv. High Energy Phys.* **2016**, *2016*, 1847620, doi:10.1155/2016/1847620.
21. Berestetskii, V.B.; Pitaevskii, L.P.; Lifshitz, E.M. *Quantum Electrodynamics, Volume 4 of the Course of Theoretical Physics*, 2nd ed.; Pergamon Press: Oxford, UK, 1982.
22. Gel'fand, I.M.; Minlos, R.A.; Shapiro, A. *Representations of the Rotation and Lorentz Groups and their Applications*; MacMillan: New York, NY, USA, 1963.
23. Weinberg, S. Feynman rules for any spin. *Phys. Rev.* **1964**, *133*, B1318–B1332.
24. Wigner, E. *Gruppentheorie und ihre Anwendungen auf die Quantenmechanik der Atomspektren*; Friedrich Vieweg und Sohn: Braunschweig, Germany, 1931. (In German)
25. Wigner, E.P. *Group Theory and Its Applications to the Quantum Mechanics of Atomic Spectra, Translated from the German*; Griffin, J.J., Ed.; Academic Press: New York, NY, USA, 1959.
26. Condon, E.U.; Shortley, G.H. *The Theory of Atomic Spectra*; Cambridge University Press: London, UK, 1951.
27. Hamermesh, M. *Group Theory and Application to Physical Problems*; Addison-Wesley: Reading, MA, USA, 1962.
28. Feynman, R.P.; Kislinger, M.; Ravndal, F. Current matrix elements from a relativistic quark model. *Phys. Rev. D* **1971**, *3*, 2706–2732.
29. Hussar, P.E.; Kim, Y.S.; Noz, M.E. Three-particle symmetry classifications according to the method of Dirac. *Am. J. Phys.* **1980**, *48*, 1038–1042.
30. Weinberg, S. Feynman rules for any spin II. massless particles. *Phys. Rev.* **1964**, *134*, B882–B896.
31. Weinberg, S. Photons and gravitons in S-Matrix theory: Derivation of charge conservation and equality of gravitational and inertial mass. *Phys. Rev.* **1964**, *135*, B1049–B1056.

© 2017 by the authors. Licensee MDPI, Basel, Switzerland. This article is an open access article distributed under the terms and conditions of the Creative Commons Attribution (CC BY) license (http://creativecommons.org/licenses/by/4.0/).

MDPI AG
St. Alban-Anlage 66
4052 Basel, Switzerland
Tel. +41 61 683 77 34
Fax +41 61 302 89 18
http://www.mdpi.com

Symmetry Editorial Office
E-mail: symmetry@mdpi.com
http://www.mdpi.com/journal/symmetry

MDPI AG
St. Alban-Anlage 66
4052 Basel
Switzerland

Tel: +41 61 683 77 34
Fax: +41 61 302 89 18

www.mdpi.com

MDPI

ISBN 978-3-03842-501-4

www.ingramcontent.com/pod-product-compliance
Lightning Source LLC
Chambersburg PA
CBHW051710210326
41597CB00032B/5428